World Geography

Australia, Antarctica & Pacific Islands

Regions | Physical Geography | Biogeography and Natural Resources
Human Geography | Economic Geography | Gazetteer

Second Edition

FLAGS OF THE WORLD

Ecuador
Egypt
El Salvador
Equatorial Guinea
Eritrea

Estonia
Eswatini (Swaziland)
Ethiopia
Fed. States of Micronesia
Fiji

Finland
France
Gabon
Gambia
Georgia

Germany
Ghana
Greece
Grenada
Guam

Guatemala
Guinea
Guinea-Bissau
Guyana
Haiti

Honduras
Hungary
Iceland
India
Indonesia

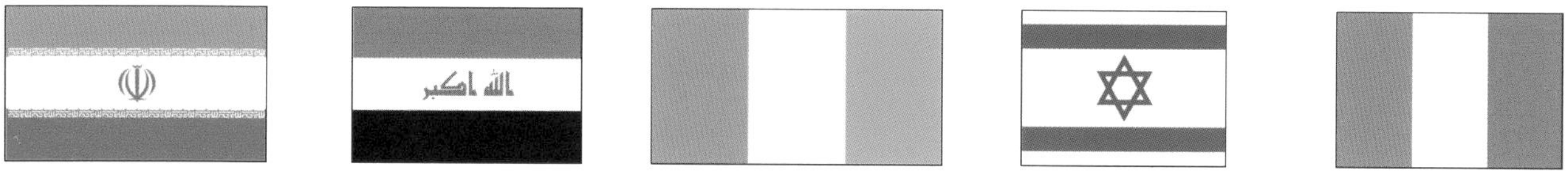

Iran
Iraq
Ireland
Israel
Italy

Jamaica
Japan
Jordan
Kazakhstan
Kenya

Kiribati
Kuwait
Kyrgyzstan
Laos
Latvia

Lebanon
Lesotho
Liberia
Libya
Liechtenstein

World Geography

Australia, Antarctica & Pacific Islands

Regions | Physical Geography | Biogeography and Natural Resources
Human Geography | Economic Geography | Gazetteer

Second Edition

Volume 6

Editor
Joseph M. Castagno
Educational Reference Publishing, LLC

SALEM PRESS
A Division of EBSCO Information Services, Inc.
Ipswich, Massachusetts

GREY HOUSE PUBLISHING

Cover photo: Australia from outer space. Image by 1xpert.

World Geography, Second Edition, published by Grey House Publishing, Inc., Amenia, NY, under exclusive license from EBSCO Information Services, Inc.

For information contact Grey House Publishing/Salem Press, 4919 Route 22, PO Box 56, Amenia, NY 12501.

∞ The paper used in these volumes conforms to the American National Standard for Permanence of Paper for Printed Library Materials, Z39.48 1992 (R2009).

Publisher's Cataloging-In-Publication Data
(Prepared by The Donohue Group, Inc.)

Names: Castagno, Joseph M., editor.
Title: World geography / editor, Joseph M. Castagno, Educational Reference Publishing, LLC.
Description: Second edition. | Ipswich, Massachusetts : Salem Press, a division of EBSCO Information Services, Inc. ; Amenia, NY : Grey House Publishing, [2020] | Interest grade level: High school. | Includes bibliographical references and index. | Summary: A six-volume geographic encyclopedia of the world, continents and countries of each continent. In addition to physical geography, the set also addresses human geography including population distribution, physiography and hydrology, biogeography and natural resources, economic geography, and political geography. | Contents: Volume 1. South & Central America — Volume 2. Asia — Volume 3. Europe — Volume 4. Africa — Volume 5. North America & the Caribbean — Volume 6. Australia, Antarctica & Pacific Islands.
Identifiers: ISBN 9781642654257 (set) | ISBN 9781642654288 (v. 1) | ISBN 9781642654318 (v. 2) | ISBN 9781642654301 (v. 3) | ISBN 9781642654295 (v. 4) | ISBN 9781642654271 (v. 5) | ISBN 9781642654325 (v. 6)
Subjects: LCSH: Geography—Encyclopedias, Juvenile. | CYAC: Geography—Encyclopedias. | LCGFT: Encyclopedias.
Classification: LCC G133 .W88 2020 | DDC 910/.3—dc23

First Printing
PRINTED IN CANADA

Contents

The Challenge of COVID-19

As *World Geography: Australia, Antarctica & Pacific Islands* goes to press in August 2020, the entire globe is grappling with the worst and most widespread pandemic in more than a century. The cause of the pandemic is a highly contagious viral condition known as Coronavirus Disease 2019, or COVID-19.

The first documented emergence of COVID-19 occurred in December 2019 as an outbreak of pneumonia in Wuhan City, in Hubai Province, China. By January 2020, Chinese health officials had reported tens of thousands of infections and dozens of deaths. That same month, COVID-19 cases were appearing across Asia, Europe, and North America, and spreading rapidly. On March 11, the World Health Organization (WHO) declared COVID-19 viral disease a pandemic.

The rapid spread of COVID-19 viral disease has strong geographical components. The virus emerged in Wuhan, a huge, densely populated city. It spread rapidly among a population living largely indoors during the cold-weather months. But the most significant geographical factor in the spread of the virus may well be the globalization of transportation. Every day, thousands of people fly to destinations near and far. Each traveler carries the potential to unknowingly spread disease.

To curtail COVID-19's spread, countries have closed their borders, air travel has been drastically reduced, and sweeping mitigation policies have been inaugurated. Some densely populated Asian countries, including South Korea, Singapore, and Taiwan, enforced these measures very early, and have, as of August 2020, managed to dampen the effect of the virus and limit the number of confirmed cases and deaths due to COVID-19. Other places, slower to act, such as China and Iran, have been very hard hit.

Many COVID-19 questions remain: Where will the disease strike next? Will the onset of warmer weather reduce the communicability of the virus? Will a vaccine be available soon? Will there be a second wave of infection? When will life be back to normal?

Geographers will continue to play a unique role in answering these questions, applying the tools and techniques of their discipline to achieving the fullest possible epidemiological understanding of the pandemic.

Publisher's Note

North Americans have long thought of the field of geography as little more than the study of the names and locations of places. This notion is not without a basis in fact: Through much of the twentieth century, geography courses forced students to memorize names of states, capitals, rivers, seas, mountains, and countries. Both students and educators eventually rebelled against that approach, geography courses gradually fell out of favor, and the future of geography as a discipline looked doubtful. Happily, however, the field has undergone a remarkable transformation, starting in the 1990s. Geography now has a bright and pivotal significance at all levels of education.

While learning the locations of places remains an important part of geography studies, educators recognize that place-name recognition is merely the beginning of geographic understanding. Geography now places much greater emphasis on understanding the characteristics of, and interconnections among, places. Modern students address such questions as how the weather in Brazil can affect the price of coffee in the United States, why global warming threatens island nations, and how preserving endangered plant and animal species can conflict with the economic development of poor nations.

World Geography, Second Edition, addresses these and many other questions. Designed and written to meet the needs of high school students, while being accessible to both middle school and undergraduate college students, these six volumes take an integrated approach to the study of geography, emphasizing the connections among world regions and peoples. The set's six volumes concentrate on major world regions: South and Central America; Asia; Europe; Africa; North America; and Australia, Antarctica and Pacific Islands. Each volume begins with common overview information related to the geography, maps and mapmaking. The core essays in the volumes begin with an overview section to provide global context and then go on to examine important geographic aspects of the regions in that area of the world: its physical geography; biogeography and natural resources; human geography (including its political geography); and economic geography. These essays range in length from three to ten pages. A gazetteer indicates major political, geographic, and manmade features throughout the region.

A robust appendix found in each volume provides further information:

- The Earth in Space (The Solar System, Earth's Moon, The Sun and the Earth, The Seasons);
- Earth's Interior (Earths Internal Structure, Plate Tectonics, Volcanoes, Geologic Time Scale);
- Earth's Surface (Internal Geological Processes, External Processes, Fluvial and Karst Processes, Glaciation, Desert Landforms, Ocean Margins);
- Earth's Climates (The Atmosphere, Global Climates; Cloud Formation, Storms);
- Earth's Biological Systems (Biomes);
- Natural Resources (Soils, Water);
- Exploration and Transportation (Exploration and Historical Trade Routes, Road Transportation, Railways, Air Transportation);
- Energy and Engineering (Energy Sources, Alternative Energies, Engineering Projects);
- Industry and Trade (Manufacturing, Globalization of Manufacturing and Trade, Modern World Trade Patterns);
- Political Geography (Forms of Government, Political Geography, Geopolitics, National Park Systems);
- Boundaries and Time Zones (International Boundaries, Global Time and Time Zones);
- Global Education (Themes and Standards in Geography Education);
- Global Data (The World Gazetteer of Oceans and Continents, The World's Oceans and Seas, Major Land Areas of the World, Major Islands of the World, Countries of the World (including population and pollution density), Past and Projected World Population Growth, 1950-2050, The World's Largest Countries by Area, The World's Smallest Countries by Area, The World's Largest Countries by Population,

The World's Smallest Countries by Population, The World's Most Densely Populated Countries, The World's Least Densely Populated Countries, The World's Most Populous Cities, Major Lakes of the World, Major Rivers of the World, The Highest Peaks in Each Continent, Major Deserts of the World, Highest Waterfalls of the World.

- A Glossary, General Bibliography, and Index complete the backmatter.

The regional divisions in the set make it possible to study specific countries or parts of the world. Pairing the specific regional information, organized by regions, physical geography, biogeography and natural resources, human geography, economic geography, and a gazetteer, with global information makes it possible for students to see the connections not only between countries and places within the region, but also between the regions and the entire global system, all within a single volume.

To make this set as easy as possible to use, all of its volumes are organized in a similar fashion, with six major divisions—Regions (organized into subregions by volume), Physical Geography, Biogeography and Natural Resources, Human Geography, Economic Geography, and Gazetteer. The number of subregions in each volume varies, depending upon the major world division being examined—this volume, for example, includes Antarctica, Australia, and the Pacific Islands.

Physical geography considers a world region's physiography, hydrology, and climatology. Biogeography and natural resources explores renewable and nonrenewable resources, flora, and fauna. Human geography addresses the people, population distribution, culture regions, urbanization, and political geography of the area. Economic geography considers the region's agriculture, industries, engineering projects, transportation, trade, and communications.

Gazetteers include descriptive entries on hundreds of important places, especially those mentioned in the volume's essays. A typical entry gives the place name and location, indicating the category into which the place falls (mountain, river, city, country, lake, etc). The entries also include statistics relevant to the categories of place (height of mountains, length of rivers, population of cities and countries).

A feature new to this edition is the discussion questions included throughout the volume. These questions are meant to foster discussion and further research into the topics related to the history, current issues, and future concerns related to physical, human, economic, and political geography.

Both a physical and a social science, geography is unique among social sciences in the demands it makes for visual support. For this reason, *World Geography* contains more than 100 maps, more than 1300 photographs, and scores of other graphical elements. In addition, essays are punctuated with more than 500 textual sidebars and tables, which amplify information in the essays and call attention to especially important or interesting points.

Both English and metric measures are used throughout this set. In most instances, English measures are given first, followed by their metric equivalents in parentheses. It should be noted that in cases of measures that are only estimates, such as the areas of deserts or average heights of mountain ranges, the metric figures are often rounded off to estimates that may not be exact equivalents of the English-measure estimates. In order to enhance clarity, units of measure are not abbreviated in the text, with these exceptions: kilometers are abbreviated as km. and square kilometers as sq. km. This exception has been made because of the frequency with which these measures appear.

Reference works such as this would be impossible without the expertise of a large team of contributing scholars. This project is no exception. Salem Press would like to thank the more than 175 people who wrote the signed essays and contributed entries to the gazetteers. A full list of contributors follows this note. We recognize the efforts of Dr. Ray Sumner, of California's Long Beach City College, for the expertise and insights that she brought to the previous edition of this book, and which have formed the strong foundation for this new edition. We also acknowledge the work of the editor of this current volume, Joseph Castagno, Educational Reference Publishing, LLC.

Introduction

When Henry Morton Stanley of the *New York Herald* shook the hand of David Livingstone on the shore of Central Africa's Lake Tanganyika in 1871, the moment represented the high point of geography to many people throughout the world. A Scottish missionary and explorer, Livingstone had been out of contact with the outside world for nearly two years, and European and American newspapers had buzzed with speculation about his disappearance. At that time, so little was known about the geography of the interior of Africa that Stanley's finding Livingstone was acclaimed as a brilliant triumph of explorations.

The field of geography in Stanley and Livingstone's time was—and to a large extent still is—synonymous with explorations. Stories of epic journeys, both historic and contemporary, continue to exert a powerful attraction on readers. Mountains, deserts, forest, caves, and glaciers still draw intrepid explorers, while even more armchair travelers are thrilled by accounts and pictures of these exploits and discoveries. We all love to travel—to the beach, into the mountains, to our great national parks, and to foreign countries. In the need and desire to explore our surroundings, we are all geographers.

Numerous geographical societies welcome both professional geographers and the general public into their membership, as they promote a greater knowledge and understanding of the earth. The National Geographic Society, founded in 1888 "for the increase and diffusion of geographical knowledge," has awarded more than 11,000 grants for scientific exploration and research. Each year, the society invests millions of dollars in expeditions and fieldwork related to environmental concerns and global geographic issues. The findings are recorded in the pages of the familiar yellow-bordered *National Geographic* magazine, now produced in 40 local-language editions in many countries around the world, publishing around 6.8 million copies monthly, with some 60 million readers. The magazine, along with the National Geographic International television network, reaches more than 135 million readers and viewers worldwide and has more than 85 million subscribers.

An even older geographical association is Great Britain's Royal Geographical Society, which grew out of the Geographical Society of London, founded in 1830 with the "sole object" of promoting "that most important and entertaining branch of knowledge—geography." Over the century that followed, the Royal Geographical Society focused on exploration of the continents of Africa and Antarctica. In the society's London headquarters adjacent to the Albert Hall, visitors can still view such historic artifacts as David Livingstone's cap and chair, as well as diaries, sketches, and maps covering the great period of the British Empire and beyond. Today the society assists more than five hundred field expeditions every year.

With the aid of satellites and remote-sensing instruments, we can now obtain images and data from almost anywhere on Earth. However, remote and inaccessible places still invite the intrepid to visit and explore them in person. Although the outlines of the continents have now been completed, and their interiors filled in with details of mountains and rivers, cities and political boundaries, remote places still exert a fascination on modern urbanites.

The enchantment of tales about strange sights and courageous journeys has been with us since the ancient voyages of Homer's *Ulysses,* Marco Polo's travels to China, and the nautical expeditions of Christopher Columbus, Ferdinand Magellan, and James Cook. While those great travelers are from the remote past, the age of exploration is far from over—a fact repeatedly demonstrated by the modern Norwegian navigator Thor Heyerdahl. Moreover, new journeys of discovery are still taking place. In 1993, after dragging a sled wearily across the frigid wastes of Antarctica for more than three months, Sir Ranulph Twisleton-Wykeham-Fiennes announced that the age of exploration is not dead. Six years later, in 1999, the long-missing body of British mountain climber George Mallory was found on the slopes of Mount Everest, near whose top he had mysteriously vanished in 1924. That discovery sparked a new wave of admiration and respect for explorers of such courage and endurance.

How many people have been enthralled by the bravery of Antarctic explorer Robert Falcon Scott and the noble sacrifice his injured colleague Lawrence Oates made in 1912, when he gave up his life in order not to slow down the rest of the expedition? There can be no doubt that the thrills and the dangers of exploring find resonance among many modern readers.

The struggle to survive in environments hostile to human beings reminds us of the power of our planet Earth. Significant books on this theme have included Jon Krakauer's *Into Thin Air* (1998), an account of a disastrous expedition climbing Mount Everest, and Sebastian Junger's *The Perfect Storm* (1997), the story of the worst gale of the twentieth century and its effect on a fishing fleet off the East Coast of North America. *Endurance* (1998), the epic of Sir Ernest Shackleton's survival and leadership for two years on the frozen Arctic, attracts the same people who avidly read *Undaunted Courage* (1996) the story of Meriwether Lewis and William Clark's epic exploration of the Louisiana Purchase territories in the early nineteenth century. In 1997 *Seven Years in Tibet* premiered, a popular film about the Austrian Heinrich Harrer, who lived in Tibet in the mid-twentieth century. The more urban people become, the greater their desire for adventurous, remote places, a least vicariously, to raise the human spirit.

There are, of course, also scientific achievements associated with modern exploration. In November 1999, the elevation of Mount Everest, the world tallest peak was raised by 7 feet (2.1 meters) to a new height of 29,035 feet (8,850 meters) above sea level; the previously accepted height had been based on surveys made during the 1950s. This new value was the result of Global Positioning System (GPS) technology enabling a more accurate measurement than had been possible with land-based earthbound surveying equipment. A team of climbers supported by the National Geographic Society and the Boston Museum of Science was equipped with GPS equipment, which enabled a fifty-minute recording of data based on satellite signals. At the same time, the expedition was able to ascertain that Mount Everest is moving northeast, atop the Indo-Australian Plate, at a rate of approximately 2.4 inches (10 centimeters) per year.

In 2000, the International Hydrographic Organization named a "new" ocean, the Southern Ocean, which encompasses all the water surrounding Antarctica up to 60° south latitude. With an area of approximately 7.8 million square miles (20.3 million sq. km.), the Southern Ocean is about twice the size of the entire United States and ranks as the world's fourth largest ocean, after the Pacific, Atlantic, and Indian Oceans, but just ahead of the Arctic Ocean.

Despite the humanistic and scientific advantage of geographic knowledge, to many people today, geography is a subject where one merely memorized longs lists of facts dealing with "where" questions. (Where is Andorra? Where is Prince Edward Island? Where is Kalamazoo?) or "what" questions (What is the highest mountain in South America? What is the capital of Costa Rica?) This approach to the study of geography has been perpetuated by the annual National Geographic Bee, conducted in the United States each year for students in grades four through eight. Participants in the competition display an astonishing recall of facts but do not have the opportunity of showing any real geographic thought. To a geographer, such factual knowledge is simply a foundation for investigating and explaining the much more important questions dealing with "why"—"Why is the Sahara a desert?"

Geographers aim to understand why environments and societies occur where and as they do, and how they change. Geography must be seen as an integrative science; the collection of factual data and evidence, as in exploration, is the empirical foundation for deductive reasoning. This leads to the creation of a range of geographical methods, models, theories, and analytical approaches that serve to unify a very broad area of knowledge—the interaction between natural and human environments. Although geography as an academic discipline became established in nineteenth century Germany, there have always been geographers, in the sense of people curious about their world. Humans have always wanted to know about day and night, the shape of the earth, the nature of climates, differences in plants and animals, as well as what lies beyond the horizon. Today, as we hear about and actually experience the sweeping effects of globalization, we need more than ever to develop our geographic skills. Not only are we connected by economic ties to the countries of the world, but we must also appreciate the consequences of North America's high standard of living.

Political boundaries are artificial human inventions, but the natural world is one biosphere. As concern over global warming escalates, national leaders meet to seek a solution to the emission of greenhouse gases, rising ocean levels, and mass extinctions. Are we connected to our environment? At a time when the rate of species extinction is a hundred times above normal, and the human population is crowding in increasing numbers into huge urban centers, we have, nevertheless, taken time each year in April to celebrate Earth Day since 1970. We need now to realize that every day is Earth Day.

Geography languished in the United States in the 1960s, as social studies was taught with a history emphasis in schools. American students became alarmingly disadvantaged in geographic knowledge, compared with most other countries. Fortunately, members of the profession acted to restore geography to the curriculum. In 1984, the National Geographic Society undertook the challenge of restoring geography in the United States. The society turned to two organizations active in geographic education: The Association of American Geographers, the professional geographers' group with more than 10,000 members, mostly in higher education in the United States; and the National Council for Geographic Education that supports geography teaching at all levels—from kindergarten through university, with members that include U.S. and international teachers, professors, students, businesses, and others who support geography education. The council administers the Geographic Alliances, found in every state of the United Sates, with a national membership of about 120,000 schoolteachers. Together, they produced the "Guidelines in Geographic Education," which introduced the Five Themes of Geography, to enhance the teaching of geography in schools. Using the themes of Location, Place, Human/Environment Interaction, Movement and Regions, teachers were able to plan and conduct lessons in which students encountered interesting real-world examples of the relevance and importance of geography. Continued research into geographic education led to the inclusion of geography in 1990 as one of the core subjects of the National Education Goals, along with English, mathematics, science, and history.

Another milestone was the publication in 1994 of "Geography for Life," the national Geography Standards. The earlier Five Themes were subsumed under the new Six Essential Elements: The World in Spatial Terms; Places and Regions; Physical Systems; Human Systems; Environment Systems; Environment and Society; and The Uses of Geography. Eighteen geography standards are included, describing what a geographically informed person knows and understands. States, schools, and individual teachers have welcomed the new prominence of geography, and enthusiastically adopted new approaches to introduce the geography standards to new learners. The rapid spread of computer technology, especially in the field of Geographical Information Science, has also meant a new importance for spatial analysis, a traditional area of geographical expertise. No longer is geography seen as an outdated mass of useless or arcane facts; instead, geography is now seen, again, to be an innovative an integrative science, which can contribute to solving complex problems associated with the human-environmental relationship in the twenty-first century.

Geographers may no longer travel across uncharted realms, but there is still much we long to explore, to learn, and seek to understand, even if it is only as "armchair" geographers. This reference work, *World Geography,* will help carry readers on their own journeys of exploration.

Ray Sumner
Long Beach City College

Joseph M. Castagno
Educational Reference Publishing, LLC

Contributors

Emily Alward
Henderson, Nevada Public Library

Earl P. Andresen
University of Texas at Arlington

Debra D. Andrist
St. Thomas University

Charles F. Bahmueller
Center for Civic Education

Timothy J. Bailey
Pittsburg State University

Irina Balakina
Writer/Editor, Educational Reference Publishing, LLC

David Barratt
Nottingham, England

Maryanne Barsotti
Warren, Michigan

Thomas F. Baucom
Jacksonville State University

Michelle Behr
Western New Mexico University

Alvin K. Benson
Brigham Young University

Cynthia Breslin Beres
Glendale, California

Nicholas Birns
New School University

Olwyn Mary Blouet
Virginia State University

Margaret F. Boorstein
C.W. Post College of Long Island University

Fred Buchstein
John Carroll University

Joseph P. Byrne
Belmont University

Laura M. Calkins
Palm Beach Gardens, Florida

Gary A. Campbell
Michigan Technological University

Byron D. Cannon
University of Utah

Steven D. Carey
University of Mobile

Roger V. Carlson
Jet Propulsion Laboratory

Robert S. Carmichael
University of Iowa

Joseph M. Castagno
Principal, Educational Reference Publishing, LLC

Habte Giorgis Churnet
University of Tennessee at Chattanooga

Richard A. Crooker
Kutztown University

William A. Dando
Indiana State University

Larry E. Davis
College of St. Benedict

Ronald W. Davis
Western Michigan University

Cyrus B. Dawsey
Auburn University

Frank Day
Clemson University

M. Casey Diana
University of Illinois at Urbana-Champaign

Stephen B. Dobrow
Farleigh Dickinson University

Steven L. Driever
University of Missouri, Kansas City

Sherry L. Eaton
San Diego City College

Femi Ferreira
Hutchinson Community College

Helen Finken
Iowa City High School

Eric J. Fournier
Samford University

Anne Galantowicz
El Camino College

Hari P. Garbharran
Middle Tennessee State University

Keith Garebian
Ontario, Canada

Laurie A. B. Garo
University of North Carolina, Charlotte

Jay D. Gatrell
Indiana State University

Carol Ann Gillespie
Grove City College

Nancy M. Gordon
Amherst, Massachusetts

Noreen A. Grice
Boston Museum of Science

Johnpeter Horst Grill
Mississippi State University

Charles F. Gritzner
South Dakota State University

C. James Haug
Mississippi State University

Douglas Heffington
Middle Tennessee State University

Thomas E. Hemmerly
Middle Tennessee State University

Jane F. Hill
Bethesda, Maryland

Carl W. Hoagstrom
Ohio Northern University

Catherine A. Hooey
Pittsburg State University

Robert M. Hordon
Rutgers University

Kelly Howard
La Jolla, California

Paul F. Hudson
University of Texas at Austin

Huia Richard Hutton
University of Hawaii/Kapiolani Community College

Raymond Pierre Hylton
Virginia Union University

Solomon A. Isiorho
Indiana University/Purdue University at Fort Wayne

Ronald A. Janke
Valparaiso University

Albert C. Jensen
Central Florida Community College

Jeffry Jensen
Altadena, California

Bruce E. Johansen
University of Nebraska at Omaha

Kenneth A. Johnson
State University of New York, Oneonta

Walter B. Jung
University of Central Oklahoma

James R. Keese
California Polytechnic State University, San Luis Obispo

Leigh Husband Kimmel
Indianapolis, Indiana

Denise Knotwell
Wayne, Nebraska

James Knotwell
Wayne State College

Grove Koger
Boise Idaho Public Library

Alvin S. Konigsberg
State University of New York at New Paltz

Doris Lechner
Principal, Educational Reference Publishing, LLC

Steven Lehman
John Abbott College

Denyse Lemaire
Rowan University

Dale R. Lightfoot
Oklahoma State University

Jose Javier Lopez
Minnesota State University

James D. Lowry, Jr.
East Central University

Jinshuang Ma
Arnold Arboretum of Harvard University Herbaria

Dana P. McDermott
Chicago, Illinois

Thomas R. MacDonald
University of San Francisco

Robert R. McKay
Clarion University of Pennsylvania

Nancy Farm Männikkö
L'Anse, Michigan

Carl Henry Marcoux
University of California, Riverside

Christopher Marshall
Unity College

Rubén A. Mazariegos-Alfaro
University of Texas/Pan American

Christopher D. Merrett
Western Illinois University

John A. Milbauer
Northeastern State University

Randall L. Milstein
Oregon State University

Judith Mimbs
Loftis Middle School

Karen A. Mulcahy
East Carolina University

B. Keith Murphy
Fort Valley State University

M. Mustoe
Omak, Washington

Bryan Ness
Pacific Union College

Kikombo Ilunga Ngoy
Vassar College

Joseph R. Oppong
University of North Texas

Richard L. Orndorff
University of Nevada, Las Vegas

Bimal K. Paul
Kansas State University

Nis Petersen
New Jersey City University

Mark Anthony Phelps
Ozarks Technical Community College

John R. Phillips
Purdue University, Calumet

Alison Philpotts
Shippensburg University

Julio César Pino
Kent State University

Timothy C. Pitts
Morehead State University

Carolyn V. Prorok
Slippery Rock University

P. S. Ramsey
Highland Michigan

Robert M. Rauber
University of Illinois at Urbana-Champaign

Ronald J. Raven
State University of New York at Buffalo

Neil Reid
University of Toledo

Susan Pommering Reynolds
Southern Oregon University

Nathaniel Richmond
Utica College

Edward A. Riedinger
Ohio State University Libraries

Mika Roinila
West Virginia University

Thomas E. Rotnem
Brenau University

Joyce Sakkal-Gastinel
Marseille, France

Helen Salmon
University of Guelph

Elizabeth D. Schafer
Loachapoka, Alabama

Kathleen Valimont Schreiber
Millersville University of Pennsylvania

Ralph C. Scott
Towson University

Guofan Shao
Purdue University

Wendy Shaw
Southern Illinois University, Edwardsville

R. Baird Shuman
University of Illinois, Champaign-Urbana

Sherman E. Silverman
Prince George's Community College

Roger Smith
Portland, Oregon

Robert J. Stewart
California Maritime Academy

Toby R. Stewart
Alamosa, Colorado

Ray Sumner
Long Beach City College

Paul Charles Sutton
University of Denver

Glenn L. Swygart
Tennessee Temple University

Sue Tarjan
Santa Cruz, California

Robert J. Tata
Florida Atlantic University

John M. Theilmann
Converse College

Virginia Thompson
Towson University

Norman J. W. Thrower
University of California, Los Angeles

Paul B. Trescott
Southern Illinois University

Robert D. Ubriaco, Jr.
Illinois Wesleyan University

Mark M. Van Steeter
Western Oregon University

Johan C. Varekamp
Wesleyan University

Anthony J. Vega
Clarion University

William T. Walker
Chestnut Hill College

William D. Walters, Jr.
Illinois State University

Linda Qingling Wang
University of South Carolina, Aiken

Annita Marie Ward
Salem-Teikyo University

Kristopher D. White
University of Connecticut

P. Gary White
Western Carolina University

Thomas A. Wikle
Oklahoma State University

Rowena Wildin
Pasadena, California

Donald Andrew Wiley
Anne Arundel Community College

Kay R. S. Williams
Shippensburg University

Lisa A. Wroble
Redford Township District Library

Bin Zhou
Southern Illinois University, Edwardsville

Regions

Overview

The History of Geography

The moment that early humans first looked around their world with inquiring minds was the moment that geography was born. The history of geography is the history of human effort to understand the nature of the world. Through the centuries, people have asked of geography three basic questions: What is Earth like? Where are things located? How can one explain these observations?

Geography in the Ancient World

In the Western world, the Greeks and the Romans were among the first to write about and study geography. Eratosthenes, a Greek scholar who lived in the third century BCE, is often called the "father of geography and is credited with first using the word geography (from the Greek words *ge*, which means "earth," and *graphe*, which means "to describe"). The ancient Greeks had contact with many older civilizations and began to gather together information about the known world. Some, such as Hecataeus, described the multitude of places and peoples with which the Greeks had contact and wrote of the adventures of mythical characters in strange and exotic lands. However, the ancient Greek scholars went beyond just describing the world. They used their knowledge of mathematics to measure and locate. The Greek scholars also used their philosophical nature to theorize about Earth's place in the universe.

One Greek scholar who used mathematics in the study of geography was Anaximander, who lived from 610 to 547 BCE. Anaximander is credited with being the first person to draw a map of the world to scale. He also invented a sundial that could be used to calculate time and direction and to distinguish the seasons. Eratosthenes is also famous for his mathematical calculations, in particular of the circumference of Earth, using observations of the Sun. Hipparchus, who lived around 140 BCE, used his mathematical skills to solve geographic problems and was the first person to introduce the idea of a latitude and longitude grid system to locate places.

Such early Greek philosophers as Plato and Aristotle were also concerned with geography. They discussed such issues as whether Earth was flat or spherical and if it was the center of the universe, and debated the nature of Earth as the home of humankind.

Whereas the Greeks were great thinkers and introduced many new ideas into geography, the Roman contribution was to compile and gather available knowledge. Although this did not add much that was new to geography, it meant that the knowledge of the ancient world was available as a base to work from and was passed down across the centuries. Geogra-

Curiosity: The Root of Geography

The earliest human beings, as they hunted and gathered food and used primitive tools in order to survive, must have had detailed knowledge of the geography of their part of the world. The environment could be a hostile place, and knowledge of the world meant the difference between life and death. Human curiosity took them one step further. As they lived in an ancient world of ice and fire, human beings looked to the horizon for new worlds, crossing continents and spreading out to all areas of the globe. They learned not only to live as a part of their environment, but also to understand it, predict it, and adapt it to their needs.

phy in the ancient world is often said to have ended with the great work of Ptolemy (Claudius Ptolemaeus), who lived from 90 to 168 CE. Ptolemy is best known for his eight-volume *Guide to Geography*, which included a gazetteer of places located by latitude and longitude, and his world map.

Geography in China

The study of geography also was important in ancient China. Chinese scholars described their resources, climate, transportation routes, and travels, and were mapping their known world at the same time as were the great Western civilizations. The study of geography in China begins in the Warring States period (fifth century BCE). It expands its scope beyond the Chinese homeland with the growth of the Chinese Empire under the Han dynasty. It enters its golden age with the invention of the compass in the eleventh century CE (Song dynasty) and peaks with fifteenth century CE (Ming dynasty) Chinese exploration of the Pacific under admiral Zheng He during the treasure voyages.

Geography in the Middle Ages

With the collapse of the Roman Empire in the fifth century CE, Europe entered into what is commonly known as the Early Middle Ages. During this time, which lasted until the fifteenth century, the geographic knowledge of the ancient world was either lost or challenged as being counter to Christian teachings. For example, the early Greeks had theorized that Earth was a sphere, but this was rejected during the Middle Ages. Scholars of the Middle Ages believed that the world was a flat disk, with the holy city of Jerusalem at its center.

The knowledge and ideas of the ancient world might have been lost if they had not been preserved by Muslim scholars. In the Islamic countries of North Africa and the Middle East, some of the scholarship of the ancient world was sheltered in libraries and universities. This knowledge was extensively added to as Muslims traveled and traded across the known world, gathering their own information.

Among the most famous Muslim geographers were Ibn Battutah, al-Idrisi, and Ibn Khaldun. Ibn Battutah traveled east to India and China in the fourteenth century. Al-Idrisi, at the command of King Roger II of Sicily, wrote *Roger's Book*, which systematically described the world. Information from *Roger's Book* was engraved on a huge planisphere (disk), crafted in silver; this once was considered a wonder of the world, but it is thought to have been destroyed. Ibn Khaldun (1332-1406) is best known for his written world history, but he also was a pioneer in focusing on the relationship of human beings to their environment.

The Age of European Exploration

Beginning in the fifteenth century, the isolation of Europe came to an end, and Europeans turned their attention to exploration. The two major goals of this sudden surge in exploration were to spread the Christian faith and to obtain needed resources. In 1418 Prince Henry the Navigator established a school for navigators and began to gather the tools and knowledge needed for exploration. He was the first of many Europeans to travel beyond the limits of the known world, mapping, describing, and cataloging all that they saw.

The great wave of European exploration brought new interest in geography, and the monumental works of the Greeks and Romans—so carefully preserved by Muslim scholars—were rediscovered and translated into Latin. The maps produced in the Middle Ages were of little use to the explorers who were traveling to, and beyond, the limits of the known world. Christopher Columbus, for example, relied on Ptolemy's work during his voyages to the Americas, but soon newer, more accurate maps were drawn and, for the first time, globes were made. A particularly famous map, which is still used as a base map, is the Mercator projection. On the world map produced by Gerardus Mercator (born Geert de Kremer) in 1569, compass directions appear as straight lines, which was a great benefit on navigational charts.

When the age of European exploration began, even the best world maps crudely depicted only a few limited areas of the world. Explorers quickly began to gather huge quantities of information, making detailed charts of coastlines, discovering new continents, and eventually filling in the maps of those continents with information about both the natural and human features they encountered. This age of exploration is often said to have ended when Roald Amundsen planted the Norwegian flag at the South Pole in 1911. At that time, the world map became complete, and human beings had mapped and explored every part of the globe. However, the beginning of modern geography is usually associated with the work of two nineteenth century German geographers: Alexander von Humboldt and Carl Ritter.

The Beginning of Modern Geography

The writings of Alexander von Humboldt and Carl Ritter mark a leap into modern geography, because these writers took an important step beyond the work of previous scholars. The explorers of the previous centuries had focused on gathering information, describing the world, and filling in the world map with as much detail as possible. Humboldt and Ritter took a more scientific and systematic approach to geography. They began not only to compile descriptive information, but also to ask why: Humboldt spent his lifetime looking for relationships among such things as climate and topography (landscape), while Ritter was intrigued by the multitude of connections and relationships he observed within human geographic patterns. Both Humboldt and Ritter died in 1859, ending a period when information-gathering had been paramount. They brought geography into a new age in which synthesis, analysis, and theory-building became central.

European Geography

After the work of Humboldt and Ritter, geography became an accepted academic discipline in Europe, particularly in Germany, France, and Great Britain. Each of these countries emphasized different aspects of geographic study. German geographers continued the tradition of the scientific view, using observable data to answer geographic questions. They also introduced the concept that geography could take a chorological view, studying all aspects, physical and human, of a region and of the interrelationships involved.

The chorological view came to dominate French geography. Paul Vidal de la Blache (1845-1918) was the most prominent French geographer. He advocated the study of small, distinct areas, and French geographers set about identifying the many regions of France. They described and analyzed the unique physical and human geographic complex that was to be found in each region. An important concept that emerged from French geography was "possibilism." German geographers had introduced the notion of environmental determinism—that human beings were largely shaped and controlled by their environments. Possibilism rejected the concept of environmental determinism, asserting that the relationship between human beings and the environment works in two directions: The environment creates both limits and opportunities for people, but people can react in different ways to a given environment, so they are not controlled by it.

British geographers, influenced by the French approach, conducted regional surveys. British regional studies were unique in their emphasis on planning and geography as an applied science. From this work came the concept of a functional region—an area that works together as a unit based on interaction and interdependence.

American Geography

Prior to World War II, only a small group of people in the United States called themselves geographers. They were mostly influenced by German

National Geographic Society and Geographic Research

In 1888 the National Geographic Society was founded to support the "increase and diffusion of geographic knowledge" of the world. In its first 110 years, the society funded more than five thousand expeditions and research projects with more than 6,500 grants. By the 1990s it was the largest such foundation in the world, and the results of its funded projects are found on television programs, video discs, video cassettes, and books, as well as in the *National Geographic* magazine, established in 1888. Its productions are cutting-edge resources for information about archaeology, ethnology, biology, and both cultural and physical geography.

ideas, but the nature of geography was hotly debated. Two schools of geographers were philosophical adversaries. The Midwestern School, led by Richard Hartshorne, believed that description of unique regions was the central task of geography.

The Western (or Berkeley) School of geography, led by Carl Sauer, agreed that regional study was important, but believed it was crucial to go beyond description. Sauer and his followers included genesis and process as important elements in any study. To understand a region and to know where it is going, they argued, one must look at its past and how it got to its present state.

In the 1930s, environmental determinism was introduced to U.S. geography but ultimately was rejected. Although geography in both Europe and the United States was essentially an all-male discipline, the United States produced the first famous woman geographer, Ellen Churchill Semple (1863-1932).

World War II illustrated the importance of geographic knowledge, and after the war came to an end in 1945, geographers began to come into their own in the United States. From the end of World War II to the early 1960s, U.S. geographers produced many descriptive regional studies.

In the early 1960s, what is often called the quantitative revolution occurred. The development of computers allowed complex mathematical analysis to be performed on all kinds of geographic data, and geographers began to analyze a wide range of problems using statistics. There was great enthusiasm for this new approach to geography at first, but beginning in the 1970s, many people considered a purely mathematical approach to be somewhat sterile and thought it left out a valuable human element.

In the 1980s and 1990s, many new ways to look at geographic issues and problems were developed, including humanism, behaviorism, Marxism, feminism, realism, structuration, phenomenology, and postmodernism, all of which bring human beings back into focus within geographical studies.

Geography in the Twenty-first Century

Geography increasingly uses technology to analyze global space and answer a wide range of questions related to a host of concerns including issues related to the environment, climate change, population, rising sea levels, and pollution. The Geographic Information System (GIS), in particular, provides a powerful way for people trained in geography to understand geographic issues, solve geographic problems, and display geographic information. Geographers continue to adopt a wide variety of philosophies, approaches, and methods in their quest to answer questions concerning all things spatial.

Wendy Shaw

Mapmaking in History

Cartography is the science or art of making maps. Although workers in many fields have a concern with cartography and its history, it is most often associated with geography.

Maps of Preliterate Peoples

The history of cartography predates the written record, and most cultures show evidence of mapping skills. The earliest surviving maps are those carved in stone or painted on the walls of caves, but modern preliterate peoples still use a variety of materials to express themselves cartographically. For example, the Marshall Islanders use palm fronds, fiber from coconut husks (coir), and shells to make sea charts for their inter-island navigation. The Inuit use animal skins and driftwood, sometimes painted, in mapping. There is a growing interest in the cartography of early and preliterate peoples, but some of their maps do not fit readily into a more traditional concept of cartography.

Mapping in Antiquity

Early literate peoples, such as those of Egypt and Mesopotamia, displayed considerable variety in their maps and charts, as shown by the few maps from these civilizations that still exist. The early Egyptians painted maps on wooden coffin bases to assist the departed in finding their way in the afterlife; they also made practical route maps for their mining operations. It is thought that geometry developed from the Egyptians' riverine surveys. The Babylonians made maps of different scales, using clay tablets with cuneiform characters and stylized symbols, to create city plans, regional maps, and "world" maps. They also divided the circle in the sexigesimal system, an idea they may have obtained from India and that is commonly used in cartography to this day.

The Greeks inherited ideas from both the Egyptians and the Mesopotamians and made signal contributions to cartography themselves. No direct evidence of early Greek maps exists, but indirect evidence in texts provides information about their cosmological ideas, culminating in the concept of a perfectly spherical earth. This they attempted to measure and divide mathematically. The idea of climatic zones was proposed and possibly mapped, and the large known landmasses were divided into first two continents, then three.

Perhaps the greatest accomplishment of the early Greeks was the remarkably accurate measurement of the circumference of Earth by Eratosthenes (276-196 BCE). Serious study of map projections began at about this time. The gnomonic, orthographic, and stereographic projections were invented before the Christian era, but their use was confined to astronomy in this period. With the possible single exception of Aristarchus of Samos, the Greeks believed in a geocentric universe. They made globes (now lost) and regional maps on metal; a few map coins from this era have survived.

Later Greeks carried on these traditions and expanded upon them. Claudius Ptolemy invented two projections for his world maps in the second century CE. These were enormously important in the European Renaissance as they were modified in the light of new overseas discoveries. Ptolemy's work is known mainly through later translations and reconstructions, but he compiled maps from Greek and Phoenician travel accounts and proposed sectional maps of different scales in his *Geographia*. Ptolemy's prime meridian (0 degrees longitude) in the Canary Islands was generally accepted for a millennium and a half after his death.

Roman cartography was greatly influenced by later Greeks such as Ptolemy, but the Romans themselves improved upon route mapping and surveying. Much of the Roman Empire was subdivided by instruments into hundredths, of which there is a cartographic record in the form of marble tablets. In Rome, a small-scale map of the world known to the Romans was made on metal by Marcus Vipsanius Agrippa, the son-in-law of Augustus Caesar, and displayed publicly. This map no longer exists, however.

Cartography in Early East Asia

As these developments were taking place in the West, a rich cartographic tradition developed in Asia, particularly China. The earliest survey of China (Yu Kung) is approximately contemporaneous with the oldest reported mapmaking activity of the Greeks. Later, maps, charts, and plans accompanied Chinese texts on various geographical themes. Early rulers of China had a high regard for cartography—the science of princes. A rectangular grid was introduced by Chang Heng, a contemporary of Ptolemy, and the south-pointing needle was used for mapmaking in China from an early date.

These traditions culminated in Chinese cartographic primacy in several areas: the earliest printed maps (about 1155 CE), early printed atlases, and terrestrial globes (now lost). Chinese cartography greatly influenced that in other parts of Asia, particularly Korea and Japan, which fostered innovations of their own. It was only after the introduction of ideas from the West, in the Renaissance and later, that Asian cartographic advances were superseded.

Islamic Cartography

A link between China and the West was provided by the Arabs, particularly after the establishment of Islam. It was probably the Arabs who brought the magnetized needle to the Mediterranean, where it was developed into the magnetic compass.

Some scholars have argued that the Arabs were better astronomers than cartographers, but the Arabs did make several clear advances in mapmaking. Both fields of study were important in Muslim science, and the astrolabe, invented by the Greeks in antiquity but developed by the Arabs, was used in both their astronomical and terrestrial surveys. They made and used many maps, as indicated by the output of their most famous cartographer, al-Idrisi (who lived about 1100–1165). Some of his work still exists, including a zonal world map and detailed charts of the Mediterranean islands.

At about the same time, the magnetic compass was invented in the coastal cities of Italy, which gave rise to advanced navigational charts, including information on ports. These remarkably accurate charts were used for navigating in the Mediterranean Sea. They were superior to the European maps of the Middle Ages, which often were concerned with religious iconography, pilgrimage, and crusade. The scene was now set for the great overseas discoveries of the Europeans, which were initiated in Portugal and Spain in the fifteenth century.

In the next four centuries, most of the coasts of the world were visited and mapped. The early, projectionless navigational charts were no longer adequate, so new projections were invented to map the enlarged world as revealed by the European overseas explorations. The culmination of this activity was the development of the projection, in 1569, of Gerardus Mercator, which bears his name and is of special value in navigation.

Early Modern Mapmaking

Europeans began mapping their own countries with greater accuracy. New surveying instruments were invented for this purpose, and a great land-mapping activity was undertaken to match the worldwide coastal surveys. For about a century, the Low Countries of Belgium, Luxembourg, and the Netherlands dominated the map and chart trades, producing beautiful hand-colored engraved sheet wall maps and atlases.

France and England established new national observatories, and by the middle of the seventeenth century, the Low Countries had been eclipsed by France in surveying and making maps and charts. The French adopted the method of triangulation of Mercator's teacher, Gemma Frisius. Under four generations of the Cassini family, a topographic survey of France more comprehensive than any previous survey was completed. Rigorous coastal surveys were undertaken, as well as the precise measurement of latitude (parallels).

The invention of the marine chronometer by John Harrison made it possible for ships at sea to determine longitude. This led to the production of charts of all the oceans, with England's Greenwich eventually being adopted as the international prime meridian.

Quantitative, thematic mapping was advanced by astronomer Edmond Halley (1656–1742) who produced a map of the trade winds; the first published magnetic variation chart, using isolines; tidal charts; and the earliest map of an eclipse. The Venetian Vincenzo Coronelli made globes of greater beauty and accuracy than any previous ones. In the German lands, the study of map projections was vigorously pursued. Johann H. Lambert and others invented a number of equal-area projections that were still in use in the twentieth century.

Ideas developed in Europe were transmitted to colonial areas, and to countries such as China and Russia, where they were grafted onto existing cartographic traditions and methods. The oceanographic explorations of the British and the French built on the earlier charting of the Pacific Ocean and its islands by native navigators and the Iberians.

Nineteenth Century Cartography

Cartography was greatly diversified and developed in the nineteenth century. Quantitative, thematic mapping was expanded to include the social as well as the physical sciences. Alexander von Humboldt used isolines to show mean air temperature, a method that later was applied to other phenomena. Contour lines gradually replaced less quantitative methods of representing terrain on topographic maps. Such maps were made of many areas, for example India, which previously had been poorly mapped.

Extraterrestrial (especially lunar) mapping, had begun seriously in the preceding two centuries with the invention of the telescope. It was expanded in the nineteenth century. In the same period, regular national censuses provided a large body of data that could be mapped. Ingenious methods were created to express the distribution of population, diseases, social problems, and other data quantitatively, using uniform symbols.

Geological mapping began in the nineteenth century with the work of William Smith in England, but soon was adopted worldwide and systematized, notably in the United States. The same is true of transportation maps, as the steamship and the railway increased mobility for many people. Faster land travel in an east-west direction, as in the United States, led to the official adoption of Greenwich as the international prime meridian at a conference held in Washington, D.C., in 1884. Time zone maps were soon published and became a feature of the many world atlases then being published for use in schools, offices, and homes.

A remarkable development in cartography in the nineteenth century was the surveying of areas newly occupied by Europeans. This occurred in such places as the South American republics, Australia, and Canada, but was most evident in the United States. The U.S. Public Land Survey covered all areas not previously subdivided for settlement. Property maps arising from surveys were widely available, and in many cases, the information was contained in county and township atlases and maps.

Modern Mapping and Imaging

Cartography was revolutionized in the twentieth century by aerial photography, sonic sounding, satellite imaging, and the computer. Before those developments, however, Albrecht Penck proposed an ambitious undertaking—an International Map of the World (IMW). Cartography historically had been a nationalistic enterprise, but Penck suggested a map of the world in multiple sheets produced cooperatively by all nations at the scale of 1:1,000,000 with uniform symbols. This was started in the first half of the twentieth century but was not completed, and was superseded by the World Aeronautical Chart (WAC) project, at the same scale, during and after World War II.

The WAC project owed its existence to flight information made available following the invention of the airplane. Both photography and balloons were developed before the twentieth century, but the new, heavier-than-air craft permitted overlapping aerial photographs to be taken, which greatly facilitated the mapping process. Aerial photography revolutionized land surveys—maps could be made at less cost, in less time, and with greater accuracy than by previous methods. Similarly, marine surveying was revolutionized by the advent of sonic sounding in the second half of the twentieth century. This enabled mapping of the floor of the oceans, essentially unknown before this time.

Satellite imaging, especially continuous surveillance by Landsat since 1972, allows temporal monitoring of Earth. The computer, through Geographical Information Systems (GIS) and other technologies, has greatly simplified and speeded up the mapping process. During the twentieth century, the most widely available cartographic product was the road map for travel by automobile.

Spatial information is typically accessed through apps on computers and mobile devices; traditional maps are becoming less common. The new media also facilitate animated presentations of geographical and extraterrestrial distributions. Cartographers remain responsive to the opportunities provided by new technologies, materials, and ideas.

Norman J. W. Thrower

Mapmaking and New Technologies

The field of geography is concerned primarily with the study of the curved surface of Earth. Earth is huge, however, with an equatorial radius of 3,963 miles (6,378 km.). How can one examine anything more than the small patch of earth that can be experienced at one time? Geographers do what scientists do all of the time: create models. The most common model of Earth is a globe—a spherical map that is usually about the size of a basketball.

A globe can show physical features such as rivers, oceans, the continents, and even the ocean floor. Political globes show the division of Earth into countries and states. Globes can even present views of the distant past of Earth, when the continents and oceans were very different than they are today. Globes are excellent for learning about the distributions, shapes, sizes, and relationships of features of Earth. However, there are limits to the use of globes.

How can the distribution of people over the entire world be described at one glance? On a globe, the human eye can see only half of Earth at one time. What if a city planner needs to map every street, building, fire hydrant, and streetlight in a town? To fit this much detail on a globe, the globe might have to be bigger than the town being mapped. Globes like these would be impossible to create and to carry around. Instead of having to hire a fleet of flatbed trucks to haul oversized globes, the curved surface of the globe can be transformed to a flat plane.

The method used to change from a curved globe surface to a flat map surface is called a map projection. There are hundreds of projections, from simple to extremely complex and dating from about two thousand years ago to projections being invented today. One of the oldest is the gnomonic projection. Imagine a clear globe with a light inside. Now imagine holding a piece of paper against the surface of the globe. The coastlines and parallels of latitude and meridians of longitude would show through the globe and be visible on the paper. Computers can do the same thing because there are mathematical formulas for nearly all map projections.

Geometric Models for Map Projections

One way to organize map projections is to imagine what kind of geometric shape might be used to create a map. Like the paper (a plane surface) against the globe described above, other useful geometric shapes include a cone and a cylinder. When the rounded surface of any object, including Earth, is flattened there must be some stretching, or tearing. Map projections help to control the amount and kinds of distortion in maps. There are always a few exceptions that cannot be described in this way, but using geometric shapes helps to classify projections into groups and to organize the hundreds of projections.

Another way to describe a map projection is to consider what it might be good for. Some map projections show all of the continents and oceans at their proper sizes relative to one another. Another type of projection can show correct distances between certain points.

Map Projection Properties

When areas are retained in the proper size relationships to one another, the map is called an equal-area map, and the map projection is called an equal-area projection. Equal-area (also called equivalent or homolographic) maps are used to measure areas or view densities such as a population density.

If true angles are retained, the shapes of islands, continents, and oceans look more correct. Maps made in this way are called conformal maps or conformal map projections. They are used for navigation, topographic mapping, or in other cases when it is important to view features with a good representation of shape. It is impossible for a map to be both equal-area and conformal at the same time. One or the other must be selected based on the needs of the map user or mapmaker.

One special property—distance—can only be true on a few parts of a map at one time. To see how far it is between places hundreds or thousands of miles apart, an equidistant projection should be used. There will be several lines along which distance is true. The azimuthal equidistant projection shows true distances from the center of the map outward. Some map projections do not retain any of these properties but are useful for showing compromise views of the world.

Modern Mapmaking

Modern mapmaking is assisted from beginning to end by digital technologies. In the past, the paper map was both the primary means for communicating information about the world and the database used to store information. Today, the database is a digital database stored in computers, and cartographic visualizations have taken the place of the paper map. Visualizations may still take the form of paper maps, but they also can appear as flashes on computer screens, animations on local television news programs, and even on screens within vehicles to help drivers navigate. Communication of information is one of the primary purposes of making maps. Mapping helps people to explore and analyze the world.

Making maps has become much easier and the capability available to many people. Desktop mapping software and Internet mapping sites can make anyone with a computer an instant cartographer. The maps, or cartographic visualizations, might be quite basic but they are easy to make. The procedures that trained cartographers use to make map products vary in the choice of data, software, and hardware, but several basic design steps should always take place.

First, the purpose and audience for whom the map is being made must be clear. Is this to be a general reference map or a thematic map? What image should be created in the mind of the map reader? Who will use the map? Will it be used to teach young children the shapes of the continents and oceans, or to show scientists the results of advanced research? What form will the cartographic visualization take? Will it be a paper map, a graphic file posted to the Internet, or a video?

The answers to these questions will guide the cartographer in the design process. The design process can be broken down into stages. In the first stage of map design, imagination rules. What map type, size and shape, basic layout, and data will be used? The second stage is more practical and consists of making a specific plan. Based on the decisions made in the first stage, the symbols, line weights, colors, and text for the map are chosen. By the end of this stage, there should be a fairly clear plan for the map. During the third stage, details and specifications are finalized to account for the production method to be used. The actual software, hardware, and methods to be used must all be taken into consideration.

What makes a good map? Working in a digital environment, a mapmaker can change and test vari-

Sliding Rocks Get Digital Treatment

Dr. Paula Messina studied the trails of rocks that slide across the surface of a flat playa in Death Valley, California. The sliding rocks have been studied in the past, but no one had been able to say for certain how or when the rocks moved. It was unclear whether the rocks were caught in ice floes during the winter, were blown by strong winds coming through the nearby mountains, or were moved by some other method.

Messina gave the mystery a totally digital treatment. She mapped the locations of the rocks and the rock trails using the global positioning system (GPS) and entered her rock trail data into a geographic information system (GIS) for analysis. She was able to determine that ice was not the moving agent by studying the pattern of the trails. She also used digital elevation models (DEM) and remotely sensed imagery to model the environment of the playa. She reported her results in the form of maps using GIS' cartographic output capabilities. While she did not solve completely the mystery of the sliding rocks, she was able to disprove that winter ice caused the rocks to slide along together in rafts and that there are wind gusts strong enough to move the biggest rock on the playa.

ous designs easily. The map is a good one when it communicates the intended information, is pleasing to look at, and encourages map readers to ask thoughtful questions.

New Technologies

Mapping technology has gone from manual to magnetic, then to mechanical, optical, photochemical, and electronic methods. All of these methods have overlapped one another and each may still be used in some map-making processes. There have been recent advances in magnetic, optical, and most of all, electronic technologies.

All components of mapping systems—data collection, hardware, software, data storage, analysis, and graphical output tools—have been changing rapidly. Collecting location data, like mapping in general, has been more accessible to more people. The development of the Global Positioning System (GPS), an array of satellites orbiting Earth, gives anyone with a GPS receiver access to location information, day or night, anywhere in the world. GPS receivers are also found in planes, passenger cars, and even in the backpacks of hikers.

Satellites also have helped people to collect data about the world from space. Orbiting satellites collect images using visible light, infrared energy, and other parts of the electromagnetic spectrum. Active sensing systems send out radar signals and create images based on the return of the signal. The entire world can be seen easily with weather satellites, and other specialized satellite imagery can be used to count the trees in a yard.

These great resources of data are all stored and maintained as binary, computer-readable information. Developments in laser technology provide large amounts of storage space on media such as optical disks and compact disks. Advances in magnetic technology also provide massive storage capability in the form of tape storage, hard drives, and cloud storage. This is especially important for saving the large databases used for mapping.

Computer hardware and software continue to become more powerful and less expensive. Software continues to be developed to serve the specialized needs that mapping requires. Just as word processing software can format a paper, check spelling and grammar, draw pictures and shapes, import tables and graphics, and perform dozens of other functions, specialized software executes maps. The most common software used for mapping is called Geographic Information System (GIS) software. These systems provide tools for data input and for analysis and modeling of real-world spatial data, and provide cartographic tools for designing and producing maps.

Karen A. Mulcahy

AUSTRALIA

The only continent occupied by a single nation, Australia is roughly the size of the forty-eight states of the continental United States. Located entirely in the Southern Hemisphere, it covers about 3 million square miles (7.8 million sq. km.). In the northwest, it is about 250 miles (400 km.) from Timor across the Timor Sea, and 500 miles (800 km.) from the tip of Indonesia across the Indian Ocean. Papua New Guinea is across the Coral Sea, 315 miles (500 hundred km.) north of Queensland's Cape York Peninsula. Due south of the Australian state of Victoria, 380 miles (610 km.) across the Bass Strait, lies Tasmania, the only Australian state not on the mainland. About the size of Scotland, Tasmania, even in its interior, is green, whereas the Outback, as the interior of mainland Australia is called, is a vast desert.

In 2020, Australia had nearly 25.5 million inhabitants. Except for Antarctica, it is Earth's least-populated and most-isolated continent. Along the coast of Queensland in the east runs the Great Barrier Reef, the largest such reef in the world. It extends more than 1,250 miles (2,000 km.) from south to north and, in places, is more than 50 miles (80 km.) wide. It has more than 350 species of coral, the

Uluru at sunset. (Riley Boughten)

AUSTRALIA

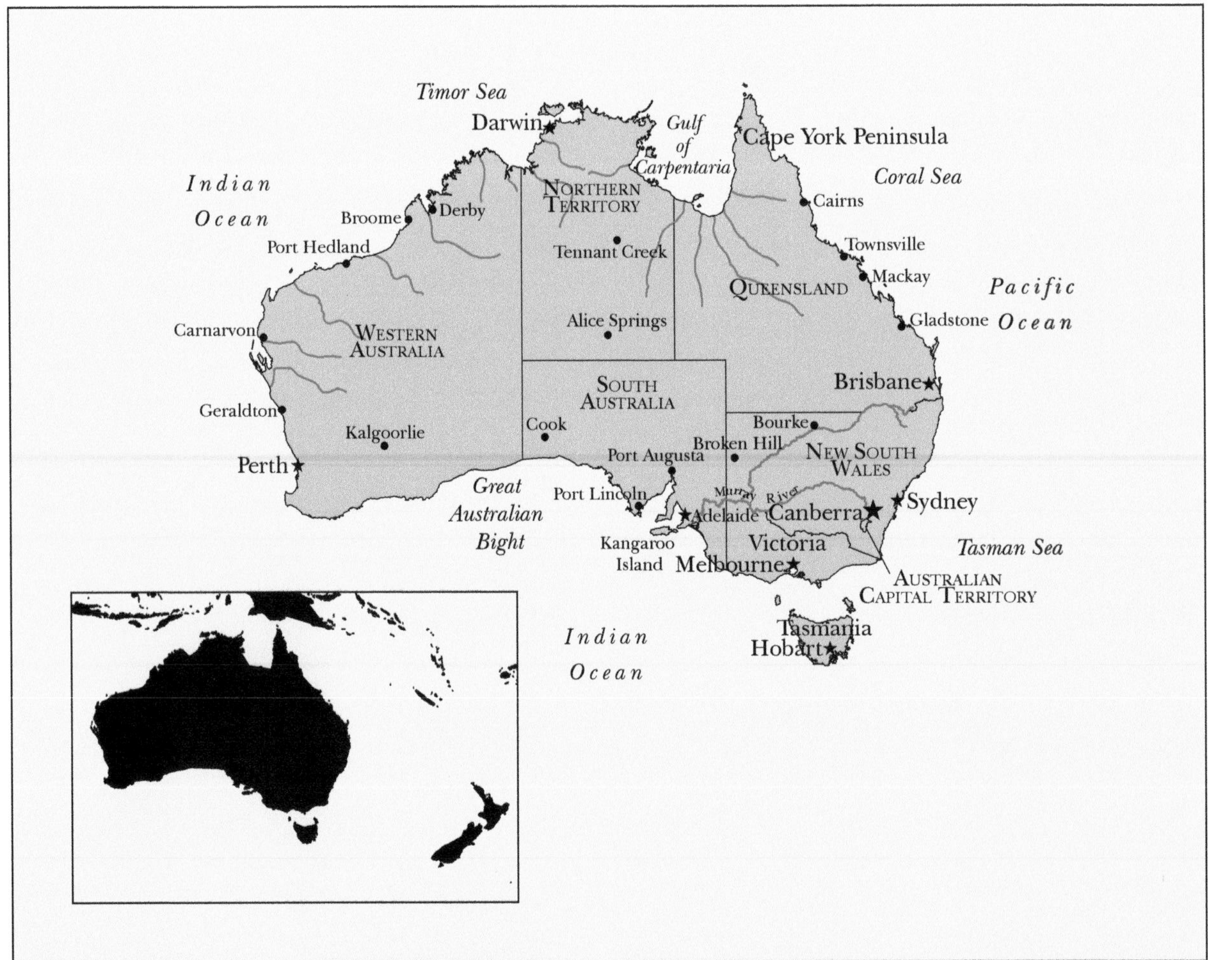

world's greatest variety, and teems with sea life, particularly shellfish.

The Geologic Past

Australia's geologic past is well documented. Long ago, Earth was one huge landmass. Over millions of years, enormous cracks developed, causing parts of it to break off. The largest piece to separate from the whole in this early break-up was Gondwanaland, which became a supercontinent that, with various nearby islands, constitutes modern Australia, geographically the sixth-largest country in the world. Despite its proximity to Asia, Australia resembles Europe more than Asia in its peoples' outlook and lifestyle.

Australia is the oldest, flattest, and driest continent. Because its interior is parched, the Outback supports only a small population. The country's east and southeast coasts are more heavily populated than those in the north and west. The uplands that separate the coastal areas from the Outback are between 20 and 200 miles (32 and 320 km.) wide.

Australia's Six States

Modern Australia consists of five contiguous states—New South Wales, Queensland, Victoria, South Australia, and Western Australia—and the island state of Tasmania. The country has two territories, the sprawling Northern Territory on the Timor and Coral seas, and the relatively small Australian Capital Territory, located within New South Wales. The national capital is Canberra.

Although all five contiguous states except Victoria have considerable land in the Outback, 80 percent of the total population lives near or along the coasts. Even in Tasmania, which has a lush, green interior, the majority of residents are centered in Hobart and other coastal cities. Most Australians live in cities, with the capital city of each state attracting two-thirds of its total population.

Notable Geographic Features

The best known geographic feature in the Outback is Uluru, also known as Ayers Rock. It is located in the Northern Territory, not far from the desert outpost of Alice Springs in Australia's red center. In this area are the MacDonnell Ranges, as well as Kings Canyon, dramatic cliffs and gorges, and the rounded sandstone forms of Kata Tjuta.

Lake Eyre in South Australia, at 49 feet (15 meters) below sea level, is Australia's lowest point and is incredibly dry. It has an area of about 3,668 square miles (9,500 sq. km.), but fills only three or four times each century. It is so flat and dry that automobile races are held in its bed.

Australia's highest point is New South Wales's Mount Kosciuszko in the Australian Alps, which reaches about 7,310 feet (2,229 meters) at its summit. In winter, between May and October, the Australian Alps are generally covered with snow, drawing skiers to them.

Other popular mountains are the Dandenongs of Victoria, less than an hour's drive south of Melbourne. Although the Dandenongs are less dramatic than the Australian Alps, Dandenong Ranges National Park attracts many tourists, particularly during the summer (November to March), when it is much cooler than the surrounding lowlands.

The Nullarbor Plain in the southern part of South Australia and Western Australia is an enormous, arid plain. It is so flat that the Indian-Pacific railroad, which runs from Sydney to Perth, a 64-hour trip, has one stretch of track that is completely straight for more than 200 miles (320 km.)!

Mount Kosciuszko. (Lupo)

Tjapukai in traditional garb. (Bgabel)

The Earliest Australians

When British explorer James Cook first landed on the Australian continent in 1770, it had been inhabited for more than 50,000 years by nomadic people. These indigenous Australians, or Australian aboriginals presumably sailed primitive boats to the continent from the nearby Indonesian islands. They were hunters and gatherers, probably numbering about 300,000 in 1788. They comprised an estimated 650 groups. These earliest inhabitants of Australia spoke more than 250 languages and some 800 dialects. About fifty the languages survived at the start of the twenty-first century, and by 2016, only thirteen traditional indigenous languages were being learned by children.

The indigenous Australians inhabited both the main continent and Tasmania. Cave drawings and other artifacts have been reliably dated to before 30,000 BCE. The early people moved from place to place in search of food. Their dwellings were usually temporary, although some wooden structures existed in the coastal regions and particularly in Tasmania, where the weather is cooler and rainier than on the Australian mainland.

These early people inhabited much of Australia. Those in the coastal regions, where fishing, hunting, and natural vegetation assured an ample food supply, had relatively comfortable lives. Because the inland areas could not support substantial populations, the coastal groups were larger than those inhabiting the arid interior. Conservation and a reverence for the land were common. They used controlled burning of brush to enrich the soil on which grass would later flourish to provide food for the animals upon which these nomads depended for sustenance.

Hydrology

From the earliest days, water has been a major problem for those in the interior. Australia has no natural, permanent lakes. Even such large bodies of water as Lake Eyre and Lake Frome are dry much of the time. Artificial reservoirs, such as Lake Argyle in Western Australia's eastern extreme, have eased water shortages in their areas. Long rivers such as the Darling (about 1,700 miles/2,740 km. long) and the Murray (about 1,600 miles/2,570 km. long), were once dry or reduced to a trickle much of the year, but now maintain a more steady flow through a series of dams and locks.

Western Queensland's semiarid plain extends over half the state. Its Great Artesian Basin, at nearly 676,000 square miles (1.75 million sq. km.), is the largest artesian basin in the world. Rain that falls into it trickles down and collects in subterranean layers through which water is naturally transported to drier regions in the west. Unlike the drier parts of this state, the Cape York Peninsula of northern Queensland contains a tropical rain forest.

Growing as a Nation

Life in Australia was rugged for the early European settlers. A wool industry grew up as shepherds developed herds to graze on the grasslands of New South Wales, Victoria, Queensland, and Tasmania. Nevertheless, by 1840, only about 150,000 Europeans lived in Australia. The population grew rapidly in 1851, when gold was discovered in Victoria's Clune and Ballarat. Prospectors flooded into Victoria. Soon eastern Australia had a population of more than 400,000 Europeans and substantial numbers of Asians, including 40,000 Chinese.

An anti-Asian sentiment soon grew among Australia's European inhabitants. This sentiment was responsible for the White Australian Policy, in effect

The Great Barrier Reef. (nickj)

The ruins of a prison settlement in Port Arthur, Tasmania. (Eli Duke)

for three-quarters of a century before its termination in 1973. This policy not only discouraged Asians from immigrating to Australia but also placed the indigenous Australians in a position of second-class citizenship.

Since 1973, Australia's government has sought to create a hospitable atmosphere for immigrants. Its stated aim is to integrate immigrants into society but not to encourage assimilation. Areas within cities reflect the ethnicity of the people living in them. Every major Australian city has sections to which Italians, Greeks, Vietnamese, Koreans, Chinese, and persons from other national groups gravitate. These enclaves preserve the cultures of their inhabitants' native countries.

Prior to World War II, 95 percent of Australian residents had British roots, and Australia was decidedly British in its outlook. Between 1945 and 1975, more than 3.5 million people immigrated to Australia. By 1990 about three-quarters of Australians were native-born. Of those born elsewhere, about 14 percent came from Europe, mostly from Italy, Greece, Yugoslavia, Germany, Poland, and the Netherlands. Another 5 percent came from Asia, largely from Vietnam, Laos, the Philippines, and Korea, and 5 percent came from North or South America. In 2019, Australia had the ninth-largest immigrant population in the world. The largest sources of immigrants were England, China, India, New Zealand, and the Philippines.

Australia has long needed outsiders to staff its industries, so an influx of foreigners creates few of the problems it might in more populous nations. The country's population has increased by nearly 62 percent since 2008. Approximately 533,500 new immigrants arrived in 2019; most found ready employment. The country's annual growth rate was about 1.4 percent in 2020.

Major Industries

Sheep and cattle stations have long existed throughout Australia. The Spanish merino sheep

A sheep rancher in Australia. (AzmanJaka)

that thrive in Australia provide high-quality wool. Because much of the grazing land gets less than 15 inches (381 millimeters) of rain annually, it often takes up to 50 acres (20 hectares) to raise one sheep. Cattle stations spawn a prosperous beef industry.

Agriculture is also important, with Australia raising some 20 million tons (18 million metric tons) of wheat annually. New South Wales and Queensland produce considerable rice, and Queensland has the most productive sugarcane fields in Australia. Because of its isolation, Australia has flora and fauna found nowhere else in the world.

Mining is a profitable enterprise in Australia. It began in 1851 with the discovery of gold in Victoria. Today, Australia is the world's second-largest producer of gold, and the world's largest producer of iron ore. Oil has been found off Australia's coasts. Queensland has deposits of silver, copper, zinc, and lead. Both the Northern Territory and South Australia have some of the world's largest deposits of uranium, much of it unmined. In 2019, Australia was the source of 95 percent of the world's opals. Valuable opal mines flourish around South Australia's Coober Pedy, a city constructed underground so that its residents can escape the intense desert heat. Australia is the world's leading exporter or coal. Deposits of lignite or brown coal are mined in Victoria. Black coal exists in enormous quantities in New South Wales and Queensland.

Australia occupies an important role in the world community. It was a signatory to the United Nations charter in 1945, and is currently a member of the Commonwealth of Nations.

R. Baird Shuman

A Torres Strait Islander. (u-tern77)

NEW ZEALAND

An isolated part of the world, New Zealand is located in the southwest Pacific Ocean, midway between the equator and the icy Antarctic. Its closest continental neighbor, Australia, lies more than 1,000 miles (1,600 km.) to the northwest. Similar in size to Great Britain or Ecuador, New Zealand consists of two main islands: North Island—44,010 square miles (113,985 sq. km.)—and South Island—58,084 square miles (150,437 sq. km.), which are separated by the Cook Strait. A third, much smaller island, Stewart Island, roughly the same size as the Hawaiian island of Oahu, sits to the south of South Island.

New Zealand is long and narrow: more than 1,000 miles (1,600 km.) in length and 280 miles (450 km.) at its widest part. If positioned along the West Coast of the United States, New Zealand would stretch from Los Angeles, California, to Seattle, Washington. No place in New Zealand is more than 75 miles (120 km.) from the ocean.

Physical Geography

Part of the Pacific Ring of Fire, New Zealand sits on a collision zone between the Indo-Australian and Pacific tectonic plates. The grinding motion of these two plates is responsible for New Zealand's mostly rugged terrain. South Island is dominated by a large mountain chain, the Southern Alps, which runs almost the entire length of the island in a north-south direction. The Alps are home to the largest mountain peak in Australasia, Aoraki/Mount Cook, at 12,218 feet (3,724 meters), as well as 3,000 glaciers.

Haupapa/Tasman Glacier, the largest glacier in New Zealand, flows down the eastern slopes of Aoraki/Mount Cook. Glaciers from previous ice ages carved out the great fjords, called "sounds" in New Zealand, that make up Fiordland National Park, located in the remote southwest corner of the island. Built by erosion, the Canterbury Plains, the country's largest lowland area, stretch along the midsection of the South Island's east coast.

The Auckland Harbour Bridge with the Auckland skyline in the background. (georgeclerk)

New Zealand

Volcanoes and thermal activity are the main features of North Island. At the heart of the island is Lake Taupo, an ancient volcanic crater. Today the country's largest natural lake, Lake Taupo, sits in an active tectonic region called the Volcanic Plateau, which runs in a line from the center of North Island northeast to Whakaari/White Island off the Bay of Plenty. To the south of Lake Taupo lie three major volcanoes: Mount Ruapehu, the highest peak in North Island at 9,177 feet (2,797 meters); Mount Ngauruhoe; and Mount Tongariro.

In 1995 Mount Ruapehu erupted, sending ash, steam, and car-sized rocks into the atmosphere. All three volcanoes are part of Tongariro National Park. The site of geysers, boiling mud pools, and hot springs, Rotorua, situated along the Volcanic Plateau between Lake Taupo and Whakaari/White Island, is the country's prime geothermal hot spot. In December 2019, the volcanic island Whakaari/White Island erupted explosively, killing twenty-one people, most of them tourists.

The Waikato, New Zealand's largest river, flows northward across central North Island. The Clutha River/Mata-Au is New Zealand's second-longest river and the longest on South Island.

Human Geography

The Maori, from Polynesia, were the first humans to occupy Aotearoa—the Maori word for New Zealand, which means "long white cloud." The Maori arrived in New Zealand from eastern Polynesia in several waves between 1320 and 1350 beginning around 700 to 1400. In 1642 the Dutch navigator Abel Tasman was the first European to reach New Zealand. After a brief offshore incident with the Maori, in which some of his men were killed, Tasman sailed off without setting foot on land. European settlement of the country, mainly by the British, occurred after 1769 with the rediscovery of New Zealand by the British naval explorer Captain James Cook. In 1840, New Zealand became a British colony with the signing of the Treaty of Waitangi on the Bay of Islands on North Island.

In 2020, of the 4.9 million people now living in New Zealand, 16.5 percent were Maori, with Europeans, mainly of British descent, making up 64 percent of the entire population. Given the country's rugged terrain, most of the population lives in communities along or near the coast. Thus, the country is highly urbanized, with 86.7 percent (2020) of the population living in cities. More than three-quarters of the New Zealand population lives on North Island. In 2020, nearly one-third of the people (1.6 million) lived in Auckland, the largest city in New Zealand, with 1.1 million people; Hamilton (169,300; 2019), located to the south of Auckland; and Wellington (215,400; 2019), the capital. Christchurch, the third-largest city in New Zealand with 377,200 people, is also the largest city on South Island.

New Zealand has a long history of promoting equal rights and social benefits for all its citizens. In 1893 New Zealand became the first country in the world to grant women the right to vote. On December 8, 1997, Jenny Shipley became the country's first female prime minister. New Zealand has established an extensive social welfare system, including public housing, universal health care, and support benefits—such as retirement pensions—which have become a model for the rest of the world. Efforts in recent years by the government to address grievances by the Maori dating back to early European settlement of the country have had mixed results.

On the world stage, New Zealand has gained international recognition for its antinuclear policy. With widespread support from the public, the fourth Labour government declared New Zealand a nuclear-free zone in 1984. Under New Zealand law, no nuclear-armed vessels or nuclear-powered ships are allowed in its waters. The country has neither nuclear weapons nor nuclear power plants. Its antinuclear policy, along with its natural beauty, low population, and rural-based economy, has earned New Zealand a reputation as a "clean and green" country.

Huia Richard Hutton

Te Ika a Maui: The Fish of Maui

Maori legend tells of how New Zealand's North Island was a great fish hauled up from the ocean floor by the Polynesian god Maui. The name given to North Island by the Maori was Te Ika a Maui—"the Fish of Maui." The canoe from which the fish was caught was South Island, and Steward Island was the anchor of Maui's canoe. The two eyes of the fish were located at the bottom of North Island, with one of its eyes at Wellington Harbor and the other at Lake Wairarapa. The fins were Mount Taranaki on the North Island's west coast and East Cape on its eastern shore. The fish's long tail was the Northland Peninsula in the far north of North Island. A map of the North Island of New Zealand turned sideways actually looks a bit like a great fish.

A scenic view from the Banks Peninsula of New Zealand's South Island. (Myriam Munoz)

The "Beehive" Parliament building in Wellington. (Nick-D.)

PACIFIC ISLANDS

To understand the geographic setting of Oceania, one must first appreciate the immensity of the Pacific Ocean itself. The world's largest ocean, the Pacific covers more than 60 million square miles (155 million sq. km.), about 22 percent of Earth's total surface. Because the continents of Asia and North America are closer to each other than Southeast Asia and South America are, the southern half of the Pacific stands out as a particularly vast body of water. Its widest zone lies along the equatorial belt.

Although the northern zones of the Pacific contain a number of important islands, it is in the southern, and particularly the southwestern, zone of the Pacific that one finds not only the largest but also the most numerous island formations. If one follows the line of the equator or the Tropic of Capricorn eastward from the heavy concentration of islands in the southwest and central Pacific, the occurrence of islands drops markedly. There are only two significant sites—the Galápagos Islands on the equator and Easter Island farther south—on the eastern fringe of Oceania.

Geographers have identified three major subregions of generally smaller to medium-size islands for concentrated study: Melanesia, Micronesia, and Polynesia. These terms mean "black islands," "small islands," and "many islands," respectively.

Melanesia

Melanesia is the area of Oceania lying north and northeast of Australia. It contains most of the island land surface in the southern Pacific Ocean. In part this is because it includes Papua New Guinea, the largest island of Oceania and the second-largest

A beach in American Samoa. (U.S. National Park Service)

Pacific Islands

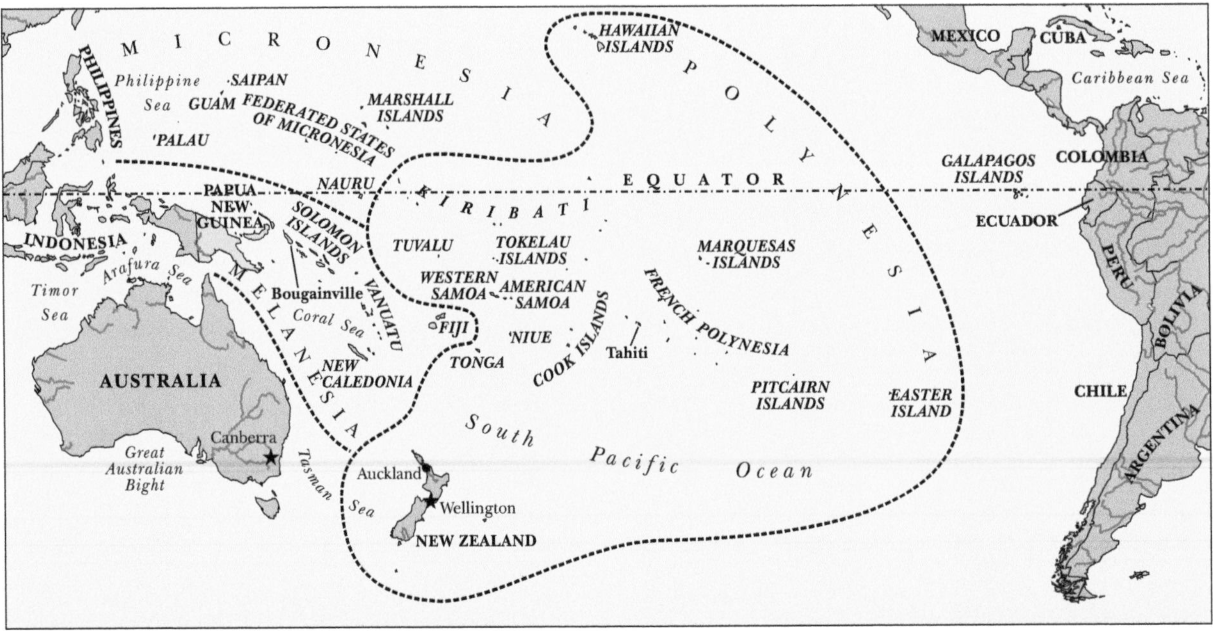

in the world, after Greenland. The two halves of New Guinea taken together cover approximately 305,000 square miles (790,000 sq. km.). Papua New Guinea is the most densely populated island in Melanesia. Its culturally diversified inhabitants are divided into hundreds of different language groups and must rely on a form of New Guinea pidgin dialect to communicate with one another. Although a diversity of cultural groups live in Melanesia, the deep pigmentation of many inhabitants prompted early observers of these islands to coin the term "Melanesia."

The modern nation state of Papua New Guinea is the eastern half of New Guinea island itself. The dominant geographical feature of Papua New Guinea is the long central mountain chain that runs almost the entire length of the country. The populations north of the main mountain ridges are culturally distinct from those living to the south of the mountains, the Papuans. Western New Guinea came under Indonesian control in the 1970s, and was named Irian Jaya.

Several islands adjacent to New Guinea are still known as the Bismarck Archipelago, since their earliest colonizers were German. After World War I, the biggest islands came under British and Australian control, gaining names that were appropriate to their new status: New Britain and New Ireland (Niu Briten and Niu Ailan in the traditional language).

Due west of the Bismarck Archipelago is Bougainville Island, named for the French navigator Louis Antoine de Bougainville who first explored it. Bougainville is the first and largest of the long chain of the Solomon Islands, stretching more than 1,000 miles (1,600 km.) to the southeast. The Solomons are geologically similar to the Bismarck Archipelago. Ten quite large islands and many small islands make up the chain. The largest of the Solomon Islands after Bougainville is Guadalcanal, whose Spanish name is a reminder of the variety of European maritime explorers who explored the South Pacific islands following Magellan's ill-fated but historic voyage westward across the Pacific early in the sixteenth century.

The Fiji islands, located on the westernmost fringe of Melanesia, number more than 300. Only about 100 of the islands are inhabited. Until they came under British colonial administration in the last quarter of the nineteenth century, they were populated by a number of rival chieftaincies. Britain chose Fiji's main city, Suva, on the island of Viti Levu, as the seat of its colonial presence in the South Pacific after World War II. Before then, the British already had begun encouraging Indian im-

The volcano Mount Otemanu in Bora Bora. (Mlenny)

migration to the Fiji Islands as part of plans to expand sugar production under a plantation system.

Most of Fiji's population is concentrated on its two largest islands, Viti Levu and Vanua Levu. Viti Levu is nearly as big as the island of Hawaii, itself a Polynesian island. Fiji's main islands are high islands whose eastern slopes receive significant quantities of rainfall, while the western slopes are considerably drier. Viti Levu's main peak, Mount Tomanivi, is more than 4,344 feet (1,324 meters) high and is not a great distance from beautiful, although limited, zones of untouched mahogany forest.

Micronesia

Micronesia lies within the broad zone due north of Melanesia beyond the equator. As its name suggests, Micronesia consists of a multitude of small islands, some so tiny that they were never, or only very recently, named. Micronesia is bounded on the west by the open seas next to the Philippine Islands and stretches northward to 20 degrees, which is about the same latitudinal location as the Hawaiian Islands. No clear border marks the end of Micronesia to the east and the beginning of the vast zone of Polynesia. Convention has drawn a line through open Pacific Ocean waters between the Republic of Kiribati (formerly the Gilbert Islands) and Tuvalu (formerly the Ellice Islands).

Micronesia contains more than 2,000 islands spread over a vast zone of more than 3 million square miles (7.8 million sq. km.). Most of these are small, low-lying atolls. Even taken together with a few larger islands, Micronesia's total land surface is only 0.30 percent of the Pacific islands' land area.

One of the longer-term results of United Nations' recognition of newly independent states in Southeast Asia was an incentive for the United States to group many of these small islands to form constitutional governments. This ended Washington's post-World War II trusteeship over the single "unit" of Micronesia (earlier controlled by Japan and, previous to that, Germany) and created four distinct constitutional regimes, all with some form of continuing dependency on the United States. These include Palau immediately north of Indonesia's Irian Jaya (western New Guinea), the Federated States of Micronesia in the center of the region, and the republics of the Marshall Islands and Kiribati to the northeast and southeast, respectively.

The island of Guam and the Commonwealth of the Northern Mariana Islands to the northwest (the latter extending almost to the Tropic of Cancer due east of Taiwan) remain more formally tied to the United States. Beyond the Northern Marianas are the Ogasawara, or Bonin Islands, the southernmost part of Japanese territory.

The largest and most heavily populated island in Micronesia is Guam, home to nearly half the people of Micronesia. In 2020, a substantial percentage of Guam's population consisted of military or military support personnel from the United States.

Polynesia

The third geographical zone of Oceania is Polynesia, representing only about 1 percent of the total land area of the Pacific island world. Nevertheless, its dozen or more island groups are heavily populated by comparison with the other areas. It is the largest subzone of Oceania, forming a vast triangle with its southwestern tip well below 40 degrees south latitude, midway through New Zealand and south of Australia. The sides of the Polynesian triangle extend some 5,000 miles (8,000 km.) through the central and southeastern Pacific. The northern tip of Polynesia reaches the Hawaiian Islands (since 1959 a state of the United States), and its eastern point includes Easter Island (part of Chile's territory) near the Tropic of Capricorn.

Although the Hawaiian Islands may be the best known of the Polynesian island groups, several other island complexes also enjoy widespread name recognition. Western Samoa, which is much larger than the half dozen islands constituting the United States territory of American Samoa, gained independence in 1962 and, in 1997, was renamed as Samoa. It was the first independent island nation of Oceania. It consists of two medium-sized and seven small islands.

To the southeast of the Samoas are the seven Cook Islands, named after the British navigator James Cook. They are nominally a self-governing unit, with New Zealand handling most external affairs. The entire region's economy remains tied to New Zealand, and recently, to a lesser extent, with China. The Cook Islands, already rather sparsely populated, have lost many immigrants to the job market in New Zealand.

Southwest of the Samoas is the complex of more than 100 islands known as Tonga. Tonga is a monarchy comprising five island divisions: Tongatapu (Sacred Tonga) with its capital, Nuku'alofa; Ha'apai; Vava'u; Onga Niua; and Eua.

Since 2004, French Polynesia, although officially a French overseas collectivity, is also referred to as an overseas country of France, owing to its considerable level of autocracy. Among its more than 130 islands, Tahiti is the largest, most populated, and best known, although Bora-Bora has the reputation of being among the most beautiful islands of the entire Pacific. Bora-Bora combines the spectacular profile of twin volcanic peaks with a surrounding coral reef structure.

There are a number of lesser-known islands in the Polynesian group, including Wallis and Futuna,

The Hawaiian Islands

The Hawaiian Islands, originally called the Sandwich Islands after the British Earl of Sandwich, include eight major islands and a number of smaller islands, with a total area of nearly 6,500 square miles (17,000 square km.). All the main islands are of volcanic origin and most are surrounded by coral reefs. Hawaii itself, the largest island, is notably younger geologically than the other islands. The capital city of Honolulu is on the island of Oahu. The islands were discovered in 1778 by the English sea captain James Cook. Apparently, growing U.S. concern over British and French colonial interest in the Hawaiian Islands—even after a local independent monarchy was recognized in 1842—led to increasing U.S. involvement. This involvement peaked in 1898 when the United States claimed the Hawaiian Islands as its territory, expanding sugarcane plantations and establishing military bases. In 1941, the Japanese attack on Pearl Harbor thrust the United States into World War II. The islands officially became the state of Hawaii in 1959.

which in 2003 gained the status of French overseas collectivity; British-controlled Pitcairn Islands, whose approximately fifty inhabitants in 2020 are descendants of mutineers from the famous ship HMS *Bounty*; and Niue, a tiny self-governing country that is closely associated with New Zealand.

The most remote of the Polynesian islands is Easter Island, 2,000 miles (3,200 km.) off the coast of Chile. Easter Island became famous after European explorers discovered its mysterious great stone figures, the origin of which is still being debated. A 2017 census counted 7,750 inhabitants.

Physical and Floral Diversity in Oceania

Islands in the different geographic subregions of Oceania show different physical characteristics, depending on how they were formed geologically. One category, referred to as continental islands, applies to only one island of Oceania as defined here, namely Papua New Guinea. Continental islands share not only geological origins with the large landmass next to them (in this case Australia), but also a notable diversity of ecological features. These features may be the result of the broad ranges of altitudes found on such large islands, or differentiations in soil composition produced by extensive erosion.

The case of Papua New Guinea in the Melanesian zone fits this observation quite well. The highlands in the interior of New Guinea contain fertile mountain valleys whose elevations range between 4,000 and 6,000 feet (1,200 to 1,800 meters), with a diversity of plant and animal life on the even higher slopes surrounding them.

The second category of islands in Oceania is referred to as "high islands." These include fairly large islands, most of which owe their geologic origins to volcanic eruptions. Because of their volcanic origins, most high islands have sharply sloping mountainous features. Some high islands are a single volcanic peak, like the visibly spectacular island of Mo'oréa in French Polynesia.

Other islands, like Bougainville in Melanesia (where most of the high islands of Oceania are found), have a series of volcanic peaks, or, in the case of New Caledonia, an extensive range of mountains forming the backbone of an elongated island formation. Although the elevation of Bougainville's peaks is impressive (reaching 10,000 feet/3,048 meters), the highest volcanic peaks in Oceania, some reaching 13,000 feet (3,962 meters), are found in the distant Hawaiian Islands at the northernmost fringe of Polynesia.

The vast majority of islands throughout Oceania are coral atolls. Coral atolls are composed of the long-term accumulation of skeletons of small marine animals (specifically, polyps) that become cemented together by the partial breakdown of their calcareous (calcium-rich) content. Atolls usually are found in relatively shallow waters close to an already existing island. The neighbor island may owe its origins to totally different geologic causes. Its function is to provide an appropriate marine environment for polyps to breed and multiply. Nearly all Pacific islands have some form of coral reef nearby offshore. All reefs do not necessarily emerge above the surface to become atolls. It sometimes happens, however, that coral atolls survive as circularly shaped reefs rising above the surface long after the original island at their core has sunk beneath the waters.

Coral formations also can become islands as a result of tectonic uplifting, or upward pressures coming from Earth's tectonic plates under the ocean floor. Uplifted coral islands are less common in Oceania than the typical flat, low-lying atoll formations. Although the mass of coral material may have been essentially a flat subsurface platform, the ef-

The Mystery of Easter Island

There are many theories concerning the origin of Easter Island's numerous carved volcanic, or tufa stone, heads and the inscriptions on them. Some have argued, unsuccessfully, that the petroglyphs found on Easter Island were derived from ancient Egyptian or Hindu writing systems, or hieroglyphs. The stone heads range in height from 10 to 40 feet (3 to 12 meters) and weigh up to fifty tons. Most scholars agree that the stones were carved by people of Polynesian stock that arrived on the island from the central Pacific. When Europeans first arrived in 1722, the sparse population of the island was visibly of Polynesian origin.

fects of tectonic uplifting can produce varied results. There may be tilting slopes (not actual mountain chains) and fissures or faults, much like the effects of uplifting of sections of the larger landmass on the continents. In such cases, medium-sized coral islands have varied topographical features, ranging from low-lying coastal fringes to fairly high mountainous elevations with valleys caused by erosion. Two of the most prominent examples of uplifted coral islands appear in the Polynesian region of Oceania: Tongatapu, the main island of Tonga, and Makatéa in the Tuamotu group of French Polynesia.

Because all three regions of Oceania are subject to high levels of precipitation and sometimes even major typhoons, one might expect plant life to be dense and varied. Both density and variation depend, however, on a host of factors that are unique to the Pacific island world.

First, the heaviest rainfall accumulates mainly on the high islands that are characteristic of the Melanesian area. These include New Guinea, New Britain, Bougainville, and Guadalcanal, where mountain slopes are covered by extensive rain forests. Rain forest conditions are otherwise rather rare, occurring on the windward slopes of Fiji and on a few islands in American Samoa. Another form of island forest in areas of high humidity, especially on the larger islands like New Guinea, occurs because of near-perpetual mist. Here the trees of the so-called moss, or cloud, forest never grow very high, and most of the misty humidity is absorbed by thick carpets of moss on the forest floor.

Savanna-like (drier and less dense) tree and scrub vegetation is the prevailing characteristic of most islands in Oceania. The nature of such vegetation depends on relative elevation and soil conditions. Larger islands, such as New Guinea, can have fairly extensive grassland zones, especially at middle elevations (4,000 to 6,000 feet/1,200 to 1,800 meters) where rainfall is seasonal and moderate by comparison to higher forested areas.

The spread of different plant species also depends on degrees of geographical isolation from other islands. Wide expanses of ocean water separate the

A resort in Tahiti. (zelg)

Coral reefs in Papua New Guinea. (Brocken Inaglory)

different island groupings, and this has served to block transmission of plant species from one region to another. In some places, the seeds of certain species may have been carried considerable distances across the water by birds. The spread of the coconut palm to almost all parts of Oceania is due in large part to the fact that coconuts have floated, sometimes for years, before washing up on distant shores where soil conditions were conducive to their growth. A similar pattern of dispersion by means of floating seeds has spread another typical Pacific island tree, the pandanus (sometimes incorrectly called pandanus palm) throughout Oceania. Several pandanus species are known as "screw pines."

When vast distances or patterns of oceanic currents and winds have prevented the spread of plants across the Pacific, the opposite phenomenon may occur—concentrations of endemic plant species that are found only in certain areas. This is the case on the Hawaiian Islands and on New Caledonia in Melanesia, where up to 70 percent of plant species exist in isolation only in those subregions of Oceania. While volcanic soils in particular possess high levels of fertility, other soils typical of Oceania do not contain adequate nutrients to support a diversity of plants. This is particularly true of the smaller coral island formations, due not only to their low elevations above sea level (and resultant low rainfall "catch"), but also to the porous nature of the subsoils underlying thin layers of topsoil. Many such islands are completely barren, or support only a few species of grasses and bushes.

Byron D. Cannon

Discussion Questions: Regions

Q1. How is Australia unique compared to the other continents? What role has its relatively remote location played in its development? Into what geopolitical divisions is modern Australia comprised?

Q2. Where is New Zealand located? With which nearby landmasses do the people of New Zealand most closely identify? How does New Zealand differ from Australia?

Q3. What criteria do geographers use to group the Pacific islands? Into what three divisions are they usually classified? What is meant by the term "Oceania"?

Physical Geography

Overview

Climate and Human Settlement

"Everyone talks about the weather," goes an old saying, "but nobody does anything about it." If everyone talks about the weather, it is because it is important to them—to how they feel and to how their bodies and minds function. There is plenty they can do about it, from going to a different location to creating an artificial indoor environment.

Climate

The term "climate" refers to average weather conditions over a long period of time and to the variations around that average from day to day or month to month. Temperature, air pressure, humidity, wind conditions, sunshine, and rainfall—all are important elements of climate and differ systematically with location. Temperatures tend to be higher near the equator and are so low in the polar regions that very few people live there. In any given region, temperatures are lower at higher altitudes. Areas close to large bodies of water have more stable temperatures. Rainfall depends on topography: The Pacific Coast of the United States receives a great deal of rain, but the nearby mountains prevent it from moving very far inland. Seasonal variations in temperature are larger in temperate zones.

Throughout human history, climate has affected where and how people live. People in technologically primitive cultures, lacking much protective clothing or housing, needed to live in mild climates, in environments favorable to hunting and gathering. As agricultural cultivation developed, populations located where soil fertility, topography, and climate were favorable to growing crops and raising livestock. Areas in the Middle East and near the Mediterranean Sea flourished before 1000 BCE. Many equatorial areas were too hot and humid for human and animal health and comfort, and too infested with insect pests and diseases.

Improvements in technology allowed settlement to range more widely north and south. Sturdy houses and stables, internal heating, and warm clothing enabled people to survive and be active in long cold winters. Some peoples developed nomadic patterns, moving with herds of animals to adapt to seasonal variations.

A major challenge in the evolution of settled agriculture was to adapt production to climate and soil conditions. In North America, such crops as cotton, tobacco, rice, and sugarcane have relatively restricted areas of cultivation. Wheat, corn, and soybeans are more widely grown, but usually further north. Winter wheat is an ingenious adaptation to climate. It is sown and germinates in autumn, then matures and is harvested the following spring. Rice, which generally grows in standing water, requires special environmental conditions.

Tropical Problems

Some scholars argue that tropical climates encourage life to flourish but do not promote quality of life. In hot climates, people do not need much caloric intake to maintain body heat. Clothing and housing do not need to protect people from the cold. Where temperatures never fall below freezing, crops can be grown all year round. Large numbers of people can survive even where productivity is not high. However, hot, humid conditions are not favorable to human exertion nor (it is claimed) to mental, spiritual, and artistic creativity. Some tropical areas, such as South India, Bangladesh, Indonesia, and Central Africa, have developed large populations living at relatively low levels of income.

Slavery

Efforts to develop tropical regions played an important part in the rise of the slave trade after 1500 CE. Black Africans were kidnapped and forceably transported to work in hot, humid regions. The West Indian islands became an important location for slave labor, particularly in sugar production. On the North American continent, slave labor was important for producing rice, indigo, and tobacco in colonial times. All these were eclipsed by the enormous growth of cotton production in the early years of U.S. independence. It has been estimated that the forced migration of Africans to the Americas involved about 1,800 Africans per year from 1450 to 1600, 13,400 per year in the seventeenth century, and 55,000 per year from 1701 to 1810. Estimates vary wildly, but at least 12 million Africans were forced to migrate in this process.

European Migration

Migration of European peoples also accelerated after the discovery of the New World. They settled mainly in temperate-zone regions, particularly North America. Although Great Britain gained colonial dominion over India, the Netherlands over present-day Indonesia, and Belgium over a vast part of central Africa, few Europeans went to those places to live. However, many Chinese migrated throughout the Nanyang (South Sea) region, becoming commercial leaders in present-day Malaysia, Thailand, Indonesia, and the Philippines, despite the heat and humidity. British emigrants settled in Australia and New Zealand, Spanish and Italians in Argentina, Dutch (Boers) in South Africa—all temperate regions.

Climate and Economics

Most of the economic progress of the world between 1492 and 2000 occurred in the temperate zones, primarily in Europe and North America. Climatic conditions favored agricultural productivity. Some scholars believe that these areas had climatic conditions that were stimulating to intellectual and technological development. They argue that people are invigorated by seasonal variation in temperature, sunshine, rain, and snow. Storms—particularly thunderstorms—can be especially stimulating, as many parents of young children have observed for themselves.

Climate has contributed to the great economic productivity of the United States. This productivity has attracted a flow of immigrants, which averaged about 1 million a year from 1905 to 1914. Immigration approached that level again in the 1990s, as large numbers of Mexicans crossed the southern border of the United States, often coming for jobs as agricultural laborers in the hot conditions of the Southwest—a climate that made such work unattractive to many others.

Unpredictable climate variability was important in the peopling of North America. During the 1870s and 1880s, unusually favorable weather encouraged a large flow of migration into the grain-producing areas just west of the one-hundredth me-

IRELAND'S POTATO FAMINE AND EUROPEAN EMIGRATION

Mass migration from Europe to North America began in the 1840s after a serious blight destroyed a large part of the potato crop in Ireland and other parts of Northern Europe. The weather played a part in the famine; during the autumns of 1845 and 1846 climatic conditions were ideal for spreading the potato blight. The major cause, however, was the blight itself, and the impact was severe on low-income farmers for whom the potato was the major food.

The famine and related political disturbances led to mass emigration from Ireland and from Germany. By 1850 there were nearly a million Irish and more than half a million Germans in the United States. Combined, these two groups made up more than two-thirds of the foreign-born U.S. population of 1850. The settlement patterns of each group were very different. Most Irish were so poor they had to work for wages in cities or in construction of canals and railroads. Many Germans took up farming in areas similar in climate and soil conditions to their homelands, moving to Wisconsin, Minnesota, and the Dakotas.

ridian. Then came severe drought and much agrarian distress. Between 1880 and 1890, the combined population of Kansas and Nebraska increased by about a million, an increase of 72 percent. During the 1890s, however, their combined population was virtually constant, indicating that a large out-migration was offsetting the natural increase. Much of the area reverted to pasture, as climate and soil conditions could not sustain the grain production that had attracted so many earlier settlers.

Climate variability can be a serious hazard. Freezing temperatures for more than a few hours during spring can seriously damage fruits and vegetables. A few days of heavy rain can produce serious flooding.

Recreation and Retirement

Whenever people have been able to separate decisions about where to live from decisions about where to work, they have gravitated toward pleasant climatic conditions. Vacationers head for Caribbean islands, Hawaii, the Crimea, the Mediterranean Coast, even the Baltic coast. "The mountains" and "the seashore" are attractive the world over. Paradoxically, some of these areas (the Caribbean, for instance) have monotonous weather year-round and thus have not attracted large inflows of permanent residents. Winter sports have created popular resorts such as Vail and Aspen in Colorado, and numerous older counterparts in New England. Large numbers of Americans have retired to the warm climates in Florida, California, and Arizona. These areas then attract working-age adults who earn a living serving vacationers and retirees. Since these locations are uncomfortably hot in summer, their attractiveness for residence had to await the coming of air conditioning in the latter half of the twentieth century.

Human Impact on Climate

Climate interacts with pollution. Bad-smelling factories and refineries have long relied on the wind to disperse atmospheric pollutants. The city of Los Angeles, California, is uniquely vulnerable to atmospheric pollution because of its topography and wind currents. Government regulations of automobile emissions have had to be much more stringent there than in other areas to keep pollution under control.

Human activities have sometimes altered the climate. Development of a large city substitutes buildings and pavements for grass and trees, raising summer temperatures and changing patterns of water evaporation. Atmospheric pollutants have contributed to acid rain, which damages vegetation and pollutes water resources. Many observers have also blamed human activities for a trend toward global warming. Much of this has been blamed on carbon dioxide generated by combustion, particularly of fossil fuels. A widespread and continuing rise in temperatures is expected to raise water levels in the oceans as polar icecaps melt and change the relative attractiveness of many locations.

Paul B. Trescott

Flood Control

Flood control presents one of the most daunting challenges humanity faces. The regions that human communities have generally found most desirable, for both agriculture and industry, have also been the lands at greatest risk of experiencing devastating floods. Early civilization developed along river valleys and in coastal floodplains because those lands contained the most fertile, most easily irrigated soils for agriculture, combined with the convenience of water transportation.

The Nile River in North Africa, the Ganges River on the Indian subcontinent, and the Yangtze River in China all witnessed the emergence of civilizations that relied on those rivers for their growth. People learned quickly that residing in such areas meant living with the regular occurrence of life-threatening floods.

Knowledge that floods would come did not lead immediately to attempts to prevent them. For thousands of years, attempts at flood control were rare. The people living along river valleys and in floodplains often developed elaborate systems of irrigation canals to take advantage of the available water for agriculture and became adept at using rivers for transportation, but they did not try to control the river itself. For millennia, people viewed periodic flooding as inevitable, a force of nature over which they had no control. In Egypt, for example, early people learned how far out over the riverbanks the annual flooding of the Nile River would spread and accommodated their society to the river's seasonal patterns. Villagers built their homes on the edge of the desert, beyond the reach of the flood waters, while the land between the towns and the river became the area where farmers planted crops or grazed livestock.

In other regions of the world, buildings were placed on high foundations or built with two stories on the assumption that the local rivers would regularly overflow their banks. In Southeast Asian countries such as Thailand and Vietnam, it is common to see houses constructed on high wooden posts above the rivers' edge. The inhabitants have learned to allow for the water levels' seasonal changes.

Flood Control Structures

Eventually, societies began to try to control floods rather than merely attempting to survive them. Levees and dikes—earthen embankments constructed to prevent water from flowing into low-lying areas—were built to force river waters to remain within their channels rather than spilling out over a floodplain. Flood channels or canals that fill with water only during times of flooding, diverting water away from populated areas, are also a common component of flood control systems. Areas that are particularly susceptible to flash floods have constructed numerous flood channels to prevent flooding in the city. For example, for much of the year, Southern California's Los Angeles River is a small stream flowing down the middle of an enormous, 20- to 30-foot-deep (6–9 meters) concrete-lined channel, but winter rains can fill its bed from bank to bank. Flood channels prevent the river from washing out neighborhoods and freeways.

Engineers designed dams with reservoirs to prevent annual rains or snowmelt entering the river upstream from running into populated areas. By the end of the twentieth century, extremely complex flood control systems of dams, dikes, levees, and flood channels were common. Patterns of flooding that had existed for thousands of years ended as civil engineers attempted to dominate natural forces.

The annual inundation of the Egyptian delta by the flood waters of the Nile River ceased in 1968 following construction of the 365-foot-high (111 meters) Aswan High Dam. The reservoir behind the 3,280-foot-long (1,000-meter) dam forms a lake almost ten miles (16 km.) wide and almost 300 miles (480 km.) long. Flood waters are now trapped behind the dam and released gradually over a year's time.

Environmental Concerns

Such high dams are increasingly being questioned as a viable solution for flood control. As human understanding of both hydrology and ecology have improved, the disruptive effects of flood control projects such as high dams, levees, and other engineering projects are being examined more closely.

Hydrologists and other scientists who study the behavior of water in rivers and soils have long known that vegetation and soil types in watersheds can have a profound effect on downstream flooding. The removal of forest cover through logging or clearing for agriculture can lead to severe flooding in the future. Often that flooding will occur many miles downstream from the logging activity. Devastating floods in the South Asian country of Bangladesh, for example, have been blamed in part on clear-cutting of forested hillsides in the Himalaya Mountains in India and Nepal. Monsoon rains that once were absorbed or slowed by forests now run quickly off mountainsides, causing rivers to reach unprecedented flood levels. Concerns about cause-and-effect relationships between logging and flood control in the mountains of the United States were one reason for the creation of the U.S. Forest Service in the nineteenth century.

In populated areas, even seemingly trivial events such as the construction of a shopping center parking lot can affect flood runoff. When thousands of square feet of land are paved, all the water from rain runs into storm drains rather than being absorbed slowly into the soil and then filtered through the watertable. Engineers have learned to include catch basins, either hidden underground or openly visible but disguised as landscaping features such as ponds, when planning a large paving project.

Wetlands and Flooding

Less well known than the influence of watersheds on flooding is the impact of wetlands along rivers. Many river systems are bordered by long stretches of marsh and bog. In the past, flood control agencies often allowed farmers to drain these areas for use in agriculture and then built levees and dikes to hold the river within a narrow channel. Scientists now know that these wetlands actually serve as giant sponges in the flood cycle. Flood waters coming down a river would spread out into wetlands and be held there, much like water is trapped in a sponge.

Draining wetlands not only removes these natural flood control areas but worsens flooding problems by allowing floodwater to precede downstream faster. Even if life-threatening or property-damaging floods do not occur, faster-flowing water significantly changes the ecology of the river system. Waterborne silt and debris will be carried farther. Trying to control floods on the Mississippi River has had the unintended consequence of causing waterborne silt to be carried farther out into the Gulf of Mexico by the river, rather than its being deposited in the delta region. This, in turn, has led to the loss of shore land as ocean wave actions washes soil away, but no new alluvial deposits arrive to replace it.

In any river system, some species of aquatic life will disappear and others replace them as the speed of flow of the water affects water temperature and the amount of dissolved oxygen available for fish. Warm-water fish such as bass will be replaced by cold-water fish such as trout, or vice versa. Biologists estimate that more than twenty species of freshwater mussels have vanished from the Tennessee River since construction of a series of flood control and hydroelectric power generation dams have turned a fast-moving river into a series of slow-moving reservoirs.

Future of Flood Control

By the end of the twentieth century, engineers increasingly recognized the limitations of human interventions in flood control. Following devastating floods in the early 1990s in the Mississippi River drainage, the U.S. Army Corps of Engineers recommended that many towns that had stood right at the river's edge be moved to higher ground. That is, rather than trying to prevent a future flood, the Corps advised citizens to recognize that one would inevitably occur, and that they should remove themselves from its path. In the United States and a number of other countries, land that has been zoned as floodplains can no longer be developed for residential use. While there are many things humanity can do to help prevent floods, such as maintaining well-forested watersheds and preserving wetlands, true flood control is probably impossible. Dams, levees, and dikes can slow the water down, but eventually, the water always wins.

Nancy Farm Männikkö

ATMOSPHERIC POLLUTION

Pollution of the Earth's atmosphere comes from many sources. Some forces are natural, such as volcanoes and lightning-caused forest fires, but most sources of pollution are byproducts of industrial society. Atmospheric pollution cannot be confined by national boundaries; pollution generated in one country often spills over into another country, as is the case for acid deposition, or acid rain, generated in the midwestern states of the United States that affects lakes in Canada.

Major Air Pollutants

Each of eight major forms of air pollution has an impact on the atmosphere. Often two or more forms of pollution have a combined impact that exceeds the impact of the two acting separately. These eight forms are:

1. Suspended particulate matter: This is a mixture of solid particles and aerosols suspended in the air. These particles can have a harmful impact on human respiratory functions.

2. Carbon monoxide (CO): An invisible, colorless gas that is highly poisonous to air-breathing animals.

3. Nitrogen oxides: These include several forms of nitrogen-oxygen compounds that are converted to nitric acid in the atmosphere and are a major source of acid deposition.

4. Sulfur oxides, mainly sulfur dioxide: This sulfur-oxygen compound is converted to sulfuric acid in the atmosphere and is another source of acid deposition.

5. Volatile organic compounds: These include such materials as gasoline and organic cleaning solvents, which evaporate and enter the air in a vapor state. VOCs are a major source of ozone formation in the lower atmosphere.

6. Ozone and other petrochemical oxidants: Ground-level ozone is highly toxic to animals and plants. Ozone in the upper atmosphere, however, helps to shield living creatures from ultraviolet radiation.

7. Lead and other heavy metals: Generated by various industrial processes, lead is harmful to human health even at very low concentrations.

8. Air toxics and radon: Examples include cancer-causing agents, radioactive materials, or asbestos. Radon is a radioactive gas produced by natural processes in the earth.

All eight forms of pollution can have adverse effects on human, animal, and plant life. Some, such as lead, can have a very harmful effect over a small range. Others, such as sulfur and nitrogen oxides, can cross national boundaries as they enter the atmosphere and are carried many miles by prevailing wind currents. For example, the radioactive discharge from the explosion of the Chernobyl nuclear plant in the former Soviet Union in 1986 had harmful impacts in many countries. Atmospheric radiation generated by the explosion rapidly spread over much of the Northern Hemisphere, especially the countries of northern Europe.

Impacts of Atmospheric Pollution

Atmospheric pollution not only has a direct impact on the health of humans, animals, and plants but also affects life in more subtle, often long-term, ways. It also affects the economic well-being of people and nations and complicates political life.

Atmospheric pollution can kill quickly, as was the case with the killer smog, brought about by a temperature inversion, that struck London in 1952 and led to more than 4,000 pollution-related deaths. In the late 1990s, the atmosphere of Mexico City was so polluted from automobile exhausts and industrial pollution that sidewalk stands selling pure oxygen to people with breathing problems became thriving businesses. Many of the heavy metals and organic constituents of air pollution can cause cancer when people are exposed to large doses or for long periods of time. Exposure to radioactivity in the atmosphere can also increase the likelihood of cancer.

In some parts of Germany and Scandinavia in the 1990s, as well as places in southern Canada and the southern Appalachians in the United States, certain types of trees began dying. There are several possible reasons for this die-off of forests, but one potential culprit is acid deposition. As noted above, one byproduct of burning fossil fuels (for example, in coal-fired electric power plants) is the sulfur and nitrous oxides emitted from the smokestacks. Once in the atmosphere, these gases can be carried for many miles and produce sulfuric and nitric acids.

These acids combine with rain and snow to produce acidic precipitation. Acid deposition harms crops and forests and can make a lake so acidic that aquatic life cannot exist in it. Forests stressed by contact with acid deposition can become more susceptible to damage by insects and other pathogens. Ozone generated from automobile emissions also kills many plants and causes human respiratory problems in urban areas.

Air pollution also has an impact on the quality of life. Acid pollutants have damaged many monuments and building facades in urban areas in Europe and the United States. By the late 1990s, the distance that people could see in some regions, such as the Appalachians, was reduced drastically because of air pollution.

The economic impact of air pollution may not be as readily apparent as dying trees or someone with a respiratory ailment, but it is just as real. Crop damage reduces agricultural yield and helps to drive up the cost of food. The costs of repairing buildings or monuments damaged by acid rain are substantial. Increased health-care claims resulting from exposure to air pollution are hard to measure but are a cost to society nevertheless.

It is impossible to predict the potential for harm from rapid global warming arising from greenhouse gases and the destruction of the ozone layer by chlorofluorocarbons (CFCs), but it could be cata-

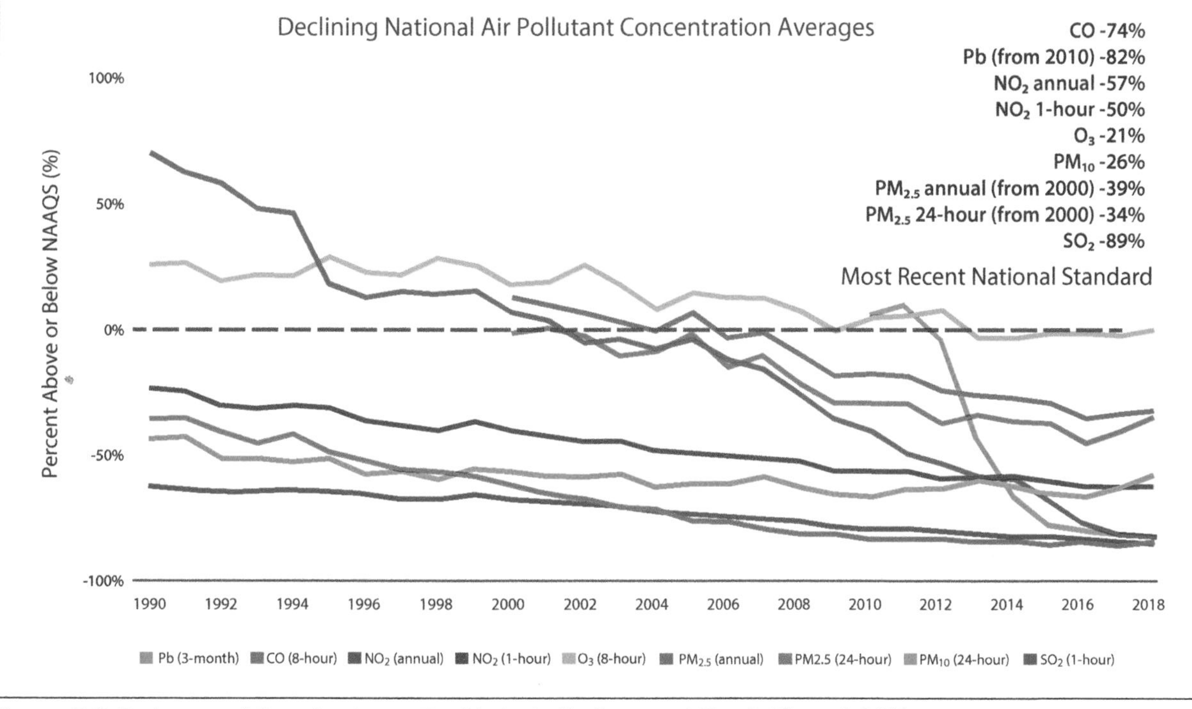

Source: U.S. Environmental Protection Agency, Our Nation's Air, Status and Trends Through 2018.

strophic. Rapid global warming would cause the sea level to rise because of the melting of the polar ice caps. Low-lying coastal areas would be flooded, or, in the case of Bangladesh, much of the country. Global warming would also change crop patterns for much of the world.

Solutions for Atmospheric Pollution

Although there is still some debate, especially among political leaders, most scientists recognize that air pollution is a problem that affects both the industrialized and less-industrialized world. In their rush to industrialize, many nations begin generating substantial amounts of air pollution; China's extensive use of coal-fired power plants is just one example.

The major industrial nations are the primary contributors to atmospheric pollution. North America, Europe, and East Asia produce 60 percent of the world's air pollution and 60 percent of its food supply. Because of their role in supplying food for many other nations, anything that damages their ability to grow crops hurts the rest of the world. In 2018, about 76 million tons of pollution were emitted into the atmosphere in the United States. These emissions mostly contribute to the formation of ozone and particles, the deposition of acids, and visibility impairment.

Many industrialized nations are making efforts to control air pollution, for example, the Clean Air Act of 1970 in the United States or the international Montreal Accord to curtail CFC production. Progress is slow and the costs of reducing air pollution are often high. Worldwide, bad outdoor air caused an estimated 4.2 million premature deaths in 2016, about 90 percent of them in low- and middle-income countries, according to the World Health Organization. Indoor smoke is an ongoing health threat to the 3 billion people who cook and heat their homes by burning biomass, kerosene, and coal.

In the year 2019 the record of the nations of the world in dealing with air pollution was a mixed one. There were some signs of progress, such as reduced automobile emissions and sulfur and nitrous oxides in industrialized nations, but acid deposition remains a problem in some areas. CFC production has been halted, but the impact of CFCs on the ozone layer will continue for many years. However, more nations are becoming aware of the health and economic impact of air pollution and are working to keep the problem from getting worse.

John M. Theilmann

DISEASE AND CLIMATE

Climate influences the spread and persistence of many diseases, such as tuberculosis and influenza, which thrive in cold climates, and malaria and encephalitis, which are limited by the warmth and humidity that sustains the mosquitoes carrying them. Because the earth is warming as a result of the generation of carbon dioxide and other "greenhouse gases" from the burning of fossil fuels, there is intensified scientific concern that warm-weather diseases will reemerge as a major health threat in the near future.

Scientific Findings

The question of whether the earth is warming as a result of human activity was settled in scientific circles in 1995, when the Second Assessment Report of the Intergovernmental Panel on Climate Change, a worldwide group of about 2,500 experts, was issued. The panel concluded that the earth's temperature had increased between 0.5 to 1.1 degrees Fahrenheit (0.3 to 0.6 degrees Celsius) since reliable worldwide records first became available in the late nineteenth century. Furthermore, the intensity of warming had increased over time. By the 1990s, the temperature was rising at the most rapid rate in at least 10,000 years.

The Intergovernmental Panel concluded that human activity—the increased generation of carbon dioxide and other "greenhouse gases"—is responsible for the accelerating rise in global temperatures. The amount of carbon dioxide in the atmosphere has risen nearly every year because of increased use of fossil fuels by ever-larger human populations experiencing higher living standards.

In 1998, Paul Epstein of the Harvard School of Public Health described the spread of malaria and dengue fever to higher altitudes in tropical areas of the earth as a result of warmer temperatures. Rising winter temperatures have allowed disease-bearing insects to survive in areas that could not support them previously. According to Epstein, frequent flooding, which is associated with warmer temperatures, also promotes the growth of fungus and provides excellent breeding grounds for large numbers of mosquitoes. Some experts cite the flooding caused by Hurricane Floyd and other storms in North Carolina during 1999 as an example of how global warming promotes conditions ideal for the spread of diseases imported from the Tropics.

Heat, Humidity, and Disease

During the middle 1990s, an explosion of termites, mosquitoes, and cockroaches hit New Orleans, following an unprecedented five years without frost. At the same time, dengue fever spread from Mexico across the border into Texas for the first time since records have been kept. Dengue fever, like malaria, is carried by a mosquito that is limited by temperature and humidity. Colombia was experiencing plagues of mosquitoes and outbreaks of the diseases they carry, including dengue fever and encephalitis, triggered by a record heat wave followed by heavy rains. In 1997 Italy also had an outbreak of malaria. An outbreak of zika in 2015–16, related to a virus spread by mosquitoes, raised concerns regarding the safety of athletes and spectators at the 2016 Summer Olympics in Rio de Janeiro and led to travel warnings and recommendations to delay getting pregnant for those living or traveling in areas where the mosquitoes are active.

The global temperature is undeniably rising. According to the National Oceanic and Atmospheric Administration, July 2019, was the hottest month since reliable worldwide records have been kept, or about 150 years. The previous record had been set in July 2017.

The rising incidence of some respiratory diseases may be related to a warmer, more humid environment. The American Lung Association reported that more than 5,600 people died of asthma in the United States during 1995, a 45.3 percent increase in mortality over ten years, and a 75 percent increase since 1980. Roughly a third of those cases occurred in children under the age of eighteen. Asthma is now one of the leading diseases among the young. Since 1980, there has been a 160 percent increase in asthma in children under the age of five.

Heat Waves and Health

A study by the Sierra Club found that air pollution, which will be enhanced by global warming, could be responsible for many human health problems, including respiratory diseases such as asthma, bronchitis, and pneumonia.

According to Joel Schwartz, an epidemiologist at Harvard University, air pollution concentrations in the late 1990s were responsible for 70,000 early deaths per year and more than 100,000 excess hospitalizations for heart and lung disease in the United States. Global warming could cause these numbers to increase 10 to 20 percent in the United States, with significantly greater increases in countries that are more polluted to begin with, according to Schwartz.

Studies indicate that global warming will directly kill hundreds of Americans from exposure to extreme heat during summer months. The U.S. Centers for Disease Control and Prevention have found that between 1979 and 2014, the death rate as a direct result of exposure to heat (underlying cause of death) generally hovered around 0.5 to 1 deaths per million people, with spikes in certain years). Overall, a total of more than 9,000 Americans have died from heat-related causes since 1979, according to death certificates. Heat waves can double or triple the overall death rates in large cities. The death toll in the United States from a heat wave during July 1999 surpassed 200 people. As many as 600 people died in Chicago alone during the 1990s due to heat waves. The elderly and very young have been most at risk.

Respiratory illness is only part of the picture. The Sierra Club study indicated that rising heat and humidity would broaden the range of tropical diseases, resulting in increasing illness and death from diseases such as malaria, cholera, and dengue fever, whose range will spread as mosquitoes and other disease vectors migrate.

The effects of El Niño in the 1990s indicate how sensitive diseases can be to changes in climate. A study conducted by Harvard University showed that warming waters in the Pacific Ocean likely contributed to the severe outbreak of cholera that led to thousands of deaths in Latin American countries. Since 1981, the number of cases of dengue fever has risen significantly in South America and has begun to spread into the United States. According to health experts cited by the Sierra Club study, the outbreak of dengue near Texas shows the risks that a warming climate might pose. Epstein and the Sierra Club study concur that if tropical weather expands, tropical diseases will expand.

In many regions of the world, malaria is already resistant to the least expensive, most widely distributed drugs. According to the World Health Organization (WHO), there were 219 million cases of malaria globally in 2017 and 435,000 malaria deaths, representing a decrease in malaria cases and deaths rates of 18 percent and 28 percent since 2010. Of those deaths, 403,000 (approximately 93 percent) were in the WHO African Region.

Bruce E. Johansen

Physiography and Hydrology of Antarctica

Antarctica is the fifth-largest of the earth's seven continents. It is an enormous island with a network of other islands along its 11,165-mile (17,968 km.) coastline and farther out in the three oceans that touch its shores: the Atlantic, Pacific, and Indian oceans. Because about 98 percent of Antarctica is covered with ice and snow that has not melted for thousands of years, most of the adjacent islands look from the air as though they belong to the mainland.

Physical Dimensions

The continent is roughly circular, with indentations for the Ross Sea on the western side and the Weddell Sea 3,000 miles (4,828 km.) to the north. Antarctica has a landmass of about 5.48 million square miles (14.2 million sq. km.)—about the size of the United States, Mexico, and Argentina combined. The Transantarctic Mountains run like a great spine from the Ross Sea to the Weddell Sea, crossing the geographic South Pole at 90 degrees south latitude. They divide Antarctica into East (Greater) Antarctica and West (Lesser) Antarctica.

The Antarctic Peninsula, consisting of Palmer Land and Graham Land, juts from the mainland and extends north from about 85 degrees south latitude to about 60 degrees south latitude and longitude 60

The McMurdo Dry Valleys in Antarctica. (U.S. Department of State)

Physical Geography of Antarctica

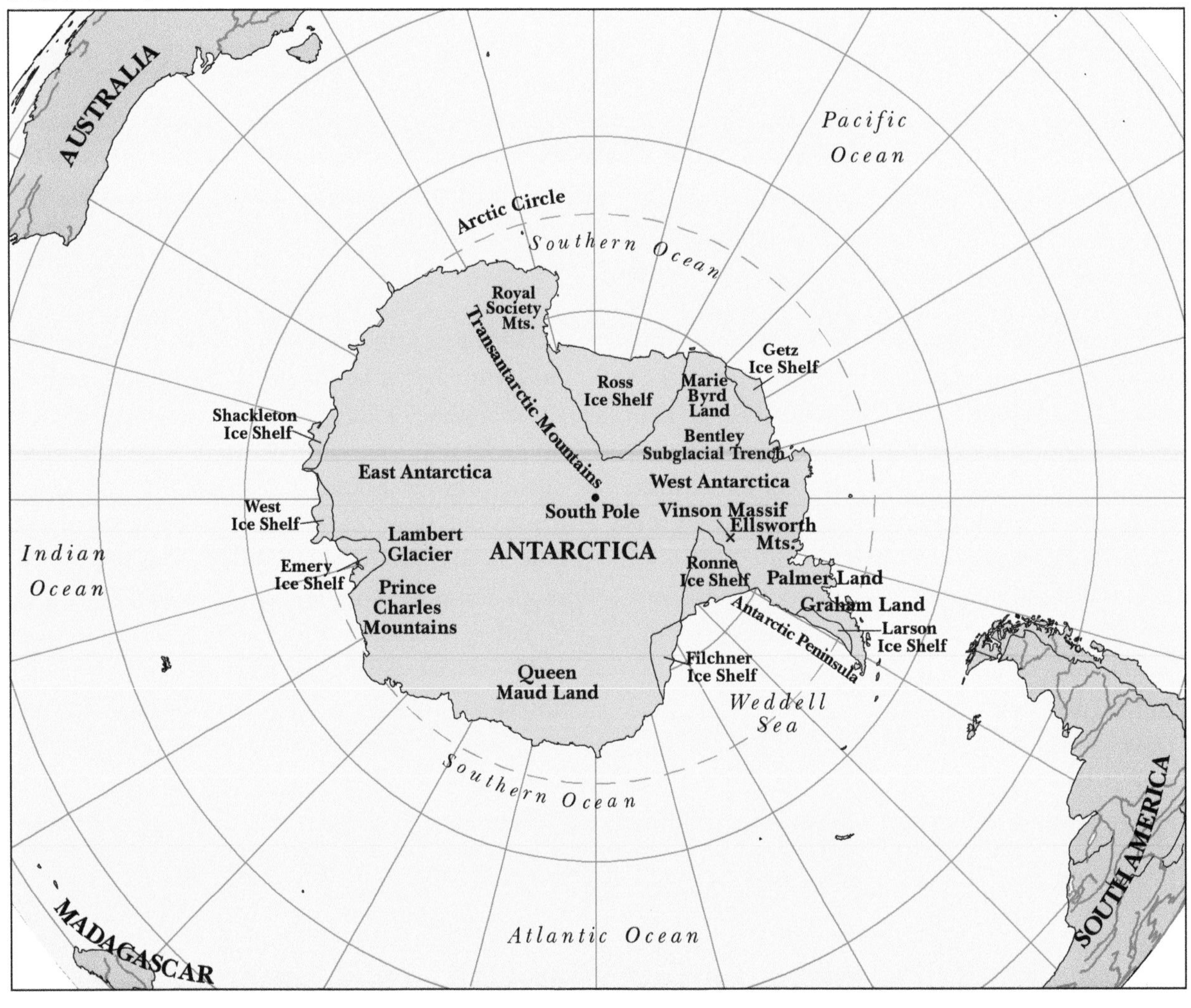

to 80 degrees west. Its mountains are part of the range that includes South America's Andes and the submarine mountains of the South Atlantic Ocean. South Orkney, South Georgia, and Elephant Island are mountain peaks of that submerged range.

Antarctica has huge ice shelves, fed largely by glaciers, off its coasts. The largest of these is the Ross Ice Shelf, about the size of France. It is fed by six glaciers, the largest of which is the Beardmore, which starts at the South Pole and, attaining altitudes of 7,200 feet (2,200 meters), flows into the Ross Sea, depositing huge quantities of ice on the Ross Ice Shelf.

The Lambert Glacier, which runs through the Prince Charles Mountains, feeds the Amery Ice Shelf. Every year it discharges vast amounts of ice into the Amery Ice Shelf.

Antarctica rises to a height of 16,050 feet (4,892 meters) at the Vinson Massif of the Ellsworth Mountains in Marie Byrd Land. Antarctica also has the lowest point on Earth: a canyon that descends some 11,500 feet (3,500 meters) beneath the Denman Glacier, nearly 2 miles (3.2 km.) lower than the Dead Sea in Israel, the lowest point on any continent. Much of Antarctica is below sea level, gradually having been sunk by the constant weight of an ice cover, in many places 2 miles (3.2 km.) thick. Nevertheless, Antarctica's average altitude is 7,546 feet (2,300 meters), excluding its ice cover, whose average thickness of 7,086 feet (2,160 meters) increases the continent's average altitude to 14,633 feet (4,460 meters).

Antarctica has three major dry valleys—Taylor Valley, Wright Valley, and Victoria Valley—sheltered from the harsh climate by the Royal Society Range. They were once filled with ice, but over time, the ice dissipated, leaving the surface exposed. The climate is extremely dry—about 3.9 inches (100 millimeters) of rainfall annually. Each of these dramatic valleys amid the great ice fields of the continent is 3 miles (5 km.) wide and about 25 miles (40 km.) long.

Besides the Ross Ice Shelf, major ice shelves off the Antarctic coast include the Amery, at longitude 70 degrees east and 70 degrees south latitude; the Filchner-Ronne, at longitude 40 degrees west and 79 degrees south latitude; and the Larsen, at longitude 65 degrees west and 65 degrees south latitude. Global warming and climate change were at least partially at play in 2017, when the calving of approximately 10 percent of the Larsen Ice Shelf produced one of the largest icebergs ever recorded, comparable in size to the US state of Delaware. Other large ice shelves are the Shackleton, Voyeykov, Getz, and West.

Geology

The geologic history of Antarctica reveals that more than 170,000 years ago, the continent had a moderate climate that sustained a broad range of plant and animal life. There is evidence that once trees grew in the area, it had swamps, and many varieties of animals roved about in it. The Ice Age that covered the continent with a coating of ice that grew to an average thickness of about 8,000 feet (2,440 meters) changed the entire face of the continent, turning it into the frozen mass that it is now.

East Antarctica south of the Indian Ocean is a single landmass composed largely of igneous rock. It appears to have been formed more than 1 billion years ago during the Precambrian period. A deep

ANTARCTICA'S GEOGRAPHICAL DIVISIONS

Unlike the other continents, Antarctica has no large cities, no towns or villages, and no permanent human residents. The Transarctic Mountains, also known as the Great Antarctic Horst, divide the continent into East (Greater) Antarctica, the continent's larger land area, and West (Lesser) Antarctica, the smaller. The Transarctic Mountains cut through the continent for a distance of more than 3,000 miles (4,800 km.).

Drilling in Antarctica. (Alexyz3d)

trough filled with 2 miles (3.2 km.) of ice runs along the Great Antarctic Horst into West Antarctica, which is a collection of islands, many below sea level because of the weight of the ice that has accumulated on them. They are covered by an ice cap that makes them look like a solid mass from the air. The northern portion of West Antarctica, which resulted from volcanic action, developed in the third period of Antarctica's geologic time.

The Ellsworth Mountains, located in West Antarctica and extending northward to the Antarctic Peninsula, are composed partly of granite and sedimentary rock. The mountains on the Antarctic Peninsula are part of the mountain chain that forms the Andes in South America.

The Antarctic Peninsula, with oceans on its coastline, has the most moderate climate in Antarctica. In summer, it can reach temperatures of 50 degrees Fahrenheit (10 degrees Celsius). The all-time high temperature for Antarctica—69.3 degrees Fahrenheit (20.72 degrees Celsius)—was recorded in February 2020. The lowest air temperature on Earth—-128.6 degrees Fahrenheit (-89.2 degrees Celsius)—was recorded in Antarctica on July 21, 1983. Some rudimentary plant life—types of grasses, pennywort, lichens, and mosses—grow on the Antarctica Peninsula, but the continent is without the plant life that characterizes the other six continents. Trees cannot grow anywhere on Antarctica.

The World's Biggest Iceberg

The largest iceberg ever seen by humans was calved from the Ross Ice Shelf and first spotted in 1956. It was 208 miles (335 km.) long and 60 miles (97 km.) wide, and its total area was six times the size of the state of Delaware.

Volcanic Activity

Major parts of Antarctica were attached at one time to the South American continent. In prehistory, large segments of the landmass detached from the mainland and drifted slowly south to what is now Antarctica. Other parts of the continent, however, appear to be of volcanic origin, particularly in the north reaches on the Antarctic Peninsula immediately below the Antarctic Circle, which lies at 65° 32′ south latitude. There also has been recent volcanic activity near the Ross Sea: Mount Erebus, at an altitude of 12,448 feet (3,794 meters), still spews smoke

An iceberg in Antarctica. (PaoMic)

and flame into the air to a height that Sir James Clark Ross estimated in 1841 to reach 2,000 feet (600 meters) above its cone.

It is indisputable that Mount Erebus underwent a major eruption around 1500. Lava and volcanic ash from such an eruption have been found by scientists and dated with reasonable accuracy. Eight of the eleven South Sandwich Islands, north of the Weddell Sea, originated from volcanic action that occurred an estimated 4 million years ago. Volcanic activity also is evidenced by major splits in rock formations found in West Antarctica that date from between 25 million and 30 million years ago.

The Marie Byrd Land volcanic area shows evidence of volcanoes from the Cenozoic period, dating back over sixty-five million years. The lower beds of volcanic rock are overlaid with other layers of volcanic rocks that suggest continued volcanic activity in the period from sixty-five million to about thirty million years ago.

When what is now Antarctica broke loose from the South American continent, vast amounts of heat and gas were released violently as blisteringly hot molten rock from Earth's core flowed into the cold waters to the south. Scientists have speculated on the effect a major contemporary eruption of either of Antarctica's two active volcanoes might have upon the surrounding environs. Some have speculated that such eruptions might, over the next millions of years, make the continent habitable, as recent archaeological evidence indicates that it once was.

Antarctica's Ice Cover

Antarctica contains 90 percent of Earth's permanent ice, which amounts to hundreds of billions of tons. This ice sheet, which has built up over at least 170,000 years without any substantial melting, is the largest accumulation of freshwater on the planet, holding 70 percent of Earth's freshwater. It is estimated that this amounts to about 6.2 million cubic miles (25 million cubic km.) of freshwater.

At times, it has been suggested that huge icebergs might be hauled to countries where water is in short supply. Icebergs, some almost unimaginably large, are constantly calving from the huge ice shelves that abound in Antarctica.

Antarctica: The Windiest Place on Earth

Antarctica has the distinction of being the windiest place on Earth, although, in some areas of the Antarctic, the wind is no stronger than in parts of the midwestern United States. Commonwealth Bay on the George V Coast east of the Ross Sea, is the windiest place in Antarctica. The wind speeds there sometimes reach 50 to 90 miles (80 to 145 km.) per hour for several consecutive days. Winds as high as 200 miles (320 km.) per hour have been clocked in the region. These winds reach their greatest intensity during storms.

About 600,000 square miles (1.55 million sq. km.) of ice shelves are found off the coastline of Antarctica. Ice shelves exist almost exclusively in Antarctica. The Arctic regions of Canada, Greenland, and Russia have much smaller ice shelves, several of which have disappeared in recent years.

In 1986 the ice shelf on the northeastern side of the Filchner-Ronne Ice Shelf calved, and the resulting iceberg drifted into the Weddell Sea, carrying with it two important research stations. This demonstrates how unstable the ice cover can be. Although the ice cover looks like a solid mass, when water is beneath it, it often is an enormous iceberg rather than an ice cover over a land base. The area that floated away from the Filchner-Ronne Ice Shelf totaled about 5,000 square miles (13,000 sq. km.), about the size of the US state of Connecticut. Then, in September 2019, climate scientists were alarmed when a huge iceberg dubbed D-28 broke off from the ice shelf. By early 2020, the iceberg, twice the size of New York City, was drifting slowly northward.

Icebergs extend far below sea level, so what one can see with the naked eye might, depending on its size and weight, be only 10 percent of the total mass. When ice cover forms on land, particularly on islands like those that lie beneath the ice of West Antarctica, their great weight, which increases every year as more snow falls and more ice accumulates, depresses the land mass on which it rests, often forcing it below sea level. As the ice cover overflows the land beneath it, it comes to rest on a less stable base than land provides. In time, fissures occur and large portions of the ice shelf break away.

The Antarctic Ocean

The line of demarcation between warm and cold water in the Antarctic Ocean is known as the Antarctic Convergence. At the Antarctic Convergence, there is a drop of about 5 degrees Fahrenheit (2.8 degrees Celsius) in the surface temperature of the water, and its salt content also drops appreciably. Ocean currents running along the ocean's floor from the Northern Hemisphere to the Antarctic Ocean carry small bits of sediment they encounter on their way south. This sediment, rich in nutrients, ends up in the upper layers of the frigid Antarctic between 50 degrees and 60 degrees south latitude. The warm water currents rise, as heat does, pushing the colder Antarctic waters deep into the sea, where they continue as cold, deep-water currents.

Hydrology and Animal Life

Antarctica's hydrology supports the animal life found on the continent. The Antarctic Ocean has an abundance of plankton and crustaceans. The Antarctic Ocean's murky, gray-green color is caused by the presence of microorganisms that make its waters nutritionally rich. Fish and mammals, particularly seals and whales, abound in the Antarctic waters, although whale hunting has brought some species of whales to the verge of extinction. The strong surface winds that blow from out of the west, often cyclonic in their speed and intensity, cause the surface water to move rapidly from west to east around the continent.

Antarctica has a variety of birds. Some, like the blue-eyed shag and various species of gulls and albatrosses, come during the long summer to mate; a few, such as snow petrels, stay in Antarctica through the long winter, sometimes continuing to sit on their eggs despite being covered by snow almost up to their beaks. The best-known animals of Antarctica are the penguins, which move about easily on its slippery surfaces, dive into its frigid waters to catch the fish that constitute the major part of their diet, and generally delight people who watch them. The penguin lacks the ability to fly, but the absence of predators and the abundance of food from the sea combine to make Antarctica an environment where these playful creatures can survive and flourish.

The four kinds of Antarctic penguins are the Adélie, chinstrap, emperor, and gentoo. Adélie and gentoo penguins seldom stray from the continent. Chinstraps are often found on the Antarctic maritime islands and on the Antarctic Peninsula. The gentoo usually breeds on the sub-Antarctic islands. The gentoos number about 600,000.

A Hydrologic Discovery

Russian scientists at the Vostok Research Station, 625 miles (1,000 km.) from the geographic South Pole, have done pioneering studies of the microorganisms found in deep ice core samples. They were retrieved by instruments developed in Russia that bring samples up with no contamination. Some of the ice in these samples is more than 400 million years old and reveals extremely interesting information about the microorganisms that existed then and the chemical composition of their environments.

Even more interesting is the discovery in 1996, while deep core ice samples were being taken, of a huge liquid lake that exists under 2 miles (3.2 km.)

ICE CORES: FROZEN TIME CAPSULES

Scientists use special drills to remove samples of ice from deep within the Antarctic ice cap. These ice cores hold clues to the earth's climate and environment, dating from the present to hundreds of thousands of years ago. As each snowflake forms and falls, it acquires a tiny sample of the atmosphere's chemistry and is compressed into ice by the weight of successive snowfalls. The ice preserves a record of the conditions that existed when it was formed, such as traces of past volcanic eruptions, lightweight pollen grains blown by winds, and air bubbles. When the sources of the pollen can be identified, knowledge about the plants' present requirements for growth give hints about past climates. When air bubbles are analyzed, chemists can determine changes in the percentages of different gases such as carbon dioxide and methane, both linked to possible climate change.

of glacial ice. This lake is presumably pristine, and efforts are being made to keep it from contamination of any sort. The ice cover directly above it will not be breached until there is evidence that the lake can be opened without the risk of contamination. Scientists involved in this project compare this hydrologic find to the possibility of finding water or ice on Mars or slush on Jupiter's moon Europa, which could be possible in the not-too-distant future. The ice in a core drilled in Antarctica in 2017 was found to be 2.7 million years old, much older than the previous oldest specimen. The ice's analysis by scientists from Princeton University discovered the presence of greenhouse-gas bubbles. These finding may help scientists determine how the ice ages came to occur.

These and other experiments and discoveries pose interesting puzzles to be solved. In the deep ice core samples, along with the elements one would expect to find—oxygen, nitrogen, and carbon—are large quantities of antimony, a highly toxic heavy element that is not currently found in our atmosphere in nearly the concentration found in the deep ice core samples.

French, Soviet, and American scientists in the Vostok team photo with unprocessed ice cores. (NOAA)

Because of the harsh climatic conditions in Antarctica, scientists still know relatively little about the continent's composition and about what lies beneath its surface. With year-around research stations being operated by several countries on Antarctica, this information gap could eventually decrease or even disappear.

R. Baird Shuman

Discussion Questions: Physiography and Hydrology of Antarctica

Q1. How do scientists describe the terrain of Antarctica? How did the continent come to be covered by ice? How much of Earth's freshwater supply is contained in Antarctic ice?

Q2. What are the dry valleys of Antarctica? Why do they lack ice?

Q3. How do geographers define the Antarctic Ocean? Why are its waters noted for their greenish color? What is the significance of the Antarctic Circle?

Physiography and Hydrology of Australia

Australia is the only country in the world that occupies an entire continent by itself. Australia's area of 2.97 million square miles (7.7 million sq. km.), which includes the island of Tasmania, is about the same size as the forty-eight contiguous states of the continental United States, making it the world's sixth-largest country.

The Tropic of Capricorn runs through the center of Australia, placing the northern half of the country within the tropics. The continent's northernmost tip, Cape York, reaches a point only 10 degrees south of the equator. Australia's southernmost point, located on the island of Tasmania, is South East Cape. It reaches 43 degrees south latitude. Australia extends farther south than the African continent, but not as far south as South America or New Zealand. The distance between Cape York and South East Cape is 2,285 miles (3,680 km.).

Australia's most easterly point is Cape Byron in New South Wales, at 153 degrees east latitude. Its

The Pinnacles Desert. (Iurii Ostakhov)

Physical Geography of Australia

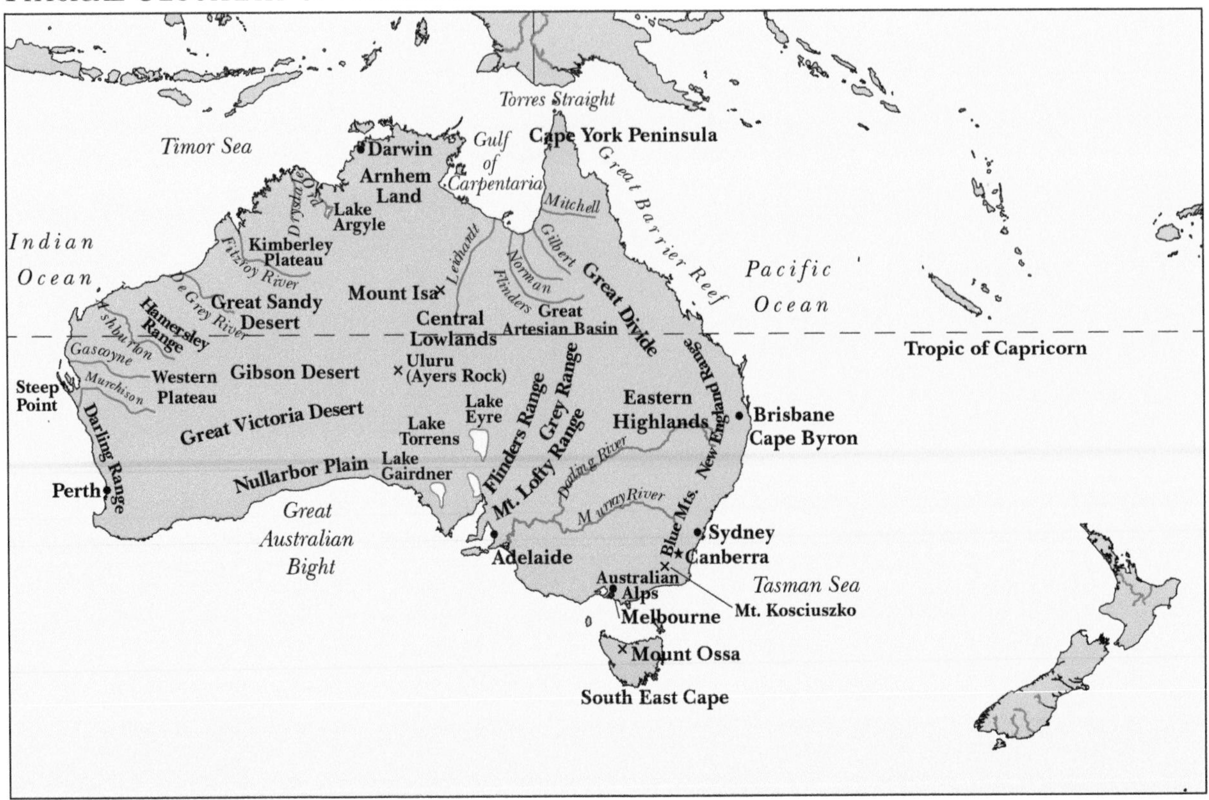

most westerly tip is Steep Point in Western Australia (113 degrees east). The east-to-west distance across Australia is 2,484 miles (4,000 km.)—roughly equivalent to the distance between the East and West Coasts of the United States.

Landforms and Physiographic Regions

The landforms of Australia are millions of years old. They have evolved as a result of unique physical circumstances and have influenced the patterns of flora and fauna on the Australian continent. They continue to affect the human geography of this part of the world. Australia is not only the world's smallest continent, it is also the flattest. It has few mountains, and those it does have are comparatively small. The continent's average elevation is 1,085 feet (330 meters) above sea level. More than a third of the continent lies below 656 feet (200 meters) in elevation. The low-lying regions are mostly narrow coastal plains and interior lowlands. Half of Australia is between 650 and 1,640 feet (200 and 500 meters) above sea level. Less than 1 percent of the continent rises more than 3,281 feet (1,000 meters) above sea level.

The history of Australian landscapes has been traced back to the Permian period, over 280 million years ago. At that time a huge ice cap covered most of the continent. After the ice melted, central parts of the continent had sunk so much that they were covered by shallow seas. There sediments were laid down that became sedimentary rocks. By the Cretaceous period, over 100 million years ago, seas rose over the flat land, separating Australia into three distinct sections of land.

Many Australian rocks contain fossils of very large animals, both land and marine creatures. In other basins, deposits developed that later formed coal, oil, and natural gas fields. These features show that the climate was warm and quite wet for long periods of geologic history.

The extreme aridity now characteristic of interior Australia is thought to be, in geologic terms, relatively recent, occurring between 10 million and 55 million years ago, during the Tertiary period, as the continent moved northward from its earlier position in the Antarctic. The Indo-Australian Plate is still moving northward, but very slowly—at a rate of about 1 inch (2.5 centimeters) per year.

Volcanism

Queensland is an area of relatively recent volcanic activity, as is the Western District of Victoria. Although there are no active volcanoes in Australia, there are lava flows and other signs of volcanic activity, dating from the Tertiary period to as recently as 5,000 years ago at Mount Gambier. Australia's island neighbors—New Zealand, New Guinea, and Indonesia—all have active volcanoes, as they are located at the boundaries of tectonic plates. Australia, by contrast, is located centrally on the Indo-Australian Plate.

Western Plateau

Australia can be divided into three fairly distinct physiographic regions: the Western Plateau, the Central Lowlands, and the Eastern Highlands. More than half of the continent is part of the great Western Plateau. This plateau covers all of Western Australia and most of South Australia and the Northern Territory and has an average elevation of 984 feet (300 meters).

The plateau contains great expanses of desert. Some of the world's oldest rocks—over 3 billion years old—are found in western Australia in the area known as the Australian Shield. The size and shape of the continent have changed greatly throughout Earth's history. Over 200 million years ago, it was part of the supercontinent Pangaea. Around 100 million years ago, it started to separate from Antarctica and to move northward.

The Great Sandy Desert, the Gibson Desert, and the Great Victoria Desert occupy much of the Western Plateau. Huge fields of longitudinal sand dunes, aligned in the direction of the prevailing winds, cover parts of these deserts. A great part of these deserts has a stony surface covering called reg, known in Australia as "gibber plains."

The Bungle Bungle Range in the Purnululu National Park in Western Australia. (W. Bulach)

Around the continent's ocean edges the Western Plateau is rimmed by escarpments. The most interesting of these is the Nullarbor Plain. The word "Nullarbor" is from the Latin meaning "no trees"—a fitting name for this treeless limestone area with no surface streams but many underground caves. Its rocks were a seafloor about 25 million years ago. Since then the area has been uplifted to form the present landscape.

At the southeastern edge of the Western Plateau are the Flinders and Mount Lofty ranges, a series of fault-block mountains. They have an unusually distinctive appearance and are sometimes called the Shatter Belt. Most of these mountains are about 3,281 feet (1,000 meters) high and contain some very old rocks. Two places in the Western Plateau are rich in minerals—the goldfield area around Kalgoorlie in Western Australia and the Mount Isa area in western Queensland, where copper and lead are mined.

A few mountains and smaller plateaus rise above the generally flat landscape of the Western Plateau. In the west is the Hamersley Range, from which a number of rivers flow west to the Indian Ocean: the De Grey, Fortescue, Ashburton, Gascoyne, and Murchison.

In the northwest is the Kimberley Plateau, from which the Fitzroy River flows west, and the Drysdale and Ord flow north to the Timor Sea. In the Northern Territory, the Victoria, Daly and Roper are important rivers. The rugged areas of Arnhem Land and the MacDonnell Ranges are prominent higher areas, as are the Musgrave Ranges in South Australia.

In the center of Australia is one of the world's famous landforms, Uluru, also known as Ayers Rock. This sandstone monolith rises 1,132 feet (345 meters) above the very flat surrounding plains, and covers 1.3 square miles (3.3 sq. km.). To the east, the Western Plateau slopes down into the Central Lowlands, a series of three basins running from the Gulf of Carpentaria to the South Australia-Victoria border.

Central Lowlands

The northernmost basin of the Central Lowlands drains to the Gulf of Carpentaria. Long rivers here include the Leichhardt, Cloncurry, Flinders, Norman, Gilbert, and Mitchell. In the center of the Lowlands is the Lake Eyre basin, where the streams of the western Queensland Channel Country drain to the lowest point on the Australian continent, Lake Eyre. That lake's surface is 50 feet (15 meters) below sea level and covers almost 3,861 square miles (10,000 sq. km.). This salt lake is normally dry, although it has filled occasionally during strong La Nina years. Other large salt pans, or salinas, in South Australia are Lake Torrens and Lake Gairdner.

Many Australian rivers are intermittent, containing water only for a short period each year. In the dry period, they shrink to a series of waterholes, or billabongs, and sometimes a river is only a sandy bed for very long periods.

The third basin of the Central Lowlands is occupied by Australia's largest and most important river system, the Murray-Darling. The Murray River is over 1,560 miles (2,500 km.) long and forms part of the New South Wales-Victoria border. The Darling and its tributaries, known as the Upper Darling, are another 1,767 miles (2,884 km.) in length. One tributary of the Darling, the Macintyre River, forms part of the New South Wales-Queensland border.

Although the Murray-Darling drainage basin covers 386,000 square miles (1 million sq. km.), or about one-seventh of the continent, its discharge of water is small. Because of climatic variability, it is also extremely variable. Nevertheless, during the nineteenth century the Murray River was used as a waterway on which a steamboat trade flourished. Today, large dams, weirs, and barrages make water available for farming throughout the Murray-Murrumbidgee agricultural area.

Artesian Water

Although the interior of Australia is extremely arid, it contains a valuable resource in the form of groundwater. This is the Great Artesian Basin, the world's largest aquifer, or reservoir of underground water. The Great Artesian Basin stretches under the Central Lowlands, from the Gulf of Carpentaria to northern New South Wales and South Australia, covering an area of 656,000 square miles (1.7 million sq. km.) and containing over 7 million acre feet

An artesian bore in Queensland, Australia. (JohnCarnemolla)

(8,700 million megaliters) of water. The intake beds for the basin are located along the western slopes of the Eastern Highlands. Because artesian water is under pressure, it rises to the surface without needing to be pumped via boreholes drilled to an average depth of about 1,640 feet (500 meters). Artesian water is warm, contains minerals, and is suitable for animals to drink.

The discovery of the Central Lowland's artesian water in the 1870s prompted a great expansion of cattle and sheep farming, especially in Queensland. These developments created two new problems. Native plant species began declining because of the overgrazing of cattle and sheep. This in turn led to a decline in the number of native animal species. The drilling of thousands of bores to tap the underground water—many of which flow without restriction—severely decreased the water's flow. Hundreds of bores have dried up. Recharge is lower than demand, so attempts are now being made to conserve the water of the Great Artesian Basin and to reduce or prevent environmental damage.

Eastern Highlands

Running the length of the east coast, from Cape York in the north to Melbourne in the southeast, are the Eastern Highlands. These uplands also continue on to the island of Tasmania. This area was uplifted during the Tertiary period, between 2 million and 65 million years ago. It is also known as the Great Dividing Range, a misleading term, as much of the highland consists only of tablelands and plateaus. Moreover, there are many different ranges, ridges, and mountains within the Eastern Highlands.

The Eastern Highlands do, however, form a true divide, separating the short rivers flowing across the coastal plains to the Pacific Ocean from the rivers that flow inland. Rainfall is plentiful, so these streams are permanent, and they have provided the initial sites for the founding of Australia's cities. Near their headwaters, waterfalls and gorges are common. Some of the longer coastal rivers are the Burdekin and Fitzroy in Queensland and the Hunter River at Newcastle, which is the heart of a wine-growing district.

The steep cliffs and slopes on the eastern side of this region form a long escarpment, running from Queensland to Victoria, with a gentle slope on the western side, down over the interior plains. The coastal plains lying between the Pacific Ocean and the Eastern Highlands are as narrow as 31 miles (50 km.) in some places and as wide as 497 miles (800 km.) in others. In Queensland, the highest peak is Mount Bartle Frere, which rises only to 5,285 feet (1,611 meters). The New England Tableland of northern New South Wales is one of Australia's coldest environments. The Atherton Tableland in north Queensland has a cool climate, enabling dairying, even though it is within a tropical region.

Exploration

The Blue Mountains, on the inland side of Sydney, are not a high escarpment, but are highly dissected and quite rugged. For this reason, they presented a formidable barrier to the early European settlers in the British colony of New South Wales. The explorers Gregory Blaxland, William Lawson, and William Charles Wentworth crossed the mountains in 1813 and discovered the wonderful grasslands of the interior, which became sheep pastures. As later explorers ventured into the strange new environment they were often deceived by misinformation and preconceived ideas.

A myth that a great river flowed to a large inland sea persisted into the 1840s, when Thomas Mitchell, the surveyor-general of New South Wales, traveled into central Queensland along what is now called the Barcoo River. In 1861 Robert Burke and William John Wills died at Cooper Creek, after crossing Australia from Melbourne to the Gulf of Carpentaria without finding any major rivers or lakes. Ludwig Leichhardt discovered the Burdekin River in 1845 during an epic journey from the Dar-

The Murray River in Western Australia. (Calistemon)

Major Watersheds of Australia

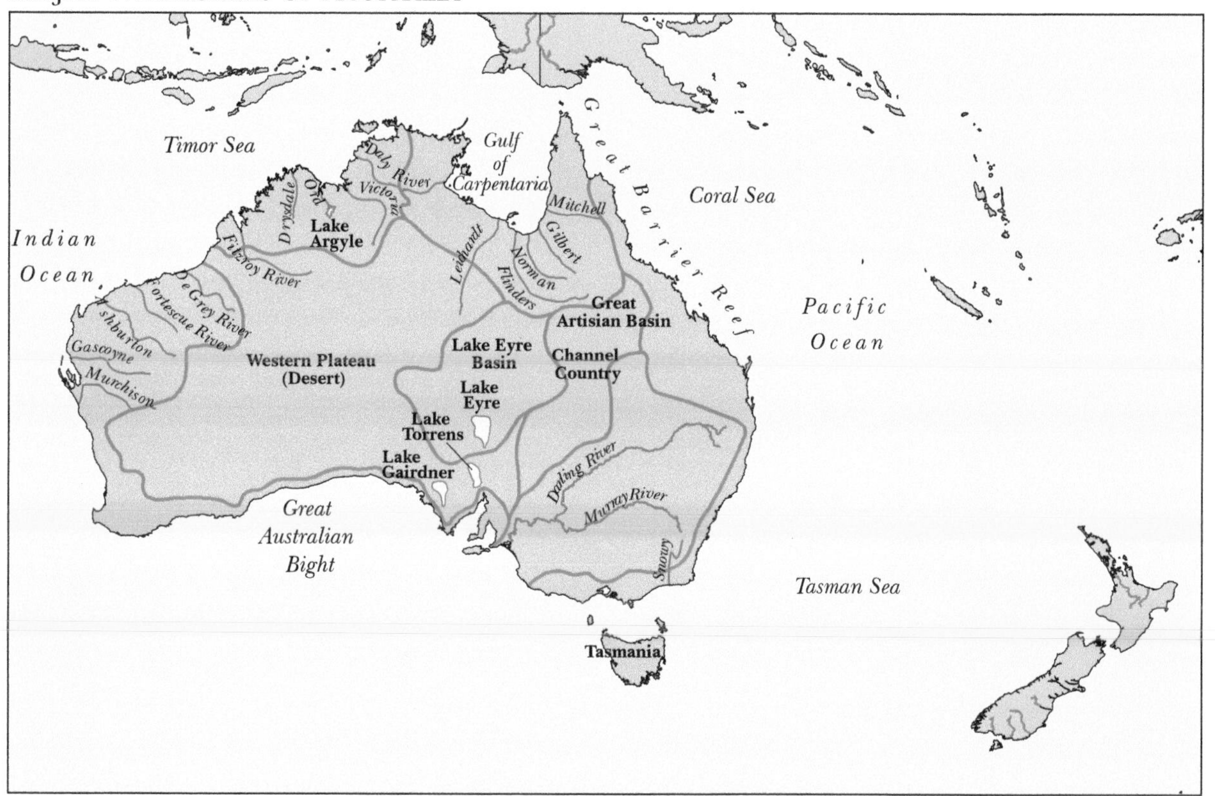

ling Downs to Port Essington, near modern Darwin in the Northern Territory. On another expedition, three years later, his party vanished in the central Queensland desert.

Mountains

Australia's ten tallest mountains all stand in an area of the Eastern Highlands known as the Australian Alps, near the border between New South Wales and Victoria. The tallest of these is Mount Kosciuszko, which is only 7,310 feet (2,229 meters) high. In comparison, the world's highest mountain, Mount Everest, is four times higher than Kosciuszko. Covered with snow during the winter months, the alps are the source of important rivers.

Eastern Highlands and Tasmania

On Australia's mainland, the Eastern Highlands end in the Grampians of western Victoria. The rugged mountains of Tasmania form the southernmost portion of the Eastern Highlands. Much of the island is a plateau, while the island's tallest peak is Mount Ossa which rises to 5,305 feet (1,618 meters).

During the Quaternary period, over a million years ago, Earth experienced its most recent ice age. Huge areas of North America, Asia, and Europe were covered by ice, which advanced and retreated numerous times. Like the northern continents, Australia also experienced glaciation, which left glacial lakes and other typical features in Tasmania. A small part of the Australian Alps also was affected by this glaciation. At this time too, sea level was some 330 feet (100 meters) lower than at present; what is now the continental shelf was part of the mainland. The Great Barrier Reef now marks this outer edge. Three parts of the Eastern Highlands have been designated World Heritage Sites—the Wet Tropics of Queensland; the Central Eastern Rainforest Reserves; and the Tasmanian Wilderness. The Great Barrier Reef is a marine World Heritage Site.

Water Resources

The flat topography of Australia affects rainfall, because there are no high mountain barriers to block winds and produce what is called orographic rainfall. Australia's drainage patterns are clearly the result of topography. The main divide of the Eastern Highlands separates the moist coastal plains from the dry basin and plains of the interior. Most permanent streams in densely populated areas have their flows regulated by dams, creating surface storage in artificial lakes and reservoirs. A drawback to this is that large amounts of water are lost to surface evaporation.

Permanent streams in the unpopulated north are too remote for development to be economically practical. The availability of water resources has been a major influence on human settlement patterns throughout Australia's history, and has resulted in the concentration of population around the moist east, southeast, and far southwest coastal areas. The percentage of the population living in rural areas declined from 43 percent in 1911 to 14 percent in 2020. Less than 4 percent of Australian workers were employed in agriculture, forestry, and fishing in 2020.

Ray Sumner

Discussion Questions: Physiography and Hydrology of Australia

Q1. How does Australia compare in size to the United States? What island lies off Australia's southeast coast? What country lies closest to Australia?

Q2. What three distinct physiographic regions make up Australia? Where are Australia's vast deserts? In what part of Australia are the highest mountains found?

Q3. What is the Great Artesian Basin? How large is it? What does it contain? Why has the Basin become a matter of concern to Australian conservationists?

Physiography of the Pacific Islands

Most islands of the Pacific Ocean are located between the Tropics of Cancer and Capricorn. While there are some large islands, most are quite small in size and population and poor in natural resources, but thanks to beautiful tropical landscapes, are conducive to travel and tourism. New Zealand is larger than all but Papua New Guinea, and is located south of the main island region, yet its indigenous population, the Maori, is similar to some island peoples in cultural and physical characteristics; thus, it is included in the Pacific Island region.

The west-to-east extent of the Pacific island region stretches from Palau in the west, at 7 degrees 30 feet north latitude and longitude 134 degrees 30 feet east, to Easter Island in the east, at 28 degrees 10 feet south latitude and longitude 109 degrees 30 feet west. The north-to-south extent is from Midway Islands in the north at 28 degrees 13 feet north latitude and longitude 177 degrees 22 feet west, to New Zealand in the south, with the southernmost tip of the southern island at 47 degrees south latitude and longitude 168 degrees east — more than 1,000 miles (1,600 km.) southeast of Australia.

Physiographic Regions

Three physiographic regions define the Pacific Islands and New Zealand. The first is Melanesia ("dark islands" in Greek), which includes the culturally rich peoples of Papua New Guinea, the Solomon Islands, Vanuatu, New Caledonia, and Fiji. These islands are diverse both physically and culturally. They are primarily mountainous, but differ

A volcanic island in Tonga, Polynesia. (Janos)

Physical Geography of the Pacific Islands

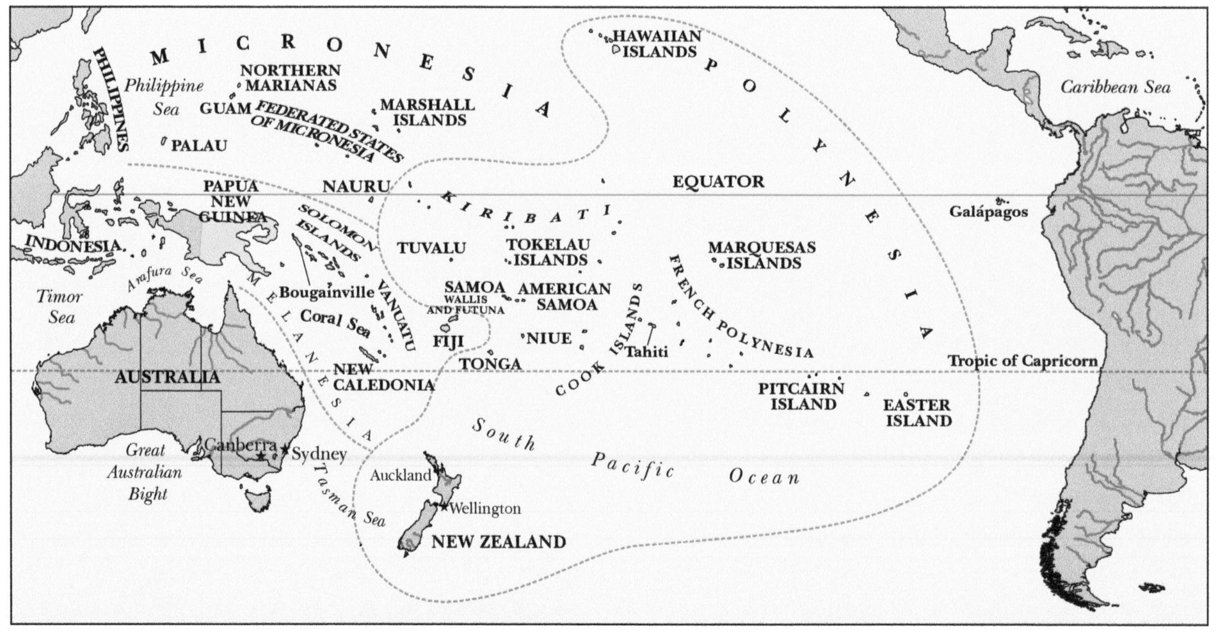

greatly in size. Papua New Guinea is by far the largest at 178,704 square miles (462,840 sq. km.). All Melanesian territories are large by Pacific standards, which gives them relatively more regional and international political influence.

The second region, Micronesia (from the Greek for “little islands”), consists of more than 2,000 coral atolls and volcanic islands scattered across the western Pacific, primarily north of the equator. The main island groups are the Federated States of Micronesia; Palau; Kiribati; the Mariana Islands, including Guam; the Marshall Islands; and Nauru. Although covering an ocean area larger than the continental United States, the islands of Micronesia have a combined land area of only 271 square miles (702 sq. km.) Fewer than 100 of these resource-poor islands are inhabited.

Polynesia (from the Greek for “many islands”), the third region, is vast in terms of sea area, covering the largest expanse of ocean. A triangular shape defines the area, with the Hawaiian Islands in the north, New Zealand in the south, and Easter Island to the east. Polynesia also includes French Polynesia, and the smaller political states of Tonga, Tuvalu, Cook Islands, and Samoa. Excluding New Zealand and the Hawaiian Islands, the total land area of these small island groups, or archipelagos, is only about 3,189 square miles (8,260 sq. km.). The largest island is Tahiti at 402 square miles (1,041 sq. km.).

Landforms

The landforms of the Pacific Islands and New Zealand are divided into high islands and low islands or atolls. They are primarily volcanic in nature and geologically fairly young. The high islands are mountainous and spectacularly scenic, with steep slopes and peaks of 3,300 to 13,000 feet (1,000 to 4,000 meters), but with great local variations in elevation, slope, soil, rainfall, and plant life. Papua New Guinea, for example, is dominated by extensive, east-west trending volcanic mountain ranges separated by rugged, elevated plateaus and bordered by coastal lowlands.

The Hawaiian Islands, on the other hand, contain several more classic-looking shield volcanoes, relatively gently sloping mountains produced from numerous fluid lava flows, and rimmed at the top by volcanic cones that reveal the inner core of the volcano. All larger islands of Melanesia and several Polynesian islands are volcanic high islands.

The low islands are generally quite small and lacking in vital resources such as fertile soils and drinking water; therefore, they are sparsely populated. Coral reefs, sandy beaches, and coconut palms, typical of a picturesque island paradise, characterize the low islands. These islands, called atolls, are made of broken pieces of coral and usually form an irregular ring around a lagoon, with the coral pieces separated by channels leading into the lagoon.

The world's largest atoll, Kwajalein, in Micronesia's Marshall Islands, is 75 miles (120 km.) long and 15 miles (24 km.) wide. Polynesia, Micronesia, and parts of Melanesia are dotted with extensive atoll systems. Some major archipelagos, such as the Marshall Islands in Micronesia and Tuamotu in Polynesia, are composed entirely of atolls.

New Zealand consists of two large islands, North Island and South Island, plus several small islands. Cook Strait separates North and South Island. The two main islands are rugged in terrain. North Island primarily comprises high volcanoes, many exceeding 5,000 feet (1,500 meters), and a central highland plateau; South Island is formed of a range of high mountains, called the Southern Alps, and an eastern lowland plain.

The Waikite Geyser in New Zealand. (Janos)

The highest and most rugged mountains are found on the western side of the South Island. Aoraki/Mount Cook is New Zealand's highest peak, cresting at more than 12,000 feet (3,660 meters). The Southern Alps/Ka Tiritiri o te Moana is one of the world's most visually spectacular mountain ranges, complete with high mountain glaciers and steep, narrow, fjord-like valleys that cut into much of South Island's isolated western coast.

Geologic History

Island geology varies across the Pacific Island region. The high islands were formed by undersea volcanic or tectonic mountain-building activity. The large islands of Melanesia, along with New Zealand, are composed of continental rock and are geologically quite complex. When the Indian continental plate collided with the Pacific continental plate, a series of mountain ranges formed in Papua New Guinea, extending across the Solomon Islands to Vanuatu and New Caledonia, and down to New Zealand to form a mountainous island chain or arc.

Most of the islands of Polynesia and Micronesia originated from volcanic activity on the ocean floor without any geologic connection to continental landmasses. These volcanic islands are formed by oceanic crust sliding over a hot spot in Earth's mantle where molten magma (hot liquid rock), relatively close to the crust, rises and cools to form new volcanic islands. The Hawaiian Islands, the largest and youngest of the Pacific's high islands, are an excellent example with several young, active, and recently active volcanoes rising to more than 13,000 feet (3,960 meters). Many French Polynesian islands are smaller examples of high volcanic islands. In tropical latitudes, most high islands are ringed by coral reefs that quickly establish themselves in the shallow waters near the shore.

High islands are limited in geologic lifespan. Volcanic activity gradually decreases, and the volcanoes eventually become extinct. These high islands slowly subside or sink, and erode away to begin formation of low island atolls. After a few hundred years, only a few low peaks, called seamounts, or underwater volcanoes, may rise out of a shallow, reef-rimmed lagoon. Eventually, even remnant peaks erode away, leaving only the reef surrounding the lagoon. Reefs tend to persist because they are composed of living organisms that constantly create new coral, even as the island base subsides. Atoll formation continues as large waves periodically break off and crush large pieces of coral, and then deposit the coral on adjacent sections of the reef to form narrow sandy islands. An atoll, therefore, comprises the combination of circular or oval-shaped barrier coral reefs, narrow sandy islands, and shallow central lagoons. The nineteenth century scientist Charles Darwin was the originator or this explanation for the geologic formation of atolls.

Volcanism

Most of the Pacific Islands have a volcanic core that developed from eruptions of lava deep beneath the ocean floor during tectonic movement of the Pacific Plate. Much of Melanesia and Polynesia, including New Zealand and the Hawaiian Islands, are part of the seismically active Pacific Rim of Fire, so named for its geologic development through undersea mountain building from "hot spots" beneath the ocean crust. The North Island of New Zealand, for

An aerial view of the Kwajalein Atoll in the Marshall Islands. (Matt Kleffer)

Recent Volcanic Eruptions in the Pacific

Volcanic eruptions and related earthquakes and tsunamis, or seismically induced sea waves, are common events across the Pacific Island region, and they impose major environmental hazards upon its inhabitants. A summary of recent notable volcanic activity in the Pacific Islands region follows.

1774: Mount Yasur on Tanna Island, Vanuatu, erupts continually for centuries, into 2020.

1832-1984: Hawaii's Mauna Loa erupts thirty-nine times.

1983-2018: Hawaii's Mount Kilauea erupts continually during this period.

1994: Eruptions and earthquakes on Papua New Guinea's island of New Britain force more than 100,000 people from their homes, destroying Rabaul and nearby villages.

1995: New Zealand's Mount Ruapehu erupts.

1995: Mariana Islands' Mount Ruby, a prominent, active submarine volcano, erupts.

1995: Tonga's Metis Shoal (also known as Lateiki) erupts, forming a new island.

1996: Mount Loihi, the youngest volcano in the Hawaiian chain, erupts, followed by a swarm of more than 4,000 earthquakes.

2000: Papua New Guinea's Mount Bagana, on Bougainville Island, begins a continual eruption that lasts into 2020.

2005: Mount Manaro, on Vanuatu's Ambae Island, erupts, leading to the relocation of 3,300 residents on the island.

2012: New Zealand's Mount Tongariro erupts twice.

2014: Mount Manam, off the northern coast of Papua New Guinea, begins a continual eruption that includes explosive eruptions in 2019.

2014: Papua New Guinea's Mount Tavurvur, which destroyed Rabaul in 1994, erupts again.

2017: Tinakula in the Solomon Islands emits an ash explosion that drifts to nearby inhabited islands, covering crops and contaminating water supplies. A small population around the volcano was previously evacuated in 1971.

2017: Vanuatu's Mount Manaro erupts again, this time leading to the total evacuation of the island.

2018: The Vanuatu island of Ambrym erupts, draining active lava lakes in both the Benbow and Marum craters.

2018: Kadovar, an island volcano north of Papua New Guinea, erupts prompting the evacuation of residents and changing the morphology of the south east side of the island.

2018: Tinakula in the Solomon Islands begins a continual eruption that lasts into 2020.

2019: Mount Ulawun in Papua New Guinea erupts, forcing the evacuation of more than 5,000 people.

2019: New Zealand's Whakaari/White Island erupts with forty seven people on the island, killing twenty one.

2019: Tonga's Metis Shoal (Lateiki) erupts again, forming a new island larger than the previous one.

example, contains several volcanic peaks and geothermal features as a result of volcanic formation.

The Hawaiian Islands were produced by the Hawaiian hot spot, presently located under the "Big Island" of Hawaii. Each island has at least one primary volcano, although many islands are composites of more than one. The Big Island, for instance, is constructed of five major volcanoes: Kilauea, Mauna Loa, Mauna Kea, Hualalai, and Kohala.

Hydrology

Many Pacific Islands are located in a tropical wet/dry climate region where abundant rains and tropical cyclones can bring heavy seasonal precipitation. High islands are particularly noted for their heavy rainfall, with frequent snows on the higher peaks (13,000 feet/3,960 meters). Drainage patterns are established in mountainous terrain according to rock type and slope steepness. Steep, volcanic slopes erode easily during heavy rains, developing deep drainage channels that transport rainwater to lands below. Major populations are located in the coastal lowland regions and urban areas of the high islands where freshwater is abundant. Inhabitants of the highlands have ample freshwater as well.

Low-lying atolls have significantly less precipitation than high islands, and limited water storage capacity as well; thus, they often experience water shortages. When it does rain, small "lenses" of freshwater often float above the salt water. In dry periods, the salt water in the center of each sandy island quickly depletes the stores of freshwater. Limited availability of fresh water limits human settlement on the low islands and atolls. Most freshwater must be imported to support inhabitants of atolls.

New Zealand's highlands lie in the Southern Hemisphere belt of westerly winds and receive abundant precipitation, with local variations within the rain shadows of east-facing slopes. Moderate marine weather keeps year-round temperatures fairly warm, causing glaciers to flow on both islands. The glaciers, lakes, and mountain streams of the highland regions provide plenty of freshwater for New Zealand's population. Rugged terrain and heavy precipitation in the Southern Alps and mountainous core of the North Island limit highland settlement. The well-populated areas are restricted to the fringing lowlands of the North Island and along the drier east and south coasts of the South Island where freshwater flows naturally from the highlands.

Island hydrology also provides transportation routes for human settlement. In some high islands, rugged terrain and lack of roads make canoe travel by navigable river a necessity. The large and powerful Fly and Sepik rivers of Papua New Guinea, for example, are widely used as travel and trade routes between villages in these southern and northern coastal regions, respectively.

Laurie A. Garo

Climatology of Antarctica

Antarctica is the coldest, windiest, driest continent on Earth. The cold is so extreme that it is known as a "heat sink," because warm air masses flowing over Antarctica quickly chill and lose altitude, affecting the entire world's weather. Most of Antarctica (98 percent) is an ice cap with a few bare, rocky coastal areas, dry valleys, and mountain peaks blown free of snow by persistent winds. Annual precipitation on the ice cap is scant—between 1.25 and 3 inches (30 and 75 millimeters). The high-altitude central plateau is a frozen desert, with annual mean temperatures between -58 and -76 degrees Fahrenheit (-50 and -60 degrees Celsius)—identical to the average surface temperature on the planet Mars. Even on a summer's day at the South Pole, temperatures hover around -22 degrees Fahrenheit (-30 degrees Celsius). By comparison, the temperature of the average home freezer is a mere +5 degrees Fahrenheit (-15 degrees Celsius).

The Antarctic Peninsula is the warmest part of Antarctica. The summer temperature reached a balmy 69.3 degrees Fahrenheit (20.72 degrees Celsius) in February 2020, though average temperatures are much lower; the area received as much as 30 inches (760 millimeters) of annual precipitation. The temperatures of the islands ringing Antarctica, warmed by the sea, range from a few degrees below freezing up to around 37 degrees Fahrenheit (3 degrees Celsius). These warmer regions are much windier than the interior, however.

An icebreaker plying through Antarctic waters. (GentooMultimediaLimited)

Antarctica holds the record for the coldest temperature ever recorded on Earth: -128 degrees Fahrenheit (-89 degrees Celsius), measured at the Russian research station of Vostok in July 1983. Even colder temperatures could be measured at higher elevations on mountain summits if recording devices were present, but Vostok is definitely the coldest inhabited place on Antarctica, partly because of its location at what is called the Pole of Inaccessibility, the point on the ice cap farthest inland from the ocean's moderating influence. In addition, both its latitude (78 degrees south) and its altitude (11,320 feet/3,450 meters) are quite high.

Latitude and altitude both contribute to the continent's extreme coldness. At such low temperatures, water sprayed into the air freezes before striking the ground. Frozen water in the form of ice crystals suspended in the frigid atmosphere scatters light rays to create spectacular optical effects unique to Antarctica. Among them are vertical sun pillars, ice and fog bows, haloes, arcs, multiple suns and moons, and mirages. During the Antarctic's fall and winter months (March through September), the sky is the scene of magnificent aurora displays (aurora australis, or southern lights).

Climate scientists have identified parts of central West Antarctica as among the fastest-warming areas on Earth. Some studies have linked ozone depletion in the stratosphere over Antarctica as playing a role in the continent's climate change. At the same time, it has been noted that in 2019, ozone depletion was at its lowest in thirty years.

THE OZONE HOLE

Ozone is a molecule composed of three atoms of oxygen instead of two. At the earth's surface, it is one of the ingredients in smog. In the upper atmosphere, however, ozone forms a protective chemical shield against deadly cosmic radiation. In 1981 atmospheric scientists became alarmed when September and October (Antarctic springtime) ozone readings over an area of Antarctica the size of the continental United States registered 20 percent below normal. This drop in ozone coincided with the increased presence in the upper atmosphere of artificial chemicals called chlorofluorocarbons (CFCs), which contain chlorine, an element that destroys ozone. The chlorine is released when ultraviolet light strikes CFC molecules in the upper atmosphere. Each chlorine atom remains in the atmosphere an average of forty years, destroying tens of thousands of ozone molecules.

The ozone hole began to develop over Antarctica because of the continent's extremely cold atmosphere. Polar stratospheric clouds, composed of ice particles containing water and nitrogen compounds, only occur when the air temperature drops below -112 degrees Fahrenheit (-80 degrees Celsius). These clouds occur during the Antarctic winter when the cold air circulates in a swirling pattern called the polar vortex at altitudes between 6 and 15 miles (10 to 24 km.). The clouds persist throughout the long winter without any influx of air from warmer areas. The surfaces of their ice crystals store chlorine compounds such as CFCs.

In the spring, the sun melts the ice crystals, freeing enormous amounts of chlorine to rapidly deplete the ozone layer, thus creating the ozone hole. In the summer, warm air breaks up the polar vortex and replenishes the ozone from other areas in the Southern Hemisphere. Bubbles of the ozone hole can break away, however, and drift north over populated areas.

The long-term trend from 1980 to the early 1990s was for the Antarctic hole to grow in both area and depth. However, by the early 21st century the annual holes stabilized, and scientists observed a 20 percent decrease in ozone depletion from 2005 to 2016. The smallest hole recorded since 1982 was observed in 2019, and models predict that the Antarctic ozone layer will largely recover by 2040.

High Latitude

The primary reason both polar regions are cold is their high latitude. Earth is tilted on its axis, so the poles receive less of the Sun's direct radiation than the lower latitudes do. Even in midsummer, the Sun never rises very high above the horizon, so the air stays cool—like early morning or late afternoon in the temperate zones. The poles never experience the heat of high noon. In addition, winters are long and dark. At the Arctic Circle (66° 32′ north latitude) and the Antarctic Circle (66° 32′ south latitude), the Sun does not rise above the horizon on midwinter's day.

As latitude increases from the Arctic and Antarctic circles to the poles, the winter night lengthens until, at the poles, the Sun sets just once a year. Once the last light fades, around March 21 at the

A dogsled team in the Antarctic. (ventdusud)

South Pole, the poles are left in darkness for half the year with no direct solar warming at all. Although the Sun stays above the horizon for six months when it finally returns in late September, the warmth Antarctica receives in summer cannot make up for the heat it loses in winter. In fact, only at the height of summer (November and December in the Southern Hemisphere) does the South Pole actually gain heat by absorbing more solar radiation than it reflects away.

Solar Reflection

Antarctica actually receives about 7 percent more solar radiation annually than the Arctic because of Earth's elliptical orbit, bringing it closest to the Sun in January, the height of the Southern Hemisphere's summer. However, the climate of Antarctica is much harsher and far colder than the climate of the Arctic. The Arctic is an ocean surrounded by land. The ice covering the Arctic Ocean melts every summer, partly because the continents that surround the Arctic Ocean heat up considerably at that time of year and then radiate warm air north. In addition, the dark waters absorb heat from the Sun, keeping them warmer than the surrounding snow-covered continents in winter. Warm currents flowing from the south also raise temperatures there.

By contrast, the Antarctic is land surrounded by ocean. Land does not retain heat as well as water. The sea reflects back only about 5 percent of incoming solar short-wave radiation into the atmosphere. Exposed land reflects between 15 and 35 percent back. In the Antarctic, reflection from the land is intensified by the permanent mass of ice that covers all but 2 percent of the surface. The ice cap reflects back up to 90 percent of the sunlight it receives in the summer, so little heat can be absorbed. Because it reflects away more total radiation than it absorbs, it would get colder and colder if not for the structure of the atmosphere above Antarctica, which is separated into three layers. The bottom and top layers transport cold air away from the continent. The middle layer is warm, moist air flowing in from temperate regions. As the water vapor in this layer condenses and freezes, it gives up its heat, providing a stabilizing influx of warm air.

Like the Arctic Ocean, the ocean that surrounds Antarctica is warmer than the land, but it cannot act to raise the land's temperature much because it, too, is blanketed most of the year by sea ice up to 6 feet (1.8 meters) thick, extending hundreds of miles out from shore. About 6.2 million square miles (16 million sq. km.) freeze and thaw every year, the biggest seasonal change to occur anywhere. When the southern ocean freezes in winter, about 8 percent of the Southern Hemisphere is covered by ice. The pack ice almost doubles the size of the continent in winter and also reflects solar energy back into the atmosphere.

Ocean Currents

Pack ice around Antarctica deepens its frigid temperatures, not just by reflecting the Sun's energy back into space but also by insulating the mainland as bags of ice do in a cooler. The sea ice also dampens wave motion and creates downward-flowing ocean currents of cold, dense water that affect ocean circulation. Powerful currents in the southern ocean flow around the entire globe unhindered by land, generating huge waves, intense winds, and powerful low-pressure weather systems. Antarctica is surrounded by the planet's stormiest seas — seas so treacherous and wild that sailors call these latitudes the Roaring Forties, the Furious Fifties, and the Screaming Sixties.

To add to the commotion, a current called the Antarctic Convergence, a belt of water about 25 miles (40 km.) wide, encircles Antarctica where the cold currents streaming away from the ice cap meet the warmer currents flowing south from the tropics, creating cold, wet, windy weather offshore. People sailing across the Antarctic Convergence can immediately feel the difference when the air and sea temperatures suddenly drop and mist appears. Its location shifts north every winter and south every summer, but it is always there somewhere between 50 degrees and 60 degrees south latitude. These ocean currents together form an effective barrier isolating Antarctica from the moderating influences of currents and air masses from warmer latitudes.

Wind and Altitude

Another reason Antarctica is so cold is its high altitude. Higher altitudes are generally colder than lower altitudes, and Antarctica has the highest average altitude of any continent—8,039 feet (2,450 meters) because of its thick ice cap. By contrast, most of the Arctic is at or near sea level.

Closer to shore, frigid winds are constantly blowing down from the high altitudes of the ice cap because the relatively warmer, lighter air flowing at high altitudes over the interior quickly chills and sinks as it becomes colder and denser. These katabatic, or gravity, winds gain power as they accelerate down the slopes of the ice cap toward the coast. Faster and stronger than any other winds on Earth, they commonly travel at more than 50 miles (80 km.) per hour and have been recorded reaching 186 miles (300 km.) per hour, more than twice the velocity of the average hurricane.

Aridity

Prevailing ocean currents and winds prevent moist air masses from reaching Antarctica while the cold temperatures allow little evaporation to occur. As a result, most of Antarctica experiences little precipitation even though more than two-thirds of the planet's freshwater is locked in its frigid embrace. The explanation for this apparent contradiction is simply that the snow, too cold to melt, has been accumulating for millions of years. What appear to be raging blizzards are actually like the sandstorms of the Sahara Desert. Most Antarctic snowfalls consist not of flakes but of pellets of loose, already fallen snow compacted into icy fragments that are then blown about by the fierce winds.

Sue Tarjan

Climatology of Australia

The dominant feature of Australia's climate is its general aridity. Indeed, the continent is the world's driest. More than a third of its area has a hot desert climate. Another third has a semiarid climate. Annual precipitation throughout the continent averages only 17.7 inches (450 millimeters). Thanks to the low rainfall, clear skies are characteristic of most Australian weather.

Precipitation Zones

Australia's precipitation zones can be visualized as rings encircling the north, east, and south portions of the continent. In the west the dry desert extends all the way to the coast. The outermost ring is the wettest, receiving more than 31 inches (780 millimeters) of rain a year. That ring encompasses the Kimberley, the northernmost region of Western Australia; Arnhem Land, the northeastern corner of Northern Territory; and the entire east coast— from Cape York to the South Australia border. It also includes most of the island of Tasmania and a small section of the southwest corner of Western Australia.

Inland, away from the imaginary outer rings, rainfall decreases sharply. A large area in the interior receives less than 12 inches (300 millimeters) of precipitation annually. Eighty percent of Australia receives less than 23.6 inches (600 millimeters) of rainfall annually.

Sunshine and Heat

The second important feature of the prevailing Australian climate is its high temperatures. Indeed, heat is such a pervasive part of Australian life that virtually every Australian is familiar with Dorothea McKellar's poem "My Country," which begins: "I love a sunburnt country."

Almost all of Australia receives at least 3,000 hours of sunshine every year. The central and western desert areas receive even more. Heat waves, or successive days of temperatures over 104 degrees Fahrenheit (40 degrees Celsius), are frequent during summer months (December through March) over much of inland Australia.

As a general rule, mean temperatures decrease as one moves from the northwest to the southeast. Between Broome and Darwin on the west, average annual temperatures are around 82 degrees Fahrenheit (28 degrees Celsius). Throughout the rest of tropical Australia (the northern half), annual average temperatures range between 70 and 79 degrees

A typically rainy scene in Tasmania. (Chris Ingham)

CLIMATE REGIONS OF AUSTRALIA

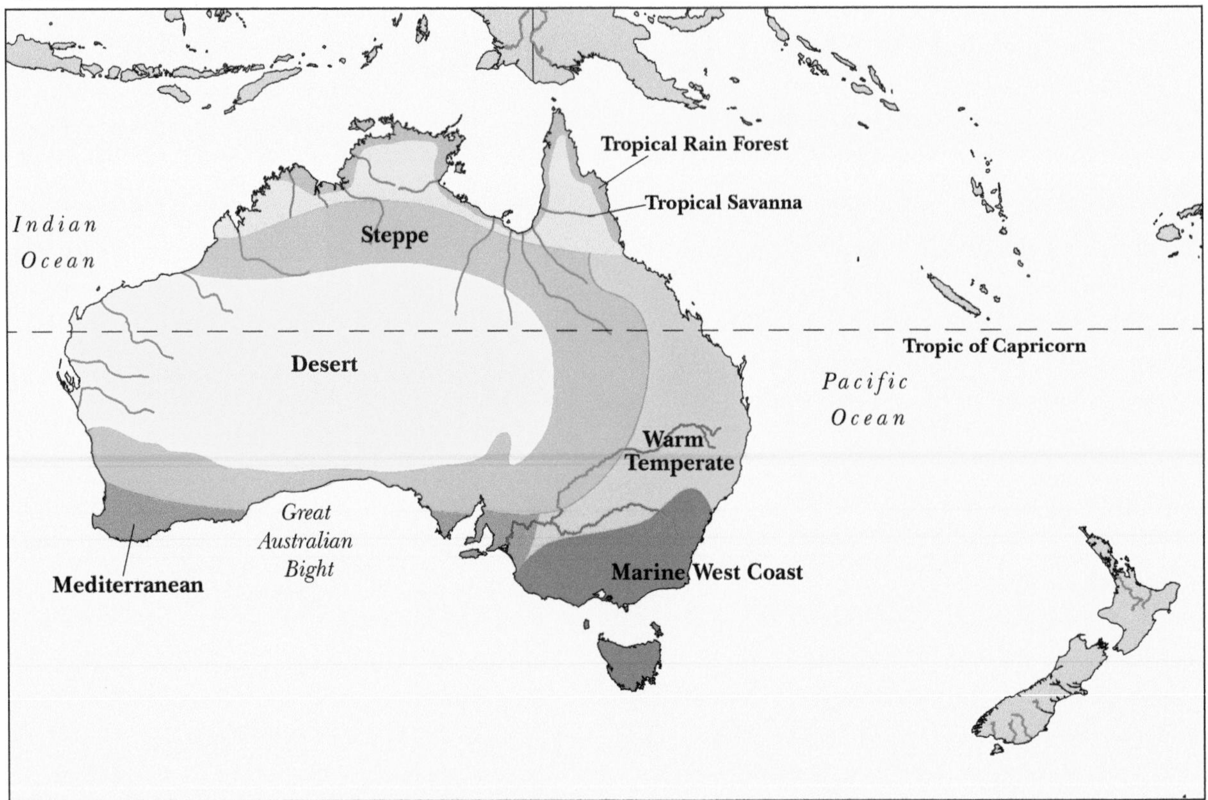

Fahrenheit (21 and 26 degrees Celsius). Between Perth and Sydney, annual averages range between 59 and 68 degrees Fahrenheit (15 and 20 degrees Celsius). Victoria and Tasmania experience annual average temperatures below 59 degrees Fahrenheit (15 degrees Celsius).

Australia's coldest regions are in its southeastern corner: the Australian Alps—the only region where snow falls regularly—and in Tasmania. Frost is common in winter in elevated areas of Australia, such as the Darling Downs and New England Tableland and throughout inland Victoria and Tasmania. At the opposite corner of the continent, in northwestern Australia, temperatures above 104 degrees Fahrenheit (40 degrees Celsius) are common throughout the summer. The highest recorded temperature in Australia, 127.4 degrees Fahrenheit (53 degrees Celsius), was recorded at Cloncurry in western Queensland, although this reading is now widely disputed. The hottest inhabited place in Australia is the small settlement of Marble Bar in Western Australia. There temperatures have risen above 100 degrees Fahrenheit (37.8 degrees Celsius) on as many as 160 days a year.

Winds

Because it straddles the Tropic of Capricorn, Australia is in the belt of subtropical highs, where dry descending air dominates the climate. During winter, large anticyclones, or highs, move across the continent, from west to east. They move at a rate of about 375 miles (600 km.) per day. Sometimes they remain stationary over the interior for several days.

Australia is in the Southern Hemisphere, where the direction of rotation of winds around subtropical cyclones is counterclockwise. This means that southeast winds—the trade winds—influence northern Australia, while the continent's southern parts are affected by strong westerly winds. These winds are called the "Roaring Forties" because they develop around 40 degrees south latitude. These cool, moist winds and accompanying cold fronts and midlatitude cyclones deliver rainfall to the southern part of Australia during the midyear winter.

The southwest corner of Western Australia receives abundant winter rain—more than 23.6 inches (600 millimeters)—followed by intense summer drought when the high pressure cells are farther south, producing dry easterly winds. This region has a typical Mediterranean climate, similar to southern California. Moving farther east, there is a similar but drier region around Adelaide in South Australia, which also has a Mediterranean climate.

In summer, when the highs move farther southward, the southern parts of Australia experience dry conditions, and heat waves are common. Southeast trade winds dominate the climate of southeast and east Australia. This is the humid subtropical climate region, where temperatures are moderate and rain falls in every month. This type of climate is also found in the eastern United States and in Western Europe, where most Australian immigrants came from in the nineteenth century. It is still the most densely populated part of Australia, containing the two largest cities, Sydney and Melbourne. The rainfall here is reliable, although there are occasional droughts, as well as occasional flooding in the tropical parts, brought on by hurricanes.

Temperatures vary greatly throughout this long region, from tropical conditions in the north to cool conditions in the south, where frosts are common and winter precipitation falls as snow in the Australian Alps and in Tasmania, and even on the Blue Mountains and the New England Tableland. Some patches of snow may remain through summer, but there are no permanent snowfields in Australia. Seasonality of precipitation also varies, with northern regions receiving more rain in summer, and the southernmost regions receiving a winter maximum.

Rain Patterns

Where the warm moist southeast trades meet mountains close to the coast, orographic rainfall occurs, making the small region of North Queensland from Ingham to Cooktown the wettest in Australia. This is a tropical wet or tropical rain forest climate. Annual rainfall totals exceed 157.2 inches (4,000 millimeters) in smaller towns like Tully and

An underground house in Coober Pedy, South Australia. (chameleonseye)

Kids wearing Santa hats play on a beach in Australia. (Imgorthand)

Innisfail, while the tiny settlement of Deeral claims the record of over 275.1 inches (7,000 millimeters) of rainfall per year. Only the mountainous western coast of Tasmania has a similar rainfall, with around 137.6 inches (3,500 millimeters) annually.

The Snowy Mountains also receive over 117.9 inches (3,000 millimeters) of precipitation a year. The record for the highest rainfall in Australia in a single year was recorded at Mount Bartle Frere in 1979: 442 inches (11,227 millimeters). The weather station there also holds the record for the highest rainfall in a twenty-four-hour period: 37.7 inches (960 millimeters). Temperature and humidity are high in this region throughout the year. Tropical cyclones are also common along all of the tropical Queensland coast, bringing heavy rains in the summer months.

The northernmost part of Australia has a tropical savanna climate and is part of the monsoon system which dominates life in India and Southeast Asia. Starting in October and continuing until April, moist northwesterly winds bring heavy rainfalls accompanied by dramatic thunderstorms to the northern coastal regions of Australia's Northern Territory, Queensland, and Western Australia.

Darwin has more than seventy thunderstorm days per year. Since temperatures are high year-round in the tropics, people of the Northern Territory describe their climate as having two seasons, the wet and the dry. Tall tropical grasses thrive in the wet summer, but dry out quickly during the rest of the year, when the wind comes from the opposite direction, bringing dry southwesterly winds from the arid heart of the continent. Evaporation is very high, and bushfires are common. These conditions are similar to those of Southern California's Santa Ana winds.

Tropical Storms

Hurricanes are referred to as tropical cyclones in Australia. These storms frequent northern and eastern tropical Australia during the summer wet season. They develop over the warm ocean waters south of 5 degrees south latitude; on average, three hurricanes affect the Queensland coast and three affect the Northern Territory coast during a season. Wind speeds exceed 46 miles (74 km.) per hour. The highest wind speed recorded in Australia was 253 miles (407 km.) per hour, during Tropical Cyclone Olivia on April 10, 1996.

Every year, tropical cyclones inflict millions of dollars' worth of damage on Australia. Fortunately, because northern Australia has a very small population, there is generally little loss of life during these storms, but severe flood damage is common. An exception to this was Tropical Cyclone Tracy, which struck the city of Darwin around midnight on December 24, 1974, killing sixty-five people and injuring 500. Because of health problems in the devastated city, most of the inhabitants were evacuated within a week. Many never returned, even though Darwin was rebuilt using strict building codes to lessen wind damage should another tropical cyclone occur.

Desert Interior

The interior of Australia is a true desert, where temperatures are high, evaporation rates are high, and rainfall is very low. About half the continent has a desert climate, a belt extending from the west Australian coast to western Queensland, and extending from the Great Australian Bight into the Northern Territory. The driest part of Australia, around Lake Eyre, receives only 4.9 inches (125 millimeters) of rainfall annually.

Surrounding the desert interior is a semiarid belt; the northern parts receive most of their scanty rainfall in the summer, while the southern parts receive their rainfall in the winter. Many towns in the south of Australia experience an interesting weather pattern in the summer: a low pressure trough causes hot dry winds from the center to bring very high temperatures that persist for several days. These winds are sometimes called brick-fielders. The situation can even affect coastal areas.

Sydney and Melbourne have recorded January temperatures above 113 degrees Fahrenheit (45 degrees Celsius). City-dwellers suffer the scorching heat and dusty conditions until a cold front brings welcome relief with thunderstorms and a very rapid drop in temperatures. This abrupt and dramatic

Lake Eyre. The striking pink coloration is from an algal bloom. (Hiltonj)

change is common in summer in Melbourne, where it is called a "cool change."

Unpredictable Weather

With rainfall and temperature data that have been collected over a long period of time, climatologists have come to understand that rainfall seasonality and variability are themselves important climatic factors in Australia. Australia is a country of climate extremes: Droughts have long alternated with great floods, and both have seemed unpredictable. Thanks to technology advances, scientists have begun to understand and even predict years of flood or drought.

The extreme variability of precipitation over much of the continent is linked to what is called the El Niño-Southern Oscillation (ENSO) phenomenon, which affects large parts of the planet. The term "El Niño" relates to times when unusually warm waters prevail in the southwestern Pacific Ocean, off the coast of Peru. "Southern Oscillation" relates to changes in atmospheric pressure in the western Pacific region, from Tahiti to Indonesia—a large region that includes Australia.

During the 1980s scientists realized, for the first time, that El Niño and Southern Oscillation are linked. Moreover, they affect large parts of the world—not just regions around the Pacific Ocean. For example, when California experiences exceptionally heavy El Niño rains, lands on the opposite side of the Pacific experience severe droughts, with added danger from fires. During periods of neutral El Niño—Southern Oscillations, as occurred in 2018-2019, rainfall in Australia is far below normal while temperatures are near or at record-breaking levels. Such were the conditions that led to Australia's catastrophic wild fires in many parts of Australia during 2019 and 2020.

Global warming, climate change, and long-term climate forecasting are areas of research with great interest and significance for Australians. Much of the Australian environment is vulnerable to significant climate variation. The Great Barrier Reef, for instance, has sustained severe damage as a result of warmer waters. A better understanding of the relevant climate processes can help protect people and resources, especially when extreme events occur.

Ray Sumner

Australian Seasons

Because Australia lies in the Southern Hemisphere, summer runs from December to February; autumn from March to May; winter from June to August; and spring from September to November. For most of the country the hottest month is January, and the coldest months are July and August. Because this pattern of seasons is opposite to that of the Northern Hemisphere, people who live in the Northern Hemisphere might find it difficult to imagine a winter with sweltering heat.

Climatology of the Pacific Islands

The Pacific Islands stretch from Palau in the west to Pitcairn in the east and from the Hawaiian Islands in the north to New Zealand in the south. The islands generally fall into two types, high and low. The high islands consist of hills and mountains, some of which are active volcanoes. They are the larger islands, such as New Zealand and New Guinea, and the main islands in groups such as the Hawaiian Islands, the Solomons, and the Marianas. The low islands are formed of coral reefs and are generally small, some rising only a few feet above sea level. When a number of low islands surround a lagoon, they are called atolls. Low islands include the Marshall and Gilbert islands. Their relatively small size and their dispersion across the ocean means that none of them has a truly continental climate; rather, their climates are governed by the ocean waters that surround them and the trade winds that blow over them.

Temperatures and Rainfall

Most of the Pacific islands are within the tropics and thus are warm, humid, and temperate the entire year. The temperature in the islands ranges from

A beach resort in Polynesia. (courtneyk)

The St. Joseph Cathedral in Neiafu, Vavau, Tonga with its whitewashed exterior. (Donyanedomam)

about 70 degrees to 80 degrees Fahrenheit (21.1 degrees to 26.7 degrees Celsius). However, the climate in parts of New Zealand, which lies on the fortieth parallel of south latitude, can be termed temperate oceanic. In New Zealand, average temperatures range from 56 degrees to 66 degrees Fahrenheit (13.3 degrees to 18.9 degrees Celsius) in January and from 42 degrees to 50 degrees Fahrenheit (5.6 degrees to 10.0 degrees Celsius) in July. The mountain areas in New Guinea, New Zealand, and some other high islands are somewhat cooler; the tallest mountains on the largest islands retain their snowcaps all year.

In general, the Pacific islands near the equator experience high humidity and abundant rainfall distributed evenly through the year; islands more distant from the equator and more affected by winds have greater seasonal fluctuations in rainfall. In the Marshall Islands, where the climate is equatorial, rain falls year-round, surpassing 117 to 156 inches (3,000 to 4,000 millimeters) annually. Rains in the interior of Pohnpei Island in the Federated States of Micronesia can reach 390 inches (10,000 millimeters) per year. The amount of rain received in a particular area also depends on whether it is exposed to the trade winds or is on the leeward side. For example, the southeastern slopes of the main Fiji islands, on which the trade winds blow, receive more than 117 inches (3,000 millimeters) of rain per year.

Many islands have wet and dry seasons. In Melanesia (the larger, mountainous islands from New Guinea to Fiji) and Polynesia (the islands within the triangle formed by Hawaii, New Zealand, and Easter Island), the wet season is from December to March and the dry season is from April to November. In Micronesia (the small islands and atolls reaching from Palau and the Marianas toward Kiribati), the wet season is from May to December and the dry season is from January to April.

Effect of the Winds

Most of the Pacific islands lie in the intertropical zone, between the trade winds—air currents that flow from the subtropical high-pressure zones found between the thirtieth and fortieth parallels both north and south of the equator. North of the equator, the trade winds, diverted by Earth's rotation from their southward path, turn to the southeast; south of the equator, the northbound winds turn to the northeast. In the eastern Pacific, the southeast trade winds lie between the twenty-fifth parallel south and slightly north of the equator, moving slightly farther north in summer. The northeast trade winds lie between the fifth and twenty-fifth parallels north.

Overall, the climatic conditions in the Pacific islands are uniform, lacking a great deal of seasonal variation, thanks to the steadiness of the trade winds. The trade winds carry relatively cool air, moving at an average wind speed of 13 knots (15 miles or 24 km. per hour). They become increasingly moisture-laden and warm as they near the equator and produce light to moderate showers, although the weather is generally fine in these areas. The windward sides of the higher islands tend to be cloudy and wet, with the leeward side relatively dry.

In the intertropical zone, the winds become generally easterly, particularly in the east Pacific, where the zone diminishes to a width of only 200 to 300 miles (322 to 483 km.). In the part of the equatorial region known as the doldrums, where the trade winds converge, the wind is generally calm or light and variable. The doldrums are over the equator, which receives the most heat from the Sun of any place on Earth. This heat causes the air to expand, creating a belt of low pressure. In the doldrums, the sky is often cloudy and humidity is particularly high. Thunderstorms or showers are common, and rainfall is abundant in the western Pacific although much scarcer in the east.

In the far western Pacific islands, the effect of the trade winds is mitigated by that of the monsoon winds. The seasonal heating and cooling of the Asian continent produces a reversal of winds that blow across the islands. From November to March, northwest monsoonal winds bring rain to Papua New Guinea, the Solomon Islands, and the western Caroline Islands; in the summer, the monsoonal winds reverse, becoming southeast winds.

Hurricanes, Typhoons, and Tropical Cyclones

Tropical cyclones are called either hurricanes or typhoons, depending on where they occur. Tropical cyclones are low-pressure systems over tropical or subtropical waters, with organized convection, such as thunderstorm activity and surface winds that rotate counterclockwise in the Northern Hemisphere and clockwise in the Southern Hemisphere. These storms are termed tropical depressions until the maximum sustained (for one minute) surface winds reach 39 miles (62.9 km.) per hour, at which point they are called tropical storms. Once the winds reach 74 miles (119.3 km.) per hour, the storm becomes either a hurricane or typhoon, depending on where it occurs. Tropical storms in the North Atlantic Ocean, in the Northeast Pacific Ocean east of the International Dateline, or in the South Pacific Ocean east of longitude 160 degrees east are called hurricanes; those in the Northwest Pacific Ocean west of the dateline are called typhoons. Those that occur in the southwest Pacific Ocean west of longitude 160 degrees east or in the Southeast Indian Ocean east of longitude 90 degrees east are called severe tropical cyclones. Those in the North Indian Ocean are severe cyclonic storms, and those in the Southwest Indian Ocean are tropical cyclones.

From July to November, typhoons frequently occur in western Micronesia. Typhoons produce gale force winds, torrential rains, and sometimes storm surges, which can damage the low islands. Tropical cyclones also occur in the southern Pacific islands.

Just below the trade winds, from the latitudes of about 30 degrees south to 40 degrees south, high-pressure zones known as anticyclones occur. They rotate and move, either toward the subpolar latitudes or in the same direction as the trade winds. In these high-pressure areas, known as the horse latitudes, the weather is generally fair.

Below the fortieth parallel south, where part of New Zealand lies, the winds are generally westerly. These warmer westerly winds converge with cold easterly winds from the polar regions, producing tropical depressions noted for their gale-force winds. This phenomenon is most pronounced in winter, when the contrast in air humidity and temperature is greatest.

The Ocean's Effect

The climate of the Pacific islands is also influenced by the movement of the waters of the ocean, particularly the tides, currents, wind-generated waves, and the rising and sinking waters produced by differences in water temperature. The temperature of the water affects that of the air above it, producing changes in atmospheric pressure. Cold surface water tends to produce high-pressure zones, and warmer water produces low-pressure zones.

The trade winds and westerlies create drifts of warm ocean water. In the northern Pacific, the northeast trade winds create a drift that splits at Hawaii to form two clockwise circulating drifts, the western of which is the larger. In the South Pacific, trade winds produce a similar, counterclockwise-moving drift that flows around many of the Pacific islands. The trade winds also produce an equatorial current that flows in a westerly direction on both sides of the doldrums. These two currents meet around Palau and the Carolines, where part of the current heads north and feeds the north Pacific drift, while the other part turns eastward through the doldrums to create the Equatorial Countercurrent.

Climate and Culture

The trade winds and currents of the Pacific islands are believed to have carried people to the islands from southeastern Asia. However, experts do not know how groups as diverse as the Asiatic Micronesians and Polynesians developed. The early islanders survived by fishing, agriculture, and hunting, but their ways of life differed greatly, reflecting the climatic differences among the islands, which produced flora and fauna that were often unique to a particular island. The Pacific islands are far from Europe, small, and spread out across the ocean; they also had varying climatic conditions, such as heat and humidity, that required adaptations in Western life-styles and few easily tapped resources. Therefore, European exploration and colonization came fairly late. Although the islands were discovered by Europeans in the sixteenth

A Polynesian man shades himself while he relaxes in a hammock. (oversnap)

AVERAGE TEMPERATURES IN SELECTED PACIFIC ISLANDS

	Average Temperature, January		*Average Temperature, July*	
City/Country	*Degrees Fahrenheit*	*Degrees Celsius*	*Degrees Fahrenheit*	*Degrees Celsius*
Suva, Fiji	80.0	26.5	73.0	23.0
Tarawa, Kiribati	83.0	28.5	83.0	28.5
Yap, Federated States of Micronesia	80.0	27.0	80.5	27.0
Auckland, New Zealand	68.0	20.0	51.5	10.5
Wellington, New Zealand	64.0	17.5	48.5	9.5
Madang, Papua New Guinea	81.5	27.5	80.0	26.5
Apia, Samoa	79.5	26.5	77.5	25.5
Honiara, Solomon Islands	80.5	27.0	79.5	26.5
Vava'u, Tonga	81.0	27.0	75.0	24.0
Fongafale, Tuvalu	83.0	28.5	83.5	28.5
Port Vila, Vanuatu	79.5	26.5	71.5	22.0

Source: National Centers for Environmental Information, National Oceanic and Atmospheric Administration (NOAA)

century, colonization did not take place on a large scale until the late nineteenth century. Fishing and agriculture still play a large part in the lives of the inhabitants of the Pacific islands.

Climate Change

For more than two decades, scientists have emphasized and re-emphasized their grave concerns about the effects of climate change on the low islands of the Pacific. Large parts of these islands are not far above sea level, and rising ocean levels caused by the melting of glacial ice may severely reduce the landmass above water, disrupting residential patterns and interfering with how people make their living. A number of Pacific islands have reported detrimental effects believed to stem from global warming. Fiji and Samoa, for instance, have seen substantial recession of their shorelines. Tonga, a group of 175 small islands, reported that the rise in sea levels had contaminated drinking water on its central and northern islands. Strong winds accompanied by salt-water spray had reduced farm production, warmer waters had reduced the supply of fish, and beaches had experienced erosion, making them less attractive to tourists. Vanuatu reported that rising seas had ruined low-lying coconut plantations, and Palau blamed global warming for the damage to its coral reefs.

In 2018, the Pacific Island Forum (PIF)adopted the Niue Declaration on Climate Change and coordinated with the United Nations to help Pacific island nations mitigate the effects of climate change on the region. In July 2019, the PIF expressed deep concern about the impact of greenhouse-gas emissions on coral atoll nations, cautioning that damage from such emissions could ultimately render the islands uninhabitable.

Rowena Wildin

Biogeography and Natural Resources

Overview

Minerals

Mineral resources make up all the nonliving matter found in the earth, its atmosphere, and its waters that are useful to humankind. The great ages of history are classified by the resources that were exploited. First came the Stone Age, when flint was used to make tools and weapons. The Bronze Age followed; it was a time when metals such as copper and tin began to be extracted and used. Finally came the Iron Age, the time of steel and other ferrous alloys that required higher temperatures and more sophisticated metallurgy.

Metals, however, are not the whole story—economic progress also requires fossil fuels such as coal, oil, natural gas, tar sands, or oil shale as energy sources. Beyond metals and fuels, there are a host of mineral resources that make modern life possible: building stone, salt, atmospheric gases (oxygen, nitrogen), fertilizer minerals (phosphates, nitrates, and potash), sulfur, quartz, clay, asbestos, and diamonds are some examples.

Mining and Prospecting

Exploitation of mineral resources begins with the discovery and recognition of the value of the deposits. To be economically viable, the mineral must be salable at a price greater than the cost of its extraction, and great care is taken to determine the probable size of a deposit and the labor involved in isolating it before operations begin. Iron, aluminum, copper, lead, and zinc occur as mineral ores that are mined, then subjected to chemical processes to separate the metal from the other elements (usually oxygen or sulfur) that are bonded to the metal in the ore.

Some deposits of gold or platinum are found in elemental (native) form as nuggets or powder and may be isolated by alluvial mining—using running water to wash away low-density impurities, leaving the dense metal behind. Most metal ores, however, are obtained only after extensive digging and blasting and the use of large-scale earthmoving equipment. Surface mining or strip mining is far simpler and safer than underground mining.

Safety and Environmental Considerations

Underground mines can extend as far as a mile into the earth and are subject to cave-ins, water leakage, and dangerous gases that can explode or suffocate miners. Safety is an overriding issue in deep mines, and there is legislation in many countries designed to regulate mine safety and to enforce practices that reduce hazards to the miners from breathing dust or gases.

In the past, mining often was conducted without regard to the effects on the environment. In economically advanced countries such as the United States, this is now seen as unacceptable. Mines are expected to be filled in, not just abandoned after they are worked out, and care must be taken that rivers and streams are not contaminated with mine wastes.

Iron, Steel, and Coal

Iron ore and coal are essential for the manufacture of steel, the most important structural metal. Both raw materials occur in many geographic regions. Before the mid-nineteenth century, iron was smelted in the eastern United States—New Jersey, New York, and Massachusetts—but then huge hematite deposits were discovered near Duluth, Minnesota, on Lake Superior. The ore traveled by ship to steel mills in northwest Indiana and northeast Illinois, and coal came from Illinois or Ohio. Steel also was made in Pittsburgh and Bethlehem in Pennsylvania, and in Birmingham, Alabama.

After World War II, the U.S. steel industry was slow to modernize its facilities, and after 1970 it had great difficulty producing steel at a price that could compete with imports from countries such as Japan, Korea, and Brazil. In Europe, the German steel industry centered in the Ruhr River valley in cities such as Essen and Düsseldorf. In Russia, iron ore is mined in the Urals, in the Crimea, and at Krivoi Rog in Ukraine. Elsewhere in Europe, the French "minette" ores of Alsace-Lorraine, the Swedish magnetite deposits near Kiruna, and the British hematite deposits in Lancashire are all significant. Hematite is also found in Labrador, Canada, near the Quebec border.

Coal is widely distributed on earth. In the United States, Kentucky, West Virginia, and Pennsylvania are known for their coal mines, but coal is also found in Illinois, Indiana, Ohio, Montana, and other states. Much of the anthracite (hard coal) is taken from underground mines, where networks of tunnels are dug through the coal seam, and the coal is loosened by blasting, use of digging machines, or human labor. A huge deposit of brown coal is mined at the Yallourn open pit mine west of Melbourne, Australia. In Germany, the mines are near Garsdorf in Nord-Rhein/Westfalen, and in the United Kingdom, coal is mined in Wales. South Africa has coal and is a leader in manufacture of liquid fuels from coal. There is coal in Antarctica, but it cannot yet be mined profitably. China and Japan both have coal mines, as does Russia.

Aluminum

Aluminum is the most important structural metal after iron. It is extremely abundant in the earth's crust, but the only readily extractable ore is bauxite, a hydrated oxide usually contaminated with iron and silica. Bauxite was originally found in France but also exists in many other places in Europe, as well as in Australia, India, China, the former Soviet Union, Indonesia, Malaysia, Suriname, and Jamaica.

Much of the bauxite in the United States comes from Arkansas. After purification, the bauxite is combined with the mineral cryolite at high temperature and subjected to electrolysis between carbon electrodes (the Hall-Héroult process), yielding pure aluminum. Because of the enormous electrical energy requirements of the Hall-Héroult method, aluminum can be made economically only where cheap power (preferably hydroelectric) is available. This means that the bauxite often must be shipped long distances—Jamaican bauxite comes to the United States for electrolysis, for example.

Copper, Silver, and Gold

These coinage metals have been known and used since antiquity. Copper came from Cyprus and takes its name from the name of the island. Copper ores include oxides or sulfides (cuprite, bornite, covellite, and others). Not enough native copper occurs to be commercially significant. Mines in Bingham, Utah, and Ely, Nevada, are major sources in the United States. The El Teniente mine in Chile is the world's largest underground copper mine, and major amounts of copper also come from Canada, the former Soviet Union, and the Katanga region mines in Congo-Kinshasa and Zambia.

Silver often occurs native, as well as in combination with other metals, including lead, copper, and gold. Famous silver mines in the United States include those near Virginia City (the Comstock lode) and Tonopah, Nevada, and Coeur d'Alene, Idaho. Silver has been mined in the past in Bolivia (Potosi mines), Peru (Cerro de Pasco mines), Mexico, and Ontario and British Columbia in Canada.

Gold occurs native as gold dust or nuggets, sometimes with silver as a natural alloy called electrum. Other gold minerals include selenides and tellurides. Small amounts of gold are present in sea water, but attempts to isolate gold economically from this source have so far failed. Famous gold rushes occurred in California and Colorado in the United States, Canada's Yukon, and Alaska's Klondike region. Major gold-producing countries include South Africa, Siberia, Ghana (once called the Gold Coast), the Philippines, Australia, and Canada.

The Exxon Valdez Oil Spill

On March 24, 1989, the tanker Exxon Valdez, with a cargo of 53 million gallons of crude oil, ran aground on Bligh Reef in Prince William Sound, Alaska. Approximately 11 million gallons of oil were released into the water, in the worst environmental disaster of this type recorded to date. Despite immediate and lengthy efforts to contain and clean up the spill, there was extensive damage to wildlife, including aquatic birds, seals, and fish. Lawsuits and calls for new regulatory legislation on tankers continued a decade later. Such regrettable incidents as these are the almost inevitable result of attempting to transport the huge oil supplies demanded in the industrialized world.

Petroleum and Natural Gas

Petroleum has been found on every continent except Antarctica, with 600,000 producing wells in 100 different countries. In the United States, petroleum was originally discovered in Pennsylvania, with more important discoveries being made later in west Texas, Oklahoma, California, and Alaska. New wells are often drilled offshore, for example in the Gulf of Mexico or the North Sea. The United States depends heavily on oil imported from Mexico, South America, Saudi Arabia and the Persian Gulf states, and Canada.

Over the years, the price of oil has varied dramatically, particularly due to the attempts of the Organization of Petroleum Exporting Countries (OPEC) to limit production and drive up prices. In Europe, oil is produced in Azerbaijan near the Caspian Sea, where a pipeline is planned to carry the crude to the Mediterranean port of Ceyhan, in Turkey. In Africa, there are oil wells in Gabon, Libya, and Nigeria; in the Persian Gulf region, oil is found in Kuwait, Qatar, Iran, and Iraq. Much crude oil travels in huge tankers to Europe, Japan, and the United States, but some supplies refineries in Saudi Arabia at Abadan. Tankers must exit the Persian Gulf through the narrow Gulf of Hormuz, which thus assumes great strategic importance.

After oil was discovered on the shores of the Beaufort Sea in northern Alaska (the so-called North Slope) in the 1960s, a pipeline was built across Alaska, ending at the port of Valdez. The pipeline is heated to keep the oil liquid in cold weather and elevated to prevent its melting through the permanently frozen ground (permafrost) that supports it. From Valdez, tankers reach Japan or California.

Drilling activities occasionally result in discovery of natural gas, which is valued as a low-pollution fuel. Vast fields of gas exist in Siberia, and gas is piped to Western Europe through a pipeline. Algerian gas is shipped in the liquid state in ships equipped with refrigeration equipment to maintain the low temperatures needed. Britain and Northern Europe benefit from gas produced in the North Sea, between Norway and Scotland.

Shale oil, a plentiful but difficult-to-exploit fossil fuel, exists in enormous amounts near Rifle, Colorado. A form of oil-bearing rock, the shale must be crushed and heated to recover the oil, a more expensive proposition than drilling conventional oil wells. In spite of ingenious schemes such as burning the shale oil in place, this resource is likely to remain largely unused until conventional petroleum is used up. A similar resource exists in Alberta, Canada, where the Athabasca tar sands are exploited for heavy oils.

John R. Phillips

Renewable Resources

Most renewable resources are living resources, such as plants, animals, and their products. With careful management, human societies can harvest such resources for their own use without imperiling future supplies. However, human history has seen many instances of resource mismanagement that has led to the virtual destruction of valuable resources.

Forests

Forests are large tracts of land supporting growths of trees and perhaps some underbrush or shrubs. Trees constitute probably the earth s most valuable, versatile, and easily grown renewable resource. When they are harvested intelligently, their natural environments continue to replace them. However, if a harvest is beyond the environment's ability to restore the resource that had been present, new and different plants and animals will take over the area. This phenomenon has been demonstrated many times in overused forests and grasslands that reverted to scrubby brushlands. In the worst cases, the abused lands degenerated into barren deserts.

The forest resources of the earth range from the tropical rainforests with their huge trees and broad diversity of species to the dry savannas featuring scattered trees separated by broad grasslands. Cold, subarctic lands support dense growths of spruces and firs, while moderate temperature regimes produce a variety of pines and hardwoods such as oak and ash. The forests of the world cover about 30 percent of the land surface, as compared with the oceans, which cover about 70 percent of the global surface.

Harvested wood, cut in the forest and hauled away to be processed, is termed roundwood. Globally, the cut of roundwood for all uses amounts to about 130.6 billion cubic feet (3.7 billion cubic meters). Slightly more than half of the harvested wood is used for fuel, including charcoal.

Roundwood that is not used for fuel is described as industrial wood and used to produce lumber, veneer for fine furniture, and pulp for paper products. Some industrial wood is chipped to produce such products as subflooring and sheathing board for home and other building construction. Most roundwood harvested in Africa, South America, and Asia is used for fuel. In contrast, roundwood harvested in North America, Europe, and the former Soviet Union generally is produced for industrial use.

It is easy to consider forests only in the sense of the useful wood they produce. However, many forests also yield valuable resources such as rubber, edible nuts, and what the U.S. Forest Service calls special forest products. These include ferns, mosses, and lichens for the florist trade, wild edible mushrooms such as morels and matsutakes for domestic markets and for export, and mistletoe and pine cones for Christmas decorations.

There is growing interest among the industrialized nations of the world in a unique group of forest products for use in the treatment of human disease. Most of them grow in the tropical rainforests. These medicinal plants have long been known and used by shamans (traditional healers). Hundreds of pharmaceutical drugs, first used by shamans, have been derived from plants, many gathered in tropical rainforests. The drugs include quinine, from the bark of the cinchona tree, long used to combat malaria, and the alkaloid drug reserpine. Reserpine, derived from the roots of a group of tropical trees and shrubs, is used to treat high blood pressure (hypertension) and as a mild tranquilizer. It has been estimated that 25 percent of all prescriptions dispensed in the United States contain ingredients derived from tropical rainforest plants. The value of the finished pharmaceuticals is estimated at US$6.25 billion per year.

Scientists screening tropical rainforest plants for additional useful medical compounds have drawn on the knowledge and experience of the shamans. In this way, the scientists seek to reduce the search time and costs involved in screening potentially useful plants. Researchers hope that somewhere in the dense tropical foliage are plant products that could treat, or perhaps cure, diseases such as cancer or AIDS.

Many as-yet undiscovered medicinal plants may be lost forever as a consequence of deforestation of large tracts of equatorial land. The trees are cut down or burned in place and the forest converted to grassland for raising cattle. The tropical soils cannot support grasses without the input of large amounts of fertilizer. The destruction of the forests also causes flooding, leaving standing pools of water and breeding areas for mosquitoes, which can spread disease.

Marine Resources

When renewable marine resources such as fish and shellfish are harvested or used, they continue to reproduce in their environment, as happens in forests and with other living natural resources. However, like overharvested forests, if the marine resource is overfished—that is, harvested beyond its ability to reproduce—new, perhaps undesirable, kinds of marine organisms will occupy the area. This has happened to a number of marine fishes, particularly the Atlantic cod.

When the first Europeans reached the shores of what is now New England in the early seventeenth century, they encountered vast schools of cod in the local ocean waters. The cod were so plentiful they could be caught in baskets lowered into the water from a boat.

At the height of the New England cod fishery, in the 1970s, efficient, motor-driven trawlers were able to catch about 32,000 tons. The catch began to decline that year, mostly as a result of the impact of fifteen different nations fishing on the cod stocks. As a result of overfishing, rough species such as dogfish and skates constitute 70 percent of the fish in the local waters. Experts on fisheries management decided that fishing for cod had to be stopped.

The decline of the cod was attributed to two causes: a worldwide demand for more fish as food and great changes in the technology of fishing. The technique of fishing progressed from a lone fisher with a baited hook and line, to small steam-powered boats towing large nets, to huge diesel-powered trawlers towing monster nets that could cover a football field. Some of the largest trawlers were floating factories. The cod could be skinned, the edible parts cut and quick-frozen for market ashore, and the skin, scales, and bones cooked and ground for animal feed and oil. A lone fisher was lucky to be able to catch 1,000 pounds (455 kilograms) in one day. In contrast, the largest trawlers were capable of catching and processing 200 tons per day.

In the 1990s, the world ocean population of swordfish had declined dramatically. With a worldwide distribution, these large members of the billfish family have been eagerly sought after as a food fish. Because swordfish have a habit of basking at the surface, fishermen learned to sneak up on the swordfish and harpoon them. Fishermen began to catch swordfish with fishing lines 25 to 40 miles (40 to 65 kilometers) long. Baited hooks hung at intervals on the main line successfully caught many swordfish, as well as tuna and large sharks. Whereas the harpoon fisher took only the largest (thus most valuable) swordfish, the longline gear was indiscriminate, catching and killing many swordfish too small for the market, as well as sea turtles and dolphins

As a result of the catching and killing of both sexually mature and immature swordfish, the reproductive capacity of the species was greatly reduced. Harpoons killed mostly the large, mature adults that had spawned several times. Longlines took all sizes of swordfish, including the small ones that had not yet reached sexual maturity and spawned. The decline of the swordfish population was quickly obvious in the reduced landings. But things have changed remarkably, thanks to a 1999 international plan that rebuilt this stock several years ahead of schedule. Today, North Atlantic swordfish is one of the most sustainable seafood choices.

Albert C. Jensen

NONRENEWABLE RESOURCES

Nonrenewable resources are useful raw materials that exist in fixed quantities in nature and cannot be replaced. They differ from renewable resources, such as trees and fish, which can be replaced if managed correctly. Most nonrenewable resources are minerals—inorganic and organic substances that exhibit consistent chemical composition and properties. Minerals are found naturally in the earth's crust or dissolved in seawater. Of roughly 2,000 different minerals, about 100 are sources of raw materials that are needed for human activities. Where useful minerals are found in sufficiently high concentrations—that is, as ores—they can be mined as profitable commercial products.

Economic nonrenewable resources can be divided into four general categories: metallic (hardrock) minerals, which are the source of metals such as iron, gold, and copper; fuel minerals, which include petroleum (oil), natural gas, coal, and uranium; industrial (soft rock) minerals, which provide materials like sulfur, talc, and potassium; and construction materials, such as sand and gravel.

Nonrenewable resources are required as direct or indirect parts of all the products that humans use. For example, metals are necessary in industrial sectors such as construction, transportation equipment, electrical equipment and electronics, and consumer durable goods—long-lasting products such as refrigerators and stoves. Fuel minerals provide energy for transportation, heating, and electrical power. Industrial minerals provide ingredients needed in products ranging from baby powder to fertilizer to the space shuttle. Construction materials are used in roads and buildings.

Location

When minerals have naturally combined together (aggregated) they are called rocks. The three general rock categories are igneous, sedimentary, and metamorphic. Igneous rocks are created by the cooling of molten material (magma). Sedimentary rocks are caused when weathering, erosion, transportation, and compaction or cementation act on existing rocks.

Metamorphic rocks are created when the other two types of rock are changed by heat and pressure. The availability of nonrenewable resources from these rocks varies greatly, because it depends not only on the natural distribution of the rocks but also on people's ability to discover and process them. It is difficult to find rock formations that are covered by the ocean, material left by glaciers, or a rainforest. As a result, nonrenewable resources are distributed unevenly throughout the world.

Some nonrenewable resources, such as construction materials, are found easily around the world and are available almost everywhere. Other nonrenewable resources can only be exploited profitably when the useful minerals have an unusually high concentration compared with their average concentration in the earth's crust. These high concentrations are caused by rare geological events and are difficult to find. For example, an exceptionally rare nonrenewable resource like platinum is produced in only a few limited areas.

No one country or region is self-sufficient in providing all the nonrenewable resources it needs, but some regions have many more nonrenewable resources than others. Minerals can be found in all types of rocks, but some types of rocks are more likely to have economic concentrations than others. Metallic minerals often are associated with shields (blocks) of old igneous (Precambrian) rocks. Important shield areas near the earth's surface are found in Canada, Siberia, Scandinavia, and Eastern Europe. Another important shield was split by the movement of the continents, and pieces of it can be found in Brazil, Africa, and Australia.

Similar rock types are in the mountain formations in Western Europe, Central Asia, the Pacific coast of the Americas, and Southeast Asia. Minerals for construction and industry are found in all three types of rocks and are widely and randomly distributed among the regions of the world.

The fuel minerals—petroleum and natural gas—are unique in that they occur in liquid and gaseous states in the rocks. These resources must be captured and collected within a rock site. Such a site needs source rock to provide the resource, a rock type that allows the resource to collect, and another surrounding rock type that traps the resource. Sedimentary rock basins are particularly good sites for fuel collection. Important fuel-producing regions are the Middle East, the Americas, and Asia.

Impact on Human Settlement

Nonrenewable resources have always provided raw materials for human economic development, from the flint used in early stone tools to the silicon used in the sophisticated chips in personal computers. Whole eras of human history and development have been linked with the nonrenewable resources that were key to the period and its events. For example, early human culture eras were called the Stone, Bronze, and Iron Ages.

Political conflicts and wars have occurred over who owns and controls nonrenewable resources and their trade. One example is the Persian Gulf War of 1991. Many nations, including the United States, fought against Iraq over control of petroleum production and reserves in the Middle East.

Since the actual production sites often are not attractive places for human settlement and the output is transportable, these sites are seldom important population centers. There are some exceptions, such as Johannesburg, South Africa, which grew up almost solely because of the gold found there. However, because it is necessary to protect and work the production sites, towns always spring up near the sites. Examples of such towns can be found near the quarries used to provide the material for the great monuments of ancient Egypt and in the Rocky Mountains of North America near gold and silver mines. These towns existed because of the nonrenewable resources nearby and the needs of the people exploiting them; once the resource was gone, the towns often were abandoned, creating "ghost towns," or had to find new purposes, such as tourism.

More important to human settlement is the control of the trade routes for nonrenewable resources. Such controlling sites often became regions of great wealth and political power as the residents taxed the products that passed through their community and provided the necessary services and protection for the traveling traders. Just one example of this type of development is the great cities of wealth and culture that arose along the trade routes of the Sahara Desert and West Africa like Timbuktu (in present-day Mali) and Kumasi (in present-day Ghana) based on the trade of resources like gold and salt.

Even with modern transportation systems, ownership of nonrenewable resources and control of their trade is still an important factor in generating national wealth and economic development. Modern examples include Saudi Arabia's oil resources, Egypt's control of the Suez Canal, South Africa's gold, Chile's copper, Turkey's control over the Bosporus Strait, Indonesia's metals and oil, and China's rare earth element.

Gary A. Campbell

NATURAL RESOURCES OF ANTARCTICA

Antarctica, the most isolated of the seven continents, generally is not considered by scientists and mineralogists as a place for productive mining and drilling. Its natural resources fall into three categories: animal, mineral, and hydrologic. The first has been overexploited. The other two have not yet been developed significantly.

What is known about the continent's mineral resources is largely speculative. The conclusions reached are based upon comparisons with mineral deposits in other parts of the world, but one cannot easily or relevantly compare Antarctica with any other place on Earth. It is the only continent that has no cities and towns, no conventional infrastructure, and no permanent human inhabitants. Its severe climate and isolation make it unique. Strides have been made, however, in assessing the hydrologic resources underlying Antarctica. It has long been known that the continent holds approximately 70 percent of the world's freshwater in its glaciers and ice shelves.

Animal Resources

The animal resources of Antarctica begin with plankton and krill in the waters around the continent's 11,165-mile (17,968-km.) coastline. These creatures at the lower end of the food chain nourish, directly or indirectly, the fish, birds, penguins, seals, and whales that constitute part of Antarctica's treasure of animal resources.

The exploitation of Antarctica's animal resources began not with whaling, which was a major, worldwide industry in the eighteenth century, but with sealing. The English navigator James Cook, who sailed into southern waters, wrote about the abundance of seals there. This launched a period of sealing activity by the British, beginning in 1778 and continuing with the arrival of American sealers into the area shortly afterward. The hunters sought seals mostly for their fur, for which ready European, Chinese, and American markets existed. They hunted the elephant seal for its oil as well. By the 1850s, seals were almost extinct in the southern polar regions.

Whaling was not a major activity in the Antarctic until the whale population in more climatically hospitable areas had been depleted almost to the point of extinction by the nineteenth century. In 1892 Christen Christensen, a Norwegian, sent a whaling expedition into Antarctic waters. By 1904 a whaling station had been established on South Georgia Island. This lucrative industry prospered and grew during World War I, when whale oil was in great demand.

The blue whale, the largest known creature on Earth, was plentiful in Antarctic waters until the 1950s. The International Whaling Commission, established in 1946, introduced zero-catch quotas for all commercial whaling so that the whale population could replenish itself. Some nations, however,

A scientist performs measurements on an ice core. (Tenedos)

Selected Resources of Antarctica

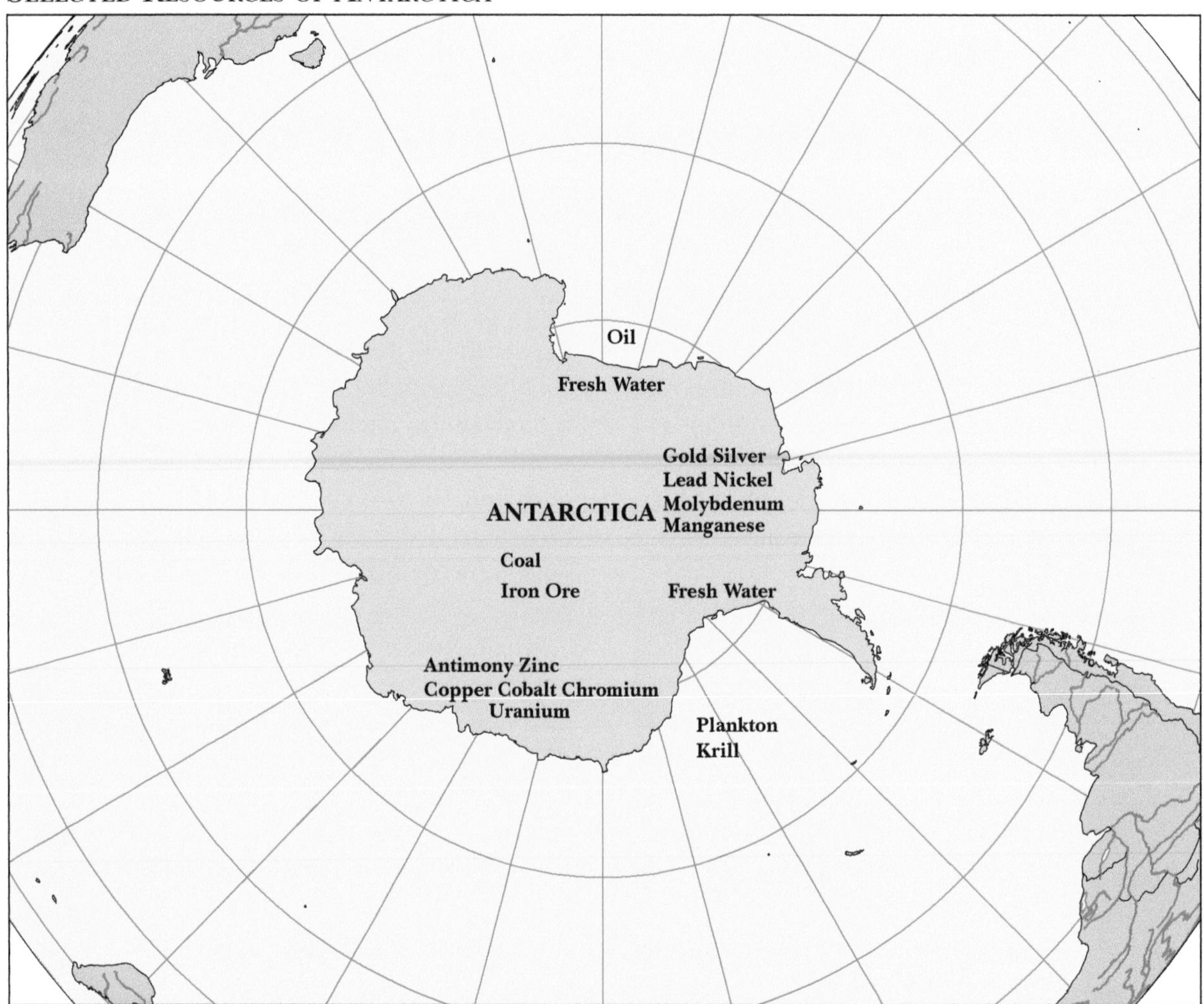

failed to honor their quota agreements and hunted the blue whale almost to extinction.

In 1912-1913, an official catch of 10,670 whales in Antarctic waters was documented. In 1930-1931, a record 40,000 whales were taken from the Ross Sea and the South Atlantic Ocean bordering Antarctica. With the depletion of the whale population, the South Georgia Whaling Station closed in 1965, giving way to ship-based processing plants. By 1990 seal hunting and whaling were no longer commercially profitable in Antarctica and its adjacent waters.

Mineral Resources

Most of what is known about the mineral resources of Antarctica is based on informed speculation by mineralogists and other scientists who must base their conclusions on comparative data from other areas where mining and drilling take place. Such conclusions are drawn at least in part by analyzing what is in the rocks that constitute that landmass. Ninety-eight percent of Antarctica is under an ice sheet whose thickness averages 8,000 feet (2,440 meters). The conjectures that have been made are based upon what has been found in a sampling of rocks from the 2 percent of the continent that is exposed and accessible.

From this small sample, it has been established that the continent probably has coal and iron ore in significant quantities. Among the minerals found in much smaller quantities are gold, silver, lead, nickel, molybdenum, antimony, zinc, copper, cobalt, and chromium. Nodules on the seafloor off Antarctica contain iron, manganese, copper, nickel, and cobalt, but in quantities too small to consider them viable mineral resources.

In 1909, during the Ernest Shackleton expedition to the Ross Sea, one of Shackleton's party found in the sedimentary rock of the Beardmore Glacier that terminates in the Ross Sea the first coal deposits ever discovered in Antarctica. Some scientists cite geologic evidence suggesting that Antarctica has no significant petroleum resources, although the waters off the continent apparently have considerable reserves of oil beneath them. Getting to such reserves is hindered by three major factors: Antarctica's lack of cities, towns, and conventional infrastructure; its lack of permanent inhabitants; and the rough seas with their frigid cyclonic winds that surround the continent. As long as petroleum reserves exist in parts of the world that are more accessible, reserves that lie beneath Antarctica's surface or beneath its adjacent waters will likely remain untapped.

Technology does not currently exist for making mining and drilling in Antarctica and off its shores

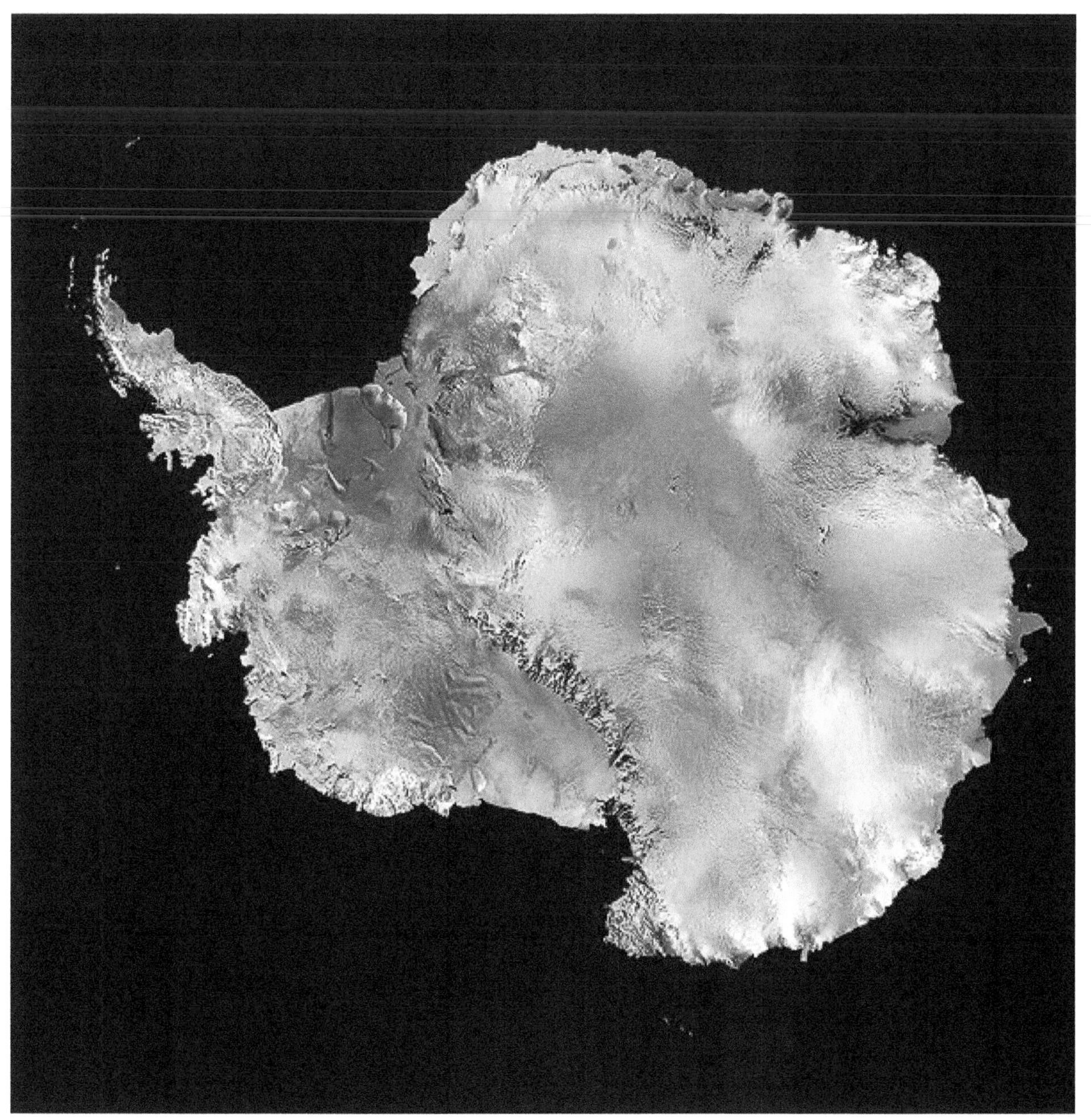

A satellite image of Antarctica. (Dave Pape)

a profitable enterprise. From a long-term perspective, it is likely that many of the problems currently related to climate and lack of a permanent population will be overcome. As shortages of various minerals develop, the technology needed to unlock Antarctica's mineral resources will be developed. When the necessity arises, it is even possible that the means will be found for controlling the climate to the point that workers can spend extended periods of time in Antarctica and possibly even live there.

One restraint on exploiting Antarctica's mineral resources in the immediate future is the Antarctic Treaty's Protocol on Environmental Protection, enacted in 1991 and enforceable as of 1998. This legally binding international act, which can be modified by the Consultative Parties, imposes an indefinite ban on all activities involving mining or drilling for minerals in Antarctica's fragile ecosystem. The act will be reviewed in 2048. It is unlikely that any substantial mining or drilling activities will occur in Antarctica before that review.

Hydrological Resources

With 70 percent of the world's total supply of fresh water, Antarctica could supply other parts of the world with water for agriculture, drinking, bathing, and industry. Recent developments at Russia's Vostok Research Station have suggested that even more fresh, liquid water may lie beneath Antarctica's surface than was originally suspected.

In 1996, while drilling through an ice cover 2 miles (3.2 km.) deep, scientists uncovered evidence of a huge underground lake, perhaps one of many, resting beneath the surface and the deep ice cover above it. To date, this newly discovered lake has not been breached because the scientific community does not wish to risk contaminating it in any way. Research is now under way to determine how this resource can best be studied without compromising its pristine qualities.

It has been suggested that huge icebergs that are calved from Antarctica's many ice shelves might, in time, be hauled to parts of the world that have water shortages. Some speculate that it would be possible not only to do this but also to create electrical power in the process by converting the ocean's thermal energy to electrical power. At present, the technology to accomplish this does not exist, but once the idea was articulated, creative minds began to work on its possible implementation.

Among the problems to be solved before such an undertaking could occur is that of keeping the icebergs from melting as they move into warmer waters and from breaking apart as they are subjected to the relentless pounding of waves in rough seas before they reach the processing plants. The hydrologic resources of Antarctica are greater than those of any other continent. It is inconceivable that they would not one day be put to use, regardless of the difficulties involved in accomplishing this objective.

R. Baird Shuman

Natural Resources of Australia

Although the smallest continent, Australia is rich in mineral and hydrocarbon (petroleum and natural gas) resources. Since the gold rushes in the middle of the nineteenth century, Australia has maintained its proud tradition as one of the world's major mining nations. It is in the top six producing countries for bauxite, gold, iron ore, lead, zinc, mineral sands, uranium, copper, diamond, lithium, manganese, nickel, silver, tantalum, tin, and vanadium. New discoveries of mineral resources and hydrocarbon resources happen frequently, but the new finds sometimes are located in remote areas and thus are difficult to exploit.

Australia is not so blessed when it comes to water resources. The continent is dry except for the tropical north. True rivers are few. Australia's water supply consists mostly of fossil water that has lain underground for millions of years and is seldom or never replaced. Problems with salinity—the salt content of the water and soil—are becoming common. More than 6 million acres (2.4 million hectares) of soil have been damaged by salinity.

Increasing population, urbanization, agricultural production, and mining have placed pressure on limited resources, causing land and water degradation and loss of biodiversity. Australia's twenty

A hillside stripped from mountaintop removal. (Ryan Hagerty)

SELECTED RESOURCES OF AUSTRALIA

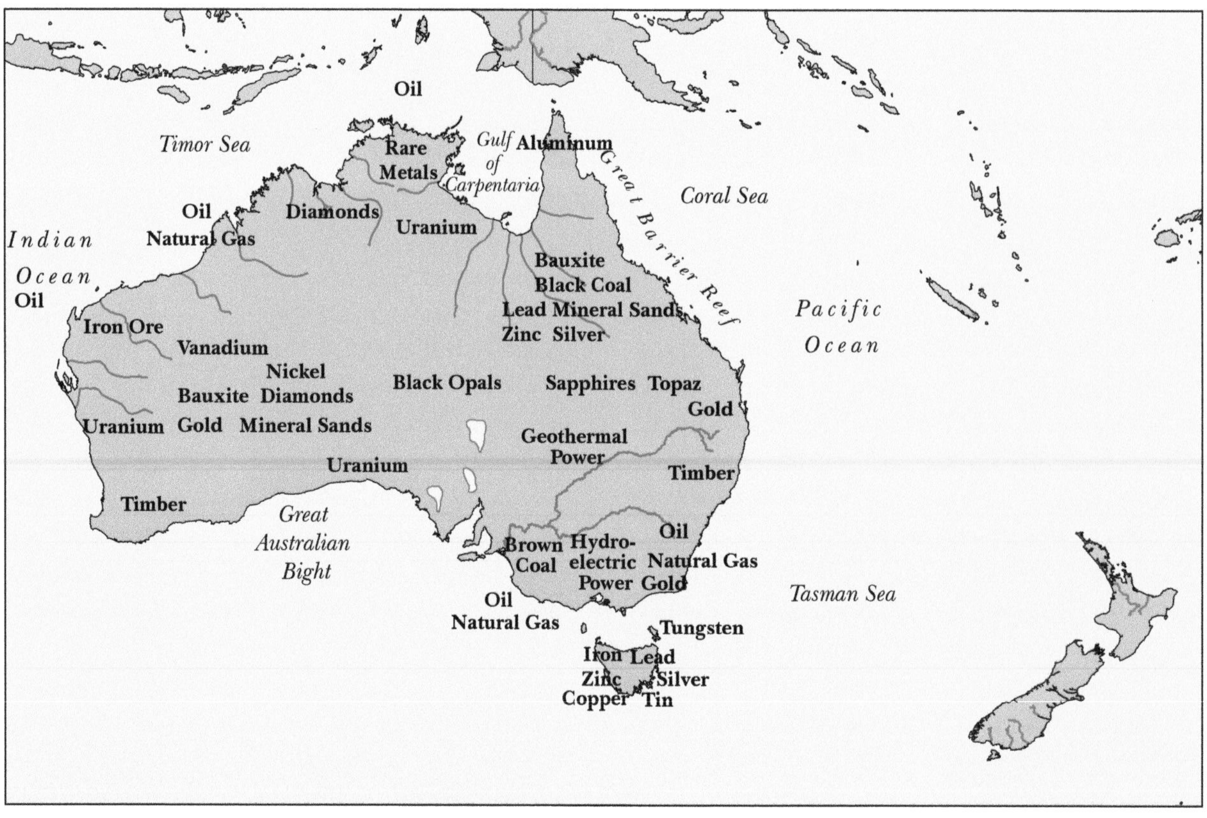

largest cities have about 86 percent of the population of 22.5 million. About 54 percent of Australia's vegetation has been cleared or disturbed. Farmers use 50-70 percent of Australia's water, 90 percent of it for irrigation. In 2016, approximately 17,300 square miles (44,800 sq. km.) of land were irrigated.

Hydrocarbon Resources

In 2017-2018, mineral fuels and lubricants accounted for 24.3 percent of the value of Australia's exports. Commercial oil reserves in 2020 were 1,821 million barrels; gas reserves (2019) were 70.2 trillion cubic feet (1.989 trillion cubic meters). Australia's oil production peaked in 2000 and has been in decline ever since, although recent discoveries have improved the production outlook. For example, the Dorado-1 well, discovered in 2018 off the coast of Western Australia, is expected to yield approximately 450 million barrels of oil.

Australia's rich natural gas reserves ensure the country's energy independence for at least the next fifty years. Today, approximately 22 percent of Australia's electricity is generated by natural gas. This percentage is expected to rise in the decades to come as greater plant capacity comes on-line. In the near term, Australia is expected to replace Qatar as the world's leading exporter of liquefied natural gas (LNG). Already, Australia is a vital source of LNG for many countries, especially Japan (the largest importer of Australian gas), China, and South Korea, as well as Taiwan, Malaysia, and India.

Renewable Energy Resources

In 2019, Australia achieved its goal of deriving 23.5 percent of its electricity generation from renewable energy sources by 2020. Leading the way is hydropower, which accounted for 32.5 percent of renewable electricity generation, or 7.5 percent of the country's overall electricity generation. Some of this energy is produced by Tasmania's thirty hydropower stations and fifteen dams. A more substantial source is the Snowy Mountain Scheme, a massive hydroelectric and irrigation complex in New South Wales, and Australia's single largest engineering project to date. A proposed expansion of the complex, dubbed "Snowy 2.0," would increase the hydropower output by upwards of 50 percent.

A close second to hydro in Australia is wind power, which in 2018 supplied about a third of Australia's renewable electricity generation and 7.1 percent of the country's total electricity. Australia has dozens of wind farms, with most facilities in the states of South Australia and Victoria. Solar power capacity is growing rapidly, and given Australia's sunny weather, holds great potential going forward. In 2019, Australia's more than 2 million solar-power installations accounted for approximately 5.2 percent of the country's electricity.

Geothermal energy holds great potential for Australia, possibly providing up to 7 percent of the country's electricity generation by 2030. Particularly promising are areas of South Australia where exploratory wells have yielded encouraging results. But currently, apart from a small plant in Birdsville, Queensland, Australia has barely begun to exploit its geothermal potential.

Mineral Reserves

The mineral industry generates a high proportion of Australia's export income. In 2016, mineral exports and scrap metal accounted for 35 percent of the value of Australia's exports. Australia's greatest mineral reserves are located in Western Australia (iron ore, nickel, bauxite, diamonds, gold, and mineral sands), Queensland (bauxite, bituminous or black coal, lead, mineral sands, zinc, and silver), New South Wales (bituminous coal, lead, zinc, silver, and mineral sands), and Victoria (brown coal or lignite and offshore oil and natural gas).

In 2016, Australia was home to 29 percent of the world's proven uranium reserves. The largest reserves are in northern and northwestern Queens-

The Snowy Mountains Scheme hydroelectricity and irrigation complex. (Cmh)

land, the Northern Territory, Western Australia, and South Australia. Australia was the world's largest producer of iron ore in 2019. Some is destined for domestic iron and steel production, but most production is for export, mainly to China, Japan, Taiwan, and South Korea. Western Australia's Hamersley iron province contains billions of tons of iron ore, but the ore's location in an isolated area is an impediment to full-scale exploitation. Tungsten has been mined in Queensland since colonial times and is a major export. The rich Kambalda nickel deposits, 35 miles (56 km.) southeast of Kalgoorlie, were discovered in 1964. Other nickel deposits are at Greenvale (Queensland) and in the Musgrave region on the borders of Western Australia, South Australia, and the Northern Territory. In 2019, Australia was the source of 6.7 percent of the world's nickel.

Enormous deposits of bauxite have made Australia the world's leading producer of bauxite and the second-largest producer of alumina (after China). Australia ranks as the world's second-largest gold producer and sixth-largest silver producer. Small amounts of platinum and palladium have also been located.

Guano in Nauru

The tiny island of Nauru is an extreme example of how the Pacific nations tend to rely for major export earnings on a few primary natural resource products that support their economies. Nauru based its industry and wealth on the mining and processing of phosphate, the product of thousands of years of accumulation of seabird excrement, called guano. Raw phosphate was mined and processed into fertilizer on the island, and then exported. Because of the phosphate industry, Nauru had one of the highest per capita incomes in the developing world, peaking in the mid-1970s. However, by the turn of the century phosphate reserves had been exhausted and the nation found itself under crushing debt, leaving its economy to rely largely on foreign aid, income from an offshore Australian detention center, and its status as a tax haven. On top of that, phosphate mining stripped the island of soil and vegetation, making the interior of the island uninhabitable. New mining initiatives began in 2005, tapping into leftover phosphate from previously inefficient techniques, but it is estimated that only thirty years' worth of revenue will be generated.

Australian opals. (oxign)

Australia has rich deposits of industrial minerals such as clays, mica, salt, dolomite (limestone), refractories, abrasives, talc, and asbestos. Gem minerals include unique black opals, sapphires and topaz. Australia is the world's third-leading source of diamonds, mostly of industrial quality. The island state of Tasmania has some important mineral deposits, including iron, high-grade lead-zinc-silver, copper, tin and tungsten.

Water

The average annual rainfall is approximately 18 inches (450 millimeters), and more than one-third of the mainland, principally the interior, receives less than 10 inches (25 centimeters) per year. Aridity or semiaridity prevails over most of Australia, and evaporation rates are extremely high. Due to the high evaporation rates, less than 2 inches (50 millimeters) of the average annual rainfall actually contributes to usable water for natural and artificial systems.

Permanently flowing rivers are found only in eastern Australia, southwestern Australia, and Tasmania. The major exception is the Murray River, a stream that rises in the Mount Kosciuszko region in the Eastern Uplands and is fed by melting snows. It acquires a volume sufficient to survive the passage across the arid and semiarid plains that bear its name and to reach the Indian Ocean southeast of Adelaide. Nevertheless, the combined discharge from all Australian rivers, including the principal river system, the Murray-Darling, is only about one-quarter that of the Mississippi River.

There are wide regional disparities in the availability of water. In the sparsely populated northern sector, runoff draining into the Timor Sea and the Gulf of Carpentaria accounts for half the national total. The tropical north as a whole contributes about two-thirds of the national total water runoff. Much of Australia's population, however, is concentrated in New South Wales and Victoria in the southeastern sector of the country.

Many areas, such as the Nullarbor Plain, which is underlain by limestone, and the sand ridge deserts are without surface drainage. There are, however, underground streams. A map of Australia can be misleading. Many of the "lakes" depicted in the interior areas are now salt lakes that often contain no water for years at a time.

The Great Artesian Basin, a huge underground aquifer, is the largest of its type in the world and provides water security to one-fifth of mainland Australia's population. Groundwater systems extend over 40 percent of Australia's land and in many cases, more water is being extracted than is replenished naturally. In some areas of Australia, people are using three to four times more water than is being recharged, so the aquifer is in danger of running out.

Scientists have warned that the quality of Australia's inland water is increasingly at risk. The two main causes are toxins produced by blue-green algae and agricultural chemical residues. Some of the algae in Australia produce unique toxins, which can be long-lived and potentially toxic to humans and animals. The algal toxins, known as paralytic shell-

Tuckers Creek in Scottsdale, Tasmania. (Thouny)

fish poisons, could accumulate in the food chain, affecting human health, the rural industries, and Australia's plant and animal life.

Soil and dryland salinity has become a serious problem in Australia. Much of the dryland salinity is due to low rainfall and high evaporation rates. Salts in groundwater originate either from minute quantities dissolved in rainwater, from the chemical breakdown of rocks, or from direct contact with seawater. The salinity issues are greatly exacerbated by land clearance to accommodate agriculture and worsened further by irrigation. Western Australia, one of the most affected states, has more than 5 million acres (2 million hectares) of salt-affected land. Biologists estimate that 75 percent of the water-bird species are in decline as a result.

In South Australia, at least 20 percent of surface water resources were sufficiently saline to be above desirable limits for human consumption. In the Northern Territory, which is relatively well watered, the groundwater in the southern half of the state is too saline for most uses. In the northern half of the Northern Territory, groundwaters are mainly fresh because the high rainfall tends to flush away salt before it is concentrated by evaporation. It has been estimated that soil salinity in Australia could expand to 95 percent of total irrigated lands and to one-third of dryland arable area by 2050.

Tasmania is well watered and thus the opposite of mainland Australia. It has two major river systems, the Derwent in the southeast and the South Esk in the northeast. There are many small systems flowing to the west coast and innumerable lakes. The Central Plateau is studded with more than 4,000 lakes in a landscape similar to northern Canada and Finland.

Australia's Share of World Mineral Resources

Mineral	*Percentage of World Resources*
Bauxite	17
Black coal	10
Brown coal	24
Copper	11
Diamond	2
Gold	19
Ilmenite (oxide of iron and titanium)	19
Iron ore	29
Lead	38
Lithium	30
Manganese Ore	13
Nickel	23
Rutile	50
Silver	16
Tantalum	67
Uranium	34
Vanadium	20
Zinc	29
Zircon	63

Forests

Approximately 16 percent of Australia's land is forested. The chief commercial forests are in high-rainfall areas on the coast or in the coastal highlands of Tasmania and the southeastern and eastern mainland, and along the southwestern coast of Western Australia. Approximately 80 percent of the industry's plantation and native log timber supplies are derived from sustainable production activities in forests throughout the country. Forestry and its products contributed approximately 0.5 percent to Australia's economy and employed some 52,000 people in 2019. Australia is the world's largest exporter of woodchips and the eighth-largest exporter of logs.

In Tasmania, there are softwood plantations in the Fingal and Scottsdale areas and inland from the northwest coast. The main types of trees are the eucalyptus, an evergreen genus providing timbers of great strength and durability, and a great variety of rain forest trees such as coachwood, crab apple, and yellow carabeen. Several regions both on the mainland and in Tasmania are intensively exploited for wood pulp, partly for export to China and Japan.

Dana P. McDermott

Natural Resources of the Pacific Islands

Resources are the part of the natural environment that is given value by human knowledge and perception. The Pacific Islands and New Zealand vary greatly in the availability of renewable and nonrenewable natural resources. These, in turn, influence settlement patterns and local economies, including exports, industry, and tourism. The Pacific region is rich in a number of natural resources that often form a basis for the countries' economies. These resources include minerals, water, forests, cash crops and plantation agriculture, and ocean fishing. The tourism industry has developed on several islands where the natural resources provide activities and attractions of interest to visitors. Some of the smaller islands, however, have virtually no natural resources apart from those in the ocean surrounding them.

The Ok Tedi Mine in southwestern Papua New Guinea. (Ok Tedi Mine CMCA Review)

Minerals

The Melanesian countries located along the Pacific and Indian tectonic plate collision zone are well endowed with mineral resources; the extraction of minerals that are desired by industrial nations has provided significantly for the national wealth of these countries. In Papua New Guinea, one of the most resource-rich countries in the region, mineral resources account for two-thirds of the country's total exports. Two of the world's largest copper and gold mines are found in Papua New Guinea. The copper deposits on Bougainville Island, an offshore island that is part of the country, contain one of the world's largest reserves and were intensively mined until terrorist activities closed the mine in the early 1990s; the mines remained closed as of 2020. Gold and petroleum are found in Papua New Guinea as well, especially along the Fly River. Gold is also mined in Fiji, Palau, and the Solomon Islands.

New Caledonia is the world's fourth-largest producer of nickel ore, and also contains chrome, iron, cobalt, manganese, copper, silver, and lead. The Solomon Islands also have bauxite, lead, zinc, and nickel. Manganese was mined in Vanuatu, but is now depleted. Phosphate, useful in producing fertilizer, was formerly contained on five great phosphate rock islands in the Pacific region: Banaba (part of Kiribati), Makatea in French Polynesia, the Solomon Islands, the Marshall Islands, and Nauru. The phosphate deposits on all these islands have been either seriously depleted or are no longer economically feasible to exploit.

Folk Medicine in the Samoan Rain Forest

The Samoan tropical rain forest is useful for more than timber: The indigenous people of Samoa use many rain forest plants to treat disease. Samoan medicine women use dozens of plant species to treat human ailments and pass down their knowledge of this folk medicine from mother to daughter.

There are some seventy remedies or cures using rain forest plants, including the polo leaf, used for fighting infections; the ulu ma'afala root, for diarrhea; and the bark of a local tree, used to combat hepatitis. Another plant species is being tested for its effectiveness against AIDS. Although these plants are termed "folk medicines," they are not merely local cures. At least twenty-five percent of the prescription drugs made in the United States are extracted from flowering plants, many of which grow in tropical rain forests.

However, during the 1990s the Samoan rain forest reached a crisis point. By 1999, commercial logging operations had cut down more than eighty percent of the coastal rain forest, and logging continued despite bans and designated conservation areas. Ethnobotanist Paul Alan Cox, noted for his research on the use of tropical plants for health remedies, won the Goldman Environmental Prize in 1997 for his efforts to save the Samoan rain forest during this time. In honor of his work, the Samoan Islanders gave him the name Nafanuna, after one of their traditional protective gods. A 2014 report to the Convention on Biological Diversity noted that Samoa's native merchantable forests were no longer profitable for logging due to depletion, and therefore only small-scale operations remained. That same report also emphasized the importance of preserving native vegetation for herbal medicine.

Water

Fresh water is a necessary resource for any area with human settlement. The Pacific islands are split in terms of water resources. Because of their location in a tropical wet/dry climatic region, heavy seasonal precipitation is common. The high islands are richly endowed with water, as the mountainous terrain provides conditions conducive to heavy, frequent rainfall, and snow on the higher peaks. Water in these countries can be tapped for energy as well. The steep terrain and heavy rains in the highlands of Papua New Guinea, for example, enable the development of reservoirs and hydroelectric power plants that supply city dwellers and much of the rural population with water and electricity. Low-lying atolls are deficient in precipitation and limited in water-storage capacity; thus, they must import most of their water to support their populations.

New Zealand's highland glaciers, lakes, and mountain streams supply its people with abundant freshwater as well as with hydroelectric power. Increasing urbanization, however, continues to place greater demands on domestic water use. To deal with this problem, storage lakes and treatment

A logger unloads timber taken from New Zealand's plantation forests. (Benchill)

plants have been built to increase the country's water storage capacity, and water conservation measures are in effect in Wellington and Auckland, New Zealand's two largest cities.

Tropical Rain Forests

Tropical rain forests are botanically complex. At lower elevations, they consist of several layers of canopy (different heights of plants) and a wide variety of plant and tree species. At higher elevations, the lower canopy is absent but several tree species can be found. Sizable expanses of tropical rain forest still exist on most of the larger islands of Melanesia, but some of the drier, flatter islands in the Pacific have only sparse vegetation. Rain forests still cover about 60 percent of Papua New Guinea (the third-largest natural tropical rain forest in the world), although deforestation as the result of illegal logging is destroying about 1.4 percent of the forest per year.

Some of the world's most biologically diverse environments are being threatened in these operations, but landowners see the quick cash sales to loggers as attractive in the short term, even though the long-term practice will damage their traditional lifestyles. The global expansion of commercial lumbering also threatens nearby portions of the Solomon Islands and many unique environmental settings in Polynesia. Much of the subalpine mountainous region of New Zealand is covered with rain forest. This country, too, has a significant forest products industry that involves the logging and processing of its rain forest.

Tropical rain forests contribute greatly to the world's oxygen supply; thus, the cutting of the rain forests depletes the supply of this vital element of air that helps humans breathe. Many plants from rain forests are used for medicinal purposes as well. The cutting of forest, therefore, reduces the availability of medicinal plants.

Cash Crops and Plantation Agriculture

The tropical and subtropical latitudes of the Pacific islands provide year-round warm temperatures, fertile soils, and adequate rainfall, making these islands ideal for tropical agriculture for both subsistence use and for export. As with minerals, specific crops have become the primary export products in certain Pacific island countries. Coconut, sugar, pineapple, and bananas are grown on small local farms and on large, foreign-owned plantations for mass production and export to industrial nations. Other tropical crops and livestock provide additional export income.

Coconuts are the most widespread plantation crop in the region. Coconuts require year-round warm temperature, well-drained soil, and plenty of moisture and sunlight; they grow best in low altitudes near the coast. The coastal lowland areas of the tropical Pacific islands are therefore ideal for large and small-scale coconut plantations. Products from the coconut include copra, the dried meat extracted from the nut; desiccated (finely grated) coconut for cooking and baking; milk and cream to drink or use in cooking; oil used in cooking and for producing margarine, fine soap, and cosmetics; and fiber from the husk of the nut, used to manufacture rugs and cord. In addition, the timber from the tree is used as a building material and to make furniture and utensils.

Coconuts are grown in Tonga, Samoa, American Samoa, Papua New Guinea, Guam, the Cook Islands, French Polynesia, New Caledonia, Fiji, Vanuatu, Tuvalu, the Solomon Islands, and the Marshall Islands. Copra, extracted from the meat of coconuts, is processed in Samoa, Papua New Guinea, the Solomon Islands (where it is the major source of income for small-scale farmers), Vanuatu, Fiji (where coconuts are the second-largest crop), Kiribati, Tuvalu, the Marshall Islands, and Tonga. Coconut cream and oil are processed in Samoa, American Samoa, and the Marshall Islands; coconut oil is processed in Tonga.

Sugar is the main plantation crop processed in and exported from the Hawaiian Islands, Fiji, Easter Island, and the Marianas. While sugarcane grows naturally in these countries, the processing industry is far more complex and expensive than that for coconut; therefore, these operations are limited to large-scale, often foreign-owned, operations. Small farmers cannot compete as they do with coconuts.

Pineapples are grown, processed, and exported from the Hawaiian Islands as well. Bananas grow in the Hawaiian Islands, Western Samoa, Tonga, and other islands of Micronesia. Other cash crops exported from the region include rubber, tea, and coffee in Papua New Guinea; coffee in Vanuatu; cocoa in Samoa, Papua New Guinea, the Solomon Islands, and Vanuatu; palm oil in the Solomon Islands and Papua New Guinea; rice in Fiji and the Solomon Islands; and vanilla in Tonga. Samoa also produces taro and passion fruit.

New Zealand is too far south for tropical agriculture but its climate and terrain are ideal for raising livestock. Sheep are the main animal raised in the country. Export products from sheep farming include wool, lamb, mutton, cheese, and other dairy products. Cattle are raised in New Zealand, Vanuatu, and Samoa, for export of beef. Poultry and pork are produced in Samoa and Papua New Guinea.

Ocean Fishing

Saltwater fish are another important resource for much of the Pacific islands. This should be no surprise, given the huge expanse of ocean within the Pacific region. Fishing of tuna and other ocean fish occurs from Papua New Guinea, Fiji, the Solomon Islands, Vanuatu, Kiribati (sent for canning to American Samoa), Tonga, Samoa (though largely untapped),

Statue of "Charlie the Tuna" at the Starkist cannery in Atu'u, American Samoa. (Eric Guinther)

New Zealand, Vanuatu, the Federated States of Micronesia, Tuvalu, the Marshall Islands, Palau, Pitcairn, French Polynesia, Niue, and the Midway Islands.

As with plantation agriculture, fishing influences industrial development in some Pacific island countries. There are fish canneries, for example, in Papua New Guinea, Fiji, the Solomon Islands, Vanuatu, and American Samoa.

Tourism

Tourism can be considered a resource in that the tourist industry uses a country's natural resources to attract visitors. This builds the economy as tourist-related infrastructure (the communication structures that make it possible for the activity to take place, such as roads and public transportation), facilities (hotels, resorts, restaurants, sporting centers, museums), and activities (swimming, camping, and other recreation) are set up to make tourism possible. The Hawaiian Islands and New Zealand have the most well-developed tourist industries, including urban and environmental attractions. Papua New Guinea, Fiji, Tahiti, the Cook Islands, New Caledonia, Tonga, Kiribati, Vanuatu, and the Samoas have invested in airport and hotel facilities and tourist activities to attract visitors to their islands. Overall, tourism revenue in the Pacific islands has increased markedly in the twenty-first century ($1.4 billion in 2013 alone), although the COVID-19 pandemic in 2020 put at least a temporary end to all nonessential travel to the region.

Resource-Poor Areas

The low islands and atolls are all poor in natural resources. Tuvalu, Wake Atoll, and the Marshall Islands, for example, have no streams or rivers, and groundwater is not potable. All water needs must be met by catchment systems with storage facilities or by desalination plants. Nauru has limited natural freshwater resources, with roof storage tanks to collect rainwater. Intensive phosphate mining during most of the twentieth century left the central 90 percent of Nauru a wasteland and threatens the limited remaining land resources. Few other resources exist in Nauru and most necessities, including freshwater, must be imported from Australia. Kiribati is another resource-poor area. With the exception of the island of Banaba, which is of limestone origin, these islands are coral atolls with shallow topsoil and low water-absorption capacity that prevent cultivation of most crops.

Laurie A. Garo

Discussion Questions: Natural Resources

Q1. Why do scientists dismiss Antarctica as a potential locale for productive mining? How might the continent's water resources be exploited? What challenges would be inherent in such an endeavor?

Q2. How much of Australia's energy is derived from renewable sources? What factors make Australia a prime candidate for solar power? What role might geothermal resources play in Australia's energy future?

Q3. Compare and contrast the so-called high Pacific islands and the low Pacific islands. Which type of island is likely to have a greater abundance of mineral resources? How are the water supplies between the two types of islands likely to differ?

Flora and Fauna of Antarctica

The harsh climate of Antarctica makes it one of the most inhospitable places on Earth, allowing only a relatively small number of organisms to live there. Most animals associated with Antarctica, such as penguins, seals, and whales, are actually seasonal visitors or live on the ice and in the ocean surrounding the continent. Permanent terrestrial (land) animals and plants are few and small. There are no trees, shrubs, or vertebrate land animals. Native organisms are hardy, yet the ecosystem is fragile and easily disturbed by human activity, pollution, global warming, and ozone-layer depletion.

The Antarctic continent has never had a native or permanent population of humans. In 1998 the United States, Russia, Belgium, Australia, and several other countries signed another in an ongoing series of treaties to preserve Antarctica and keep it for peaceful international uses such as scientific research and ecotourism. The agreement, known as the Protocol on Environmental Protection to the Antarctic Treaty, will remain in effect at least through 2048.

Terrestrial Flora

There are only two types of flowering plants in Antarctica, a grass and a small pearlwort. These are restricted to the more temperate Antarctic Peninsula. Antarctic hairgrass forms dense mats and

Antarctic pearlwort (Colobanthus quitensis) *is one of only two flowering plants found in Antarctica. (Liam Quinn; CC BY-SA 2.0)*

grows fairly rapidly in the Southern Hemisphere summer (December, January, and February). At the end of summer, the hairgrass's nutrients move underground and the leaves die. Pearlwort forms cushion-shaped clusters and grows only 0.08 to 0.25 inch (2 to 6 millimeters) per year.

Numerous species of primitive plants, such as lichens, mosses, fungi, algae, and diatoms, live in Antarctica. Lichens are made up of an alga and a fungus in a symbiotic (interdependent) relationship. They can use water in the form of vapor, liquid, snow, or ice. Lichens grow as little as 0.04 inch (1 millimeter) every 100 years, and some patches may be more than 5,000 years old. Mosses are not as hardy as lichens but also grow slowly; a bootprint in a moss carpet may be visible for years. Fungi are found on the more temperate peninsula and most are microscopic.

Algae grow in Antarctic lakes, runoff near bird colonies, moist soil, and snowfields. During the summer, algae form spectacular red, yellow, or green patches on the snow.

Bacteria are not plants, but they are also found in lakes, meltwater, and soils. As elsewhere on the Earth, bacteria play a role in decomposition. Because of the extreme cold, they are not always as efficient in Antarctica, and animal carcasses may lay preserved for hundreds of years.

Terrestrial Fauna

Terrestrial animals in Antarctica are limited to invertebrates. There are more than 100 species, including insects, mites, nematode worms, copepods (tiny crustaceans), and microscopic protozoans. The largest is an insect, a long wingless midge only 0.47 inch (1.2 centimeters) long. The invertebrates have adopted various methods of dealing with the cold: Some dry out; others avoid freezing by making antifreeze chemicals in their blood; others do freeze but ice crystals form between cells, not inside them. Many avoid the cold by attaching themselves as parasites to a warm-blooded host.

A whale breaching the Antarctic waters. (heckepics)

Antarctica's Dry Valleys

Found nowhere else on earth, Antarctica's dry valleys are cold, ice-free areas in which nearly all the snow sublimates, or vaporizes, without melting. They are some of the driest areas in the world, but some organisms—such as pockets of bacteria living underground and lichens and algae living inside rocks—manage to survive. Tiny nematode worms can tolerate being frozen solid and even survive the complete loss of body water: They can be freeze-dried and reconstituted. The nematodes lie dormant for long periods of time, then breed wildly when water and warmth become available. Some researchers believe that life in dry valleys may resemble what life could be like on other, very cold astral bodies such as Europa, a moon of Jupiter.

Marine Life

Unique in the world's oceans, the marine ecosystem of Antarctica is based on shrimplike crustaceans called krill, tiny animals that live in open water and under sea ice and grow up to 2 inches (5 centimeters) long. Nearly all Antarctic vertebrates, such as whales, seals, birds, and fish, depend directly or indirectly on krill. Masses of krill covering 300 square miles (775 sq. km.) and 328 feet (100 meters) thick have been seen. Other protozoans, small crustaceans, and larval fish float with the krill to form plankton.

Whales

Among the creatures that feed on plankton are the largest animals in the world, the baleen whales, which have sievelike plates instead of teeth. Seven species feed in the southern ocean in the austral summer. A whale can eat thousands of pounds of

Two elephant seals fighting. (Michael L. Baird)

krill per day. All the larger baleen whales have been hunted by whalers, and most species are still recovering despite decades of protection.

The blue whale is the largest animal ever to live on Earth, growing up to 100 feet (30 meters) long and weighing 150 tons (136,000 kilograms). The most common large whale, the fin whale, grows up to 85 feet (27 meters) long. The fastest whale, the sei whale, grows up to 65 feet (20 meters) long. The southern right whale was named by whalers because it was the "right" whale to catch: They are fat, very slow, float when killed, and grow up to 56 feet (17 meters) long. Their numbers were reduced almost to extinction by whaling. Humpback whales, up to 62 feet (19 meters) long, have recovered most successfully from whaling. They are known for their haunting songs and unique feeding habits. The minke is the smallest baleen whale, at 26 feet (8 meters) long, and was not hunted extensively.

There also are toothed whales in the Southern Ocean. Sperm whales are occasionally found south of 70 degrees latitude. Killer whales are found up to the edge of the Antarctic ice. They travel in groups called pods and prey on seals, penguins, and other whales. Since 1986, all commercial whaling has been prohibited in Antarctic waters.

Seals

Five species of true seals live and breed on or near Antarctica. Weddell seals and crabeater seals are the most abundant. Weddell seals are found on fast ice (sea ice that is attached to land) and eat squid and fish. They spend much of the winter under the ice through which they must chew breathing holes. Their teeth can become so severely worn that they are unable to catch food or keep their breathing holes open, and some die from being trapped under the ice. Despite their name, crabeater seals feed almost exclusively on krill and live on the pack ice (sea ice not attached to shore).

Leopard seals, named for their spotted coats, usually appear on the continent only in summer. About half their diet is krill, but they are impressive predators and will eat penguins, fish, squid, and the

A pair of wandering albatrosses. (mzphoto11)

A group of penguins in Antarctica. (NicoElNino)

young of other seals. The Ross seal is one of the least known of the Antarctic seals. It lives on the pack ice and eats mainly squid. The elephant seal is one of the most recognizable seals, and is found on the continent in summer. It is named for the large proboscis (enlarged snout) of the male seals. Pups suckle for about three weeks, during which time the mother does not feed and may lose over 660 pounds (300 kilograms) in weight.

Birds

The most familiar creatures associated with Antarctica are penguins, flightless birds well-adapted to the cold. As many as 35 million to 40 million penguins live on Antarctica and the sub-Antarctic islands. They are excellent swimmers and, depending on the species, eat fish, squid, and krill.

The Human Impact on Antarctic Life

Human activities have major impacts on Antarctica. Some are as obvious as whaling, some are less so. Adélie penguins nest on low earth mounds, often less than one foot (thirty centimeters) high. If the mounds are bulldozed flat, the penguin colony may not recover, even after the humans have gone. Even so-called green activities such as eco-tourism can introduce foreign bacteria and cause wear and tear. Strict regulations require tourists' clothing to be cleaned before they set foot on the continent, and there are high fines for littering. Research stations must burn or carry away all trash. Some impacts are invisible: climate change may raise the temperature of Antarctica. Warming such a frozen land might seem like a good thing, but native organisms are adapted to the cold environment, and warming could impact them in negative ways.

Emperor penguins and Adélie penguins are the only species that breed on the continent. Emperors are the largest penguin, at 4 feet (1.2 meters) tall. Adélie penguins are very sensitive to physical and environmental disturbance, and may serve as biological indicators of the health of the Antarctic ecosystem. Other penguin species, the king, gentoo, rockhopper, royal, and macaroni, breed on sub-Antarctic islands and occasionally on the peninsula itself.

The majority of the flighted seabirds are albatrosses and petrels. The wandering albatross has the largest wingspan in the world (11.5 feet/3.5 meters), and actually raises its chick during the Antarctic winter. Other albatrosses, storm petrels, diving petrels, terns, skuas, cormorants, and gulls also live on or visit Antarctica. Virtually all seabirds feed on fish, squid, or krill. A few land birds are found on the Antarctic Peninsula and sub-Antarctic islands, including sheathbills, pintails, and pipits.

Kelly Howard

Flora of Australia

Many species of plants in Australia are found nowhere else on Earth, except where they have been introduced by humans. Such species are known as endemic species. This distinctiveness is the result of the long isolation of the Australian continent from other landmasses. Australia broke off from the supercontinent Pangaea more than 50 million years ago, and the species of plants and animals living at that time continued to change and adapt to conditions on the isolated island. This led to distinctive plants and animals, differing from those of the interconnected Eurasian-African-American landmass, where new immigrant species changed the ecology. Australian fauna (animals) are mostly marsupials, and Australian flora (vegetation) is dominated by two types of plants—the eucalyptus and the acacia. Nevertheless, a great deal of botanical diversity exists throughout this large continent, with more than 800 known species of eucalypts and more than 900 species of acacias.

Climate is a major influence on Australian flora, and the most striking feature of the Australian environment as a whole is its aridity. Nutrient-poor soils also affect the nature of Australia's vegetation, espe-

E. camaldulensis, *immature woodland trees, showing collective crown habit, Murray River, Tocumwal, New South Wales. (Mattinbgn)*

cially in arid areas. Half of the continent receives less than 12 inches (300 millimeters) of rainfall per year; small parts of Australia receive annual rainfall of 75 inches (1,900 millimeters). Therefore, forests cover only a small percentage of Australia. A close correlation exists between rainfall and vegetation type throughout Australia. Small regions of tropical rain forest grow in mountainous areas of the northeast, in Queensland. In the cooler mountains of New South Wales, Victoria, and Tasmania, extensive temperate rain forests thrive.

More extensive than rain forest, however, is a more open forest known as sclerophyllous forest, which grows in the southern part of the Eastern Highlands in New South Wales and Victoria, in most of Tasmania, and in southwest Western Australia. Elsewhere—throughout northern Australia, the eastern half of Queensland and the inland plains of New South Wales, and to the north of the Western Australian sclerophyllous forest—extends a huge crescent-shaped region of woodland vegetation—an open forest of trees of varying height with an open canopy. Beyond this region, the climate is arid, and nontree vegetation is dominant. The tropical north of inland Queensland, Northern Territory, and a smaller part of Western Australia have extensive areas of grassland. Much of Western Australia and South Australia, as well as interior parts of New South Wales and Queensland, are shrubland, where grasses and small trees grow sparsely. In the center of the continent is the desert, which has little vegetation, except along watercourses.

AUSTRALIA'S OFFICIAL FLOWERS

In 1988 Australia proclaimed the golden wattle to be its official national floral emblem. Appropriately, the colors worn by Australian sporting teams—green and gold—are reminiscent of the wattle's green leaves and gold flowers. Australia has many beautiful wildflowers, and seven of the states and territories chose a flower as their state floral emblem; Tasmania chose a flowering tree, the Tasmanian blue gum. The waratah, a gorgeous red flower chosen by New South Wales, was once favored to be Australia's national flower, before the wattle was selected.

Queensland's floral emblem, the striking purple Cooktown orchid, is a reminder of that state's tropical rain forests. South Australia's flower, Sturt's desert pea, is a bright red and black plant found throughout a wide area of central Australia. Victoria's emblem is the common heath, a tiny pink flower. The distinctive kangaroo paw of Western Australia grows along the southwestern coasts of that state. The Tasmanian blue gum is a tall tree, 230 feet (70 meters) in height, making it difficult to see up close the tree's beautiful cream-colored flowers. The tree grows naturally in eastern Tasmania and a small area of Victoria and is also abundant in California.

Sturt's desert rose, the flower of the Northern Territory, is found in central Australia. The small shrub produces flowers with mauve petals, a stylized representation of which is on the territory's flag. The Australian Capital Territory was the last administrative division to choose a floral emblem—the royal bluebell, a small perennial herb with violet flowers. Its natural habitat is restricted to the region of the ACT.

The Eucalypts

Many people familiar with the song "Kookaburra Sits in an Old Gum Tree" may not realize that "gum tree" is the common Australian term for a eucalyptus tree. When the bark of a eucalyptus tree is cut, sticky drops of a transparent, reddish substance called "kino" ooze out. The British William Dampier noticed kino coming from trees in Western Australia in 1688, and called it "gum dragon," as he thought it was the same as commercial resin. Kino is technically not gum, as it is not water-soluble.

The scientific name *Eucalyptus* was chosen by the first botanist to study the dried leaves and flowers of a tree collected in Tasmania during Captain James Cook's third voyage in 1777. The French botanist chose the Greek name because he thought that the bud with its cap (operculum) made the flower "well" (eu) "covered" (kalyptos). The hard cases are commonly called gum nuts. Early in the twentieth century, a series of children's books was written about Gumnut Babies, tiny imaginary beings who lived among the eucalyptus leaves and wore the operculum as a cap.

The more than 800 species of eucalyptus in Australia range from tropical species in the north to al-

Eucalyptus tereticornis *buds, capsules, flowers and foliage, Rockhampton, Queensland, Australia. (Ethel Aardvark)*

pine species in the southern mountains. Rainfall, temperature, and soil type determine which particular eucalypt will be found in any area. Large eucalyptus trees dominate the Australian forests of the east and south, while smaller species of eucalypts grow in the drier woodland or shrubland areas. It is easier to mention parts of Australia where eucalypts do not grow: the icy peaks of the Australian Alps, the interior deserts, the Nullarbor Plain, and the rain forests of the Eastern Highlands, both tropical and temperate.

The scientific classification of the eucalypts proved difficult to European botanists. Various experts used flowers, leaves, or other criteria in their attempts to arrange the different species into a meaningful and useful taxonomy, or classification scheme. British botanist George Bentham eventually chose the shape of the anthers—the tiny stems that hold pollen—together with fruit, flowers, and nuts as differentiating factors.

A simpler classification of the eucalypts, commonly used by foresters, gardeners, and naturalists, arranges them into six groups based on their bark: Gums have smooth bark, which is sometimes shed; bloodwoods have rough, flaky bark; ironbarks have very hard bark with deep furrows between large pieces; stringybarks have fibrous bark that can be peeled off in long strips; peppermints have mixed but loose bark; and boxes have furrowed bark, firmly attached. This system was devised in 1859 by Sir Ferdinand von Mueller, the first Government Botanist of the Colony of Victoria, and the father of Australian botany. After emigrating from Germany in 1847 for his health, he was appointed to the Victorian position in 1853 and served until his death in 1896. He was an eccentric and controversial figure, but devoted to botanical exploration and research.

Many of the native plants of Australia show typical adaptations to the arid climate, such as deep taproots that can reach down to the water table. Another common feature is small, shiny leaves, to reduce transpiration. Eucalyptus leaves are tough or leathery, described as sclerophyllous. Sclerophyllus forests of eucalypts cover the wetter parts of Australia, the Eastern Highlands, or Great Dividing Range, and southwest Western Australia. The hardwood from these forests is generally not of a quality suitable for building, so areas are cleared and the trees made into woodchips that are exported for manufacture of newsprint paper. This has been a controversial use of Australian forests, especially where the native forest has been cleared and replaced with pine plantations.

The southwest corner of Western Australia has magnificent forests featuring two exceptional species with Aboriginal names, karri (*Eucalyptus diversicolor*) and jarrah (*Eucalyptus marginata*). Karri is one of the world's tallest trees, growing to 295 feet (90 meters) tall. This excellent hardwood tree is widely used for construction. The long straight trunks are covered in smooth bark that is shed each year, making a colorful display of pink and gray. These forests are now protected. Jarrah grows to 120 feet (37 meters) in height and is a heavy, durable timber. It was used for road construction in the nineteenth century, but now the deep red timber is prized for furniture, flooring, and paneling.

A giant cycad. (Bruno Dupont)

During the nineteenth century, Australia could also claim to be home to the world's tallest tree—a mountain ash that purportedly stood 433 feet (132 meters). Today, the height of the tallest ash tree is confirmed to be only 374 feet (114 meters).

The most widely distributed of all Australian eucalypts is the beautiful river red gum. These trees grow along riverbanks and watercourses throughout Australia, especially in inland areas; their spreading branches provide wide shade and habitat for many animals. Koalas eat leaves from this tree. In the song "Waltzing Matilda," Australia's unofficial national anthem, a man camps "under the shade of a coolabah tree." This word might apply to any eucalypt, but is most likely a river red gum.

Red river gums in Australia. (Ken Griffiths)

In drier interior areas, and in some mountain areas, there are more than 100 hundred smaller species of eucalypts that are known by the Aboriginal name "mallee." These many-trunked shrubs have an underground root that stores water. Much of this marginal country was cleared for farming, creating a situation similar to the Dust Bowl in the United States. In the dry Australian summers bushfires are a great danger, in the mallee and in any eucalyptus forest. The volatile oils of the eucalyptus trees can lead to rapid spread in the tree crowns, jumping across human-made firebreaks. On the other hand, several Australian trees can not only can survive fires but actually require fire to germinate.

Eucalypts have been introduced to many countries, including Italy, Egypt, Ethiopia, India, China, and Brazil, and they are common in California, where they were introduced more than 170 years ago.

The Acacias

These plants are usually called "wattles" in Australia, because the early European convicts and settlers used the flexible twigs of the plant for wattling, in which twigs are woven together, making a firm foundation for a thatched roof or for walls, which then are covered inside and out with mud. This style of building, known as wattle-and-daub, was common throughout Australia in pioneering days.

Wattles frequently have masses of colorful flowers, usually bright yellow. One species is the national flower of Australia. Other interesting acacias include the mulga (Acacia aneura), which has an attractive wood.

The Rain Forests

Although rain forest vegetation covers only a small area of Australia, it is exceptionally varied and of great scientific interest. Neither of the two general types of rain forest found in Australia has eucalypts. Rain forests are located along the Eastern Highlands, or Great Dividing Range, where rainfall is heavy. In small areas of tropical Queensland, where rainfall is also heavy, true tropical rain forest is found. The flora are similar to that in Indonesian and Malaysian rain forests. The tropical rain forest

Spinifex grass clumps in South Australia. (Galumphing Galah)

Acacias as seen at Royal National Park in Sydney, Australia. (KarenHBlack)

contains thousands of species of trees, as well as lianas, lawyer vine, and the fierce "stinging tree" (Dendrocnide moroides) whose touch could kill an unwary explorer. Toward the north of New South Wales, a kind of subtropical to temperate rain forest grows. Cool, wet Victoria and Tasmania have extensive areas of temperate rain forest, where only a few species dominate the forests. Arctic beech are found, as well as sassafras and tall tree ferns.

Sclerophyllus Forest

This is the typical Australian bush, which grows close to the coast of New South Wales, Victoria, and Tasmania. The bright Australian sunshine streams down through the sparse crowns and narrow leaves of the eucalypts. As the climate becomes drier, farther inland from the coast, the open forest slowly changes to a shrubbier woodland vegetation.

Grasslands

Moving farther inland, to still drier regions, woodlands give way to grasslands, where cattle are raised for beef in the tropics, and sheep are raised for wool in the temperate areas. Before Europeans came to Australia, there were native grasslands in the interior—tropical grasslands in the monsoonal north, and temperate grasslands in the south and southwest. Kangaroo grass and wallaby grass grew in the temperate interior of New South Wales, but much of this has been cleared for agriculture, especially for wheat farming. Mitchell grass is another tussock grass, which grows in western Queensland and into the

A row of baobab trees. (madmatrix)

Northern Territory; cattle and sheep graze extensively on this excellent native pasture.

The most-common grassland type in Australia is spinifex, a spiky grass clump that grows in the arid interior and west. Even cattle cannot feed on spinifex grass, so this environment is less threatened than most other grasslands. The northern grasslands are dotted with tall red termite mounds; those that are aligned north-south for protection from the hot Sun are built by so-called "magnetic" termites.

Other Trees and Plants

Many people think that macadamia nuts are native to Hawaii, which produces 90 percent of the world's crop, but in fact, the tree is native to Australia. It was discovered by Ferdinand von Mueller in 1857 on an expedition in northern Australia. Mueller named the tree after his Scottish friend, John Macadam. The trees were introduced to Hawaii in 1882.

Cycads belong to an ancient plant species they still thrive in Australia. The *Macrozamia* of North Queensland is a giant fernlike plant. Similarly ancient are the *Xanthorrhoea* grass-trees, a fire-resistant species in which a single spearlike stem rises from a delicate green skirt.

In northwest Australia, a single species of baobab tree (*Adansonia gregorii*), locally known as the boab, is found in the Kimberley region of Western Australia and in parts of Northern Territory. This fat-trunked tree collects water in its tissue. The only other baobabs are found in Africa, a reminder that these continents were once joined. Bottlebrush is an Australian shrub with colorful flowers that has become popular with gardeners in many parts of the world.

Extinct and Endangered Plant Species

Human activities on Australia have led to the extinction of more than ninety species of plants, and the list of threatened and endangered plants contains more than 1,000 species. Many non-native species have been introduced to Australia by Europeans; some have become pests, such as the blackberry in Victoria, the lantana in north Queensland, and the water hyacinth throughout the continent. In 2020, there were 681 national parks in Australia, as well as other conservation areas where native flora are protected.

Aboriginal Plant Use

The indigenous Australians used plants as sources of food and for medicinal purposes. Food plants included nuts, seeds, berries, roots, and tubers. Nectar from flowering plants, the pithy center of tree ferns, and stems and roots of reeds were eaten. Fibrous plants were made into string for weaving nets or making baskets. Weapons such as spears, clubs, and shields, as well as boomerangs, were made from hardwoods such as eucalyptus. The bunya pine (*Araucaria bidwillii*) forests of southeast Queensland were a place of great feasting when the rich bunya nuts fell.

Ray Sumner

FAUNA OF AUSTRALIA

Many scientists theorize that until about 300 million years ago, Earth had one supercontinent or landmass, called Pangaea, that covered about 40 percent of the planet's surface. During that time, animal and plant species multiplied and intermixed freely across the great land. Then, nearly 200 million years ago, many factors caused Pangaea to split into two smaller continents, Laurasia and Gondwanaland. Laurasia evolved from what had been the northern part of Pangaea and included North America, Europe, and Asia. Gondwanaland, formed from the southern part of Pangaea, comprised Antarctica, South America, Africa, India, and Australia. Over millions of years, Laurasia and Gondwanaland continued to fragment and drift apart. Once-related flora and fauna began to evolve separately.

When Gondwanaland broke apart some 60 million years ago, Australia became Earth's only island continent—completely surrounded by oceans and seas. Isolated from the other major landmasses, Australian plants and animals evolved on their own. Faunal adaptations took place as special habitats demanded change for survival, especially habitats such as the tropical rain forests of North Queens- land; the mangrove swamps along the coast; the mountainous Alps of southeast Australia; and the gorges, caves, and deserts of central Australia. Frogs, for ex-

Kangaroo with young in pouch. (Stephen Lynn)

HABITATS AND SELECTED VERTEBRATES OF AUSTRALIA

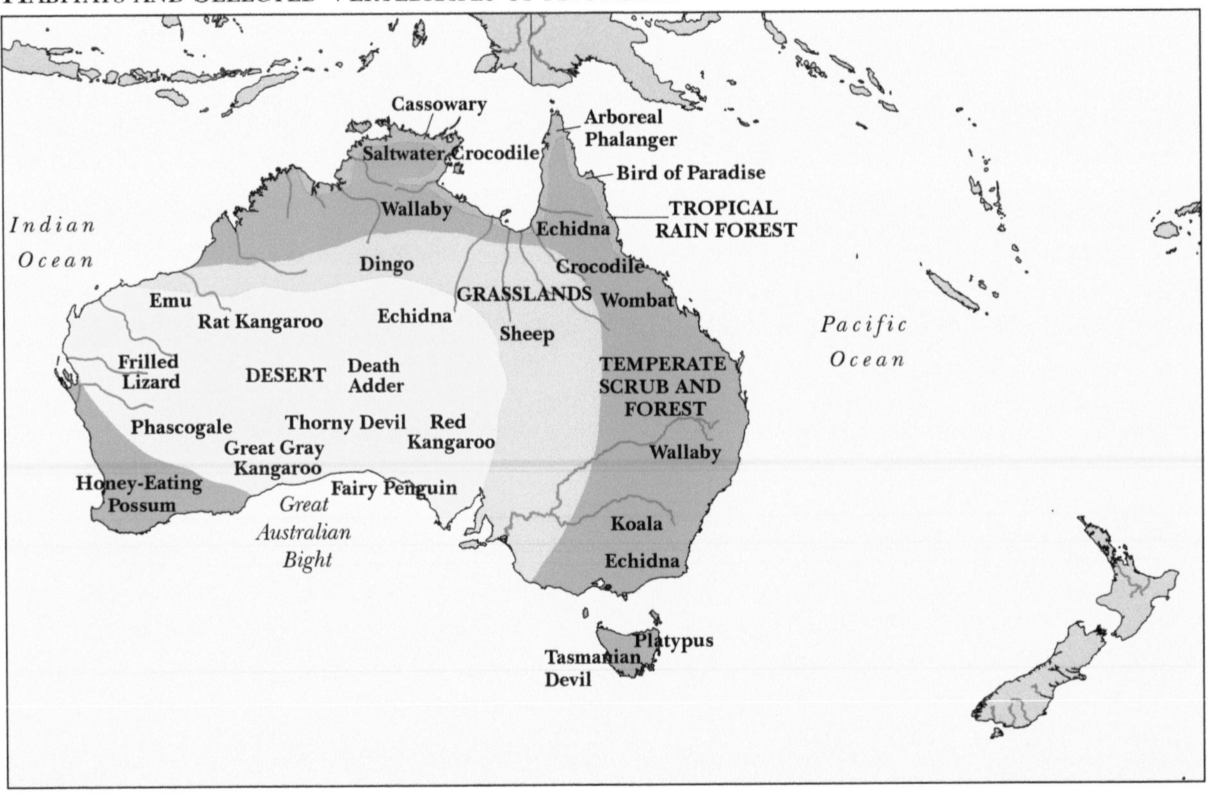

ample, learned to burrow for water. Female kangaroos can "pause" a pregnancy during times of drought. The platypus developed electroreception, a sixth sense that helps it detect food.

Some mammals adapted to the lack of drinking water by metabolizing food differently. All successful species made adjustments in order to survive. As a result, the Australian continent is now home to a rich, varied, and often unique flora and fauna. Nearly 95 percent of its mammals, 70 percent of its birds, 88 percent of its reptiles, and 94 percent of its frogs are found nowhere else in the world in the wild.

Australian Mammals

Australia's distinctive wildlife includes more than 350 species of native mammals. All three groups of modern mammals—marsupials, monotremes, and placentals—are represented. Monotremes (egg-laying mammals) are perhaps the most unusual; two of the three known species (the platypus and the echidna) exist on the island continent. Australia also has several sea mammals, including the humpback and other whale species, the dugong (or sea cow), and a variety of Australian fur seal species.

Marsupials are the most common mammals native to Australia. They give birth to live immature (fetal) young that are then carried in the mother's pouch until fully developed. Marsupials date back thousands of years and have adapted to many habitats related to their preferred foods. These include plants, eaten by the kangaroo and the koala; nectar, eaten by the pygmy possum; and insects and meat, eaten by the quoll and Tasmanian devil. Some marsupials, such as bandicoots, are omnivorous, eating both plants and small animals.

The second group of mammals, the monotremes, are the most puzzling of nature's animal species. Sometimes referred to as "living fossils," these mammals lay eggs, then nurse their hatched young. Scientists believe that monotremes may represent the stage of evolution between reptilian and placental mammals. Australia has two types of monotremes: the platypus and the echidna. Best-known is the platypus, with its duck bill, beaver-like tail, webbed feet, and furry body. Once hunted for its rich pelt, the platypus is now protected from poaching by law. The echidna, also known as the spiny anteater, resembles the porcupine. This rabbit-sized mammal has no teeth and uses its long sticky tongue to penetrate ant and termite nests that it has torn open with its sharp claws. When in danger, it burrows straight into the ground. Echidnas are common in Australia and are not a protected species.

The third group of mammals, the placentals, are those species that give birth to fully developed live young. Many Australian species fall into this group, including varieties of native mice.

Most Australian land mammals are nocturnal (active only at night). Australia abounds with wildlife parks and zoos that allow visitors to observe these animals in their own habitats.

Birds

Australia has more than 800 bird species of many types. Some major categories include birds of prey (owls, eagles, hawks/kestrels), songbirds, and especially parrots and cockatoos. Australia is also home to the smallest penguin species, the little blue or fairy penguin found on Phillip Island in Victoria.

Australia's emu and cassowary are the largest birds in the world; both are flightless. The emu stands nearly 6 feet (2 meters) tall and weighs more than 100 pounds (45 kilograms). Covered with feathers and having a ferocious beak, its wings are only prehistoric remnants. Its long legs can cover the ground at nearly 30 miles (50 km.) per hour. The emu is also a good swimmer. Its image (along with that of the kangaroo) is found on the Australian coat-of-arms. The cassowary is also distinctive in appearance. Smaller than the emu (around 5 feet/1.5 meters), it has coarse, hairlike feathers, red fleshy wattles, a bony helmet, and three-toed feet designed for running. Both the cassowary and emu have distinctive, throaty calls.

Parrot species abound in Australia. The budgerigar (often called budgies) is a common pet. The galah (rose-breasted cockatoo), crimson rosella, and rainbow lorikeet are among the most colorful

Emu. (tracielouise)

and commonly seen species. There are ten native species of kingfishers and kookaburras, famous for taking plunging dives to catch prey. The laughing kookaburra has a distinct call and is the largest representative of Australia's kingfisher families. Particularly unusual is the lyrebird, the largest Australian songbird. Sooty brown and the size of a hen, the lyrebird's tail feathers are nearly 2 feet (0.6 meter) long. When raised, the tail feathers take the shape of the lyre, a stringed instrument from antiquity.

Reptiles and Amphibians

There are 869 reptile species in Australia; 93 percent do not exist anywhere else in the world. Of the 2,400 species of snakes recognized in the world, 172 of them are found on the island continent. Australian snakes are grouped into five families: pythons and boas; blind snakes; colubrid snakes; file snakes; and elapids and sea snakes. Many of the land snakes are venomous, such as the taipan, brown snake, and red-bellied black snake.

Australian reptiles include hundreds of species of lizards. The main families are geckos; legless or snake lizards; dragon lizards (such as the frilled-neck lizard found on Australia's old two-cent coin); iguana-like monitor lizards or goannas; and skinks. Australia's critically endangered western swamp tortoise saw its numbers drop to a few dozen in the wild, but thanks to a highly successful captive-breeding program at the Perth Zoo, more than 700 of the tiny tortoises have been bred and released into the wild.

Australia is also home to the freshwater crocodile and the saltwater (estuarine) crocodile. The latter is the biggest reptile in the world. Found along the country's northern coastline, from Rockhampton in Queensland to Broome in Western Australia, "saltys" live mainly where tidal rivers meet the sea. They are a protected species and are often caught and relocated for their own survival.

Australia's large population of amphibians includes about 230 frog species. One of the most striking features of Australian frog fauna is their lack of dependence on permanent bodies of water for reproduction. Species have adapted to breeding in temporary pools of water by shortening their periods of larval development.

Sea Mammals and Marine Life

Australia is famous for its 1,250-mile (2,000-km.)-long Great Barrier Reef and the countless species of fish and other sea creatures that coexist in that habitat. These include anemones, sponges, worms, gastropods, lobsters, crayfish, shrimp and prawns,

Saltwater crocodile. (DianaLynne)

Frilled-neck lizard. (Kaan Sezer)

crabs, and starfish. One starfish species in particular, the crown-of-thorns, has caused much destruction to the reef by eating most of the living coral. Sharks, such as the fearsome great white, also live along the Australian coast. These waters are home to many sea mammal species, including whales (humpback, sei, and fin, among others); dolphins; Australian fur seals, and dugongs. Australia has the largest dugong population in the world, most living in the Torres Strait and in the northern Great Barrier Reef. A slow-moving mammal with little protection against sharks, saltwater crocodiles, or human predators, it is now a protected species in Australia.

Invertebrates

Australia's invertebrate fauna is estimated at between 275,000 and 300,000 species, most of them endemic to Australia. These include families of insects (beetles, wasps and bees, stick insects, praying mantis, grasshoppers, termites) and nearly 4,000 species of ants. Snails, earthworms (including some of the world's largest worms), moths and butterflies, dragonflies, and spiders are included in the count. Many invertebrate species have Pangean origins or can be traced back to Gondwanan times.

Introduction of Destructive Nonnative Fauna

With the passage of time and the mobility of human populations, foreign plants and animals have been introduced into Australia, and many of them thrived. The indigenous Australians, the first humans to inhabit the Australian continent, arrived there nearly 40,000 years ago. With them they brought the dingo, a wild dog originally from Asia, possibly Thailand. When Europeans began settling in Australia in the late eighteenth century, hoofed species such as horses, cattle, goats, sheep, and deer were introduced into the faunal mix. Domestic cats were imported as well.

Many of these introduced animals had a disastrous impact on the Australian ecology, causing much damage and loss of native species. Without native predators or diseases to control their growing populations, imported species have often

Koala. (Freder)

Red-belly black snake. (Lakeview_Images)

Echidna. (Andrew Haysom)

Cassowary. (BirdImages)

Endangered and Threatened Mammals of Australia

Common Name	*Scientific Name*	*Range*	*Status*
Bandicoot, desert	*Perameles eremiana*	Australia	endangered
Kangaroo, Tasmanian forester	*Macropus giganteus tasmaniensis*	Australia (Tasmania)	endangered
Koala	*Phascolarctos cinerus*	Australia	threatened
Marsupial, eastern jerboa	*Antechinomys laniger*	Australia	endangered
Mouse, Australian native	*Notomys aquilo*	Australia	endangered
Numbat	*Myrmecobius fasciatus*	Australia	endangered
Possum, mountain pygmy	*Burramys parvus*	Australia	endangered
Rat-kangaroo, desert	*Caloprymnus campestris*	Australia	endangered
Tiger, Tasmanian	*Thylacinus cynocephalus*	Australia	endangered
Wallaby, western hare	*Lagorchestes hirsutus*	Australia	endangered
Wombat, Queensland hairy-nosed	*Lasiorhinus krefftii*	Australia	endangered

Source: U.S. Fish and Wildlife Service, U.S. Department of the Interior

caused great destruction by preying on native animals and devouring native plants. Examples of harmful species include the rabbit, camel, cat, dingo, water buffalo, and cane toad. The dingo, a carnivore, competed for the same food sources as the native Tasmanian tiger, leading to the extinction of that species. Feral domestic cats have killed off native rodent species.

Camels are another example. Originally brought to Australia from Afghanistan, they were well-suited to carrying cargo across central Australia's vast arid deserts. With the advent of railroads and automobiles, however, their use ended. Now, descendants of the original camels roam the Outback.

Cane toads were first introduced by farmers in 1935, despite warnings from scientists, to protect Queensland's sugarcane fields from the cane beetle. The experiment failed; today they are a major nuisance. The highly poisonous amphibians multiply rapidly, often live up to twenty years, and have no natural enemies in Australia. They are especially deadly to native animal populations that attempt to feed on them. Rabbits are also considered pests. In 1859, Thomas Austin imported twenty-four rabbits from England and released them on his property in Victoria for sport hunting. Today the total wild rabbit population exceeds 200 million and is highly destructive to crops and native plants.

The loss of so many of Australia's native animal species cannot be blamed entirely on competitive interaction with imported species. Humans also have caused losses through poaching practices, use of poisonous herbicides, destruction of habitats through increased building and urbanization, and the commercial use of indigenous animal species.

Issues concerning the preservation of native plants, animals, and habitats remain of the utmost importance to Australians. They have great pride in the country's unique faunal heritage. To this end, Australia continues to establish safeguards and pass legislation aimed at protecting the natural environment, the destruction of which would be tragic and apt to have wide-reaching economic and other consequences.

Cynthia Beres

Flora and Fauna of the Pacific Islands

The vast region of the Pacific, collectively called Oceania, comprises thousands of islands. Oceania spreads across the Pacific Ocean from 20 degrees north latitude to 50 degrees south latitude and from longitude 125 degrees east to 130 degrees west. The major groupings are Melanesia, Micronesia, Polynesia, and New Zealand. Melanesia is a group of large islands immediately north and east of Australia, from New Guinea to New Caledonia. Micronesia is made up of hundreds of tiny atolls in the western Pacific. Polynesia covers a huge region in the central Pacific. New Zealand lies east and south of Australia. For biological purposes, these islands can be categorized by climate and formation type. Climates range from tropical to sub-Antarctic, dry to very rainy. Types include volcanic (Fiji, Guam, and Hawaii), tectonic (New Zealand and New Guinea), and low coral atolls (nearly all of Micronesia's islands).

Endemic and Exotic Plants and Animals

Organisms have difficulty reaching the islands across the broad expanses of the Pacific Ocean. This isolation leads to trends in the number of species found on any given island: Bigger islands have more species; those farthest from continents have fewer species. To reach the islands, plants must be carried by animals or rely on wind or water currents. Animals must fly (birds and bats), float on logs, or be carried in by humans. Birds are usually the first visitors, bringing with them hitchhiking insects and plant seeds in their digestive tracts. Bats are the only mammals to reach many islands without human help.

Island plants and animals evolve together, affected by difficult conditions; soil is often poor and food limited. Harsh environments and isolation contribute to the formation of new and unique species; some of the strangest creatures on Earth are endemic (found nowhere else) to particular islands. Most of the world's flightless birds developed on islands, where there originally were no large land predators. Island ecosystems are sensitive to disturbances, whether from natural causes such as severe storms, or human activities such as construction, agriculture, logging, and introduced species.

Introduced species (exotics), both accidental and deliberate, are a serious problem. Rats and feral animals (domestic animals that have gone wild) can devastate island ecologies. Pigs, cats, rats, and goats are particularly devastating—goats devour vegetation, cats eat birds and small animals, and rats and pigs eat anything. Introduced plants may overgrow native ones. Exotics also carry diseases to which native plants and animals have no resistance. Humans have tried to deal with exotics, with mixed results. For example, the mongoose was intentionally brought to Fiji to control accidentally introduced rats. No one considered that rats are active at night and mongooses during the day. The mongoose did not control the rats, but it did prey on seven native Fijian bird species, causing their extinction.

A mangrove swamp. (Fanny Schertzer)

Another strange example occurred in Hawaii. A Hawaiian bee crawls headfirst into native, barrel-shaped flowers, gathers the nectar, and then backs out. A plant that was introduced by landscapers attracted bees, but the flowers were smaller than the native plants. Once a bee crawled in, it became stuck like a cork in a bottle. There are thousands of these plants, each with hundreds of flowers stoppered with dead bees.

Many tropical and temperate islands have coastal wetlands and mangrove swamps growing at the edge of the sea. Mangroves are low-growing, salt-tolerant trees whose dense tangles are virtually impenetrable to humans. Wetlands and mangrove swamps are important breeding grounds for many types of fish and crabs, and also trap sediment, stabilize shorelines, and protect coastlines from storms. Humans often fill in the wetlands and cut down the mangroves, causing coastal erosion and the loss of food fish.

Fiji Islands

The Fiji Islands are mostly volcanic in origin and lie in the South Pacific Ocean between longitudes 175 degrees east and 178 degrees west and 15 degrees and 22 degrees south latitudes, about 1,300 miles (2,100 km.) north of Auckland, New Zealand. Some parts of the islands receive up to 13 feet (4 meters) of rain per year, while other parts remain dry. A range of volcanic peaks divides the islands; the highest, Mount Tomanivi (formerly Mount Victoria), stands 4,341 feet (1,322 meters). These differences in weather and elevation create a variety of habitats—dense rain forests, grassy savanna, and mangrove swamps—and a large diversity of species.

Human disruption on Fiji has been moderate. About half the total area is still forested, and less than one-fourth of the land is suitable for agriculture. Trees include mahogany, pine, pandanus, coconut palms, mangoes, guava, and figs. Banyan figs are difficult to cut down and are responsible for some of the lack of forest clearing. The figs are an important food for many birds and animals. Other rain- forest plants include orchids, ferns, and epiphytes (plants that grow upon other plants).

Easter Island

Easter Island (Rapa Nui) lies in the South Pacific 2,400 miles (3,862 km.) from Chile, South America. It is best known for its huge, mysterious stone statues. Archeological evidence proves that Easter Island once was densely forested and supported many animal and bird species. The inhabitants burned or cut down the forests before Europeans discovered the island in the 1700's. The only plants now living on the island are sparse grasses, a few shrubs, and two species of trees. There are almost no land animals larger than insects. The approximately 8,000 people who live on the island must import nearly all their food. In 2018, the island took the step of limiting tourist visits from 90 to 30 days, citing social and environmental concerns. The devastation of Easter Island is an extreme example of what can happen to island ecologies when human development exceeds natural capacity.

There are nearly 1,500 endemic Fijian plant species, including ten species of palm tree on the island of Viti Levu alone. Endemic animals include birds such as lorikeets (parrots that eat flowers or nectar), the Fiji goshawk, spectacularly colored pigeons and parrots, and a pigeon that barks like a dog. The Vanikoro flycatcher, which holds its mossy nest together with spider webs, is one of the few native birds that has adapted to forest clearing. It is commonly seen in town gardens and suburbs.

Other creatures native to Fiji are snakes, including a rare, tiny cobra, two species of frog, and several species of geckoes and skinks (small lizards). The critically endangered Fijian crested iguana was not discovered until 1978 and lives only on the island of Yadua Taba. Yadua Taba is now a preserve, and the crested iguana is more likely to survive since feral goats were removed from the island.

Grassy savannas are found higher on the volcanic slopes and in the dry zones. They are often planted with coconut palms and taro, a plant with potato-like roots that is grown on many Pacific islands.

Several of the reserve areas in the Fijian islands are being logged and provide little sanctuary to native plants and animals. The government is interested in increased logging of mahogany and pine. The Fiji Pine Commission hopes to remove pressure from logging native forest trees such as mahogany by encouraging the development of pine forests. Pines grow quickly and could form a sustainable logging industry, unlike the valuable but slow-growing mahogany trees. Increased world interest in herbal remedies has created a market for Fiji's traditional crop, kava root, and for ginger processing.

The University of the South Pacific is located in Fiji and is a source of important research into South Pacific species. Tourism is important to Fiji's economy and, with management, could represent a source of income to Fijians while still preserving native wildlife.

New Guinea

The world's largest tropical island, New Guinea is located north of Australia and just south of the equator. It is tectonic in origin, with great changes in elevation and many different habitats. Because of its size and varied terrain, New Guinea has a greater variety of habitats than any similar-sized land area in the world. In fact, New Guinea is so rugged that it is one of the least explored and least developed places on Earth. It provides the best remaining example of the types of organisms that can develop in island isolation.

Biological Mass Murder

The brown tree snake (*Boiga irregularis*) vividly illustrates the dangers of introducing an exotic species to an island ecosystem. The brown tree snake, a native of Australia, was accidentally introduced to Guam after World War II. It reproduced in incredible numbers and reached the highest density of any snake population on Earth, with up to sixteen to twenty thousand snakes per square mile. In the late 1970's biologists noticed that birds were disappearing from Guam. Of the fourteen bird species endemic to the island, at least nine had become extinct by the 1990's. Many small animal species, as well as chickens raised by the island's inhabitants, also disappeared. The snakes harm humans as well; they bite viciously and cause frequent power outages by climbing on power lines. Millions of dollars have been spent on attempts to control the brown tree snake, and by 2017 officials reported a slight decrease in the snake population. Measures include traps, the use of trained dogs to search for and remove snakes before vessels leave the island, and air-dropping poisoned bait for snakes to consume, in the form of dead mice filled with acetaminophen.

A klinki tree in Papua New Guinea. (Phil Markey)

New Guinea habitats include cold tundra, tropical rain forests, grassy savannas, coastal zones, montane rain forests, cloud forests, and bogs. There are at least 20,000 species of flowering plants, including more than 2,500 species of orchids, and hundreds of birds and animals. Many New Guinea species are unusual. Endemic Klinki pines (*Araucaria hunsteinii*) are the tallest tropical trees in the world, reaching 295 feet (90 meters) tall.

Native birds range from the beautiful to the bizarre. There are several species of birds of paradise, some of the most beautiful birds in the world. They have long, brilliantly colored feathers with metallic-looking feather disks which they wave and tremble to attract mates. The bowerbird builds large, complicated structures and decorates them with

A mahogany forest. (aldarinho)

colorful flowers, feathers, or trash, and even uses berry juice as paint. The megapode tunnels into the soil near volcanic hot springs with huge, powerful feet. It lays its eggs in the warm tunnel so that it does not have to sit on them.

New Guinea has several parrots, including the endangered Pesquet's parrot, whose face is completely devoid of feathers, allowing it to stick its head into fruit without getting its plumage sticky. Lorikeets are colorful, nectar-eating parrots. New Guinea is also home to the flightless cassowary, a large bird up to 6.6 feet (2 meters) tall and weighing up to 130 pounds (60 kilograms). Other bird species include feathery crowned pigeons, kingfishers (twenty-two species), ducks, herons, hawks, and egrets.

As in Australia, primates and large mammals never arrived in New Guinea. Marsupials (animals with pouches) took their ecological places. There are several species of tree-living kangaroos. The basic kangaroo shape is modified in the tree species; they have larger forearms and smaller hind legs. Other marsupials include striped and feather-tailed possums, ringtails, and land wallabies that look like small, dainty kangaroos.

Other odd New Guinea animals are the echidnas, perhaps the most primitive mammal in the world. Echidnas lay eggs and are related to the duck-billed platypus. There are several echidna species; the largest is the giant spiny anteater, which has long spines, dense fur, thick claws, and a long, slender snout. Despite its name, it eats mostly earthworms, which it reels into its mouth using spines on its tongue.

New Guinea has many types of fruit bats. One large species, the flying fox, roosts in colonies. In Madang and several other New Guinean communities, thousands of bats roost overhead, causing significant problems in street cleaning and for pedestrians without hats.

Reptiles and amphibians on New Guinea consist of snakes, lizards, frogs, and toads. The Salvadori monitor lizard is one of the longest in the world, growing up to 16 feet (5 meters) long, most of which is tail. Crocodiles live in many rivers and are endangered from hunting and demand for their hides. Some locals have begun crocodile ranches, where they breed and raise the crocodiles for meat and hides without damaging wild populations. Snakes include adders and pythons (eight species), and the deadly taipan and Papuan black snakes. There are many species of frogs, including several odd rain-forest species, whose metamorphosis no longer includes a tadpole stage.

New Guinea's insects include the largest moth in the world and the largest butterfly, the Queen Alexandra's birdwing. This species is endangered due to collecting and loss of habitat. In recent years, special butterfly museums have stepped in to raise the insects, and sell them to collectors and museums, and in doing so help preserve the species. New Guinea is also home to giant millipedes, stick insects 12 inches (30 centimeters) long, and a fly with its eyes on stalks. Several species of flies have antlers on their heads, which they use to fight in defense of egg-laying territory.

Many forests host "ant-plants," warty looking epiphytes that have hollow mazes inside their tissues. Ants live in the maze, safe from predators. The ants provide nutrition for the plant in the form of droppings, scraps of food, and dead ants.

Even though New Guinea is rugged and isolated, human impact is increasing. The population is rising, leading to more forests being cut and grasslands being plowed for agriculture, roads, and development. Humans have brought in food plants such as the sago palm, which can be cultivated in areas where more traditional crops will not survive. They also cultivate pandanus trees, several varieties of fruiting vine, breadfruit, fungi, tubers, sugarcane, bananas, taro, and yams. Gold, silver, and copper have been discovered, which encourages destructive mining.

It has proven difficult to develop New Guinea economically without destroying the unique flora and fauna of the island. It is hoped that lessons learned on other islands, such as Guam and New

Tree kangaroo. (CraigRJD)

Flying foxes roosting in Madang, Papua New Guinea. (Ianm35)

Zealand, may be applied to New Guinea. Incentives have been introduced to keep wild areas wild, such as encouraging ecologically friendly businesses like crocodile and butterfly farms and ecotourism. The National Park reserve that includes Puncak Jays (Mount Jaya), the highest point between the Himalayas and the Andes, is the only place in the world where it is possible to visit a glacier and a coral reef in the same park.

New Zealand

Located off the eastern edge of Australia, New Zealand has a fairly moderate climate that comes from conflicting warm, humid Pacific air interacting with colder Antarctic weather. It is similar to New Guinea, with rugged terrain, high mountains, and habitats from grassy open plains to dense forests, wet areas to near-deserts. Unlike New Guinea, however, New Zealand has been occupied and developed by humans for hundreds of years. Before large-scale agriculture, about half of New Zealand was covered with forests and one-third with grassland communities. Now, half is pasture for grazing and a quarter is forest, mostly comprised of introduced species. Much of the remaining native forest is maintained as national parks and reserves. Pastureland usually consists of a single species of grass and does not support the wide variety of bird and animal life of the original grassland communities.

It is sometimes said that the dominant mammal in New Zealand is the sheep. Sheep are vital to the economy, but their grazing lands have displaced native plants and animals from their natural habitats. Today, more than 600 species native to New Zealand are threatened and several are extinct, including 43 percent of New Zealand's frogs and more than 40 percent of its native birds.

Like New Guinea and Australia, New Zealand did not originally have any large mammals or carnivores. Fourteen flightless or weak-flying birds developed there. The native Maori people hunted the large, flightless moa to extinction before the Europeans arrived, one of the few known instances in recent history when humans not from a Western culture were responsible for a species' extinction.

Flightless birds that still survive in New Zealand are the kiwi and the kakapo parrot. The introduction of cats, dogs, rats, and pigs has severely endangered both species, although by 2019, intensive captive-breeding efforts had restored the population to approximately 200 juvenile or older birds. All had been moved to predator-free, offshore islands. The kiwi is New Zealand's state bird and efforts are being made to save it as well. Two species of another flightless bird, the penguin, breed on the south and east coasts. There are more than 200 species of flying birds, at least forty of them introduced, including tropic birds, gulls, hawks, harriers, skuas, spoonbills, and pheasants.

The only original endemic mammals were two species of bats. In the past 200 years, many exotic marsupials and mammals have appeared, including rabbits, rats, mice, weasels, otters, cats, pigs, cattle, deer, goats, and sheep. Domestic cattle, sheep, and pigs are economically important.

Amphibians and reptiles are interesting but not numerous. Among them is a frog that retains its tail as an adult and one that bypasses the tadpole stage. Some frog species have been introduced, but have had little impact. Reptiles include geckoes, skinks, and the extremely primitive, lizard-like tuatara.

Invertebrate species include such oddities as a giant carnivorous land snail. There are more than 4,000 species of beetles, 2,000 species of flies, and 1,500 species of butterflies and moths. In a reversal of the usual ecological scenario, some native insects are destroying introduced pasture grasses.

Coral Atolls

The Federated States of Micronesia consists mainly of small atolls. Coral atolls are found only in tropical latitudes because coral (small, colonial animals) grows only in warm water. Coral reefs support a tremendous variety of fish, crabs, and mollusks. Atolls tend to have porous, infertile soil and to be very low in elevation; the inhabited state of Tokelau, three small islands located at 9 degrees south latitude, longitude 172 degrees west, has a maximum elevation of only 16 feet (5 meters). Due to the low profile, poor soil, and occasional scouring by typhoons, flora are mostly limited to hardy root crops and

Penguin in New Zealand. (TerryJLawrence)

fast-growing trees such as coconut and pandanus. Other vegetation may include native and introduced species such as papaya, banana, arrowroot, taro, lime, breadfruit, and pumpkin. Fauna usually consists of lizards, rodents, crabs, and other small creatures. Pigs, ducks, and chickens are raised for food, and dogs and cats are kept as pets.

Human disturbances on coral atolls often have been particularly impactful; several nuclear test bombs were exploded on Bikini Atoll and other islands in the 1940s and 1950s. Kwajalein, the largest atoll in the world, is leased by the US military from the Marshall Islands as a ballistic missile defense test site. Johnston Atoll, a US territory about 820 miles (1,320 km.) southwest of Honolulu, Hawaii, was formerly a military base and storage facility for radioactive and toxic substances; today, it is undergoing environmental remediation. The atoll is administered by the US Fish and Wildlife Service as a protected area and bird-breeding ground, and is closed to visitors.

Future Prospects

Island ecologies are unique and fragile. Some, like those in Guam and New Zealand, can never be returned to their original state, but with extensive wildlife management, many native species can be saved from extinction. Ironically, people in the developed world who visit zoos see more native Pacific island species than those who live on those islands. Some endemic species have been wiped out or had their habitat destroyed, making zoos the final refuge for many creatures.

New Guinea, Fiji, and many smaller islands are in earlier stages of modern development during which wildlife destruction can still be controlled. Conservationists sometimes do not realize that islands are not small, geographical zoos; people live there and want to improve their lives. Development cannot be stopped, but it can be managed so that the humans can improve their standard of living as they wish, and the original, amazing island dwellers can still survive.

Kelly Howard

New Zealand's National Parks

New Zealand's strikingly beautiful and diverse landscape contains thirteen national parks: four on North Island and nine on South Island. These parks, comprising thirty percent of New Zealand's land area, are set aside to preserve spectacular scenery, rare and endangered flora and fauna, and archeological sites. These parks offer various outdoor activities, such as golfing, walking, mountaineering and climbing, snow sports, exploring archeological and mining sites and old volcanic landforms, geyser watching, swimming, diving, sailing, and camping.

Many parks contain features of historic and spiritual significance to the Maori. In 1887 the Maori chiefs of the Tuwharetoa tribe dedicated their ancestral North Island volcanic peaks of Tongariro, Ruapehu, and Nguaruhoe to all people of New Zealand. This area, named Tongariro National Park, is the country's first—and the world's fourth—national park. It is one of three World Heritage sites within the country. Te Wahipounamu, on South Island, includes parts of Westland, Fiordland, Mount Cook, and Mount Aspiring National Parks. The New Zealand Sub-Antarctic Islands, in the Southern Ocean, consist of five island groups (the Snares, Bounty Islands, Antipodes Islands, Auckland Islands and Campbell Island) and are nesting grounds for 126 species of birds.

Discussion Questions: Flora and Fauna

Q1. What types of flowering plants grow in Antarctica? Why is Antarctica's mammal life limited to sea-dwelling creatures? What characteristics of the Antarctic environment make it ideal for penguins?

Q2. Why are Australia's and New Zealand's flora and fauna so different from those of the rest of the world? What are monotremes? How are they unique among mammals? Which marsupial is most closely associated with Australia?

Q3. What impact have introduced species had on the flora and fauna of the Pacific islands? Why are island ecologies considered fragile?

Human Geography

Overview

The Human Environment

No person lives in a vacuum. Every human being and community is surrounded by a world of external influences with which it interacts and by which it is affected. In turn, humans influence and change their environments: sometimes intentionally, sometimes not, and sometimes with effects that are harmful to these environments, and, in turn, to humans themselves. Humans have always shaped the world in which they live, but developments over the past few centuries have greatly enhanced this capacity.

Many people feel a sense of alarm about the consequences of widespread adoption of modern technology, including artificial intelligence (AI) and accelerating human population growth in the world. Travel and transportation among the world's regions have been made surer, safer, and faster, and global communication is virtually instantaneous. The human environment is no longer a matter of local physical, biological, or social conditions, or even of merely national or regional concerns—the postmodern world has become a true global community.

Students of human geography divide the human environment into three broad areas: the physical, biological, and social environments. The study of ecology describes and analyzes the interactions of biological forms (mainly plants and animals) and seeks to uncover the optimal means of species cooperation, or symbiosis. Everything that humans do affects life and the physical world around them, and this world provides potentials for and constraints on how humans can live.

As people acquired and shared ever-more knowledge about the world, their abilities to alter and shape it increased. Humans have always had a direct impact on Earth. Even 10,000 years ago, Neolithic people cut down trees, scratched the earth's surface with simple plows, and replaced diverse plant forms with single crops. From this basic agricultural technology grew more complex human communities, and people were freed from the need to hunt and gather. The alteration of the local ecosystems could have deleterious effects, however, as gardens turned eventually to deserts in places like North Africa and what later became Iraq. Those who kept herds of animals grazed them in areas rich in grasses, and animal fertilizer helped keep them rich. If the area was overgrazed, however, destroying important ground cover, the herders moved on, leaving a perfect setup for erosion and even desertification. Today, people have an even greater ability to alter their environments than did Neolithic people, and ecologists and other scientists as well as citizens and politicians are increasingly concerned about the negative effects of modern alterations.

The Physical Environment

The earth's biosphere is made up of the atmosphere—the mass of air surrounding the earth; the hydrosphere—bodies of water; and the lithosphere—the outer portion of the earth's crust. Each of these, alone and working together, affect human life and human communities.

Climate and weather at their most extreme can make human habitation impossible, or at least extremely uncomfortable. Desert and polar climates do not have the liquid water, vegetation, and animal life necessary to sustain human existence. Humans can adapt to a range of climates, however. Mild variations can be addressed simply, with clothing and shelter. Local droughts, tornadoes, hurricanes, heavy winds, lightning, and hail can have devastating effects even in the most comfortable of climates. Excess rain can be drained away to make habitable land, and arid areas can be irrigated. Most people live in temperate zones where weather extremes are rare or dealt with by technological adaptation. Heating and, more recently, air conditioning can create healthy microclimates, whatever the external conditions. Food can be grown and then transported across long distances to supply populations throughout the year.

The hydrosphere affects the atmosphere in countless ways, and provides the water necessary for human and other life. Bodies of water provide plants and animals for food, transportation routes, and aesthetic pleasure to people, and often serve to flush away waste products. People locate near water sources for all of these reasons, but sometimes suffer from sudden shifts in the water level, as in tidal waves (tsunamis) or flooding. Encroachment of salt water into freshwater bodies (salination) is a problem that can have natural or human causes.

The lithosphere provides the solid, generally dry surface on which people usually live. It has been shaped by the atmosphere (especially wind and rain that erode rocks into soil) and the hydrosphere (for example, alluvial deposits and beach erosion). It serves as the base for much plant life and for most agriculture. People have tapped its mineral deposits and reshaped it in many places; it also reshapes itself through, for example, earthquakes and volcanic eruption. Its great variations—including vegetation—draw or repel people, who exploit or enjoy them for reasons as varied as recreation, military defense, or farming.

The Biological Environment

Humans share the earth with over 8 million different species of plants, animals, and microorganisms—of which only about 2 million have been identified and named. As part of the natural food chain, people rely upon other life-forms for nourishment.

Through perhaps the first 99 percent of human history, people harvested the bounty of nature in its native setting, by hunting and gathering. Domestication of plants and animals, beginning about 10,000 years ago, provided humans a more stable and reliable food supply, revolutionizing human communities. Being omnivores, people can use a wide variety of plants and animals for food, and they have come to control or manage most important food sources through herding, agriculture, or mechanized harvesting. Which plants and animals are chosen as food, and thus which are cultivated, bred, or exploited, are matters of human culture, not, at least in the modern world, of necessity.

Huge increases in human population worldwide have, however, put tremendous strains on provision of adequate nourishment. Areas poorly endowed with foodstuffs or that suffer disastrous droughts or blights may benefit from the importation of food in the short run, but cannot sustain high populations fostered by medical advances and cultural considerations.

Human beings themselves are also hosts to myriad organisms, such as fungi, viruses, bacteria, eyelash mites, worms, and lice. While people usually can coexist with these organisms, at times they are destructive and even fatal to the human organism. Public health and medical efforts have eradicated some of humankind's biological enemies, but others remain, or are evolving, and continue to baffle modern science.

The presence of these enemies to health once played a major role in locating human habitations to avoid so-called "bad air" (*mal-aria*) and the breeding grounds of tsetse flies or other pests. The use of pesticides and draining of marshy grounds have alleviated a good deal of human suffering. Human efforts can also control or eliminate biological threats to the plants and animals used for food, clothing, and other purposes.

Social Environments

Human reproduction and the nurturing of young require cooperation among people. Over time, people gathered in groups that were diverse in age if not in other qualities, and the development of towns and cities eventually created an environment in which otherwise unrelated people interacted on intimate and constructive levels. Specialization, or division of labor, created a higher level of material wealth and culture and ensured interpersonal reliance.

The pooling of labor—both voluntary and forced—allowed for the creation of artificial living environments that defied the elements and met human needs for sustenance. Some seemingly basic human drives of exclusivity and territoriality may be responsible for interpersonal friction, violence and, at the extreme, war. Physical differences, such as size, skin, or hair color, and cultural differences, including language, religion, and customs, have often divided humans or communities. Even within close quarters such as cities, people often separate themselves along lines of perceived differences. Human social identity comes from shared characteristics, but which things are seen as shared, and which as differentiating, is arbitrary.

People can affect their social environment for good and ill through trade and war, cooperation and bigotry, altruism and greed. While people still are somewhat at the mercy of the biological and physical environments, technological developments have balanced the human relationship with these. Negative effects of human interaction, however, often offset the positive gains. People can seed clouds for rain, but also pollute the atmosphere around large cities, create acid rain, and perhaps contribute to global warming.

Human actions can direct water to where it is needed, but people also drain freshwater bodies and increase salination, pollute streams, lakes, and oceans, and encourage flooding by modifying riverbeds. People have terraced mountainsides and irrigated them to create gardens in mountains and deserts, but also lose about 24 billion metric tons of soil to erosion and 30 million acres (12 million hectares) of grazing land to desertification each year. These negative effects not only jeopardize other species of terrestrial life, but also humans' ability to live comfortably, or perhaps at all.

Globalization

Humankind's ability to affect its natural environments has increased enormously in the wake of the Industrial Revolution. The harnessing of steam, chemical, electrical, and atomic energy has enabled people to transform life on a global scale. Economically, the Western world still to dominates global markets despite effort of China to capture the crown, and computer and satellite technology have made even remote parts of the globe reliant on Western information and products. Efficient transportation of goods and people over huge distances has eliminated physical barriers to travel and commerce. The power and influence of multinational corporations and national corporations in international markets continues to grow.

Human environmental problems also have a global scope: Extreme weather, changes in ocean temperatures and sea level rise, global warming, and the spread of disease by travelers have become planetary concerns. International agencies seek to deal with such matters, and also social and political concerns once left to nations or colonial powers, such as population growth, the provision of justice, or environmental destruction within a country. Pessimists warn of horrendous trends in population and ecological damage, and further deterioration of human life and its environments. Optimists dismiss negative reports as exaggerated and alarmist, or expect further technological advances to mitigate the negative effects of human action.

Joseph P. Byrne

Population Growth and Distribution

The population of the world has been growing steadily for thousands of years and has grown more in some places than in others. On November 2019, the total population of the earth had reached 7.7 billion people. The population of the United States in August 2019 was approximately 329.45 million. India's population in November 2019 was 1.37 billion, making it the world's second most populous country. China's population was about 1.45 billion—about 1 in 5 people on the planet.

How Populations Are Counted

The U.S. Constitution requires that a census, or enumeration, of the population of the United States be conducted every ten years. The U.S. Census Bureau mails out millions of census forms and pays thousands of people (enumerators) to count people that did not fill out their census forms. This task cost about US$5.6 billion in the year 2010, and estimates for the 2020 census have risen to over US$15 billion. Despite this great effort, millions of people are probably not counted in every U.S. census. Moreover, many countries have much less money to spend on censuses and more people to count. Therefore, information about the population of many poor or less-developed countries is even less accurate than that for the population of the United States.

Counting how many people were alive a hundred, a thousand, or hundreds of thousands of years ago is even more difficult. Estimates are made from archaeological findings, which include human skeletons, ruins of ancient buildings, and evidence of ancient agricultural practices. Historical records of births, deaths, taxes paid, and other information are also used. Although it is not possible to estimate the global population 1,000 years ago with great accuracy, it is a fascinating topic, and many people have participated in estimating the total population of the planet through the ages.

History of Human Population Growth

Ancient ancestors of humans, known as hominids, were alive in Africa and Europe around 1 million years ago. It is believed that modern humans *(Homo sapiens sapiens)* coexisted with the Neanderthals *(Homo sapiens neandertalensis)* about 100,000 years ago. By 8000 BCE (10,000 years ago) fully modern humans numbered around 8 million. If the presence of archaic *Homo sapiens* is accepted as the beginning of the human population 1 million years ago, then the first 990,000 years of human existence are characterized by a very low population growth rate (15 persons per million per year).

Around 10,000 years ago, humans began a practice that dramatically changed their growth rate: planting food crops. This shift in human history, called the Agricultural Revolution, paved the way for the development of cities, government, and civilizations. Before the Agricultural Revolution, there were no governments to count people. The earliest censuses were conducted less than 10,000 years ago in the ancient civilizations of Egypt, Babylon, China, Palestine, and Rome. For this reason, historical estimates of the earth's total population are difficult to make. However, there is no argument that human numbers have increased dramatically in the past 10,000 years. The dramatic changes in the growth rates of the human population are typically attributed to three significant epochs of human cultural evolution: the Agricultural, Industrial, and Green Revolutions.

Before the Agricultural Revolution, the size of the human population was probably fewer than 10 million people, who survived primarily by hunting and gathering. After plant and animal species were domesticated, the human population increased its growth rate. By about 5000 BCE, gains in food production caused by the Agricultural Revolution meant that the planet could support about 50 million people. For the next several thousand years, the human population continued to grow at a rate of about 0.03 percent per year. By the first year of the common era, the planet's population numbered about 300 million.

At the end of the Middle Ages, the human population numbered about 400 million. As people lived in densely populated cities, the effects of disease increased. Starting in 1348 and continuing to 1650, the human population was subjected to massive declines caused by the bubonic plague—the Black Death. At its peak in about 1400, the Black Death may have killed 25 percent of Europe's population in just over fifty years. By the end of the last great plague in 1650, the human population numbered 600 million.

The Industrial Revolution began between 1650 and 1750. Since then, the growth of the human population has increased greatly. In just under 300 years, the earth's population went from 0.5 billion to 7.7 billion people, and the annual rate of increase went from 0.1 percent to 1.1 percent. This population growth was not because people were having more babies, but because more babies lived to become adults and the average adult lived a longer life.

The Green Revolution occurred in the 1960s. The development of various vaccines and antibiotics in the twentieth century and the spread of their use to most of the world after World War II caused big drops in the death rate, increasing population growth rates. Feeding this growing population has presented a challenge. This third revolution is called the Green Revolution because of the technology used to increase the amount of food produced by farms. However, the Green Revolution was really a combination of improvements in health care, medicine, and sanitation, in addition to an increase in food production.

Geography of Human Population Growth

The present-day human race traces its lineage to Africa. Humans migrated from Africa to the Middle East, Europe, Asia, and eventually to Australia, North and South America, and the Pacific Islands. It is believed that during the last Ice Age, the world's sea levels were lower because much of the world's water was trapped in ice sheets. This lower sea level created land bridges that facilitated many of the major human migrations across the world.

Patterns of human settlement are not random. People generally avoid living in deserts because they lack water. Few humans are found above the Arctic Circle because of that region's severely cold climate. Environmental factors, such as the availability of water and food and the livability of climate, influence where humans choose to live. How much these factors influence the evolution and development of human societies is a subject of debate.

The domestication of plants and animals that resulted from the Agricultural Revolution did not take place everywhere on the earth. In many parts of the world, humans remained as hunter-gatherers while agriculture developed in other parts of the world. Eventually, the agriculturalists outbred the hunter-gatherers, and few hunter-gatherers remain in the twenty-first century. Early agricultural sites have been found in many places, including Central and South America, Southeast Asia and China, and along the Tigris and Euphrates Rivers in what is now Iraq. The practice of agriculture spread from these areas throughout most of the world.

By the time Christopher Columbus reached the Americas in the late fifteenth century, there were millions of Native Americans living in towns and villages and practicing agriculture. Most of them died from diseases that were brought by European colonists. Colonization, disease, and war are major mechanisms that have changed the composition and distribution of the world's population in the last 300 years.

The last few centuries also produced another change in the geography of the human population. During this period, the concentration of industry in urban areas and the efficiency gains of modern agricultural machinery caused large numbers of people to move from rural areas to cities to find jobs. From 1900 to 2020 the percentage of people living in cities went from 14 percent to just about 55 percent. Demographers estimate that by the year 2025, more than 68 percent of the earth's population will live in cities. Scientists estimate that the human population will continue to increase until the year

2050, at which time it will level out at between 8 and 15 billion.

Earth's Carrying Capacity

Many people are concerned that the earth cannot grow enough food or provide enough other resources to support 15 billion people. There is great debate about the concept of the earth's carrying capacity—the maximum human population that the earth can support indefinitely. Answers to questions about the earth's carrying capacity must account for variations in human behavior. For example, the earth could support more bicycle-riding vegetarians than car-driving carnivores. Questions about carrying capacity and the environmental impacts of the human race on the planet are fundamental to the United Nations' goals of sustainable development. Dealing with these questions will be one of the major challenges of the twenty-first century.

Paul C. Sutton

Global Urbanization

Urbanization is the process of building and living in cities. Although the human impulse to live in groups, sharing a "home base" probably dates back to cave-dweller times or before. The creation of towns and cities with a few hundred to many thousands to millions of inhabitants required several other developments.

Foremost of these was the invention of agriculture. Tilling crops requires a permanent living place near the cultivated land. The first agricultural villages were small. Jarmo, a village site from c. 7000 BCE, located in the Zagros Mountains of present-day Iran, appears to have had only twenty to twenty-five houses. Still, farmers' crops and livestock provided a food surplus that could be stored in the village or traded for other goods. Surplus food also meant surplus time, enabling some people to specialize in producing other useful items, or to engage in less tangible things like religious rituals or recordkeeping.

Given these conditions, it took people with foresight and political talents to lead the process of city formation. Once in cities, however, the inhabitants found many benefits. Walls and guards provided more security than the open country. Cities had regular markets where local craftspeople and traveling merchants displayed a variety of goods. City governments often provided amenities like primitive street lighting and sanitary facilities. The faster pace of life, and the exchange of ideas from diverse people interacting, made city life more interesting and speeded up the processes of social change and invention. Writing, law, and money all evolved in the earliest cities.

Ancient and Medieval Cities

Cities seem to have appeared almost simultaneously, around 3500 BCE, in three separate regions. In the Fertile Crescent, a wide curve of land stretching from the Persian gulf to the northwest Mediterranean Sea, the cities of Ur, Akkad, and Babylon rose, flourished, and succeeded one another. In Egypt, a connected chain of cities grew, soon unified by a ruler using Memphis, just south of the Nile River's delta, as his strategic and ceremonial base. On the Indian subcontinent, Mohenjo-Daro and Harappa oversaw about a hundred smaller towns in the Indus River valley. Similar developments took place about a thousand years later in northern China.

These first city sites were in the valleys of great river systems, where rich alluvial soil boosted large-scale food production. The rivers served as a "water highway" for ships carrying commodities and luxury items to and from the cities. They also furnished water for drinking, irrigation, and waste disposal. Even the rivers' rampages promoted civilization, as making flood control and irrigation systems required practical engineering, an organized workforce, and ongoing political authority to direct them.

Eurasia was still full of peoples who were not urbanized, however, and who lived by herding, pirating, or raiding. Early cities declined or disappeared, in some cases destroyed by invasions from such forces around 1200 BCE. Afterward, the cities of Greece became newly important. Their surrounding land was poor, but their access to the sea was an advantage. Greek cities prospered from fishing and trade. They also developed a new idea, the city-state, run by and for its citizens.

Rome, the Greek cities' successor to power, reached a new level of urbanization. Its rise owed more to historical accident and its citizens' political and military talents than to location, but some geographical features are salient. In some ways, the fertile coastal plain of Latium was an ideal site for a great city, central to both the Italian peninsula and the Mediterranean Sea. There, the Tiber River becomes navigable and crossable.

In other ways, Rome's site was far from ideal. Its lower areas were swampy and mosquito-ridden. The seven hills, with their sacred sites later filled with public buildings and luxury houses, imposed a crazy-quilt pattern on the city's growth. Romans built cities with a simple rectangular plan all over Europe and the Middle East, but their home city grew in a less rational way.

At its peak, Rome had a million residents, a population no other city reached before nineteenth century London. It provided facilities found in modern cities: a piped water supply, a sewage disposal system, a police force, public buildings, entertainment districts, shops, inns, restaurants, and taverns. The streets were crowded and noisy; to control traffic, wheeled wagons could make deliveries only at night. Fire and building collapse were constant risks in the cheaply built apartment structures that housed the city's poorer residents. Still, few wanted to live anywhere but in Rome, their world's preeminent city.

In the Early Middle Ages after the western Roman Empire collapsed, feudalism, based on land holdings, eclipsed urban life. Cities never disappeared, but their populations and services declined drastically. Urban life still flourished for another millenium in the eastern capital of Constantinople. When Islam spread across the Middle East, it caused the growth of new cities, centered around a mosque and a marketplace.

In the twelfth and thirteenth centuries, life revived in Western Europe. As in the Islamic cities, the driving forces were both religious—the building of cathedrals—and commercial—merchants and artisans expanding the reach of their activities. Medieval cities were usually walled, with narrow, twisting streets and a lack of basic sanitary measures, but they drew ambitious people and innovative forces together. Italy's cities revived the concept of the city-state with its outward reach. Venice sent its merchant fleet all over the known world. Farther north, Paris and Bologna hosted the first universities. The feudal system slowly gave way to nation-states ruled by one king.

Modern Cities

Modern cities differ from earlier ones because of changes wrought by technology, but most of today's cities arose before the Industrial Revolution. Until the early nineteenth century, travel within a city was by foot or on horse, which limited street widths and city sizes. The first effect of railroads was to shorten travel time between cities. This helped country residents moving to the cities, and speeded raw materials going into and manufactured goods coming out of the factories that increasingly dotted urban areas. Rail transit soon caused the growth of a suburban ring. Prosperous city workers could live in more spacious homes outside the city and ride rail lines to work every day. This pattern was common in London and New York City.

Factories, the lifeblood of the Industrial Revolution, were built in pockets of existing cities. Smaller cities like Glasgow, Scotland, and Pittsburgh, Pennsylvania, grew as ironworking industries, using nearby or easily transported coal and ore resources, built large foundries there. Neither industrialists nor city authorities worried about where the people working there would live. Workers took whatever housing they could find in tenements or subdivided old mansions.

Beginning in the 1880s, metal-framed construction made taller buildings possible. These skyscrapers towered over stately three- to eight-story structures of an earlier period. Because this technology enabled expensive central-city ground space to house many profitable office suites, up through the 1930s, city cores became quite compacted. Many people believed such skyward growth was the wave of the future and warned that city streets were becoming sunless, dangerous canyons.

Automobiles kept these predictions from fully coming true. As car ownership became widespread, more roads were built or widened to carry the traffic. Urban areas began to decentralize. The car, like rail transit before it, allowed people to flee the urban core for suburban living. Because roads could be built almost anywhere, built-up areas around cities came to resemble large patches filling a circle, rather than the spokes-of-a-wheel pattern introduced by rail lines. Cities born during the automotive age tend to have an indistinct city center, surrounded by large areas of diffuse urban development. The prime example is Los Angeles: It has a small downtown area, but a consolidated metropolitan area of about 34,000 square miles (88,000 sq. km.).

Almost everywhere, urban sprawl has created satellite cities with major manufacturing, office, and shopping nodes. These cause an increasing portion of daily travel within metropolitan areas to be between one edge city and another, rather than to and from downtown. Since these journeys have an almost limitless variety of start points and destinations within the urban region, mass transit is only a partial solution to highway crowding and air pollution problems.

The above trends typify the so-called developed world, especially the United States. Many cities in poor nations have grown even more rapidly but with a different mix of patterns and problems. However, the basic pattern can be detected around the globe, as urban dwellers seek to better their own circumstances. Today, 55 percent of the world's population lives in urban areas, and that percentage is expected to rise to 68 percent by 2050. Projections show that urbanization combined with the overall growth of the world's population could add another 2.5 billion people to urban areas by 2050, with close to 90 percent of this increase taking place in Asia and Africa, according to a United Nations data set published in May 2018.

URBANIZATION AND DEVELOPING NATIONS

The urban population, or number of people living in cities, in North America accounts for about 75 percent of its total population. In Europe, about 90 percent of the population lives in cities. In developing countries, the urban population is often less than 30 percent. The term "urbanization" refers to the rate of population growth of cities. Urbanization mainly results from people moving to cities from elsewhere. In developing countries, the urbanization rate is very high compared to those of North America or Europe. The high rate of urbanization of these countries makes it difficult for their governments to provide housing, water, sewers, jobs, schools, and other services for their fast-growing urban populations.

Megacities and the Future

In the year 2019 the world had thirty-three megacities, defined as urban areas with a population of 10 million or more. The largest was Tokyo, with an estimated 37.5 million people in 2018, predicted to grow to around 37 million by 2030. Second-largest was Delhi, with more than 28.5 million in 2018 and predicted to grow to around 38.94 million by 2030. Megacities in the United States include New York-Newark with a population of 18.8 million and Los Angeles at 12.5.

Megacities profoundly affect the air, weather, and terrain of their surrounding territory. Smog is a feature of urban life almost everywhere, but is worse where the exhaust from millions of cars mixes with industrial pollution. Some megacities have slowed the problem by regulating combustion technology; none have solved it. Huge expanses of soil preempted by buildings and pavements can turn heavy rains into floods almost instantly, and the ambient heat in large cities stays several degrees higher than in comparable rural areas. Recent engineering studies suggest that megacities create instability in the ground beneath, compressing and undermining it.

How will cities evolve? Barring an unforeseen technological or social breakthrough, the current growth and problems will probably continue. The process of megapolis—metropolitan areas blending together along the corridors between them—is well underway in many areas. Predictions that the computer will so change the nature of work as to cause massive population shifts away from cities have not been proven correct. Despite its drawbacks, increasing numbers of people are drawn to urban life, seeking the economic opportunities and wider social world that cities offer.

Emily Alward

Urban Heat Islands

Large cities have distinctly different climates from the rural areas that surround them. The most important climatic characteristic of a city is the urban heat island, a concentration of relatively warmer temperatures, especially at nighttime. Large cities are frequently at least 11 degrees Fahrenheit (6 degrees Celsius) warmer than the surrounding countryside.

The urban heat island results from several factors. Primary among these are human activities, such as heating homes and operating factories and vehicles, that produce and release large quantities of energy to the atmosphere. Most of these activities involve the burning of fossil fuels such as oil, gas, and coal. A second factor is the abundance of heat-absorbing urban materials, such as brick, concrete, and asphalt. A third factor is the surface dryness of a city. Urban surface materials normally absorb little water and therefore quickly dry out after a storm. In contrast, the evaporation of moisture from wet soil and vegetation in rural areas uses a large quantity of solar energy—often more than is converted directly to heat—resulting in cooler air temperatures and higher relative humidities.

People of Australia

In 2020, Australia had about 25.5 million people, the smallest population of any continent except Antarctica. Although it is about the same size as the continental United States, Australia has only about one-fourteenth the number of inhabitants. The size of the population is the result of two major processes: migration from outside of the country and natural increase.

Migration

Australia is a multiethnic country. Its population, like those of the United States and Canada, is composed of descendants of the original native peoples plus migrants from many other regions of the world. Even the indigenous Australians (also called Aboriginal peoples) were immigrants long ago. About 55 percent of Australia's people are of British and Irish ancestry, reflecting colonial ties to the United Kingdom. The number of migrants from other regions—especially Asia—has increased as Australia weaves closer bonds with its Pacific Rim neighbors.

The ancestors of the indigenous Australians came to the continent at least 60,000 years ago and perhaps as early as 120,000 years ago. Apparently, they traveled from Southeast Asia through the Indonesian islands during periods of lower sea levels. In the Pleistocene geologic epoch, much of Earth's water was locked in ice caps, which reduced sea levels by 600 feet (180 meters) or more below current

An indigenous Australian festival. (chameleonseye)

depths. The shallow straits between New Guinea and Australia were often dry during this period, and the seas that separated the continent from Indonesia were narrower than they are now. These conditions reduced Australia's isolation and allowed humans to enter its pristine environments for the first time.

It is not known how many of these early settlers arrived in northern Australia or how they traveled. They may have paddled simple rafts or canoes. Much of the evidence of their earliest settlement is presumed to be submerged on the drowned continental shelf. In any event, their descendants gradually spread over the entire continent, even reaching Tasmania before it became an island isolated by rising sea levels.

The new arrivals probably distributed themselves in patterns somewhat resembling the modern population, with larger numbers along the coastlines and in the wetter environments of the east, south, and north, and fewer in the arid interior. The central interior may not have been settled until as recently as 10,000 years ago. The groups learned to use the varied plant and animal resources of Australia to support their hunting and gathering economies.

By the time Europeans reached Australia, the indigenous population was probably about 500,000, although estimates range from 300,000 to as high as 1 million. These estimates are based on archaeological remains and evidence of food supplies, but no one knows exactly how many people descended from those initial immigrants. The arrival of European colonists proved devastating; their numbers shrank rapidly as they suffered from warfare, diseases for which they possessed no immunity, and disruption of their traditional cultures and economies. Until the mid-twentieth century, the indigenous Australian people were denied citizenship rights and many other Australians believed that they were a dying race. In 2016, people descended from indigenous Australians comprised about 2.8 percent of the Australian population, or nearly 650,000 people.

Convict Colonies

The first British settlers in Australia were mostly deported convicts and their guards. The English judicial system typically sent prisoners to serve their sentences in the colonies instead of building expensive prisons in the British Isles. North America stopped accepting British prisoners after the American Revolution, so a new penal colony was needed. The east coast of Australia, explored by Captain James Cook in 1770, was selected by the British government for this purpose.

The present city of Sydney was the site of the initial prison camp established by the First Fleet in 1788. From that year until 1858, over 160,000 convicts were transported to Australia. Most were English, but Scottish, Welsh, and Irish prisoners also ar-

A cricket match in Melbourne, Australia. (danieldep)

rived; roughly 20 percent were women. The convicts were usually sentenced to serve seven- to fourteen-year terms for crimes such as robbery, burglary, or fraud. Prison colonies were established in Tasmania, Queensland, Victoria, and Western Australia, as well as in New South Wales. When their terms expired, most of the convicts stayed in Australia as emancipated settlers.

The convicts were joined by a trickle of free settlers who wanted to take advantage of agricultural and business opportunities in the new colonies. The British government began to pay the costs of passage for some migrants, mainly single women seeking marriage partners and skilled workers needed in specific industries. A few thousand Germans arrived in the 1830s and 1840s, fleeing political upheaval in their homeland. By 1850 the total European population of Australia was about half a million. These people clustered in the growing port towns of Sydney, Melbourne, Hobart, Brisbane, Adelaide, and Perth, and in nearby agricultural areas, displacing the indigenous Australians who had once lived along the coastlines.

Other Settlers

From 1850 to 1860, Australia's population doubled from about 500,000 to 1 million as a result of the discovery of gold in the colonies of Victoria and New South Wales. Fortune seekers arrived from the British Isles, other European countries, the United States, and China. Although some miners returned to their homelands at the end of the gold rush decade, most stayed to take up new occupations in the booming cities and towns. Since most of the new migrants were relatively unskilled male laborers, the British government continued to pay moving costs for young women and skilled workers who wished to make the long journey to Australia. By assisting potential brides and small business owners, the government helped create a more balanced colonial economy and society.

Chinese gold miners formed the first large group of Asian migrants to Australia. By 1860 there were at least 40,000 Chinese in the gold fields of Victoria. Discrimination against the Chinese was legalized through state laws that prevented non-Europeans from becoming citizens, thus effectively barring them from property ownership and professional occupations. Despite these restrictions and the Immigration Restriction Act of 1901, Chinese communities persisted in Melbourne, Sydney, Brisbane, and Perth.

The other major non-European migration in the nineteenth century was the movement of South Pacific islanders called Kanakas into Queensland. Young men were recruited from Vanuatu, the Solomon Islands, and other western Pacific islands to work for low wages in the sugar and cotton plantations of the tropical Queensland coast. Humanitarian groups and labor unions protested that this arrangement was little better than slavery; others deplored the fact that a large number of nonwhites had been admitted to Australia. Importation of Kanaka laborers was prohibited in 1901, and most were sent back to their homelands by 1908.

Later Migrations

Australia became an independent state within the British Commonwealth in 1901. Isolated by distance from Great Britain and fearing the growing

The Chinatown neighborhood in Melbourne, Australia. (brightsea)

A diverse classroom in Australia. (JohnnyGreig)

power of Asian countries, the new Australian federal government developed a "White Australia" policy aimed at excluding non-European immigrants. This policy, implemented by the aforementioned Immigration Restriction Act of 1901, gave the government the ability to prohibit "undesirable" migrants without specifying which groups were to be rejected or how this goal would be accomplished. In practice, Asian migrants usually could be rejected by an English literacy test. This law impeded Asian migration into the 1970s.

Although migration from the British Isles continued, the twentieth century brought people from other areas as well. Political and economic turmoil caused by World War I created a major surge of newcomers. At least 200,000 migrants, including former British soldiers and other Europeans, received assisted passages to Australia in the aftermath of World War I. In 1938, Australia agreed to accept 15,000 refugees from Nazi persecution, many of them Jewish. These humane acts marked the opening of a new era of migration from Europe.

After World War II ended in 1945, about 250,000 European displaced persons moved to Australia. In the 1950s and 1960s, the refugees were joined by thousands of people from the less-prosperous Mediterranean countries, which suffered from high unemployment and lack of economic opportunities. Groups from countries such as Italy, Greece, the former Yugoslavia, and Lebanon created ethnic neighborhoods that have become important features of the cultural landscape of Melbourne, Sydney, and other cities.

The number of Asian migrants increased in the last quarter of the twentieth century. Australia's involvement in the Vietnam War resulted in a strong flow of refugees from Vietnam and Cambodia in the 1970s and 1980s. When Australia discarded the Immigration Restriction Act in 1973 and reoriented its economic ties to Asia, migrants began to arrive from Malaysia, the Philippines, Hong Kong, India, and China, among other places. Large numbers of the Asian migrants were well-educated people with professional or skilled occupations.

Now, in the third decade of the twenty-first century, more than 30 percent of Australia's population is made up of immigrants. Proportionally, this ranks Australia as having the eighth-largest immigrant population in the world. The majority of the newcomers bring with them specific skills, though a significant number are admitted as refugees or family members. In 2019, England was the leading source of immigrants, followed by China, India, New Zealand, the Philippines, and Vietnam.

Natural Increase

Although a simple timeline shows that the population has grown steadily since Europeans arrived in the region, the rate of increase has varied substantially. Natural increase is the surplus of births over deaths in an existing population. In general, both migration and natural increase rates have helped Australia grow during periods of economic prosperity and social stability. Birth rates rose after both world wars, for example, as returning soldiers married and the peacetime economies stabilized. In periods of war and economic depression, however, population growth leveled off. For a few short periods, the population actually declined. Marriage rates and birth rates plummeted during the world wars and the major depressions of the 1890s and 1930s, when few could afford to start a family. Since both depressions also prevented potential migrants from reaching Australia and caused some people to seek work in other countries, the population declined slightly during these decades.

Australia's Population Future

In 2020, Australia's rate of natural increase averaged about 5.5 per 1,000 per year. Assuming that this rate stays about the same and immigration levels also do not change, it can be extrapolated that another million people will be added to the population every three years for the next few decades.

Australia's Immigration Policies

The White Australia Policy and the Immigration Restriction Act were based on attitudes of racial superiority that were common in Australia, as well as in Europe and North America, during the late nineteenth and early twentieth centuries. These policies also reflected concern about Japan's growing economic and military power and China's large, land-hungry population. In 1901 many Australians believed that their new country must "populate or perish." Their fear was that if there were not enough British or other European settlers to occupy the entire continent, they would risk losing it to the rising Asian forces. In the second half of the twentieth century, however, Australia realized that its closest neighbors were Asian-Pacific countries, not Great Britain and Europe. Trade, investment, and tourism now tightly link Australia to Japan and other Pacific Rim states. Anti-Asian immigration policies were largely abandoned by the Australian government in the 1970s.

Current immigration policies focus on individual migrants' qualifications, such as their educational level and specific skills, rather than their country of origin. Family reunification and refugee status also are considered. Ongoing debate about migration policies centers on environmental, economic, and political concerns, including the mandatory immigration detention for all arrivals deemed unlawful.

Although rates of natural increase are comparable to those of other developed nations, Australia historically has not been able to attract as many migrants as North America or South America. Isolated from potential migrants by vast distances and expensive transportation Australia's economic opportunities generally could not compete with those in many other regions. Still, by 2020, Australia offered a stable democratic society with opportunities to participate in the global economy.

Susan P. Reynolds

People of the Pacific Islands

The Pacific is the world's largest ocean, covering about 64 million square miles (165 million sq. km.) between Singapore and Panama—more than 30 percent of Earth's surface. While most of the Pacific is devoid of land, it does have a large number of islands stretching from Southeast Asia to the coast of South America.

Prehistory

Several hundred million years ago, the world's continents were linked together in one landmass that geologists have called Pangaea. After the Ice Age, rising waters separated the land into two sections, Laurasia in the north and Gondwanaland in the south. Water spaces between these new continents were narrow, making travel from one land to the other possible.

The Pacific islands were probably the last place on Earth to be inhabited by humans. It is also the last major region of the world, except for the North and South Poles, to be explored and settled by Europeans. The majority of islands in the South Pacific are grouped together under the name Oceania. The three groups within Oceania are Melanesia, Micronesia, and Polynesia.

The Voyage

Prevailing winds and currents in the South Pacific follow east-to-west patterns. Therefore, voyagers from Southeast Asia to the Pacific islands would

Native Hawaiian schoolchildren. (Henry Wetherbee Henshaw)

have had to sail hundreds, even thousands, of miles against the currents to their new homes. The immigrants had no navigational tools and depended solely upon skills they learned from their ancestors. They followed the Sun, clouds, the Moon and stars, the waves and winds, and flights of migratory birds. It is probable that some vessels drifted off course during cloudy days or storms.

The canoes within which the voyagers traveled were made of hollowed-out tree trunks. The sides were built up with planks and lashed together with coconut fibers. They were made watertight with vegetable gum or a mix of charcoal and soil. To provide stabilization, a smaller canoe often was connected to the side by a wooden boom. This canoe was called an outrigger. Some double canoes had small shelters built over the connections, which the travelers used for protection during longer voyages. These vessels were paddled by oars or powered by triangular sails.

Melanesia

Situated northeast of Australia, the island group known as Melanesia was settled as early as 50,000 years ago by Austronesian tribes who left their homeland of Southeast Asia, traveling south and east in dugout canoes. These hunters and gatherers are the direct ancestors of the indigenous peoples of Australia, the highland people of Papua New Guinea, and peoples still living in the interior of the Philippines and Malaysia.

About 5,000 years ago, another wave of immigrants left Southeast Asia. These people, known as Proto-Malay, spoke a language far different than that of the Austronesians. Later groups of Proto-Malay sailed from Southeast Asia directly to western Melanesia.

About 8,000 years ago, more settlers arrived. These people, were of a different racial background than the first group and are known to anthropologists as Papuans. They eventually traveled north and east into Micronesia.

Micronesia

At about the same time the Proto-Malay were migrating to Melanesia, another group from Southeast Asia headed for the islands of Micronesia.

THOR HEYERDAHL AND THE PEOPLING OF POLYNESIA

Norwegian ethnologist Thor Heyerdahl developed a theory that Polynesia may have been settled by indigenous people from South America, specifically Bolivia. He reasoned that most of the European voyagers took routes from east to west. He believed west-to-east travel would have been too difficult against the prevailing winds and currents. Heyerdahl also reasoned that the first settlers did not come from Indonesia because Sanskrit words, found in Indonesian languages, was absent from Polynesian languages. There are, however, some similarities between the Polynesian languages and the Inca language. Heyerdahl also pointed out that the sweet potato, native to South America, was also found in the South Pacific. Heyerdahl believed that the original settlers traveled from Asia to North America through the Bering Strait, down to South America, then across to Polynesia. In 1947, to prove his theory, Heyerdahl and a crew constructed a balsa raft called the Kon-Tiki and set sail from Peru, South America. They sailed 101 days, following winds and currents until they crash-landed on the atoll of Raroia in the Tuamotus, thus proving that South American Indians could have reached Polynesia by this route.

THE PEOPLING OF EASTER ISLAND

Easter Island lies 2,400 miles (3,862 km.) from Chile and 1,500 miles (2,414 km.) from the nearest inhabited island. It was discovered by Dutch explorer Jacob Roggeveen on Easter Sunday, April 15, 1722. It is not known exactly how the first inhabitants reached Easter Island, but it is generally agreed that people of Polynesian stock arrived on the island from the central Pacific. When Europeans first arrived in 1722, the sparse population of the island was visibly of Polynesian origin. The inhabitants of Easter Island carved huge stone statues, weighing more than fifty tons and standing as tall as 40 feet (12 meters) on stone platforms, line the shore with their backs to the sea. Some of the statues have "hats." Although Roggeveen noted that the islanders often bowed to these statues, he did not ask them how they had managed to move these massive monuments.

A Maori family in New Zealand. (MollyNZ)

Predominately farmers, the immigrants found life on these atolls—small islands composed of coral reefs—very different from what they were used to. They had to change their methods of farming, losing old skills but learning new ones. They became expert boat builders and navigators and developed their outrigger canoes into the fastest and most maneuverable long-distance vessels in the Pacific.

Migration from one island group to another was slow, usually being spread over several generations. After a group had settled on an island for some length of time, a new sailing party would continue eastward to the next group of islands. This migration accounts for the similarity in languages, cultures, traditions, and appearance of the peoples of the Pacific. Some islanders stayed at home, intermarrying with Melanesians and other Asian travelers.

Polynesia

The largest island group of the region, Polynesia forms a triangle bounded by Hawaii in the north, New Zealand in the southwest, and Easter Island, in the southeast. It is believed to be the last region to have been settled. The first Polynesians were tall, and comparatively advanced culturally and technologically. Since most Polynesians speak closely related languages, it is believed they must have come from one direction in a short period of time, possibly through Micronesia or Melanesia, or both. Polynesians eventually traveled farther east to New Zealand.

Uninhabited longer than any major landmass, New Zealand is the southernmost point of the Polynesian triangle and consists of three major islands: North, South and Stewart. The first human inhabitants were Polynesian. The first phase of migrants, who arrived approximately 1,200 years ago, can be identified by their archaeological remains and are known as Archaic Maori. The second phase, Classic Maori, is probably the culture encountered by European explorers. British explorer Captain James Cook noted the similarity in the languages of New Zealand Maori and Polynesians. It is not yet clear how the first phase evolved into the second. It is believed that the first Maori may have reached New Zealand by accident, being blown off course on fishing expeditions.

A self-sufficient people, the Maori easily adapted to the temperate climate of their new home. They had no concept of culture, nationhood, or race. They were a tribal people, protective of their ancestry, often living in the same place from birth to death. Their culture was the same throughout the islands, with slight regional differences. Languages spoken on North Island differed in dialect from that spoken on South Island. By the end of the eighteenth century, the Maori had to make room for new immigrants: explorers, whalers, escaped convicts, and traders. The British began their colonization of the islands in the 1830s.

Languages and Cultures

Language, folklore, and culture were, and still are, individual to each region. This indicates that immigrants probably did not originate from the general place. Migrations were probably made as a result of changes on their home islands, including changes in environment, economics, or war. Languages spoken in Melanesia, Micronesia, and Polynesia all derive from one basic language group, the Austronesian family. Melanesia alone has more than 1,200 different languages. In Micronesia, the people in the west speak differently than those in the east. Languages from eastern Micronesia have similarities to those in eastern Melanesia. Polynesia is the only region whose languages, nearly thirty, are alike enough that one islander can understand what another from a different area is saying.

A boy on the beach in Kiribati (Vladimir Lysenko)

A New Zealander of Anglo descent eats breakfast in Auckland with the Sky Tower in the background. (chameleonseye)

European Explorers

Additional migrations to the South Pacific were initiated by Europeans, beginning with the Spanish and Portuguese in the sixteenth century, followed by the Dutch in the seventeenth century and the English and French in the eighteenth century. The explorers were followed by whalers and traders, missionaries, planters and merchants, soldiers and sailors, and slave traders. With the appearance of European explorers, the populations of the islands were threatened by diseases and social and political issues that were introduced. By the early nineteenth century, missionaries were spreading through the area converting a proud people who were now regarded as savages. Respect for the Polynesian culture was gone. Bans were put on dancing and singing, and European-style dress was enforced.

Maryanne Barsotti

Discussion Questions: People

Q1. Who were the ancestors of the indigenous people of Australia? Where did they come from? When do scientists think the first humans arrived in Australia?

Q2. How do anthropologists describe the populating of the Pacific islands? Over how long a period did it last? How did early peoples travel from one island to the next?

Q3. How did the Maori come to inhabit New Zealand? What have anthropologists surmised from studying the Maori language and its dialects?

Population Distribution of Australia

The vast majority of Australia's population of 25.5 million people is concentrated in a core region that stretches along the eastern and southeastern coastline of the continent. From Cairns in northern Queensland through the states of New South Wales and Victoria and beyond to Adelaide in the state of South Australia, this core region contains the largest cities and the industrial and agricultural heartland of the country. Much smaller secondary concentrations of population lie in Western Australia around the city of Perth and on the northern and southern coasts of the island state of Tasmania. Beyond these core regions, a scattering of small communities holds the rest of the Australian people.

Physical and Historical Factors

The map of population distribution clearly reveals the peripheral location of Australia's settlement. This spatial pattern reflects the constraints of the physical environment. The eastern plains and highlands, including Tasmania, contain the best agricultural lands and receive sufficient rainfall year-round for crop production.

The western slopes of the Great Dividing Range in New South Wales and Victoria are drier, but provide enough moisture for grain farming. Along the

The Australian Outback. (Timothy Ball)

boundary between these two states, the Murray River basin offers irrigated lands for specialty crops. The southern and western margins of the continent around Adelaide and Perth also have good soils and a Mediterranean-type climate with adequate winter rainfall. The agricultural population of these regions is fairly dense. Beyond these areas, the arid interior is suitable mainly for grazing sheep and cattle on huge pastures; the human population is sparse. The driest areas are virtually uninhabited, because they lack enough surface water and vegetation for even a grazing economy.

Other natural resources, especially minerals, attract people to specific locations. Beginning in the gold rush days of the 1850s, the settlement frontier advanced into the harsher environments of the Australian interior in response to new mineral discoveries. Rich deposits of silver, copper, lead, and zinc brought miners to places such as Broken Hill, New South Wales, and Mount Isa, Queensland. Recent developments are even more remote from the population core: The bauxite mines at Weipa in northern Queensland and the Hamersley Range iron mines in Western Australia are good examples.

Climate and Population Distribution

The first British immigrants who settled in Australia during the late eighteenth and early nineteenth centuries had no idea how dry their new land was. When Captain James Cook first reached Australia during the 1770's, he landed at Botany Bay on the temperate and comparatively moist east coast. As his voyage of discovery continued northward, Cook crossed the Tropic of Capricorn, which neatly divides the northern and southern portions of the continent, and found a region where rainfall was abundant.

It was not until after the British began settling the region that the continent's dry interior was penetrated by explorers, some of whom perished in the harsh desert interior. After that time, climate played a major role in determining the pattern of the continent's population distribution. The vast majority of Australia's people now live in coastal cities in the temperate midlatitudes, particularly in the southeast.

Bondi, Australia, a suburb of Sydney. (Searsie)

Australian State and City Populations in 2016

State	*Population (millions)*	*Capital City*	*Population (millions)*
New South Wales	7.4	Sydney	4.8
Victoria	5.9	Melbourne	4.4
Queensland	4.7	Brisbane	2.2
Western Australia	2.4	Perth	1.9
South Australia	1.6	Adelaide	1.2
Tasmania	0.5	Hobart	0.2
Australian Capital Territory	0.3	Canberra	0.3
Northern Territory	0.2	Darwin	0.1

Modern mechanized mining requires few workers, and the extreme climates of these newer resource regions are not attractive to the average Australian. Thus, these mining towns have remained relatively small.

Each of the largest cities was the focus of early colonial settlement in its region. The initial settlements were located on the coast to facilitate transportation and communication with Great Britain and the other Australian colonies. These small port towns grew by consolidating government functions, trade, industry, and services, and developed into Australia's major cities. Sydney, for example, was the first convict settlement and from the beginning was the administrative capital of New South Wales. Its beautiful bay offered deep water to accommodate ships, and railroads were built there to bring agricultural products and natural resources to the port. Once Sydney established a large enough population base, it attracted manufacturing industries and consumer services. The first British settlement in Australia, Sydney has always remained the largest.

The national capital city, Canberra, is the only exception to the rule that all large Australian cities are located on the coast. Canberra was established after the individual Australian colonies federated as an independent state within the British Commonwealth in 1901. The new city was deliberately situated in the Eastern Highlands midway between Sydney and its major rival Melbourne, capital of the state of Victoria. Canberra represents a geographic compromise between the two largest cities, each of which had hoped to become the national capital.

Urbanization

Australia is a highly urbanized society. About 86 percent of the population is urban and more than 40 percent live in Sydney or Melbourne, the two largest metropolitan centers. In several states, the majority of people live in the capital city and its suburbs. For example, about 71 percent of the New South Wales population lives in Sydney and 83 percent of Victorians live in the Melbourne metropolitan area. About 80 percent of South Australians cluster in Adelaide, while Perth and its suburbs contain 83 percent of Western Australia's inhabitants.

As in North America and other economically developed countries, Australia's cities attract those who are searching for new opportunities. Urban manufacturing and service businesses, universities and cultural organizations, and entertainment facilities offer jobs and leisure activities. Former rural residents and new immigrants alike are drawn to join those who were born and raised in the cities. Recent migrants, especially those from non-English speaking homelands, tend to cluster in ethnic neighborhoods. Approximately 79 percent of the indigenous population also lives in urban settings.

Within the larger cities, sprawling residential areas surround the skyscraper-studded central business districts. Single-family houses and apartment buildings are linked by multilane highways and

A shopping district in Melbourne, Australia. (kokkai)

served by shopping malls in a suburban landscape that resembles the cities of the western United States.

Rural Population

About 14 percent of the Australian population lives in small towns and rural areas. Most farmers live on individual farmsteads at some distance from their closest neighbors; livestock grazers often live on huge pastoral stations (ranches) that may be hundreds of miles from the nearest large city. Approximately 20 percent of the indigenous Australians are rural dwellers, some of whom live and work on livestock stations, while a few maintain their traditional hunting and gathering lifestyles in isolated regions of the Outback.

Although many agricultural districts face declining numbers as economic reorganization continues, some small towns and rural areas attract new residents. Growth areas include pleasant towns within commuting distance of urban jobs and coastal beach communities that draw retired persons. The expansion of international tourism has also created resort settlements on Great Barrier Reef islands, in the Red Center near Uluru (Ayers Rock), and in other remote places.

Susan P. Reynolds

Population Distribution of the Pacific Islands

The Pacific islands region, including New Zealand, has approximately 16 million people, which is roughly equivalent to the combined populations of the US states of Michigan and Wisconsin. The obvious difference is that Oceania's population is distributed on islands that are spread out over a vast area of ocean. Hawaii and New Zealand have highly urbanized populations and advanced economies. Most of the remaining islands have large rural populations, because many of their inhabitants depend on farming, fishing, and, until recently, mining for their livelihoods.

On less-developed islands, rising population and loss of soil fertility and forestland in rural areas are spurring a migration from the countryside to urban areas. Consequently, the population of urban areas is growing faster than that of rural areas, urban housing and health and transportation services are not keeping pace with the rate of population growth, and squatter populations have become permanent fixtures around the edges of many cities and large towns. Migration to the United States (especially Hawaii and California), Canada, Australia, and New Zealand serves as a safety-valve to lessen

A rural village in Papua New Guinea. (dane-mo)

A busy street in Auckland, New Zealand. (AsianDream)

the impact of overpopulation occurring on many islands.

There are two true metropolitan centers: Honolulu, Hawaii, with a metro population of about 950,000 (2019 estimate), and Auckland, New Zealand, with about 1.47 million people (2020). New Zealand has six other cities with populations exceeding 100,000: Christchurch, Wellington, Hamilton, Tauranga, Lower Hutt, and Dunedin. Other cities with populations of more than 100,000 include Port Moresby in Papua New Guinea (364,000;2011), Suva in Fiji (185,913; 2017 metro), Noumea in New Caledonia (182,341; 2019 metro), and Papeete in French Polynesia (136,771; 2017 metro).

Melanesia

This subregion has about 9.47 million inhabitants (2020), about 55 percent of the total population of the Pacific islands. More than three-fourths of the population is in Papua New Guinea. Melanesia has one of the highest proportions of rural population in the world.

Fiji (2020 population, 935,974) has an area 1.5 times the size of Washington, D.C. Its capital is Suva. Most of the land (86 percent) and people (87 percent) are on the two largest islands—Viti Levu and Vanua Levu. Cities have grown rapidly on these two islands, and today, 57 percent of Fiji's population has become urbanized. Fiji's economy is relatively advanced, and its annual growth rate was 3 percent in 2017.

New Caledonia (2020 population, 290,009), a special collectivity of France, has an area slightly smaller than that of New Jersey. The capital, Noumea, is the principal center on the island. Slightly more than 71 percent of its population is urbanized. New Caledonia Island, the largest oceanic island in the Pacific basin, is home to a majority of the territory's population.

Papua New Guinea (2020 population, 7.25 million) has an area slightly larger than California. It is the most populous nation in the Pacific islands. The capital, Port Moresby, located on the southern coast, is one of the fastest growing urban centers in the region. The highest proportion of population

Land Areas and Population Densities of Pacific Island Countries, Mid-2020 Estimates

Country	*Population*	*Area in sq. mi.*	*Persons/sq. mi.*
Fiji	935,974	7,055	132.67
Kiribati	111,796	313	357.18
Marshall Islands	77,917	70	1,113.10
Micronesia, Federated States of	102,436	271	377.99
Nauru	11,000	8	1,375.00
New Zealand	4,925,477	102,138	48.22
Palau	21,685	177	122.51
Papua New Guinea	7,259,456	174,850	41.52
Samoa	203,774	1,089	187.12
Solomon Islands	685,097	10,805	63.41
Tonga	106,095	277	383.01
Tuvalu	11,342	10	1,134.20
Vanuatu	298,333	4,706	63.39

Source: The World Factbook.

lives in the five central highland provinces and the coastal region; less than 14 percent of the population lives in urban areas.

The Solomon Islands (2020 population, 685,097) has an area slightly smaller than Maryland. The capital, Honiara, is located on the island of Guadalcanal and had a population of 84,520 in 2017. Only 17 percent of the inhabitants live in urban areas, and of those, two-thirds reside in Honiara.

Vanuatu (2020 population, 298,333) has an area slightly smaller than Connecticut. About 25 percent of the population is urbanized. The main urban center is Port Vila (2011 population, 51,437) on the island of Efate. A scarce supply of water and deforestation are forcing rural people to move to cities. Consequently, Vanuatu has a high urban growth rate (about 2.55 percent).

Polynesia

This subregion's 7.05 million inhabitants account for 41 percent of the total population of the Pacific islands. Most of the people of Polynesia live in New Zealand and Hawaii. Together, these countries account for 90 percent of this subregion's total population and most of the urban population.

Most of the people in American Samoa (2020 population, 49,437) live on the island of Tutuila, where the capital, Pago Pago, (2010 population, 3,656) is located. Thanks to many years of a high urban growth rate, more than 87 percent of the population is urbanized. The population has dropped by 25 percent in the past two decades, largely owing to the migration of American Samoans to Hawaii and California.

The Cook Islands (2020 population, 8,574) has an area 1.3 times the size of Washington, D.C. Fifty-nine percent of the population is urbanized. Most of the people live on the island of Raratonga, where the capital, Avaru (2016 population, 4,906), is located. More than half of Cook Island residents have migrated to New Zealand in the past twenty years.

Easter Island has an area three-fourths the size of Washington, D.C. About 7,750 people lived on the island in 2017, nearly all in the town of Hanga Roa. The island has one of the lowest population densities in Polynesia. Nevertheless, there is some out-migration. A small number of Easter Islanders —known as Rapa Nuis—live in Chile, French Polynesia, and the United States.

French Polynesia (2020 population, 295,121) has an area slightly less than one-third the size of Connecticut. Sixty-two percent of the people live in cities. The island of Tahiti makes up about one-fourth the total land area and is home to about half of the total population. Tahiti includes the capital city of Papeete, with a metro population of 136,771 (2017).

Hawaii (2018 population, 1.42 million) is organized into five counties: Honolulu, Hawaii, Kauai, Maui, and Kalawao. Honolulu, the state capital, is the largest city (2019 metro population, 953,207), with two-thirds of the state's population. Honolulu County's population density is far greater than that of the next most populous county, and the highest population density for a single island in the entire Pacific islands region.

New Zealand (2020 population, 4.92 million) is the second-most populous country in the Pacific islands. Eighty-seven percent of its population is urbanized. More than three-fourths of all New Zealanders live on North Island, primarily in the Auckland region. Due to its historical ties to other Pacific Islands, New Zealand is an important destination of emigrants from Samoa, the Cook Islands, Tonga, Niue, Fiji, and Tokelau.

Niue (2019 population, 2,000) consists of a single island 1.5 times the size of Washington, D.C. Alofi, the capital, had only 597 residents in 2017. The island is losing population due to migration of Niueans to New Zealand.

Pitcairn Islands has an area twice the size of Washington, D.C. In 2020, the entire population of about four dozen people lives in the settlement of Adamstown. This sparse population, which has been stable for years, depends on a small tourist trade, the sale of limited-edition postcards and stamps, and honey production.

Most of the population of Samoa (2020 population 203,774) lives on the islands of Savai'i and Upolu. Apia, the capital, is the largest town (2016 population, 37,391). Nearly 18 percent of the population is urbanized. The minuscule (0.6 percent) annual growth rate would be higher were it not for a steady flow of emigrants to New Zealand and the United States.

Tokelau has an area about three times the size of New York City's Central Park. The population of 1,647 (2019) has been decreasing slightly as a result of emigration to New Zealand.

Tonga (2020 population, 106,095) has an area four times the size of Washington, D.C. Twenty-three percent of the population is urbanized. Approximately two-thirds of the people live on the main island of Tongatapu, where the capital, Nuku'alofa (2016 population, 23,221), is located. The negative annual population growth rate (-0.16 percent in 2020) is due in part to a steady emigration of Tongans to New Zealand and Australia.

Tuvalu (2020 population, 11,342) has an area one-tenth the size of Washington, D.C. Sixty-four percent of the population is urbanized. Its capital, Funafuti, had 6,320 residents in 2017. The annual population growth rate is fairly low (0.87 percent).

Wallis and Futuna (2020 population, 15,854) has an area 1.5 times the size of Washington, D.C. The principal town, Mata-Utu, had 1,029 residents in 2018. A low population growth rate (0.28 percent in 2020) is partly caused by the fact that, as French citizens, its inhabitants can legally migrate to France to find employment.

Micronesia

This subregion had 544,752 inhabitants in 2020. All the islands in this region are small and consequently the population density (persons per unit area) is substantially greater than ten times greater here than in Melanesia and Polynesia. The percentage of rural population (59 percent) is similar to many countries in Africa and Latin America. Population pressures are relieved somewhat because citizens of most of these islands can travel to the United States to seek employment.

Guam (2020 population, 168,485) has an area three times the size of Washington, D.C. Nearly 95 percent of the population is urbanized. Agana, the capital, has 1,051 residents. The island is the site of a large US Air Force base, and more than one-third of the island's population is from the US mainland, principally military personnel and dependents.

The Republic of Kiribati (2020 population, 111,796) has an area four times the size of Washington, D.C. Fifty-five percent of Kiribatians live in cities, most in the city of Tarawa (2015 population, 63,017), located on the island of Tarawa. The government of Kiribati had planned to move its surplus population from Tarawa Island to Kiritimati Island, but it was decided that such a move would devastate the ecosystem of the latter.

The Republic of the Marshall Islands (2020 population, 77,917) has an area the size of Washington, D.C. Seventy-eight percent of the people live in cities. Majuro Island and Kwajalein atoll are important population centers.

The Federated States of Micronesia (2020 population, 102,436) has an area four times the size of Washington, D.C. Twenty-three percent of this country's population is urbanized. The capital city of Palikir, has 6,647 residents. Approximately half the population is in the state of Chuuk. About 15,000 Micronesians reside in the United States.

The Republic of Nauru (2019 population, 102,436) consists of just a single island, which is about one-tenth the size of Washington, D.C. Nearly all its workforce was employed in phosphate mines until 2006, when the phosphate mines were exhausted. A deeper layer of phosphates is being exploited, but the economy is holding together barely, and then only thanks to Australian support. Farming is impossible because most of the island is an environmental wasteland due to intensive mining and limited water supply.

The Commonwealth of the Northern Mariana Islands (2020 population, 51,433) has an area 2.5 times the size of Washington, D.C. Nearly 92 percent of the total population resides in Saipan, the largest city. The commonwealth had a negative population growth rate (-0.55 percent) in 2020.

The Republic of Palau (2020 population, 21,685) has an area 2.5 times the size of Washington, D.C. It has an urban population of 81 percent. The population is concentrated on two islands: Babeldaob, the larger island and site of Ngerulmud, the capital since 2006 and smallest nation capital city by population in the world, and Koror, the smaller island but with a far larger population.

An arts and crafts fair in Papua New Guinea. (Kirsten Walla)

Wake Island has an area eleven times the size of the Mall in Washington, D.C. There is no indigenous population; it is a United States military and commercial base.

Future Projections

Population growth and urbanization in this region are expected to continue, albeit at a slower pace than during the first two decades of the twenty-first century. Overall, by 2050,the growth rate in the Pacific islands is projected to drop to no more than 0.80 percent. At the same time, the urbanization rate should average 72 percent. The US territory of Guam, the Solomon Islands, and Vanuatu are predicted to have the greatest population growths at least through 2050.

Richard A. Crooker

Culture Regions of Australia

Most Australians live in the coastal regions. The center part of the continent is largely desert unable to support many people, although there are cattle and sheep stations in the so-called Outback. Eastern Australia, from Cairns in Queensland to Adelaide in South Australia, is more heavily populated than the west coast. The southern coast west of Adelaide has few residents. Although the north coast has a concentration of residents in and around Darwin, this coast also is sparsely populated.

New South Wales has approximately one-third of the country's population, with more than 8 million residents. Victoria has about 6.6 million residents, and Queensland, north of New South Wales, has just over 5 million people. Western Australia has about 2.6 million residents. South Australia has slightly more than 1.75 million people. Tasmania has just over half a million. The Northern Territory has a population of about 244,000, most of them concentrated in the capital, Darwin.

Protestors march for indigenous rights in Brisbane, Australia. (David Jackmanson)

Regionality

Various regions of the United States are typified by speech patterns, cuisine, music, and other such characteristics, but such is not the case in Australia. Australian dialect differences are largely based on socioeconomic status and educational level rather than location. A university graduate from rural Georgia or coastal South Carolina usually has different speech patterns from a university graduate from Maine or Massachusetts. In Australia, university graduates from Sydney or Melbourne have speech patterns similar to those of university graduates from Perth or Darwin.

Cuisine from one end of Australia to the other is quite similar. Musical tastes and cultural offerings also tend to be very much alike across the country.

MULTICULTURALISM IN AUSTRALIA

Australia's high rate of immigration, mainly from Europe and Asia, has the potential to change voting patterns. Asian immigration, particularly, has been used by right-wing groups to arouse nativist resentment, as in the racially tinged late-1990's One Nation movement that flourished briefly in Queensland. Most Australian political parties welcome all races, affirming multiculturalism and reconciliation with Australia's often-oppressed aboriginal population, which represent 3.3% of the country's total population. Indigenous Australian land rights and cultural recognition continue to be a dominant issue. As of 2020, 40 per cent of Australia's land mass formally recognizes Aboriginal and Torres Strait Islander peoples' rights and interests.

Earliest Inhabitants

When the first Europeans reached Australia in 1788, indigenous peoples had lived in the area for more than 50,000 years. Cave paintings discovered throughout mainland Australia and in the island state of Tasmania are more than 30,000 years old.

Although estimates vary, most scholars conclude that about 300,000 indigenous Australians (also called Australian aboriginals) lived on the continent when the first Europeans arrived. They were divided into tribes and clans (extended families). Some 650 groups were formed by these people, who spoke about 250 languages, many mutually understandable to people from different tribes or clans. Currently, about fifty of these indigenous languages survive.

The campus of the University of Western Australia in Perth. (CUHRIG)

The indigenous Australians were hunters and gatherers who lived a nomadic existence. Tribes lived throughout Australia, with the largest concentrations in the coastal regions, particularly in the north and east. The Outback, mostly desert, could support few people. Hunting in the Outback was much less productive than it was closer to the coast, whose oceans were important sources of food. Fruits, vegetables, and nuts were also more plentiful in the coastal regions.

Early European immigrants drove the indigenous Australians from their property and demeaned them as primitive, despite their sophisticated art forms and notable conservation measures. These included the controlled burning of brush in the Outback to enrich the soil and to provide grasses to feed livestock in the unforgiving environment. Their reverence for nature was similar to that of the Anasazi in the southwestern United States.

Driven from their land, the indigenous Australians endured for more than two centuries on the fringes of society, an impoverished group, who, in some instances, assimilated into mainstream Australian society through intermarriage. Those with indigenous roots often sought to obscure their backgrounds, denying any indigenous connections. The 1991 census recorded 257,333 Aboriginal Australians; this number had increased to nearly 650,000 by 2016, an indication that many, who had been reluctant to openly recognize their origins, were beginning to acknowledge their indigenous heritage.

Legislation in 1967 and afterward recognized Aboriginal property claims. Although they and the Torres Strait islanders off Australia's northeastern tip comprise about 3 percent of Australia's population, they now hold claim to 9.6 percent of the land, most of it in Western Australia, the Northern Territory, and South Australia, where the largest concentrations of indigenous Australians live. Substantial numbers also live in all of Australia's major cities.

Population Shortage

From the time of its earliest European settlement, Australia had a shortage of people. Originally colonized by prisoners from Britain's overcrowded jails, Australia at first had a Caucasian population in which men outnumbered women by about six to one. Although the transportation of criminals to New South Wales and Tasmania ended officially in the 1840s, it resumed in Western Australia in 1850 and continued for an additional eighteen years.

By 1868, 137,000 men but only 25,000 women had been transported to Australia, 80 percent of both sexes having been incarcerated in Britain for crimes against property—theft, larceny, burglary—rather than crimes against persons. Most of the convicts worked for free settlers who had voluntarily relocated in Australia in search of a better life there. These convicts usually had a seven-year work obligation, after which they were freed. As unemployment and poverty engulfed Britain between 1830 and 1850, the British government agreed to pay the passage of those who wished to emigrate to Australia.

An Arab neighborhood in Sydney, Australia. (Australian Lifestyle Images)

An Anglican church in Tasmania. (Ruben Ramos)

With the discovery of gold in Victoria during the 1850s, many Asians flocked into the area. About 40,000 Chinese lived in eastern Australia during the 1880s. Melanesians were imported as field workers, but in 1904, three years after the formation of the first Australian parliament, the Melanesians were expelled to their native countries. A considerable anti-Asian bias had become apparent among Australians.

Modern Immigration Patterns

At the beginning of World War II in 1939, 98 percent of Australia's 7 million people were of British origin. Between the end of that war and 1970, some 3 million people immigrated to Australia from Britain, Italy, Greece, Germany, the Netherlands, and the Middle East. The White Australian Policy, which precluded Asians from immigrating, was repealed in 1973, after which many refugees from Vietnam, Laos, and Cambodia settled permanently in the country. By 2019, the country was 76 percent Caucasian and 12 percent Asian.

The official aim of the Australian government is to build a multicultural society in which those from other cultures will be integrated but not assimilated. In other words, people from other cultures are discouraged from abandoning the cultures from which they came. As a result, ethnic neighborhoods have grown up in all of Australia's cities as well as in many of its small towns.

Data from 2016 revealed that 240 different languages, about fifty of which are Aboriginal, are currently spoken in Australian homes. After English, the most common languages are, in descending order, Mandarin, Arabic, Cantonese, Vietnamese, Italian, Greek, and Hindi. The languages declining the most are Italian, Greek, Croatian, Polish, and Maltese. These figures indicate the direction in which population growth through immigration is taking place.

R. Baird Shuman

Culture Regions of the Pacific Islands

The Pacific islands traditionally are divided into three cultural and geographic areas: Melanesia, Micronesia, and Polynesia. Melanesia contains the larger, hilly, mountainous islands, including New Guinea, the Solomon Islands, New Caledonia, Vanuatu, and Fiji. Micronesia comprises more than 2,000 small islands, many of which are atolls formed from coral reefs that rise just a few feet above sea level. These islands include Palau, the Northern Mariana Islands, Guam, Kiribati, the Federated States of Micronesia, the Marshall Islands, and Nauru. Polynesia is the largest of these geographic areas, covering a triangular area defined by Hawaii, New Zealand, and Easter Island. It includes Midway Island, the Cook Islands, Samoa, American Samoa, Tonga, Tuvalu, French Polynesia, and Pitcairn.

The people who populate the Pacific islands are believed to have arrived from southeastern Asia in a series of migratory waves. However, the pattern of their distribution across the islands and their subsequent development into such highly differentiated groups as the dark-skinned Melanesians, Asiatic-featured Micronesians, and lighter-skinned Polynesians is unknown. Many experts believe that the newcomers gradually spread across the Pacific islands, traveling from west to east using prevailing winds and currents. After settling in separated and physically distant islands, they gradually came to differ in appearance and culture. The three cultural and geographic groupings were evident when the first Europeans arrived in the sixteenth century and are described in the writings of European explorers in the eighteenth and nineteenth centuries. These explorers noted that the regions had distinct religions, languages, and customs. Europeans brought their own way of living, which many islanders, particularly those who live in the larger cities, have adopted.

New Zealand and Hawaii

Two of the larger islands—New Zealand, an independent nation, and Hawaii, part of the United States—are highly developed economies and reflect strong European influences. New Zealand was populated originally by people who arrived by canoe from eastern Polynesia. These people became known as the Maori. They fished, hunted, farmed, and engaged in warfare. Individuals specialized in

A woman weaves a basket in Fiji. (Nnehring)

arts such as creating poetry, tattooing, and carving wood, bone, and stone. Their religion had a number of nature-based deities, as well as spirits that performed magical spells and punished those who broke taboos. Many religious concepts, including taboos and personal prestige, were similar to other Polynesian beliefs.

The Maori fiercely resisted settlement by Europeans, engaging in a number of wars after being colonized by the British in 1840. The wars and European diseases reduced the numbers of the Maori from about 120,000 in 1769 to 42,000 in 1896. By 1996 the number of Maori had grown to 500,000, about 14 percent of New Zealand's population. By 2018, the Maori population had increased by more than 50 percent, or 16.5 percent of New Zealander. Some Maori in rural areas maintain their traditions; however, a large number in urban areas have become assimilated into the dominant, British-derived culture.

The Hawaiians, like the Maori, are a minority presence in their native lands. The first Europeans arrived in the Hawaiian Islands in the late eighteenth century. In the nineteenth century, the islands developed a plantation economy, and thousands of Japanese and Filipino immigrants came seeking work. According to 2015 estimates, about 10 percent of Hawaii's population consisted of native Hawaiians, and other Pacific islanders, 37.3 percent were of Asian origin, and 26.7 percent were white. Although most Pacific islanders live in rural areas, about 89 percent of all Hawaii residents live in urban areas. Native Hawaiians retain some traditional practices, but the culture in Hawaii is basically American, with considerable Asian influence. The predominant religion in Hawaii is Christianity, although Buddhism also has a large following. The native Hawaiian religion was a form of nature worship, which was abolished by Kamehameha II in 1819.

Melanesia

Melanesians exhibit a wide cultural diversity: those who dwell along the beaches tend to be less traditional because of prolonged contact with other cultures; those who live inland in relative isolation have retained their traditions. A definite Polynesian influence marks the culture of Fiji and the outlying islands to the northwest. Many languages are spoken in Melanesia, with 740 spoken in Papua New Guinea alone.

A religious ceremony in Papua New Guinea. (Joel Carillet)

Many islanders (although not those in New Caledonia or Fiji) use Pidgin English (Neo-Melanesian), a combination of indigenous and English words, as their common language. Although Christianity is the main religion in many of the Pacific islands, in the Melanesian islands of New Guinea, the Solomons, and Vanuatu, many island religions—predominantly animistic beliefs combining magic, sorcery, and ancestor worship—still survive, and many islanders continue to believe in magic and witchcraft. Cannibalism, once practiced in a few Pacific islands, is thought to persist only in remote areas of New Guinea.

Although considerable variation exists among groups, Melanesian traditional society relies on kinship systems, usually with clan membership dependent on descent from some common ancestor. Traditional society has been described as class-free because power and status are not inherited or attained by membership in a group, but must be earned by the individual. To gain a position of leadership (to be a "big man"), the Melanesian must accumulate wealth or acquire a skill such as oratory. Traditional arts include the production of bark cloth, basket weaving, and pottery.

Micronesia

In comparison with Melanesians, Micronesians are somewhat fair-complexioned and are taller, and many have Asiatic features such as high cheekbones and straight hair. Micronesians speak about thirteen languages, including English, French, and Japanese, the languages of their colonizers. Micronesians traditionally believed in the stability of society and culture. Living in small communities on small islands, they valued harmony and good manners, preferring to settle disagreements by negotiating rather than fighting. Micronesians often lived in extended family groups, separate from other families, and often on the lagoon side of the atolls, making fishing and canoeing easier. Low-island cultures (Marshalls, Kiribati) tended to venture farther away from shore, while high-island (Northern Marianas, Palau) residents tended to stay near their own islands. Islands with higher levels of social stratification included Yap (an island of the Federated States of Micronesia), Palau, the Marshalls, and Kiribati.

The indigenous polytheistic religions of Micronesia continued to be practiced until the mid-twentieth century on Yap and parts of Palau, Christianity largely supplanted these beliefs. Micronesian traditional culture has been influenced greatly by European colonizers; several traditional crafts, such as weaving and pottery, have disappeared. Other traditional arts and crafts are matwork, tattooing, shell ornaments, body painting, and the recital of poetry and tales. The only island with substantial numbers of non-Micronesians is Guam, part of the United States since 1898. Despite the presence of many Americans, the Chamorros (the indigenous people of the Mariana Islands) have maintained political influence.

Fiji: An Uneasy Mix

In 2018, the Fiji Islands were home to about 926,276 people, spread out over more than 330 islands covering 7,095 square miles (18,375 sq. km.). About 57 percent of the population live in urban areas, including the capital, Suva.

Native Fijians (legally known as iTaukei since 2010), who comprise about 57 percent of the population, are ethnically Melanesian but culturally are more similar to the Polynesians. The next largest group of people, 37.5 percent of the population, is Indian (Indo-Fijian). From 1879, five years after Fiji became a British colony, until 1916, the British brought more than sixty thousand Indians to work as indentured laborers on the sugarcane and cotton plantations. After their five years of servitude were up, about 60 percent of the Indians remained in Fiji, farming leased land or opening shops in urban areas. The remaining 5 percent of Fiji's population consists of Europeans, Chinese, other Pacific Islanders, and people of mixed ethnicity. The population of the Fiji Islands is 64 percent Christian (Anglican, Assembly of God, Catholic, Methodist, 7th Day Adventist, and Other), 28 percent Hindu, and six percent Muslim. The iTaukei are primarily Christian, and Indo-Fijians are Hindu or Muslim. The official language is English, although the iTaukei also speak Fijian and Indo-Fijians speak Hindi.

The iTaukei tend to discriminate against Indo-Fijians, whose sheer numbers at one time threatened the iTaukei traditional way of life. The Indo-Fijians' economic success in sugarcane farming and other key industries also produced tension between the two groups. Racial clashes between the iTaukei and Indo-Fijians first erupted in 1959 and broke out again in 1970, just before Fiji gained its independence from Great Britain. From 1970 to 1987 Fiji was part of the Commonwealth, ruled by Prime Minister Ratu Sir Kamisese Mara. Mara lost the elections of April, 1987, to a coalition of iTaukei and Indo-Fijians. A perception that the new government was dominated by Indo-Fijians led to racial violence and two military coups by Lieutenant Colonel Sitiveni Ligamamada Rabuka in 1987, after which Rabuka declared Fiji to be a republic.

Fiji's 1990 constitution increased native Fijian representation in the government, mandated that the prime minister be iTaukei, and gave hereditary clan chiefs a role in the government. In 1997 the government eliminated preferential treatment for iTaukei. In May, 1999, Mahendra Chaudhry, an ethnic Indian, became prime minister. However, the rift between the iTaukei and Indo-Fijians had not healed. In May of 2000, iTaukei businessman George Speight took Chaudhry and his entire cabinet hostage, demanding iTaukei control of the country. Speight released the hostages after fifty six days and later was charged with treason. A new government excluding Indo-Fijians from positions of power was installed, with Ratu Josefa Iloilo as president and Laisenia Qarase as prime minister. Tensions continued to simmer until late 2005, when then-Commodore Frank Bainimarama escalated his verbal attacks on the government for, among other grievances, showing leniency towards the perpetrators of the 2000 coup. By late 2006, tensions boiled over and another coup took place when the government failed to meet a list of demands issued by Bainimarama. His military forces overthrew the government and, in early 2007, he became interim prime minister, with Iloilo reinstated as president. Bainimarama and the FijiFirst Party went on to win democratic elections held in 2014 and 2018.

Polynesia

Of the Pacific islanders, Polynesians are the tallest and have the lightest pigmentation and straight or wavy hair. The Tongans, Hawaiians, and Samoans tend to be large-boned and heavier. Because of intermarriages with Europeans and Asians, many Polynesians have characteristics typical of those groups. Polynesians speak about twenty languages, most of which are related.

English is the official language of Hawaii, Samoa, and Tonga, as well as those islands controlled by English-speaking nations. Polynesians are more homogeneous in custom than the Melanesians or Micronesians. Traditionally, Polynesians were gardeners and fishers whose lifestyles were remarkably in tune with the ocean environment. Polynesian languages contain many words for winds, currents, stars, and directions, all useful for life on and near the ocean.

Traditional Polynesian culture was conservative, dedicated to preserving its complicated social etiquette and rituals. Polynesians defined a person's relation to society and nature, creating elaborate mythologies and genealogies. Society was arranged by hierarchical, branching descent groups based on kinship. The main lines in the hierarchy were believed to be closest to the gods. Alternatively, societies were arranged around a direct descent line from a mythological ancestor. These descent lines handed down titles that gave their holder societal positions such as chief. Polynesians lived in extended family groupings, and their kinship patterns made distinctions among the grandparents', parents', one's own, and children's generations, as might be expected in a highly stratified society.

Traditional Polynesian religion was polytheistic, with sometimes malevolent gods that required worship through sacrifices, feasting, sexual acts, chants, and various elaborate rituals. Animals, people, and even inanimate objects were endowed with *mana*, a supernatural power that had to be guarded so that it would not be diminished through various acts. For example, stepping over a person's head would destroy the *mana* of that person. This belief created many prohibitions (*tapu*), violation of which meant harsh and violent punishment, even death. Lesser violations received spiritual punishment such as illness.

Polynesians also believed in magic; people practiced rituals designed to guarantee success in various endeavors or to bring bad luck upon others, sometimes visiting practitioners of magic. Although many indigenous religious beliefs have faded and been replaced by Christianity, many Polynesians still believe in magic. Introduction of Europeans and European culture to Polynesia brought disease and social and political instability to the islands. Missionaries also altered and destroyed the indigenous cultures. Contemporary Polynesia can sometimes be described as an uneasy combination of European and indigenous culture.

Rowena Wildin

Discussion Questions: Culture Regions

Q1. How do scholars characterize the relationship between the indigenous Australians and the European settlers? How does the indigenous Australian experience differ from that of Native Americans?

Q2. What factors led to an increase in Australia's population in the late 1800s? How were Asians and Pacific islanders treated during that period? When did Australia embrace a liberal immigration policy?

Q3. How might the cultural aspects of traditional Melanesians, Micronesians, and Polynesians be described? What role did Western missionaries play in the prevailing indigenous cultures? To what degree have traditional cultures survived in the Pacific islands?

Exploration

The aboriginal Australians were not only the first residents of Australia, but also its first explorers. They crossed into Australia when there was still a land bridge connecting Australia with the Eurasian landmass. These early people then branched out over every area of the continent. They established complex systems of communication—the "songlines"—that helped map the continent. Similarly, most of Polynesia was settled by voyagers plunging eastward and, in the case of the Maoris who settled New Zealand, southward. They paddled in large canoes (*waka*) in search of new lands.

In classical European culture, ancient geographers such as Ptolemy had postulated the existence of a great southern landmass to counterbalance the known areas of land in the Northern Hemisphere. These ideas of a Great Southern Land (*terra australis* in Latin, thus eventually "Australia") were revived after Christopher Columbus and Ferdinand Magellan empirically proved the roundness of the world in the fifteenth and sixteenth centuries.

Captain Pedro Fernandes de Queiros, a Portuguese navigator who captained a Spanish ship, is thought to be the first European to sight the Australian mainland, but he did not set foot ashore. Queiros wanted to establish a Spanish empire in the South Pacific. He thought New Guinea was the Great Southern Land, but his lieutenant, Luis Vaz de Torres, established that it was not by discovering the strait to the south of New Guinea. This body of water is now called the Torres Strait.

In 1642 Dutch navigator Abel Tasman became the first explorer to circumnavigate Australia. Ironically, during the entire voyage he never actually sighted the continent. However, his expedition did glimpse a small island to the south of the continent that was later called Tasmania in Tasman's honor. Another Dutch expedition headed by Willem Schouten had sailed east of Fiji in 1616.

William Dampier, a crew member of an English expedition that briefly touched the Australian coast in 1688, returned there eleven years later as captain of his own ship. Dampier had little formal education, but he had instinctive scientific curiosity. His expeditions made accurate geographical observation a top priority in future Pacific voyaging. Dampier became the first Englishman to set foot in Australia, landing on the north coast. Although equipment damage to his ship prevented Dampier from exploring the east coast of Australia, his narrative of his travels, published in 1703, helped spur interest in further South Pacific exploration.

British explorer Captain James Cook brought Australia, New Zealand, and Melanesia fully within

Abel Tasman. (National Library of Australia)

the orbit of Western exploration. Cook commanded ships built for the rough North Sea currents off his native Yorkshire, sailing them tens of thousands of miles away from home. In a short ten-year sequence, from 1769 to 1779, Cook vastly expanded Europe's knowledge of the globe.

Australia

By Cook's time, the motives for geographic expeditions had become less economic and more scientific in nature. This was to the benefit of South Pacific exploration, because the trading possibilities in these areas were less financially lucrative than in other areas of the globe. Nevertheless, there were extraordinary possibilities for the gain of scientific knowledge, both of these regions and the globe itself.

Cook's expedition included a botanical scientist, Joseph Banks, who named many of the new species he encountered, classifying them within the systems recently developed as a result of the European Enlightenment. Banks was so enthusiastic about the expedition that he helped finance it, contributing a considerable amount of his own funds to assist in sponsoring the voyage. Another scientist on board,

Joseph Banks. (National Portrait Gallery)

James Cook. (National Maritime Museum)

Daniel Carl Solander, had actually studied under Carl Linnaeus, who established the taxonomic classifications of genera and species still used for animals and plants today.

Cook was not a trained scientist, although he was knowledgeable about astronomy. Nevertheless, his voyages were the first in which exploration ceased to be primarily a search for spoils and instead became a search for knowledge. Cook's expeditions not only brought back geographic data, but also made advances in geographic practice. Cook perfected precise calculations of longitude, which overcame the final barrier to reliable mapping of what were, to Europeans, distant, and unknown areas. Cook's explorations marked what has been termed "the closing of the global circle." Before Cook's travels, there were still mysterious, unexplored places on Earth. After his voyages, every significant inhabited portion of the world had been mapped and surveyed, and only hidden corners remained for future explorers to discover.

This scientific orientation made it no accident that, when Cook's ship, the *Endeavor*, landed near what is now Sydney in 1770, the inlet the ship came

A replica of James Cook's ship, the HMS Endeavor. *(Dennis4trigger)*

upon was named Botany Bay after the many notable plant specimens Banks observed and classified along shore. Botany Bay is one of the few places on Earth named after a scientific field. The British followed up Cook's expedition in 1788 by sending a fleetload of convicts (transported prisoners) to a place near Botany Bay that was to become known as Sydney Harbor

At nearly the same time, a French expedition headed by Jean-François La Perouse had reached Australia. Its goal was to research the feasibility of whale and fur trading in the South Pacific. A relief voyage headed by Bruni d'Entrecasteaux was later sent out to look for La Perouse, who, with his entire expedition, mysteriously vanished. La Perouse's expedition occurred contemporaneously with the French Revolution and at the height of the influence of the French Enlightenment. It has always fascinated Australians.

Along with the much earlier voyage of Captain Pedro Fernandes de Queiros, it seemed to be an alternative, non-British genesis for European Australia. The Spanish and French voyages were never followed up by actual colonization. The British expedition, led by Arthur Phillip, appeared to be just a solution to prison overcrowding, and the loss of another site for the relocation of prisoners after the successful American rebellion. Australia, however, quickly developed into a full-fledged British colony. Australia was now not just a point on the globe to be explored, but a land to be settled.

Into the Interior

The first quarter-century of European settlement of Australia was confined to the coastline, whose lush vegetation and superb harbors offered little challenge to the continent's new residents. Soon, however, they realized that an entire continent was before them. In 1813 Gregory Blaxland became the first European to probe beyond the Blue Mountains, breaking out of the confines between the mountains and the Pacific coast and plunging into the interior. This opened up much of the long but, in terms of water resources, sparse Murray-Darling river system to exploration by men such as Charles Sturt.

The fringes of this inland area proved hospitable to agriculture and, especially, cattle and sheep grazing. It soon became clear that most of the Australian interior was barren and contained, at best, only mineral resources. Settlement could not continue along the same lines as in the lands near Botany Bay.

The goal then became to find what was at the center of Australia, as the coasts had by now been well charted. The goal once again became scientific and geographic, this time tinged with a kind of romantic quest for knowledge and great deeds. As with the eighteenth century explorers, the trekkers across nineteenth-century Australia, whatever their formal education, were intensely knowledgeable about the landscape, and what they did not know they sought to learn. The seat of exploration now became Adelaide in South Australia, and efforts were made to find a giant inland lake or a sizable river in the center of the continent, some form of oasis in the inhospitable, arid land. Instead, all the explorer Edward John Eyre found was the Nullarbor Plain, whose name is Latin for "no trees."

Although explorers of Australia showed great courage, their stories do not have the heroic ring associated with explorers of North America such as Vasco Nuñez de Balboa or Henry Hudson. In fact, some of the most prominent figures in Australian exploration lore are men who failed, such as Ludwig Leichhardt, who died trying to traverse the continent, or Robert Burke and William Wills, who tried to fulfill Leichhardt's vain quest. Burke and Wills had little geographic knowledge or wilderness know-how, and died of hunger, not having neared their goal.

One of the possible reasons for Burke and Wills' failure was their lack of indigenous Australian guides. Although the indigenous people were often victimized by the explorations, they also played a crucial role in aiding and guiding the European-led expeditions. They not only knew the lay of the land but also how to survive upon it. Without the help of the indigenous people, many of the most famous explorations of the Australian interior, particularly the passage beyond the Blue Mountains, would never have been viable.

LUDWIG LEICHHARDT AND THE EXPLORATION OF AUSTRALIA

The German explorer Ludwig Leichhardt came to Australia in 1841, his scientific curiosity stirred by the exotic land and its strange flora, fauna, and geology. He achieved celebrity when he crossed from Brisbane, in Queensland, over thousands of miles of barren land to discover the Burdekin River and reach the area of the present-day Northern Territory. In his next, ill-fated, expedition, Leichhardt attempted to cross the entire Australian continent, but he and his entire party disappeared without a trace. Their remains have never been found. Leichhardt's earlier success had made him famous, but his failure turned him into an Australian legend. An echo of Leichhardt's story is found in Patrick White's novel Voss (1957), which is considered the single greatest work of Australian fiction. The novel raised the tale of the doomed explorer in the desolate continent to the status of myth.

New Zealand

New Zealand first was sighted by Europeans in 1642, when a part of its coast was observed by Tasman in his two ships, the *Heemskerck* and the *Zeehaen*. Tasman named the islands after the Dutch province of Zeeland; the Maori name for the country is *Aotearoa*, meaning Land of the Long White Cloud. Tasman's crew did not land, but they did engage a group of Maori in battle at sea, in which four of Tasman's men were killed.

The events that led to New Zealand's permanent European settlement began on October 7, 1769, when Cook and his crew arrived at what they later named Poverty Bay. The crew of the *Endeavor* was thrilled to see such a sizable landmass after a long voyage filled with only specks of islands, but they soon realized that New Zealand was not the Great

Southern Land. The crew remained in the area through March, 1770, and circumnavigated both the North and South Islands, making accurate charts and establishing that the islands were not peninsular outgrowths of an interior landmass. The hungry crew often waded ashore to pilfer the tasty sweet potato (*kumara*), which led to several skirmishes with angry Maori. Expeditions after Cook's were largely mounted by whalers; along with a wave of British missionaries, the whalers laid the foundations, in the early nineteenth century, for New Zealand's European settlement.

Other Pacific Islands

The first Pacific island group to gain a foothold on Western consciousness was Tahiti, or, as the early explorers spelled it, Otaheite. When Captain Cook landed there in the *Endeavor* in April, 1769, he initiated an era of cultural contact that would see Polynesian and Melanesian people become conscious of the wider globe. Cook was well-received by the Tahitians and dealt judiciously with them.

Cook's main agenda there was neither economic nor military, but scientific: The Royal Society (a scientific group in Britain) had commanded him to observe the transit of Venus on June 3, 1769, and the British Admiralty (navy department) had advised him to watch for any promising areas of land he might encounter on his voyage. (It was in recognition of the astronomical purpose to Cook's voyage that one of the US space shuttles was named *Endeavor*). On this and his two subsequent voyages, Cook sailed around numerous Polynesian islands. He made informed Pacific navigation possible. Englishman John Harrison's invention of the chronometer, a mode of measuring longitude, was crucial to Cook in this task. When native peoples, apparently as a joke, stole Cook's chronometer on one island he visited, it was a moment of great crisis. Fortunately, the instrument was returned. Also important was Cook's solution to the problem of shipboard disease. He encouraged his sailors to eat fruit, which kept them healthy.

Cook made friends with a chief called "Tioony" on Tonga. Therefore, he called that island group the Friendly Islands. Cook's interaction with the native island peoples did not always go so smoothly, however. Cook himself was killed by Hawaiians in 1779, in circumstances still hotly debated among historians. Also, his expedition spread disease that often decimated indigenous communities.

As with the indigenous peoples in Australia and the Maoris in New Zealand, the Pacific islanders were subjugated and displaced by the European colonists. Sometimes the cultural interaction was happier. When one of Cook's subordinates, William Bligh, led his own Pacific expedition on the ship HMS *Bounty* in 1789, his men mutinied. This was the famous "Mutiny on the *Bounty*." The mutineers married Tahitian women and set up their own domain on remote Pitcairn Island.

Cook affected geography as much as any other individual since Columbus. By showing that the Southern Hemisphere did not have the same amount of landmass as the Northern Hemisphere, he paved the way for accurate measurement of the world.

Antarctica

The exploration of Antarctica—Earth's only uninhabited continent—does not include the scenes of cultural encounter that made exploration in other areas so dramatic. Indeed, one might say that the exploration of Antarctica is still continuing. As in other areas still unknown in the eighteenth century, Cook was the pioneering figure. Cook had established that neither Australia nor New Zealand con-

Antarctic Exploration in the International Geophysical Year

In 1958 British scholar Sir Vivian Fuchs led the first successful overland crossing of Antarctica. Fuchs' expedition took place in the context of the International Geophysical Year, a worldwide research effort extending through 1957 and 1958, which determined the mineral composition of Antarctica and surveyed the thickness of the Antarctic ice cap. This international cooperation led to the Antarctic Treaty in 1961. This treaty, renewed in 1991, deferred all national claims to Antarctica and in effect made the continent into a giant world research station. The Antarctic Treaty System, which includes related agreements, currently has 54 parties.

stituted the Great Southern Land for which explorers had been searching for the past two centuries, and that the only remaining possibility was the uttermost south. Cook's voyage on the ship *Resolution* in 1773 was the first human expedition known to cross the Antarctic Circle. Its progress was, unsurprisingly, blocked by ice. Cook was the first known explorer to set foot on the sub-Antarctic South Sandwich Islands; although he turned back north, he was able to verify that any southern continent would be completely icebound.

In 1821 US explorer John Davis became the first human being known to have set foot on Antarctica. Two years later, British explorer James Weddell sailed into a large bay in the outer fringes of West Antarctica, which was named the Weddell Sea in his honor. Unspurred by any hunger for resources beyond offshore whaling and sealing, further nautical explorations continued at a leisurely pace until a Norwegian expedition brought people to the continent again in 1895. Once it was determined that humans could survive on Antarctica and that the South Pole rested on this giant landmass, the main goal of exploration was to reach the Pole.

From 1908 to 1912, the famous expeditions to the South Pole of Norwegian Roald Amundsen and British explorers Ernest Shackleton and Robert Falcon Scott took place. Shackleton narrowly failed to reach the South Pole as a result of a shortage of supplies. Scott fell short once but reached it later on his second attempt. However, he and all his party perished on the journey. Amundsen became the first to reach the South Pole in 1911. Although there were initially hopes that mineral deposits would be found in the area, the race to the pole became its own reward. The Norwegian Borge Ousland became the first person to travel across all of Antarctica alone in January 1997.

Robert Scott. (Wikimedia Commons)

The continent today serves as a base for research into weather, climate, oceanography, and especially fossil research, as the ice preserved prehistoric remains not found on other continents. For some, the most exciting area of Antarctic research is astrophysics. Indeed, many in the scientific community believe that the conditions in Antarctica are the closest to what can be expected if humans seek to settle permanently in outer space.

Nicholas Birns

Discussion Questions: Exploration

Q1. What were the main accomplishments of British explorer Captain James Cook? How did the goals of his voyages differ from those of previous explorers? What contributions did he make to the science of navigation?

Q2. Why were the Pacific islands relatively late to be discovered by European explorers? What impact did the Europeans have on the indigenous populations?

Q3. When was Antarctica discovered? When and by whom was the South Pole reached? Why are there no permanent human settlements on Antarctica?

Urbanization of Australia

To many people, including Australians themselves, typical "Aussies" are strong, lanky, suntanned persons from the Outback, wearing a broad-brimmed hats and speaking slowly, with a distinctive accent and a dry sense of humor. Their world is thought to be one of kangaroos, horses, and sheep traveling on dusty and deserted plains. In reality, Australians have been city dwellers throughout most of the history of European settlement of the country, and Australia has long been one of the world's most urbanized populations.

Immigration History

The history of immigration to Australia explains this pattern of settlement. After the American Revolution, the British, needing a place to send their criminals, turned their attention to the southern continent discovered and mapped by Captain James Cook in 1770. A new penal colony was established in New South Wales, at Botany Bay, in 1788. Thus, immigrants to Australia were initially convicts, or prisoners, sent into exile as punishment for their crimes. A small number of penal settlements soon were located around the coast of the continent. These grew into the colonial centers that are today the capital cities of Australia's states.

Some free citizens came from Britain to the Australian colonies as settlers, but their numbers were small, and they wanted to become pastoralists, rather than agriculturalists. Not until 1830 were

River views of Brisbane CBD seen from Kangaroo Point, Queensland. (Kgbo)

MAJOR URBAN CENTERS IN AUSTRALIA

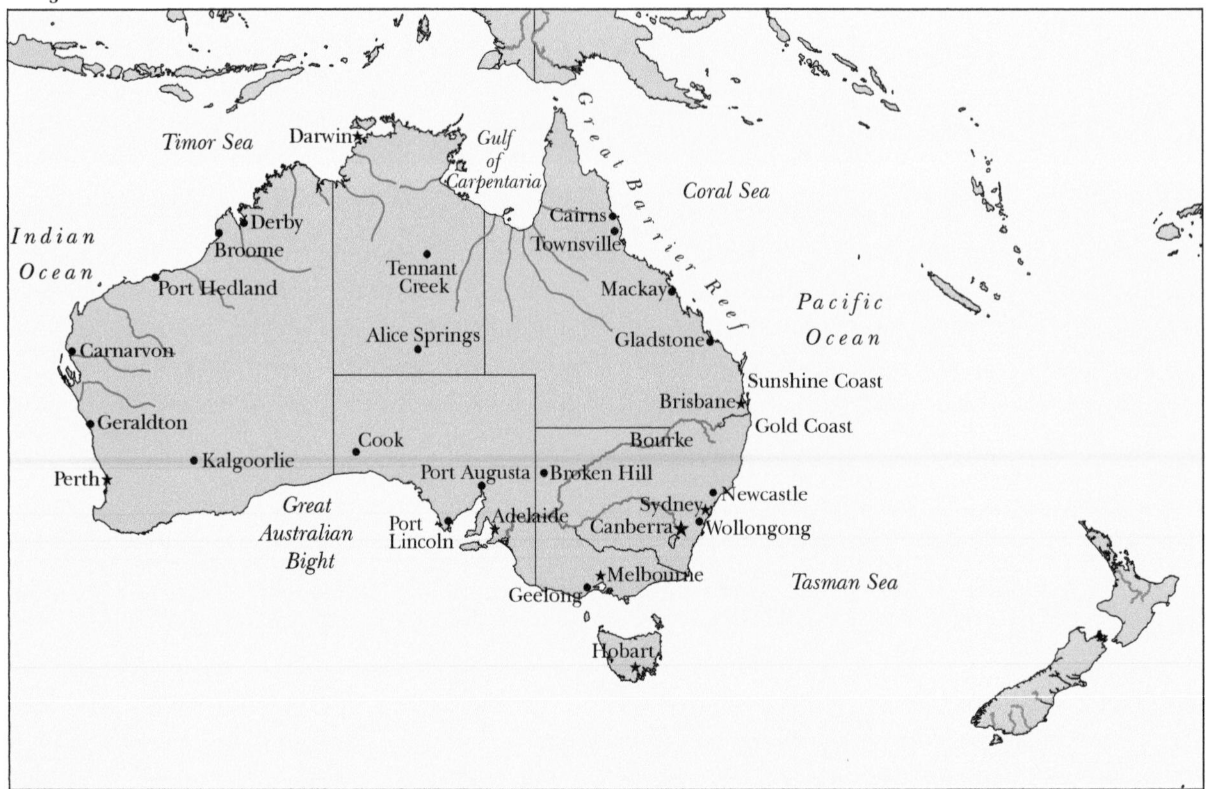

there more free settlers than convicts in Australia. The discovery of gold in Victoria in 1851 attracted thousands of miners from around the world and led to the establishment of inland towns, which persist today as Bendigo, Ballarat, and Castlemaine.

Later gold discoveries in other colonies led to population increases in New South Wales, Queensland, and Western Australia. Unsuccessful miners often stayed in Australia and established businesses. Overall, immigrants to Australia came to the cities and towns—a contrast to the United States, where the majority of immigrants in the eighteenth and nineteenth centuries wanted to be farmers. At the time of World War I, more than 50 percent of the US population still lived in rural areas, while only 40 percent of Australians lived in rural areas then. The percentage continued to decline throughout the twentieth century and into the twenty-first: by 2020, approximately 14 percent of Australians and 17 percent of people in the United States lived in rural areas.

An outdoor café in Sydney. (imamember)

Modern Population Trends

The pattern of population distribution in Australia reflects environmental constraints as well as history. The driest parts of the continent are almost empty; the arid 50 percent of the continent in the center contains less than 5 percent of Australia's population. The well-watered temperate parts have a moderate population density; the coastal areas have high-density urban settlements. The largest population concentration is in the southeastern part, in the more temperate regions. Ninety percent of Australian residents live east of a line drawn from

Adelaide to Brisbane, and 84 percent of the people are crowded into a densely populated 1 percent of the area of Australia. The southwest corner, around Perth, has the other significant population concentration. The island state of Tasmania, with a population of 537,000 in 2020, has the lowest growth rate in Australia.

In 2020, Australia had a population of 25.5 million people. New South Wales was the most populous state, with more than 8 million people, followed by Victoria with 6.6 million. Queensland's population is increasing most rapidly, due to interstate migration. The population of Western Australia has also increased rapidly, reaching 2.6 million in 2020.

Australia is one of the world's most urbanized countries. Forty percent of all Australians live in either Sydney or Melbourne, which, despite ranking only fifty-sixth and fifty-eighth, respectively, in comparative world city population size, are nevertheless cosmopolitan cities. Three other Australian capital cities have a population of more than 1 million—Brisbane, Perth, and Adelaide. Seven more urban areas have a population greater than 200,000—the conurbations of Gold Coast-Tweed Heads, Newcastle-Maitland, Canberra-Queanbeyan, and Sunshine Coast, as well as the cities of Wollongong, Geelong, and Hobart. Australia's largest inland towns include Albury-Wodonga, Toowoomba, Ballarat, and Bendigo. Suburban life is typical in Australia, with Sydney and Melbourne, each having more than 1 million privately occupied homes.

Slightly more than two-thirds of Australians live in either one of the six capital cities, the national capital, Canberra, or the administrative center of Darwin. This makes the comparative size of cities in Australia noteworthy; the typical rank-size pattern of a primate city, some middle-sized cities, and many smaller towns does not apply. The state capitals contain the greatest proportion of each state's population. In South Australia, Western Australia, and Victoria, almost three-quarters of the people live in the capital city. In Northern Territory, about 60 percent of the people live in the capital city of Darwin. In Queensland and Tasmania, 48 and 43 percent, respectively, live in the capital city.

A modern apartment complex in Canberra, Australia. (Daniiielc)

Australia's Largest Cities in 2016

City	*Population (Millions)*
Sydney	4,823.9
Melbourne	4,485.2
Brisbane	2,270.8
Perth	1,943.8
Adelaide	1,295.7
Gold Coast-Tweed Heads	624.2
Newcastle-Maitland	463.0
Canberra-Queanbeyan	432.3
Sunshine Coast	307.5
Wollongong	285.6
Hobart	222.3
Geelong	247.4
Townsville	173.8
Cairns	144.7

Source: 2016 Census QuickStats, Australia Bureau of Statistics.

This dominance of the capital cities is a demonstration of primacy, reflecting both the history and the relative location of the cities. Those states with a more dispersed population also have a less centrally located capital city. Queensland is the best example: its capital, Brisbane, is located in the far southeast of the state. Such a location has, on occasion, led to separatist movements initiated by people in areas remote from the seat of government, which has happened several times in Queensland.

Immigration led to the growth of Australian cities, and recent immigration has changed the ethnic profile of Australian cities. In the first two decades

Australia's Poetic Town Names

Mark Twain spent several months in Australia in 1895-1896 and was enchanted with the names of Australian towns—many of which were adapted from indigenous Australian words. He called these names "good words for poetry. Among the best I have ever seen," and used sixty-six of them to write the poem that follows.

His favorite name was "Woolloomooloo," which he called "the most musical and gurgly." He made no pretense of the poem's having any meaning other than the beauty of its words, but it can stand as a commentary on Australia's climate. (Some of these names are actually from New Zealand.)

A Sweltering Day in Australia.

(To be read soft and low, with the lights turned down.)

The Bombola faints in the hot Bowral tree,
Where fierce Mullengudgery's smothering fires
Far from the breezes of Coolgardie
Burn ghastly and blue as the day expires;

And Murriwillumba complaineth in song
For the garlanded bowers of Woolloomooloo,
And the Ballarat Fly and the lone Wollongong
They dream of the gardens of Jamberoo;

The wallabi sighs for the Murrumbidgee,
For the velvety sod of the Munno Parah,
Where the waters of healing from Muloowurtie
Flow dim in the gloaming by Yaranyackah;

The Koppio sorrows for lost Wolloway,
And sigheth in secret for Murrurundi,
The Whangaroa wombat lamenteth the day
That made him an exile from Jerrilderie;

The Teawamute Tumut from Wirrega's glade,
The Nangkita swallow, the Wallaroo swan,
They long for the peace of the Timaru shade
And thy balmy soft airs, O sweet Mittagong!

The Kooringa buffalo pants in the sun,
The Kondoparinga lies gaping for breath,
The Kongorong Comaum to the shadow has won
But the Goomeroo sinks in the slumber of dead

In the weltering hell of the Moorooroo plain
The Yatala Wangary withers and dies,
And the Worrow Wanilla, demented with pain,
To the Woolgoolga woodlands despairingly flies;

Sweet Nangwarry's desolate, Coonamble wails,
And Tungkillo Kuitpo in sables is drest,
For the Whangarei winds fall asleep in the sails
And the Booleroo life breeze is dead in the west.

Myponga, Kapunda, O slumber no more!
Yankalilla, Parawirra, be warned!
There's death in the air! Killanoola, wherefore
Shall the prayer of Penola be scorned?

Cootamundra, and Takee, and Wakatipu,
Toowoomba, Kaikoura are lost!
From Oukaparinga to far Oamaru
All burn in this hell holocaust!

Parramatta and Binnum are gone to their rest
In the vale of Tapanni Taroom,
Kawakawa, Deniliquin—all that was best
In the earth are but graves and a tomb!

Narrandera mourns, Cameroo answers not
When the roll of the scathless we cry:
Tongariro, Goondiwindi, Woolundunga, the spot
Is mute and forlorn where ye lie.

Source: Mark Twain, *Following the Equator* (1897), chapter 36.

of the twenty-first century, there was a significant change in the source countries of Australian immigrants. In the 1960s, the main countries of origin were the United Kingdom and Ireland, which accounted for just over half of all immigrants, followed by Greece, Italy, Yugoslavia, Malta, and Germany. In the 1990s, the major source countries were New Zealand, with over half of all immigrants, the United Kingdom, Ireland, China, Vietnam, Hong Kong, and the Philippines. In 2019, the greatest number of immigrants came from the United Kingdom, India, New Zealand, and the Philippines. Today, approximately 30 percent of the Australian population is foreign-born.

Major Cities

Sydney is located in one of the world's most beautiful settings, around a huge protected harbor, so that the inner parts of the city slope down to the water. The Sydney Opera House was designed with sail-like roof shells to reflect the water-loving characteristic of the city's people. The site at Sydney Cove was chosen in 1788 for the first penal settlement, so there was some controversy when Australians decided to celebrate the Bicentennial in 1988, since other colonies were settled later. The indigenous Australians also noted that they had little reason to celebrate the event. Sydney was the site of the Summer Olympic Games of 2000. It is the country's financial center, manufacturing hub, and Australia's tourism gateway. Sydney consistently ranks among the world's top-ten cities for quality of life.

Melbourne began as an unofficial settlement to serve pastoralists in 1853. Gold discoveries inland from Melbourne led to the rapid growth of a rich and beautiful city, full of impressive Victorian-period buildings and large public parks. Melbourne is the most English of Australia's cities in appearance. It was larger than Sydney during the latter half of the nineteenth century and was Australia's capital city until Canberra was built. The Olympic Games were held in Melbourne in 1956—the first time the event took place in the Southern Hemisphere; in 2006, the Commonwealth Games were held there. Each year, Melbourne hosts the Grand Slam Australian Open tennis tournament.

Brisbane was first settled in 1824 as the penal outpost of Moreton Bay, a place of harsh punishment for hardened criminals. Free settlement was permitted in 1841, and in 1859 the separate colony of Queensland was created. Brisbane is located in a hilly area upstream from the mouth of the Brisbane River. The subtropical climate of Brisbane gives it a different character from Australia's other state capitals, and wooden houses raised on rough wooden piers called stumps are an interesting feature of what is internationally known as "Queenlander architecture."

Adelaide was founded in 1836. The man behind this city was Edward Gibbon Wakefield, a severe critic of convict colonies, who envisioned a community of small farmers who purchased relatively expensive land; the proceeds were to be used to bring laborers from overcrowded England. The location and elegant design of the city were the work of the surveyor-general Colonel Light. Prosperity did not ensue, however, as settlers stayed in the town rather than begin farming. Today, Adelaide is known for its beauty, as a manufacturing center, and as being Aus-

Sydney Opera House, and the Harbour Bridge, two of Sydney's most famous landmarks. (Benh Lieu Song)

tralia's "city of churches" in recognition of its diversity of faiths as well as its many houses of worship.

Perth is one of the most isolated cities in the world. It was founded in 1829 as a free settlement known as the Swan River Colony, but it fared so badly that in 1846, the governor requested that it become a penal settlement. Despite opposition from the eastern colonies, almost 1,000 convicts, all men, were sent to Western Australia from 1850 to 1868. In 1890 Western Australia became a self-governing colony, and gold discoveries at Coolgardie and Kalgoorlie brought new prosperity and a rapid population increase. Perth has grown most rapidly, partly as a result of prosperity from mining in Western Australia.

Hobart was settled in 1803 as a penal colony. It prospered as a free settlement, as agriculture and mining created new wealth, but in the late twentieth century it declined in comparative importance. Hobart has emerged as the center for Antarctic research, being the closest city to that continent.

Darwin, the administrative center of the Northern Territory, relies heavily on government employment. Its port is well located for trade with Asia. It was the only Australian city bombed by the Japanese during World War II. The city experienced a 23-percent growth rate between 2011 and 2018.

Canberra, the national capital of Australia, is located within the Australian Capital Territory. Part of the agreement that led to the union of the former colonies into the Commonwealth of Australia was the construction of a new capital city, located in New South Wales. After the site was chosen, an international competition was held to select the architect. Walter Burley Griffin, an American who had worked for Frank Lloyd Wright, won, and construction was completed in 1927. Canberra remains a well-planned city with many important modern buildings, including the High Court of Australia, National Library, National Art Gallery, National War Museum, and the impressive Parliament House.

Alice Springs is a small town of fewer than 30,000 residents (2018), but is located close to the center of the continent and is visited by many tourists. About 200 miles (320 km.) is Uluru (Ayers Rock), a well-known sandstone formation, and a center of indigenous Australian art.

Cairns is located on the northeast coast of Queensland. A city of more than 150,000 residents, it is an important port and popular with tourists because of its tropical climate and its proximity to such attractions as the Great Barrier Reef, rain forests, beaches, and Outback experiences.

Ray Sumner

Australia's Administrative Divisions

When considering the names of Australian regions, one must be careful not to confuse phrases such as "southern Australia" with actual place names. The country is administratively divided into six states and three internal territories. One of these states is called "South Australia," and it occupies the center portion of the continent's southern coast, immediately west of two other states, New South Wales and Victoria. Western Australia, the largest state, covers most of the west half of the continent. Occupying most of northeastern Australia is Queensland. The Northern Territory occupies the central part of the northern coast. Jervis Bay Territory is the smallest of the states and territories at roughly 25 square miles, and is located on the southeastern coast of New South Wales. Tasmania, the sixth state, is an island directly south of Victoria. Australia's capital city, Canberra, is located in the Australian Capital Territory—much the way the US capital, Washington, is in the District of Columbia.

Urbanization of the Pacific Islands

The Pacific islands, also known as Oceania, include approximately 30,000 islands, divided into three distinct groups: Melanesia, Micronesia, and Polynesia. The Pacific Ocean has more islands than all other oceans combined; most are located in the southwest quadrant of the Pacific Ocean. It is estimated that when the first Europeans landed in the area during the sixteenth century, more than 3 million people lived there. Anthropologists believe that the region has been inhabited for 50,000 years and that the first inhabitants came from Southeast Asia.

In the nineteenth century, the Pacific islands began to become truly integrated into the world economy. Since World War II, the Pacific island countries and territories have had to cope with increasing urban, social, and economic problems that had not affected them before. Urban growth has increased air, water, and land pollution in the Pacific islands.

The market in downtown Port Vila, Vanuatu. (Torbenbrinker)

City dwellers in Saipan try to stay dry. (aksyBH)

The Pacific island countries and territories are small and isolated in the vast Pacific Ocean. Most rely on agriculture and tourism for economic survival. The population density varies throughout the islands: In New Caledonia, the density was 37 persons per square mile (14.5 per sq. km.), while Nauru had a density of 1,375 persons per square mile (523.8 per sq. km.) in 2020. It appears that family-planning programs and economic conditions have not slowed the growth of population. In recent years, there has been an increase in the number of islanders who live in towns or cities. For some of the poorer islands, the growing urban population has led to poor sanitation, scarcity of land for adequate housing, underemployment, and the breakup of the family structure.

Melanesia

While some islands have an area that is at least a few thousand square miles, many islands are no more than small mounds of rock or sand. Melanesia, which includes Fiji, New Caledonia, Papua New Guinea, the Solomon Islands, and Vanuatu, had a population in 2020 of 9.47 million—a 50 percent increase in population in twenty years. More than three-quarters of Melanesia's population live in Papua New Guinea, which had a population of almost 7.25 million in 2020.

In 2020 the overall urban population in Melanesia was 18.3 percent, a drop of nearly 3 percent since 2000. While the Solomon Islands at 24.7 percent, Papua New Guinea at 13.3 percent, and Vanuatu at 25.5 percent had relatively small urban populations percentage-wise, Fiji at 57.2 percent and New Caledonia at 71.5 percent had comparatively large urban populations. The capital of Fiji is Suva, located on the island of Viti Levu; in 2017 it had an urban population of about 186,000. New Caledonia's urban population grew at an annual rate of 1.89 percent. It is an overseas collectivity of France and is made up of the island of New Caledonia, where most of the population lives, and several outlying islands. The capital of New Caledonia, Noumea, had a population of 94,285 in 2019.

The urban population in Papua New Guinea was only 13.3 percent in 2020. Its capital, Port Moresby, had a population of almost 364,000 in 2019. While the percentage may be low for the urban population in Papua New Guinea, the annual urban growth rate in 2020 was rather robust at 2.5 percent. On the large islands of Melanesia, such as Papua New Guinea, people began leaving the interior highlands to settle in the coastal areas. As expectations were not met, some urban settlers became frustrated with city life; house burglaries, carjackings, and other destructive patterns of behavior became widespread. Social problems such as drug abuse and alcoholism have also increased.

The highest annual urban growth rate in the Pacific islands was in the Solomon Islands, with a rate of 3.91 percent. While only 13 percent of the population was urban in the year 2000, that percentage had nearly doubled by 2020 to 24.7 percent. The capital of the Solomon Islands is Honiara, located on the island of Guadalcanal; in 2017 it had a population of more than 84,000.

Vanuatu, formerly known as the New Hebrides, had a rural population that made up nearly 75 percent of its total in 2020; its annual urban growth rate was 2.55 percent. The capital of Vanuatu, Port Vila, had a population of 51,437 in 2016. Islands with rapidly growing urban areas faced the pressing issues of providing adequate housing, a safe water supply, proper sanitation, and steady employment.

Micronesia

Including the Federated States of Micronesia, Guam, Kiribati, the Marshall Islands, Nauru, the Northern Mariana Islands, and Palau, Micronesia had a population in 2020 of approximately 545,100, an increase of 9.3 percent since 2000. The overall percentage for the Micronesian urban population in 2020 was 70 percent—more than 3.5 times that of Melanesia. The Federated States of Micronesia, comprising 607 islands, had a relatively low urban population at 22.9 percent, with an annual urban growth rate of 1.05 percent. The capital of the Federated States of Micronesia, located on the island of Pohnpei, is Palikir, which had a 2010 population of 6,647.

Dilapidated housing in Papeete, Tahiti. (jfbenning)

Guam is a territory of the United States. Because of its strategic importance, US air force and naval personnel occupy part of the island. The urban population on Guam reached 94.9 percent in 2020.

Kiribati, formerly known as the Gilbert Islands, had an estimated total population of nearly 112,000 in 2020. The urban population was 55.6 percent, with an annual urban growth rate of 3.19 percent. The Kiribati capital, Tarawa, had a population in 2015 of more than 63,000.

The Marshall Islands became independent in 1986. With a population of approximately 78,000 in 2020, its urban population was 77.8 percent. The capital of the Marshall Islands is Majuro; in 2011 its population was nearly 28,000.

Nauru is the only Pacific island country with a 100 percent urban population. In 1983, Nauru's population was 8,100. By 2020 the population had grown to 11,000.

The Northern Mariana Islands had the next highest percentage of urban dwellers of any Pacific island country or territory, with 91.8 percent in 2020 — a huge increase since 1995, when the percentage of urban dwellers had been 53.6 percent. Today, the Northern Mariana Islands' annual urban growth rate has dropped to 0.45 percent. In 1994, Palau became an independent republic. With a 2020 population of little more than 21,000, its urban population had grown to 81 percent. The annual urban growth rate was 1.77 percent.

The number of urban Pacific Islanders having an adequate sewage system is small. In the Marshall Islands, surface pollution was widespread in the

1990s because of faulty septic tanks, pit latrines, and household waste.

Polynesia

The Polynesian nations and territories of American Samoa, the Cook Islands, French Polynesia, Niue, Pitcairn Island, Samoa, Tokelau, Tonga, Tuvalu, and Wallis and Futuna had a combined population of 701,644 in 2020. With Hawaii and New Zealand included as part of Polynesia, the total population was 7,047,612 (2020).

American Samoa is a territory of the United States. With a population of approximately 50,000 in 2020, the urban population hovered around 87 percent. The annual urban growth rate was 0.07 percent. The capital of American Samoa is Pago Pago, located on the island of Tutuila, which had a population of 3,656 in 2010. American Samoa's economy was not large enough to adequately support its native population, and thousands of American Samoans have emigrated to the United States.

Both the Cook Islands and French Polynesia had urban populations above 60 percent in 2020. The Cook Islands had a "self-governing in free association" relationship with New Zealand. Because of this relationship, many Cook Island citizens emigrated to New Zealand in search of a better life. French Polynesia is an overseas collectivity of France. Tahiti, one of the most important tourist locations in the South Pacific, is part of French Polynesia. The capital of French Polynesia, Papeete, is located on Tahiti. Because of the influx of people moving to Papeete, shantytowns have sprouted up on the outskirts of the city. At 64 percent, Tuvalu was the only other French Polynesian country or territory to have an urban population above 40 percent in 2020. It also had the highest annual urban growth rate (2.27 percent) in all of Polynesia.

While arguably not part of Polynesia, Hawaii has close ties to the region. Hawaii's native population originally came from Polynesia. The fiftieth state of the United States, Hawaii has a high standard of living. The capital of the state is Honolulu, located on the island of Oahu. Hawaii's largest city, in 2019 Honolulu had a population of approximately 345,000, and a metropolitan area population of more than 950,000. Honolulu is one of the primary destinations for tourists from the United States mainland and a primary destination for Polynesians looking to resettle.

New Zealand

New Zealand has an advanced economy. In 2020 the population was more than 4.9 million, an increase of more than 1 million since 2000; 86.7 percent live in urban areas. The annual urban growth rate was a modest 1.01 percent in 2020. Pacific islanders made up 7.4 percent of the total population, with newcomers arriving from the Cook Islands, Fiji, Samoa, and Tonga. Compared to the problems associated with urban growth in the other Pacific island countries or territories, New Zealand seemed to offer greater possibilities for a better life. Economic downturns have impacted the Pacific islander immigrants more than the New Zealand-born population. In 2020, New Zealand had seven cities with populations of more than 100,000. The capital of New Zealand, Wellington, had a population of 215,000 in the city proper and more than 424,000 in the metropolitan area in 2019.

Conclusion

Because land space is so limited in the Pacific islands, population density in urban areas can be high. Pacific island cities are faced with such issues as urban waste, rapid economic change, and expanding squatter encampments. In the twenty-first century, Pacific island government institutions both local and national are working with more traditional indigenous leadership to help smooth the growth of urban centers. Effective urban planning, appropriate government oversight, and vigorous public feedback are helping to make the transition from life in rural villages to swirling urban cities more efficient and less traumatic for all Pacific islanders.

Jeffry Jensen

Political Geography

The vast Pacific Ocean region, or Oceania, encompasses a variety of landforms, from the continent of Australia to tiny coral atolls. The main geographical divisions are Australia and three groups of Pacific Islands—Melanesia, Micronesia, and Polynesia. Oceania's many countries and non-sovereign territories exist under a variety of governmental systems—from monarchies to democratic republics. The nations of Australia and New Zealand share a British colonial heritage as antipodean (situated on the opposite side of the globe from Britain) settler colonies. They are closely integrated historically, culturally, economically, and politically.

Australia

While Australia is the world's sixth-largest country by landmass, it has a relatively small population (25.5 million in 2020) with most people living along the eastern and southeastern coastlines. When the Constitution of Australia came into force on January 1, 1901, the six separate self-governing British colonies on the Australian mainland became states of the Commonwealth of Australia, a dominion of the British Empire. It achieved independence after the British Parliament adopted the Statute of Westminster (1931), granting full sovereignty to all dominions, and in 1942, when the Parliament of Australia passed the Statute of Westminster Adoption Act. Modern Australia is a federation, with six states—Queensland, Western Australia, South Australia, New South Wales, Victoria, Tasmania—and ten territories—three internal ones, located on the Australian mainland, and seven external ones, located on Australian islands and in Eastern Antarctica. Two of the internal territories—Northern Territory and Australian Capital Territory—are self-governed and function essentially as states.

The House of Representatives in the Parliament Building in Canberra, Australia. (ai_yoshi)

One underappreciated aspect of Australian politics is how decentralized the country is. Each state inherited and kept the system of government it had developed as a separate colony. It has a bicameral legislature and its own premier, who is a figure of national scope, although state politicians rarely go on to assume significant roles in the federal government. The federal Australian parliament reflects this decentralization.

In Australia's parliamentary system, the head of government (the prime minister on the federal level, the premier on the state level) comes out of the legislature and must command a legislative majority. Members of the lower chamber, the House of Representatives, represent single-member electoral districts, determined according to population. As in the United States, each state elects the same number of senators to the upper chamber of the national parliament—in the case of Australia, twelve. Two senators are also elected from each of the two

Flag of the Dominion of Westralia, the Western Australia secessionist movement.

self-governing territories. That makes a total number of senators in the Australian Senate of seventy-six. The remainder of Australia's ten territories are either very sparsely populated or uninhabited; they are not represented in the Parliament of Australia and are administered directly by the federal government. (The one exception is the Jervis Bay Territory, which is administered by the Australian Capital Territory.)

The official name of the country is the Commonwealth of Australia. The country's three main parties are the Australian Labor Party, the Liberal Party of Australia, and the Nationals. Since the general election of 1949, the Liberal Party and the Nationals (under the latter party's various names) have acted as a coalition; that is why Australia is often regarded as having a two-party system.

While the British colonization of the continent was characterized by violent conflicts between indigenous Australians and settlers, Australia's post-colonial history has been notably peaceful, with few of the sectional rivalries or ideological divisions such as the ones that led to civil war in the United States. However, secessionist sentiments have been a constant feature on the political landscape of Western Australia, which was the last British colony to agree to join the Australian federation.

Western Australia is the largest of the Australian states and the farthest removed from the urban centers of the eastern states, where some 80 percent of the Australian population lives. At the same time, Western Australia, with its wealth of natural resources and a large mining sector, contributes a significant part of Australia's GDP. There is a belief among Western Australians that the eastern states live off the hard-earned wealth of western taxpayers. In 1933, the people of Western Australia voted to secede from the Commonwealth of Australia by a margin of more than two to one. The British Parliament, however, ruled the decision invalid as the constitution of Australia had no provision for states to secede from the federation. The 1974 Westralian Secession Movement also stagnated. Nevertheless, in 2020, secessionist groups were still active in rural Western Australia.

Even the individual states have internal divisions. In New South Wales, the conservative political culture of the city of Sydney has remained perceptively different from the left-leaning party politics of the city of Newcastle located some 100 miles (160 km.) from Sydney. Queensland has a reputation of being the most conservative state in Australia, especially with regard to social and racial issues. As such, it has been deemed Australia's

"Deep North," in reference to the "Deep South" of the United States. Although most residents of the capital city of Brisbane have become relatively liberal on social issues in recent decades, the same cannot be said for many of the people of central Queensland.

Australia is formally a constitutional monarchy. While the king or queen of Great Britain is Australia's head of state, Australia is not part of the British monarchy but constitutes a separate Australian monarchy. The monarch appoints his or her official representatives, the governor-general of Australia and the state governors, on the advice, respectively, of the federal government and each state government. A powerful movement launched in the 1990s for Australia to become a republic proved unsuccessful in a referendum on November 6, 1999, receiving only 45 percent of the vote. Many observers felt this was because the model for a republic on the ballot did not provide for the direct election of the president (who would replace the queen and her representative, the governor-general, as head of state). A more democratic model for the presidency might have garnered approval for a republic. In 2020, Britain's Queen Elizabeth II was still Queen of Australia.

Australia is known worldwide as the originator of the secret ballot, often termed the Australian ballot. Voting is compulsory in Australia, with only a small minority of "informal" votes (those voters who have not come out to the polls or have not followed the electoral rules) being disqualified. Until 1984, indigenous Australians were exempt from compulsory voting provisions.

Australia's treatment of its indigenous peoples has been mired in tragedy and controversy. The Aboriginal and Torres Strait Islander peoples inhabited the continent and nearby islands for over 65,000 years. The populations collapsed following the arrival of British settlers in 1788, as a result of disease epidemics and frontier conflicts. Between the 1870s and 1970s, up to 30 percent of indigenous and mixed-race children were forcibly removed from their families for the purpose of "resocialization," or improving the integration of indigenous populations into the European-Australian society. Until about the age of eighteen, they were placed in specialized institutions under the care of government-appointed social workers or church missionaries. Besides being a tragedy for thousands of families, the policy did not generate any tangible improvement in the social position of "removed" individuals as compared to "non-removed." The decades of the policy became known as the Stolen Generations. In 2008, an official apology was issued by the Australian government for "the pain, suffering, and hurt of these Stolen Generations."

Paul Eagle, right, an MP (member of Parliament) of Maori descent, addresses a colleague. (Nevada Halbert)

During the second half of the twentieth century, indigenous population numbers began to recover. During that period and in later decades the federal government has enacted a number of significant policy initiatives aimed at reconciliation between non-indigenous and indigenous Australians and addressing the key issues such as unemployment, school enrollment, child protection, alcoholism, food security, and housing.

Until recently, Australia always supported high levels of immigration. It also resettled many asylum seekers. However, the government's policy of mandatory-and since 1992, indefinite-detention of unauthorized arrivals has generated controversy, as has the offshore processing arrangements with Nauru and Papua New Guinea. In recent decades, the flow of unauthorized arrivals from Asia has increased dramatically. For instance, the number from India has grown from 3,000 migrants in 1996 to more than 40,000 by 2013. This led the government to pass Migration Amendment Acts, which initially removed Australia's outlying islands and later, the entire mainland from the country's migration zone. Public opinion has shifted from supporting immigration toward opposing the explosive population growth caused by massive migration.

New Zealand

Like Australia, New Zealand achieved independence after the British Parliament adopted the Statute of Westminster (1931), granting full sovereignty to all dominions of the British Empire, and in 1947, when the Parliament of New Zealand passed the Statute of Westminster Adoption Act. An island nation with two large landmasses-the North Island and the South Island-and some 600 small islands, New Zealand is a unitary parliamentary democracy under a constitutional monarchy.

New Zealand's parliamentary system, modeled on Britain's, was once considered an ideal form of the Westminster system. In this system, each member of parliament was elected as a representative from his or her own constituency, and elections were governed by the first-past-the-post principle, by which the party that won the most seats formed the government. In 1996, New Zealand switched to a mixed-member proportional (MMP) system. Under New Zealand's MMP rules, each voter gets two votes: one for a member of parliament from their local electoral district and one for a political party of the voter's choice.

New Zealand has a multiparty system, but only two parties have dominated the country's political life

Three immigrant women in Auckland, New Zealand. (nazar_ab)

since the 1930s. The Labour Party, traditionally left-leaning, underwent a radical reversal in the mid-1980s in favor of free markets and deregulation. The National Party, traditionally conservative, has become more open to the viewpoints of women and minorities. The party produced New Zealand's first woman prime minister, Jenny Shipley, in 1997; when Shipley was defeated in 1999, she was replaced by another woman, her Labour opponent, Helen Clark.

New Zealand is a unitary, not a federal, state. Although the country is divided into sixteen regions, regions do not have the influence there that they do in Australia or the United States. The national legislature is unicameral (having only one house). A group with a distinct electoral role is the Maori, the indigenous Polynesian people of mainland New Zealand. There have been parliamentary seats reserved for the Maori in the New Zealand Parliament since 1868. Many Maori have left their traditional lands to work in large urban centers such as Auckland and Wellington; these cities also host many new immigrants from Europe and Polynesia. This characterization is mainly true of North Island; South Island, with fewer Maori and immigrants and a more rural economy, tends to be more conservative politically. Secession movements, which surfaced several times in South Island history, have reemerged as the twenty-first century began.

Like Australia, New Zealand has retained the constitutional role of the British monarchy. The monarchy in New Zealand functions the same way as it functions in Australia and other former British dominions. There has been less of a republican movement in New Zealand than in Australia. This is partially because of the role of the Crown in signing the Treaty of Waitangi, the 1840 document that is the foundation for the relationship between the Maori and the Pakeha (New Zealanders of European descent). Although the Maori dispute several provisions of the treaty and the treaty has not been incorporated as part of the New Zealand constitution, it continues to play a major role in the system of government, politics, and many aspects of life in the country. The Maori name for New Zealand—Aotearoa—has been increasingly used by all New Zealanders since the late twentieth century.

The Pacific Islands

Polynesia, in the east-central Pacific, is bound by the US state of Hawaii in the north, New Zealand (Aotearoa) in the west, and Easter Island (Rapa Nui), a special territory of Chile, in the east. It includes the independent nations of New Zealand, Samoa (formerly Western Samoa), Tuvalu, Tonga, Niue, and the Cook Islands (the latter two are self-governing countries in free association with New Zealand). Polynesia also has several territorial dependencies of different countries-Wallis and Futuna (a French island collectivity), Tokelau (a New Zealand dependency), American Samoa (an unincorporated US territory), Pitcairn Island (a British overseas territory), Rotuma Island (a Fijian dependency), Nukuoro and Kapingamarangi (atolls belonging to the Federated States of Micronesia), and French Polynesia. The collectivity of French Polynesia includes Tahiti and the other Society Islands, the Marquesas Islands, the Austral Islands, the Tuamotu Archipelago, and the Gambier Archipelago, including Mangareva Island.

Melanesia forms an arc that begins with New Guinea (the western half of which is called Papua

Biman Prasad, the leader of Fiji's National Federation Party, one of the rival parties to the ruling FijiFirst party. (Vatuniveivukefj)

The Tongan Royal Palace. (Donyanedomam)

and is part of Indonesia, and the eastern half of which is the independent country of Papua New Guinea) and continues through the independent island nations of Solomon Islands, Vanuatu (formerly New Hebrides), and Fiji as well as the French-administered archipelago of New Caledonia, Norfolk Island (one of Australia's external territories), and numerous smaller islands belonging to Papua New Guinea and Indonesia. The Andesite Line, a chain of active volcanoes, separates Melanesia from Polynesia in the east and from Micronesia in the north, along the equator; in the south, Melanesia is bounded by Australia.

Micronesia consists of the independent microstates of Kiribati, Nauru, and Palau; the Federated States of Micronesia; the Republic of the Marshall Islands; and insular areas of the United States-Guam and the Northern Mariana Islands. The United States also has several minor outlying islands in the Pacific.

The independent Pacific island countries, inhabited largely by Polynesians, Melanesians, and other Austronesian peoples, have not seen the dictatorships that have often characterized newly independent countries in Africa and mainland Asia. The Pacific islands were decolonized later than the rest of the non-European world, and the lessons were learned from the mistakes made earlier when other colonies had been transitioning to sovereignty. Still, colonial rule has left an indelible mark on the island nations.

Papua New Guinea, for example, has been plagued by land tenure issues and a strong secessionist movement on the island of Bougainville (in a 2019 referendum, over 98 percent of the island voters chose independence). Parts of what in 1975 became an independent country of Papua New Guinea had been colonized and ruled, at different times, by different countries-namely, Germany, Britain, and Australia-which resulted in its dual name and, more importantly, had contributed to the complexity of organizing a uniform post-independence administration. Fiji saw several (albeit bloodless) coups d'état since it gained independence from Britain in 1970. At the root of much of the political upheaval has been the rivalry between indigenous Fijians and the Indo-Fijians, descendants of indentured laborers brought from India during colonial rule.

Most Pacific nations are republics, operating with a president and a legislature of one or two chambers. Several, however, are constitutional monarchies. Tonga is the only sovereign indigenous monarchy in Oceania. It has been ruled by kings and queens since the tenth century. In 2008, the reigning king relinquished most of his powers to the prime minister. Papua New Guinea, Tuvalu, and the Solomon Islands have retained the British monarch as the formal head of state. Like Australia and New Zealand, these countries are Commonwealth realms, that is, they are realms (kingdoms) in personal union with the United Kingdom. Fiji, which disestablished the monarchy during its 1987 coup, briefly considered bringing the monarchy back in the late 1990s, although this plan was rejected as unworkable. The Union Jack in the corner of Fiji's flag still testifies to how much Fijians value their British connections. Samoa became a republic in 2007, when its last monarch, who had been appointed king for life at independence in 1962, died at the age of 94.

In many island countries, traditional chiefs (known as alii, ariki, matai, iroji, and by other titles, depending on the local chiefly system) still retain substantial power. In some they have a formal constitutional role, usually comprising an upper or advisory authority; in others, they exercise informal power. On some islands, traditional chiefs and the clans they represent are major landowners. There have been many complaints about the concentration of power in the chiefly ranks, which infringes upon the ideals of democracy found in these states' constitutions. However, often the traditional chiefs serve as mediators of potentially difficult social or moral problems.

Since Pacific island nations are usually composed of archipelagoes containing several small islands, the usual administrative subdivision is by island, although subdivisions also occur within islands and by grouping a number of islands. Only on the large island of New Guinea and on New Zealand's North and South Islands are there sizable regional subdivisions of the sort typical of most continental countries.

The United States has been a direct influence on governmental forms in the Pacific islands as much as anywhere in the world. The Federated States of Micronesia and the Marshall Islands, for example, were under US trusteeship from 1945 to 1991, as was Palau from 1945 to 1994. Upon independence, all three countries adopted constitutions similar to the US model, with minor variations. American Samoa, Guam, and the Northern Marianas remain under United States control with internal self-government.

As of 2020, France was the only European power maintaining its colonial presence in the Pacific. Although French Polynesia and New Caledonia were represented in the French parliament, there was considerable discontent among the indigenous peoples of those lands. In 2003, the status of French Polynesia was changed from "overseas territory" to that of "overseas collectivity of the French Republic." The territory thus acquired more autonomy in legislative matters, including the power to oppose laws voted by the French Parliament. It also established French Polynesian citizenship. In New Caledonia, after a long conflict between the French authorities and the indigenous Melanesian inhabitants, the Kanaks, the Nouméa Accord of 1998 provided for a gradual preparation for a referendum on full independence. The referendum was held in 2018, and independence was rejected by 57 percent of voters. Another independence referendum was planned to be held in several years.

French nuclear testing was a source of great concern in the Pacific region. This was not only because of the political overtones of continued European presence but also the ecological danger nuclear explosions pose to an already fragile natural environment. Regional organizations such as the South Pacific Forum, including all states in the area, sharply condemned nuclear testing. The test site at French Polynesia's Mururoa atoll was dismantled after France conducted its last nuclear test there on January 27, 1996.

The South Pacific Forum and the Pacific Community are the two main regional organizations of the island countries and territories of Oceania. The Pacific Community is focused on developmental issues from fisheries and food security to education and climate change mitigation. It helps implement specific projects and programs, especially in small island states with limited resources or expertise. The South Pacific Forum changed its name to Pacific Islands Forum in 1999 to reflect the organization's Oceania-spanning membership. It has played an important role in harmonizing regional positions on various political issues as well as in peacekeeping, transportation, and regional trade.

Until recently, the Pacific island states, with their low populations and limited natural resources, tended not to receive much attention in the world media or to exercise great influence in international debates. However, with the growing threat of rising sea levels, the low-lying atoll countries of Oceania have become an international symbol for the potentially catastrophic consequences of climate change. Countries such as New Zealand and India as well as the United Nations have pledged assistance to the threatened islands. Meanwhile, the island nations worldwide have created the Global Island Partnership to help each other adapt to climate change and build resilient and sustainable island communities. Considering the obstacles in their way, their political achievements have been notable.

Nicholas Birns

KH-7 satellite reconnaissance image of the Mururoa Atomic Test Site in French Polynesia, May 26, 1967. (National Security Archive Electronic Briefing Book No. 184)

Antarctica: Political Status and Related Issues

Antarctica does not belong to any country or group of countries although there are some historic claims to its territory. It has no permanent population and therefore has neither citizenship nor government. Antarctica is a de facto condominium, that is, territory governed by multiple sovereign powers. The governance of Antarctica is carried out by the Consultative Nations of the Antarctic Treaty during their annual Antarctic Treaty Consultative Meeting.

The world's southernmost and coldest continent was the last region on Earth to be discovered in recorded history. It was only in 1820, when an expedition commanded by Fabian von Bellingshausen and Mikhail Lazarev, officers (and future admirals) of the Imperial Russian Navy, sighted Antarctica's ice shelf. Numerous expeditions from different countries have visited Antarctica since then, dozens of scientific research bases have been established there, and seven countries have laid claim to parts of the continent. These include the United Kingdom (British Antarctic Territory), France (Adélie Land), Norway (Peter I Island and Queen Maud Land), Australia (Australian Antarctic Territory), New Zealand (Ross Dependency), Chile (Chilean Antarctic Territory), and Argentina (Argentine Antarctica). Some of these claims overlap and have caused friction. Marie Byrd Land, a 620,000-square-mile (1,606,000 sq. km.) area in western Antarctica, is the only major land on Earth not claimed by any country.

The Antarctic Treaty was signed in 1959 by the twelve countries whose scientists had been active in and around Antarctica at the time. It entered into force in 1961 and has since been acceded to by some forty other nations. The primary purpose of the treaty was to ensure "in the interests of all mankind that Antarctica shall continue forever to be used exclusively for peaceful purposes and shall not become the scene or object of international discord." The treaty essentially preserved the status quo with respect to previously made territorial claims by neither explicitly denying these claims nor providing their de jure recognition. Apart from the seven original claimants, most other parties to the treaty do not recognize the existing territorial claims; some maintain that they reserve the right to make a claim. More than twenty countries without claims-such as the United States, Russia, China, and India-have constructed one or more research facilities within the areas claimed by other countries. Many other nations intend to build their bases on the Antarctic mainland, attracted by the continent's unique scientific research potential as well as its abundant natural resources. There has never been any commercial mining in Antarctica thanks to the Environmental Protocol, part of the Antarctic Treaty System of agreements and conventions. The ban, however, can be subject to review starting in 2048.

There is a growing number of other issues that the Antarctic Treaty System, which has kept peace and order on the continent for almost six decades, has been struggling to deal with in the twenty-first century. The treaty provides that Antarctica will be used for peaceful purposes only. It specifically prohibits "any measures of a military nature, such as the establishment of military bases and fortifications, the carrying out of military maneuvers, as well as the testing of any type of weapons." Antarctica's almost seventy bases are all professed to be peaceful research stations, but the ban on militarization has been hard to monitor under the current inspection provisions of the treaty. Some countries may have used civilian contractors for dual-use scientific research that has utility for military purposes, including possibly for controlling offensive weapon systems.

Other new geopolitical tests-from the effects of climate change to the regulation of tourism and fishing-are facing Antarctica that are increasingly difficult for a consensus-based group to address. There is a need for new regulation, stringent enforcement, and universal agreement on the problems posed by cumulative impacts. The latter is one of the greatest challenges not only for Antarctica but for the entire planet.

Irina Balakina

Economic Geography

Overview

Traditional Agriculture

Two agricultural practices that are widespread among the world's traditional cultures, slash-and-burn and nomadism, share several common features. Both are ancient forms of agriculture, both involve farmers not remaining in a fixed location, and both can pose serious environmental threats if practiced in a nonsustainable fashion. The most significant difference between the two forms is that slash-and-burn generally is associated with raising field crops, while nomadism as a rule involves herding livestock.

Slash-and-Burn Agriculture

Farmers have practiced slash-and-burn agriculture, which is also referred to as shifting cultivation or swidden agriculture, in almost every region of the world where the climate makes farming possible. Humans have practiced this method for about 12,000 years, ever since the Neolithic Revolution. Swidden agriculture once dominated agriculture in more temperate regions, such as northern Europe. It was, in fact, common in Finland and northern Russia well into the early decades of the twentieth century. Today, between 200 and 500 million people use slash-and-burn agriculture, roughly 7 percent of the world's population. It is most commonly practiced in areas where open land for farming is not readily available because of dense vegetation. These regions include central Africa, northern South America, and Southeast Asia

Slash-and-burn acquired its name from the practice of farmers who cleared land for planting crops by cutting down the trees or brush on the land and then burning the fallen timber on the site. The farmers literally slash and burn. The ashes of the burnt wood add minerals to the soil, which temporarily improves its fertility. Crops the first year following clearing and burning are generally the best crops the site will provide. Each year after that, the yield diminishes slightly as the fertility of the soil is depleted.

Farmers who practice swidden cultivation do not attempt to improve fertility by adding fertilizers such as animal manures but instead rely on the soil to replenish itself over time. When the yield from one site drops below acceptable levels, the farmers then clear another piece of land, burn the brush and other vegetation, and cultivate that site while leaving their previous field to lie fallow and its natural vegetation to return. This cycle will be repeated over and over, with some sites being allowed to lie fallow indefinitely while others may be revisited and farmed again in five, ten, or twenty years.

Farmers who practice shifting cultivation do not necessarily move their dwelling places as they change the fields they cultivate. In some geographic regions, farmers live in a central village and farm cooperatively, with the fields being alternately allowed to remain fallow, and the fields being farmed making a gradual circuit around the central village. In other cases, the village itself may move as new fields are cultivated. Anthropologists studying indigenous peoples in Amazonia, discovered that village garden sites were on a hundred-year cycle. Villagers farmed cooperatively, with the entire village working together to clear a garden site. That garden would be used for about five years, then a new site was cleared. When the garden moved an inconvenient distance from the village, about once every twenty years, the entire village would move to be closer to the new garden. Over a period of approximately 100 years, a village would make a circle through the forest, eventually ending up close to where it had been located long before any of the present villagers had been born.

In more temperate climates, individual farmers often owned and lived on the land on which they practiced swidden agriculture. Farmers in Finland, for example, would clear a portion of their land, burn the brush and other covering vegetation, grow grains for several years, and then allow that land to remain fallow for from five to twenty years. The individual farmer rotated cultivation around the land in a fashion similar to that practiced by whole villages in other areas, but did so as an individual rather than as part of a communal society.

Although slash-and-burn is frequently denounced as a cause of environmental degradation in tropical areas, the problem with shifting cultivation is not the practice itself but the length of the cycle. If the cycle of shifting cultivation is long enough, forests will grow back, the soil will regain its fertility, and minimal adverse effects will occur. In some regions, a piece of land may require as little as five years to regain its maximum fertility; in others, it may take 100 years. Problems arise when growing populations put pressure on traditional farmers to return to fallow land too soon. Crops are smaller than needed, leading to a vicious cycle in which the next strip of land is also farmed too soon, and each site yields less and less. As a result, more and more land must be cleared.

Nomadism

Nomadic peoples have no permanent homes. They earn their livings by raising herd animals, such as sheep, cattle, or horses, and they spend their lives following their herds from pasture to pasture with the seasons. Most nomadic animals tend to be hardy breeds of goats, sheep, or cattle that can withstand hardship and live on marginal lands. Traditional nomads rely on natural pasturage to support their herds and grow no grains or hay for themselves. If a drought occurs or a traditional pasturing site is unavailable, they can lose most of their herds to starvation.

The Heritage Seed Movement

Modern hybrid seeds have increased yields and enabled the tremendous productivity of the modern mechanized farm. However, the widespread use of a few hybrid varieties has meant that almost all plants of a given species in a wide area are almost identical genetically. This loss of biodiversity, or, the range of genetic difference in a given species, means that a blight could wipe out an entire season's crop. Historical examples of blight include the nineteenth century Great Potato Famine of Ireland and the 1971 corn blight in the United States.

In response to the concern for biodiversity, there has been a movement in North America to preserve older forms of crops with different genes that would otherwise be lost to the gene pool. Nostalgia also motivates many people to keep alive the varieties of fruits and vegetables that their grandparents raised. Many older recipes do not taste the same with modern varieties of vegetables that have been optimized for commercial considerations such as transportability. Thus, raising heritage varieties also can be a way of continuing to enjoy the foods one's ancestors ate.

In many nomadic societies, the herd animal is almost the entire basis for sustaining the people. The animals are slaughtered for food, clothing is woven from the fibers of their hair, and cheese and yogurt may be made from milk. The animals may also be used for sustenance without being slaughtered. Nomads in Mongolia, for example, occasionally drink horses' blood, removing only a cup or two at a time from the animal. Nomads go where there is sufficient vegetation to feed their animals.

In mountainous regions, nomads often spend the summers high up on mountain meadows, returning to lower altitudes in the autumn when snow begins to fall. In desert regions, they move from oasis to oasis, going to the places where sufficient natural water exists to allow brush and grass to grow, allowing their animals to graze for a few days, weeks, or months, then moving on. In some cases, the pressure to move on comes not from the depletion of food for the animals but from the depletion of a water source, such as a spring or well. At many natural desert oases, a natural water seep or spring provides only enough water to support a nomadic group for a few days at a time.

In addition to true nomads—people who never live in one place permanently—a number of cultures have practiced seminomadic farming: The temperate months of the year, spring through fall, are spent following the herds on a long loop, sometimes hundreds of miles long, through traditional grazing areas; then the winter is spent in a permanent village.

Nomadism has been practiced for millennia, but there is strong pressure from several sources to eliminate it. Pressures generated by industrialized society are increasingly threatening the traditional cultures of nomadic societies, such as the Bedouin of the Arabian Peninsula. Traditional grazing areas are being fenced off or developed for other purposes. Environmentalists are also concerned about the ecological damage caused by nomadism.

Nomads generally measure their wealth by the number of animals they own and so will try to develop their herds to be as large as possible, well beyond the numbers required for simple sustainability. The herd animals eat increasingly large amounts of vegetation, which then has no opportunity to regenerate, and desertification may occur. Nomadism based on herding goats and sheep, for example, has been blamed for the expansion of the Sahara Desert in Africa. For this reason, many environmental policymakers have been attempting to persuade nomads to give up their roaming lifestyle and become sedentary farmers.

Nancy Farm Männikkö

Desertification

Desertification is the extension of desert conditions into new areas. Typically, this term refers to the expansion of deserts into adjacent nondesert areas, but it can also refer to the creation of a new desert. Land that is susceptible to prolonged drought is always in danger of losing its vegetative ground cover, thereby exposing its soil to wind. The wind carries away the smaller silt particles and leaves behind the larger sand particles, stripping the land of its fertility. This naturally occurring process is assisted in many areas by overgrazing.

In the African Sahel, south of the Sahara, the impact of desertification is acute. Recurring drought has reduced the vegetation available for cattle, but the need for cattle remains high to feed populations that continue to grow. The cattle eat the grass, the soil is exposed, and the area becomes less fertile and less able to support the population. The desert slowly encroaches, and the people must either move or die.

Commercial Agriculture

Commercial farmers are those who sell substantial portions of their output of crops, livestock, and dairy products for cash. In some regions, commercial agriculture is as old as recorded history, but only in the twentieth century did the majority of farmers come to participate in it. For individual farmers, this has offered the prospect of larger income and the opportunity to buy a wider range of products. For society, commercial agriculture has been associated with specialization and increased productivity. Commercial agriculture has enabled world food production to increase more rapidly than world population, improving nutrition levels for millions of people.

Steps in Commercial Agriculture

In order for commercial agriculture to exist, products must move from farmer to ultimate consumer, usually through six stages:

1. Processing, packaging, and preserving to protect the products and reduce their bulk to facilitate shipping.

2. Transport to specialized processing facilities and to final consumers.

3. Networks of merchant middlemen who buy products in bulk from farmers and processors and sell them to final consumers.

4. Specialized suppliers of inputs to farmers, such as seed, livestock feed, chemical inputs (fertilizers, insecticides, pesticides, soil conditioners), and equipment.

5. A market for land, so that farmers can buy or lease the land they need.

6. Specialized financial services, especially loans to enable farmers to buy land and other inputs before they receive sales revenues.

Improvements in agricultural science and technology have resulted from extensive research programs by government, business firms, and universities.

International Trade

Products such as grain, olive oil, and wine moved by ship across the Mediterranean Sea in ancient times. Trade in spices, tea, coffee, and cocoa provided powerful stimulus for exploration and colonization around 1500 CE. The coming of steam locomotives and steamships in the nineteenth century greatly aided in the shipment of farm products and spurred the spread of population into potentially productive farmland all over the world. Beginning with Great Britain in the 1840s, countries were willing to relinquish agricultural self-sufficiency to obtain cheap imported food, paid for by exporting manufactured goods.

Most of the leaders in agricultural trade were highly developed countries, which typically had large amounts of both imports and exports. These countries are highly productive both in agriculture and in other commercial activities. Much of their trade is in high-value packaged and processed goods. Although the vast majority of China's labor force works in agriculture, their average productivity is low and the country showed an import surplus in agricultural products. The same was true for Russia. India, similar to China in size, development, and population, had relatively little agricultural trade. Australia and Argentina are examples of countries with large export surpluses, while Japan and South Korea had large import surpluses. Judged by volume, trade is dominated by grains, sugar, and soybeans. In contrast, meat, tobacco, cotton, and coffee reflect much higher values per unit of weight.

The United States

Blessed with advantageous soil, topography, and climate, the United States has become one of the most productive agricultural countries in the world. Technological advances have enabled the United States to feed its own residents and export substantial quantities with only 3 percent of its labor force engaged directly in farming. In the 2020s there are about 2 million farms cultivating about 1 billion acres. They produced about US$133 billion worth of products. After expenses, this yielded about US$92.5 billion of net farm income—an average of only about US$25,000 per farm. However, most farm families derive substantial income from nonfarm employment.

There is a great deal of agricultural specialization by region. Corn, soybeans, and wheat are grown in many parts of the United States (outside New England). Some other crops have much more limited growing areas. Cotton, rice, and sugarcane require warmer temperatures. Significant production of cotton occurred in seventeen states, rice in six, and sugarcane in four. In 2018, the top 10 agricultural producing states in terms of cash receipts were (in descending order): California, Iowa, Texas, Nebraska, Minnesota, Illinois, Kansas, North Carolina, Wisconsin, and Indiana Typically the top two states in a category account for about 30 percent of sales. Fruits and vegetables are the main exception; the great size, diversity, and mild climate of California gives it a dominant 45 percent.

Socialist Experiments

Under the dictatorship of Joseph Stalin, the communist government of the Soviet Union established a program of compulsory collectivized agriculture in 1929. Private ownership of land, buildings, and other assets was abolished. There were some state farms, "factories in the fields," operated on a large scale with many hired workers. Most, however, were collective farms, theoretically run as cooperative ventures of all residents of a village, but in practice directed by government functionaries. The arrangements had disastrous effects on productivity and kept the rural residents in poverty. Nevertheless, similar arrangements were established in China in 1950 under the rule of Mao Zedong. A restoration of commercial agriculture after Mao's death in 1976 enabled China to achieve greater farm output and farm incomes.

Most Western countries, including the United States, subsidize agriculture and restrict imports of competing farm products. Objectives are to support farm incomes, reduce rural discontent, and slow the downward trend in the number of farmers. Farmers in the European Union will see aid shrink in the 2021–2027 period to 365 billion euros (US$438 billion), down 5 percent from the current Common Agricultural Policy (CAP). Japan's Ministry of Agriculture, Forestry and Fisheries (MAFF) has requested 2.65 trillion yen (roughly US$24 billion) for the Japan Fiscal Year (JFY) 2018 budget, a 15 percent increase over last year. The budget request eliminates the direct payment subsidy for table rice production, but requests significant funding for a new income insurance program, agricultural export promotion, and underwriting goals to expand domestic potato production. In 2019, trade wars with China and punishing tariffs have led to increased subsidies by the U.S. government, totaling US$10 billion in 2018 and US$14.5 billion in 2019.

Problems for Farmers

Farmers in a system of commercial agriculture are vulnerable to changes in market prices as well as the universal problems of fluctuating weather. Congress tried to reduce farm subsidies through the Freedom to Farm Act of 1996, but serious price declines in 1997-1999 led to backtracking. Efforts to increase productivity by genetic alterations, radiation, and feeding synthetic hormones to livestock have drawn critical responses from some consumer groups. Environmentalists have been concerned about soil depletion and water pollution resulting from chemical inputs.

Productivity and World Hunger

Despite advances in agricultural production, the problem of world hunger persists. Even in countries that store surpluses of farm commodities, there are still people who go hungry. In less-developed countries, the prices of imported food from the West are too low for local producers to compete and too high for the poor to buy them.

Paul B. Trescott

Modern Agricultural Problems

Ever since human societies started to grow their own food, there have been problems to solve. Much of the work of nature was disrupted by the work of agriculture as many as 10,000 years ago. Nature took care of the land and made it productive in its own intricate way, through its own web of interdependent systems. Agriculture disrupts those systems with the hope of making the land even more productive, growing even more food to feed even more people. Since the first spade of soil was turned over and the first plants domesticated, farmers have been trying to figure out how to care for the land as well as nature did before.

Many modern problems in agriculture are not really modern at all. Erosion and pollution, for example, have been around as long as agriculture.

However, agriculture has changed drastically within those 10,000 years, especially since the dawn of the Industrial Revolution in the seventeenth century. Erosion and pollution are now bigger problems than before and have been joined by a host of others that are equally critical—not all related to physical deterioration. Modern farmers use many more machines than did farmers of old, and modern machines require advanced sources of energy to unleash their power. The machines do more work than could be accomplished before, so fewer farmers are needed, which causes economic problems.

Cities continue to grow bigger as land—usually the best farmland around—is converted to homes and parking lots for shopping centers. The farmers that remain on the land, needing to grow ever more food, turn to the research and engineering industries to improve their seeds. These industries have responded with recombinant technologies that move genes from one species to another; for example, genes cut from peanuts may be spliced into chickens. This creates another set of cultural problems, which are even more difficult to solve because most are still "potential"—their impact is not yet known.

Erosion

Soil loss from erosion continues to be a huge problem all over the world. As agriculture struggles to feed more millions of people, more land is plowed. The newly plowed lands usually are considered more marginal, meaning they are either too steep, too thin, or too sandy; are subject to too much rain; or suffer some other deficiency. Natural vegetative cover blankets these soils and protects them from whatever erosive agents are active in their regions: water, wind, ice, or gravity. Plant cover also increases the amount of rain that seeps downward into the soil rather than running off into rivers. The more marginal land that is turned over for crops, the faster the erosive agents will act and the more erosion will occur.

Expansion of land under cultivation is not the only factor contributing to erosion. Fragile grasslands in dry areas also are being used more intensively. Grazing more livestock than these pastures can handle decreases the amount of grass in the pasture and exposes more of the soil to wind—the primary erosive agent in dry regions.

Overgrazing can affect pastureland in tropical regions too. Thousands of acres of tropical forest have been cleared to establish cattle-grazing ranges in Latin America. Tropical soils, although thick, are not very fertile. Fertility comes from organic waste in the surface layers of the soil. Tropical soils form under constantly high temperatures and receive much more rain than soils in moderate, midlatitude climates; thus, tropical organic waste materials rot so fast they are not worked into the soil at all. After one or two growing seasons, crops grown in these soils will yield substantially less than before.

Tropical fields require fallow periods of about ten years to restore themselves after they are depleted. That is why tropical cultures using slash-and-burn methods of agriculture move to new fields every other year in a cycle that returns them to the same place about every ten years, or however long it takes those particular lands to regenerate. The heavy forest cover protects these soils from exposure to the massive amounts of rainfall and provides enough organic material for crops—as long as the forest remains in place. When the forest is cleared, however, the resulting grassland cannot provide the adequate protection, and erosion accelerates. Grasslands that are heavily grazed provide even less protection from heavy rains, and erosion accelerates even more.

The use of machines also promotes erosion, and modern agriculture relies on machinery: tractors, harvesters, trucks, balers, ditchers, and so on. In the United States, Canada, Europe, Russia, Brazil, South Africa, and other industrialized areas, machinery use is intense. Machinery use is also on the rise in countries such as India, China, Mexico, and Indonesia, where traditional nonmechanized methods are practiced widely. Farming machines, in gaining traction, loosen the topsoil and inhibit vegetative cover growth, especially when they pull behind them any of the various farm implements designed to rid the soil of weeds, that is, all vegetation except the desired crop. This leaves the soil more exposed to erosive weather, so more soil is carried away in the runoff of water to streams.

Eco-fallow farming has become more popular in the United States and Europe as a solution to reducing erosion. This method of agriculture, which leaves the crop residue in place over the fallow (nongrowing) season, does not root the soil in place, however. Dead plants do not "grab" the soil like live plants that need to extract from it the nutrients they need to live, so erosion continues, even though it is at a slower rate. Eco-fallow methods also require heavier use of chemicals, such as herbicides, to "burn down" weed growth at the start of the growing season, which contributes to accelerated erosion and increases pollution.

Pollution

Pollution, besides being a problem in general, continues to grow as an agricultural problem. With the onset of the Green Revolution, the use of herbicides, insecticides, and pesticides has increased dramatically all over the world. These chemicals are not used up completely in the growth of the crop, so the leftovers (residue) wash into, and contaminate, surface and groundwater supplies. These supplies then must be treated to become useful for other purposes, a job nature used to do on its own. Agricultural chemicals reduce nature's ability to act as a filter by inhibiting the growth of the kinds of plant life that perform that function in aquatic environments. The chemical residues that are not washed into surface supplies contaminate wells.

As chemical use increases, contamination accumulates in the soil and fertility decreases. The microorganisms and animal life in the soil, which had facilitated the breakdown of soil minerals into usable plant products, are no longer nourished because the crop residue on which they feed is depleted, or they are killed by the active ingredients in the chemical. As a result, soil fertility must be restored to maintain yield. Chemical replacement is usually the method of choice, and increased applications of chemical fertilizers intensify the toxicity of this cyclical chemical dependency.

Chemicals, although problematic, are not as difficult to contend with as the increasingly heavy silt load choking the life out of streams and rivers. Accelerated erosion from water runoff carries silt particles into streams, where they remain suspended and inhibit the growth of many beneficial forms of plant and animal life. The silt load in U.S. streams has become so heavy that the Mississippi River Delta is growing faster than it used to. The heavy silt load, combined with the increased load of chemical residues, is seriously taxing the capabilities of the ecosystems around the delta that filter out sediments, absorb nutrients, and stabilize salinity levels for ocean life, creating an expanding dead zone.

This general phenomenon is not limited to the Mississippi Delta—it is widespread. Its impact on people is high, because most of the world's population lives in coastal zones and comes in direct contact with the sea. Additionally, eighty percent of the world's fish catch comes from the coastal waters over continental shelves that are most susceptible to this form of pollution.

Monoculture

Modern agriculture emphasizes crop specialization. Farmers, especially in industrialized regions, often grow a single crop on most of their land, perhaps rotating it with a second crop in successive years: corn one year, for example, then soybeans, then back to corn. Such a strategy allows the farmer to reduce costs, but it also makes the crop, and, thus, the farmer and community, susceptible to widespread crop failure. When the crop is infested by any of an ever-changing number and variety of pests—worms, molds, bacteria, fungi, insects, or other diseases—the whole crop is likely to die quickly, unless an appropriate antidote is immediately applied. Chemical antidotes can do the job but increase pollution. Maintaining species diversity—growing several different crops instead of one or two—allows for crop failures without jeopardizing the entire income for a farm or region that specializes in a particular monoculture, such as tobacco, coffee, or bananas.

Chemicals are not the only modern methods of preventing crop loss. Genetically engineered seeds are one attempt at replacing post-infestation chemical treatments. For example, splicing genes into varieties of rice or potatoes from wholly unrelated species—say, hypothetically, a grasshopper—to prevent common forms of blight is occurring more often. Even if the new genes make the crop more resistant, however, they could trigger unknown side effects that have more serious long-term environmental and economic consequences than the problem they were used to solve. Genetically altered crops are essentially new life-forms being introduced into nature with no observable precedents to watch beforehand for clues as to what might happen.

Urban Sprawl

As more farms become mechanized, the need for farmers is being drastically reduced. There were more farmers in the United States in 1860 than there were in the year 2000. From a peak in 1935 of about 6.8 million farmers farming 1.1 billion acres, the United States at the end of the twentieth century counted fewer than 2.1 million farmers farming 950 million acres. As fewer people care for land, the potential for erosion and pollution to accelerate is likely to increase, causing land quality to decline.

As farmers are displaced and move into towns, the cities take up more space. The resulting urban sprawl converts a tremendous amount of cropland into parking lots, malls, industrial parks, or suburban neighborhoods. If cities were located in marginal areas, then the concern over the loss of farmland to commercial development would be nominal. However, the cities attracting the greatest numbers of people have too often replaced the best cropland. Taking the best cropland out of primary production imposes a severe economic penalty.

James Knotwell and Denise Knotwell

World Food Supplies

All living things need food to begin the life process and to live, grow, work, and survive. Almost all foods that humans consume come from plants and animals. Not all of Earth's people eat the same foods, however, nor do they require the same caloric intakes. The types, combinations, and amounts of food consumed by different peoples depend upon historic, socioeconomic, and environmental factors.

The History of Food Consumption

Early in human history, people ate what they could gather or scavenge. Later, people ate what they could plant and harvest and what animals they could domesticate and raise. Modern people eat what they can grow, raise, or purchase. Their diets or food composition are determined by income, local customs, religion or food biases, and advertising. There is a global food market, and many people can select what they want to eat and when they eat it according to the prices they can pay and what is available.

Historically, in places where food was plentiful, accessible, and inexpensive, humans devoted less time to basic survival needs and more time to activities that led to human progress and enjoyment of leisure. Despite a modern global food system, instant telecommunications, the United Nations, and food surpluses at places, however, the problem of providing food for everyone on Earth has not been solved.

According to the United Nations Sustainable Development Goals that were adopted by all Member States in 2015, an estimated 821 million people were undernourished in 2017. In developing countries, 12.9 percent of the population is undernourished. Sub-Saharan Africa has the highest prevalence of hunger; the number of undernourished people increased from 195 million in 2014 to 237 million in 2017. Poor nutrition causes nearly half (45 percent) of deaths in children under five—3.1 million children each year. As of 2018, 22 percent of the global under-5 population were still chronically undernourished in 2018. To meet challenge of Goal 2: Zero Hunger, significant changes both in terms of agriculture and conservation as well as in financing and social equality will be required to nourish the 821 million people who are hungry today and the additional 2 billion people expected to be undernourished by 2050.

World Food Source Regions

Agriculture and related primary food production activities, such as fishing, hunting, and gathering, continue to employ more than one-third of the world's labor force. Agriculture's relative importance in the world economic system has declined with urbanization and industrialization, but it still plays a vital role in human survival and general economic growth. Agriculture in the third millennium must supply food to an increasing world population of nonfood producers. It must also produce food and nonfood crude materials for industry, accumulate capital needed for further economic growth, and allow workers from rural areas to industrial, construction, and expanding intraurban service functions.

Soil types, topography, weather, climate, socioeconomic history, location, population pressures, dietary preferences, stages in modern agricultural development, and governmental policies combine to give a distinctive personality to regional agricultural characteristics. Two of the most productive food-producing regions of the world are North America and Asia. Countries in these regions export large amounts of food to other parts of the world.

Foods from Plants

Most basic staple foods come from a small number of plants and animals. Ranked by tonnage produced, the most important food plants throughout the world are corn (maize), wheats, rice, potatoes, cassava (manioc), barley, soybeans, sorghums and millets, beans, peas and chickpeas, and peanuts (groundnuts).

More than one-third of the world's cultivated land is planted with wheat and rice. Wheat is the dominant food staple in North America, Western and Eastern Europe, northern China, and the Middle East and North Africa. Rice is the dominant food staple in southern and eastern Asia. Corn, used primarily as animal food in developed nations, is a staple food in Latin America and Southeast Africa. Potatoes are a basic food in the highlands of South America and in Central and Eastern Europe. Cassava (manioc) is a tropical starch-producing root crop of special dietary importance in portions of lowland South America, the west coast countries of Africa, and sections of South Asia. Barley is an important component of diets in North African, Middle Eastern, and Eastern European countries. Soybeans are an integral part of the diets of those who live in eastern, southeastern, and southern Asia. Sorghums and millets are staple subsistence foods in the savanna regions of Africa and south Asia, while peanuts are a facet of dietary mixes in tropical Africa, Southeast Asia, and South America.

Food from Animals

Animals have been used as food by humans from the time the earliest people learned to hunt, trap, and fish. However, humans have domesticated only a few varieties of animals. Ranked by tonnage of meat produced, the most commonly eaten animals are cattle, pigs, chickens and turkeys, sheep, goats, water buffalo, camels, rabbits and guinea pigs, yaks, and llamas and alpacas.

Cattle, which produce milk and meat, are important food sources in North America, Western Europe, Eastern Europe, Australia and New Zealand, Argentina, and Uruguay. Pigs are bred and reared for food on a massive scale in southern and eastern Asia, North America, Western Europe, and Eastern Europe. Chickens are the most important domesticated fowl used as a human food source and are a part of the diets of most of the world's people.

Sheep and goats, as a source of meat and milk, are especially important to the diets of those who live in the Middle East and North Africa, Eastern Europe, Western Europe, and Australia and New Zealand.

Water buffalo, camels, rabbits, guinea pigs, yaks, llamas, and alpacas are food sources in regions of the world where there is low consumption of meat for religious, cultural, or socioeconomic reasons. Fish is an inexpensive and wholesome source of food. Seafood is an important component to the diets of those who live in southern and eastern Asia, Western Europe, and North America.

The World's Growing Population

The problem of feeding the world is compounded by the fact that population was increasing at a rate of nearly 82 million persons per year at the end of second decade of the twenty-first century. That rate of increase is roughly equivalent to adding a country the size of Germany to the world every single year.

Also compounding the problem of feeding the world are population redistribution patterns and changing food consumption standards. In the year 2050, the world population is projected to reach approximately 10 billion—4 billion people more than were on the earth in 2000. Most of the increase in world population is expected to occur within the developing nations.

Urbanization

Along with an increase in population in developing nations is massive urbanization. City dwellers are food consumers, not food producers. The exodus of young men and women from rural areas has given rise to a new series of megacities, most of which are in developing countries. By the year 2030, there could be as many as forty-one megacities (cities with populations of 10 million people or more).

When rural dwellers move to cities, they tend to change their dietary composition and food-consumption patterns. Qualitative changes in dietary consumption standards are positive, for the most part, and are a result of copying the diets of what is considered a more prestigious group or positive educational activities of modern nutritional scientists working in developing countries. During the last four decades of the twentieth century, a tremendous shift took place in overall dietary habits as Western foods became increasingly available and popular throughout the world. While improved nutrition has contributed to a decrease in child mortality, an increase in longevity, and a greater resistance to disease, it is also true that conditions including morbid obesity, Type II diabetes, and hypertension are on the rise.

Strategies for Increasing Food Production

To meet the food demands and the food distribution needs of the world's people in the future, several strategies have been proposed. One such strategy calls for the intensification of agriculture—improving biological, mechanical, and chemical technology and applying proven agricultural innovations to regions of the world where the physical and cultural environments are most suitable for rapid food production increases.

The second step is to expand the areas where food is produced so that areas that are empty or underused will be made productive. Reclaiming areas damaged by human mismanagement, expanding irrigation in carefully selected areas, and introducing extensive agrotechniques to areas not under cultivation could increase the production of inexpensive grains and meats.

Finally, interregional, international, and global commerce should be expanded, in most instances, increasing regional specializations and production of high-quality, high-demand agricultural products for export and importing low-cost basic foods. A disequilibrium of supply and demand for certain commodities will persist, but food producers, regional and national agricultural planners, and those who strive for regional economic integration must take advantage of local conditions and location or create the new products needed by the food-consuming public in a one-world economy.

Perspectives

Humanity is entering a time of volatility in food production and distribution. The world will produce enough food to meet the demands of those who can afford to buy food. In many developing countries, however, food production is unlikely to keep pace with increases in the demand for food by growing populations.

Factors that could lead to larger fluctuations in food availability include weather variations such as those induced by El Niño and climate change, the growing scarcity of water, civil strife and political instability, and declining food aid. In developing countries, decision makers need to ensure that policies promote broad-based economic growth—and in particular agricultural growth—so that their countries can produce enough food to feed themselves or enough income to buy the necessary food on the world market.

William A. Dando

Agriculture of Australia

Agriculture is an essential part of Australia's economy. Australia's exports were overwhelmingly agricultural until the 1960s, when mining and manufacturing grew in importance. Agricultural production has continued to increase, but so have other sectors of the economy, including the service sector. As a result, since the mid-1970s, the value of agricultural output has dropped to only 3 percent of Australia's gross domestic product (GDP). However, when value-added processing of cultivated products beyond the farm is included, this figure rises to 12 percent. In 2019, direct on-farm employment in agriculture in Australia was reported at 2.56 percent of the workforce. This was nearly twice as high as in the United States, where 1.3 percent of the workforce was directly employed in agriculture at the farm level.

In one form or another, agriculture occupies 60 percent of the total land area of Australia. Much of this land (over 85 percent) is used for open-range cattle grazing, especially huge stretches of Queensland and Western Australia. Only a small part of Australia's agricultural land is used for growing crops. Western Australia and New South Wales

A barley field in Western Australia. (Andrew Hanlon)

SELECTED AGRICULTURAL PRODUCTS OF AUSTRALIA

have the greatest areas under cultivation. The areas most suitable for growing commercial crops are limited mainly by climate, given how dry Australia is.

As a general rule, an annual rainfall of 20 inches (500 millimeters) is necessary to grow crops successfully without irrigation; less than half of Australia receives this amount, and the rainfall is often variable or unreliable. Years of drought may be followed by severe flooding. Warm temperatures throughout most of Australia also mean high evaporation rates, so rainfall figures are not the only factor that affects agricultural productivity.

Australian soils usually require the application of fertilizer to grow crops successfully. The wet tropical north of Australia is suitable for cattle raising, while sugarcane is grown on the east coast in high-rainfall and irrigated areas along coastal plains and river valleys. The cooler southern parts are suited to wheat growing and sheep raising. Dairy production is found throughout the country in selected cool, wet regions. Irrigation has opened up large areas of drier land to agriculture, especially for fruit growing. However, irrigation has contributed to the salinization of soil, which has become a major problem in some areas, especially the Murray-Darling basin in southeastern Australia, one of the continent's most significant agricultural areas. (Australia's salt-interception schemes, aimed at reducing salinity and rehabilitating irrigation areas, are discussed in the "Engineering Projects" section of this chapter.)

Listed by value, the top ten agricultural exports from Australia in 2018 were: beef, wheat, other meat, wool, wine and other alcoholic beverages, sugar and related products, vegetables, dairy, live animals, and fruits and nuts. Other important agricultural exports were cotton, rice, canola, and flowers.

Wheat and Barley

Long the most important crop of Australia, wheat is produced in several so-called wheat-belt regions in a crescent of land just west of the Eastern Highlands, or the Great Dividing Range, which extends from central Queensland through New South Wales to Victoria, as well as in the south of South Australia and southwest Western Australia. These are also the most important regions of sheep raising, so they are sometimes called the wheat-sheep belts. In the 2010s, wheat was the principal or significant grain crop on some 21,000 Australian farms. The average Australian grain farm is family-owned and has an area of over 6,440 acres (2,606 hectares). Wheat crops are rotated with legume and oilseed crops in order to improve soil fertility and reduce soil erosion, weeds, and pests.

Australian wheat is planted during the winter, which is much milder than winter on the prairies of North America. Harvesting begins in September in the warm state of Queensland and moves south to Victoria and Western Australia by January. Australian wheat is high in quality and low in moisture, so it is easy to mill. Wheat crops are frequently affected by drought; another problem is markets, since Australia competes with Russia, the United States, Canada, and several other countries in selling wheat.

When the British first came to Australia and established a penal colony near Port Jackson (Sydney Harbor), the convicts planted wheat on a government farm in what is now inner Sydney. They had difficulty growing wheat because of poor soils, unfamiliar climate, and inexperience, causing fear of widespread hunger. As settlement spread beyond the coastal plain and into the interior, wheat production rose dramatically. The rapid increase in population after the gold discoveries of the 1850s also led to increased demand for wheat. Australia began exporting wheat in 1845 and is now the world's fifth-largest exporter of the cereal grain.

Barley is Australia's second-largest commercial crop, after wheat. It is widely grown throughout a vast geographic area from southern Queensland to Western Australia. Australia is a dominant player in world barley markets, representing about 20 percent of global trade in feed barley and between 30 and 40 percent (depending on the year) of global trade in malting barley. Malting barley is barley developed and grown specifically for brewing beer and making whiskey and other distilled spirits. Australian barley is highly sought after because of its exceptional malting quality, low moisture content, and low foreign-material contamination. China and Japan are Australia's largest barley export markets.

Wool

Australia produces approximately one-fifth of the world's wool. Its wool's high-quality and high-price fleece (mostly from Merino sheep) makes Australia the largest producer of wool by value. Ninety percent of the world's fine apparel wool comes from Australia.

The Merino breed of sheep originated in Spain. In 1797, Merino sheep were imported for the first time to Australia by John Macarthur, a British officer in the New South Wales Corps. The flat grassy plains and temperate dry climate of inland New South Wales provided a favorable environment for sheep raising. Merino sheep were bred to produce new genetic variants of animals, or "strains," with the emphasis on high production of fine woolen fleece. New South Wales today has one-third of Australia's sheep. Wool is produced from the New England Tablelands to the Goulburn area. (The city of Goulburn is home to the huge Big Merino monument, the world's largest concrete-constructed sheep.) Western Australia is the second-largest wool-producing state. An average Australian fleece (the coat of wool from a single sheep) weighs 10 to 11 pounds (4.5 to 5 kilograms). To put this into perspective, 1 pound (0.45 kilogram) of wool can make up to 10 miles (16 km.) of yarn.

Wool was Australia's main export in the earliest days of British colonization. It is mostly exported as greasy (or raw) wool, meaning the wool is not cleaned or treated. Sheep and cattle farms are called "stations" in Australia. The sheep are shorn each year, usually in spring, by teams of professional shearers who travel from one station to another. The work has always been physically demanding, especially before electric shears were adopted. The Australian Labor Party traces its origins to disputes between shearers and station owners in central Queensland, which led to widespread strikes in the 1890s. Today, the working conditions, hours, and wages of contract shearers are strictly regulated. Once rare on sheep stations, women now take a large part in the shearing industry; they work as pressers, wool rollers, wool classers, and shearers. Top professional shearers (called "gun shearers") can shear, per day, more than 400 crossbred sheep or around 200 finer-wool sheep such as Merino.

Two men shearing sheep in Australia. (GaryRadler)

Britain was the traditional market for Australian wool until the 1960s, but Japan later became the main buyer. By 2020, 98 percent of Australia's wool was being exported, with the vast majority of this going to China to be processed and eventually made into clothing such as suits and sweaters. Wool ranked fourth, in terms of value, of agricultural commodities produced in Australia.

In 1970, there were 180 million sheep in Australia. By 2020, that number had fallen to about 64 million, as the Australian sheep industry has changed focus, transitioning from being predominantly wool-orientated to a combination of meat and wool. The wool industry has suffered from a lowering global demand for natural fibers. Recent years of drought have also motivated many producers to switch their emphasis from wool to highly profitable mutton and lamb. However, the overall strong international demand for Australian sheep products has endured, and the national flock has been projected to grow again in the 2020s.

Beef and Other Meat

Cattle are raised for beef throughout Australia, but predominantly in the tropical north. In the 2010s, Australia had more than 28 million head of cattle; Queensland had the largest number, more than 11 million head of cattle. Large herds are kept on the unimproved native grasses, although in southern states, improved pastures and fodder crops are more common in the cattle industry. The pastoral industries are subject to great losses through both drought and flooding. Since the early 1990s, Japan has been the largest customer for Australian beef, with the United States second. Australia exports mostly high-quality, highly marbled beef. In 2020 it was the world's third-largest beef exporter after Brazil and India.

Poultry and sheep meat (lamb and mutton) rank second in Australian meat production, followed by pork. In the decade leading up to 2020, Australia was the world's largest exporter of sheep meat. In 2018, it exported almost 57 percent of its lamb production and 92 percent of its mutton production.

The United States, China, the United Arab Emirates, and Qatar were the main markets for Australian sheep meat. Australia was the third-largest exporter of live sheep; its main markets were the Persian Gulf states.

Since the 1970s, a new type of agriculture in northern Australia has been saltwater crocodile farming. The creatures are raised for both meat and skins. Crocodile meat is sold domestically and exported. It is a healthful white meat, low in fat and high in protein. The skin of the Australian saltwater crocodile is considered one of the best leathers in the world. However, the Australian industry has been facing challenges from the recent global oversupply of lesser-quality crocodile skins from Africa and Papua New Guinea. Pressure from animal rights activists campaigning against the use of any animal leather has also increased.

Dairy Products

Dairying is Australia's third-most important type of agriculture, after beef cattle raising and wheat growing. In 2019, Australia had more than 1.5 million dairy cattle. Dairy cows are kept throughout the cooler and more closely settled areas of Australia, in a belt extending south around the east coast to Adelaide in South Australia. The southwest of Western Australia also produces milk. The milk-delivery journey from the town of Harvey in Western Australia to a bottling plant near Darwin, the capital of the Northern Territory, is believed to be the longest such run in the world. The trip by a three-trailer "road train" across 2,610 miles (4,200 km.) takes about 55 hours.

The state of Victoria produces 60 percent of Australia's milk. A dairy area of interest is the Atherton Tableland in north Queensland, where elevation enables the industry to exist in the tropics. The dominant dairy cattle breed in Australia is the Holstein, accounting for around 70 percent of all dairy cattle. This breed has the highest milk production of all dairy breeds in the world. In 2019, Australia produced almost 2.3 million gallons (8.6 million liters) of milk. While the Australian market is

A herd of cattle in Queensland, Australia. (JosuOzkaritz)

A sugar mill in Queensland, Australia. (Mastamak)

well-supplied with fresh milk and dairy products, exports are also important. Some 35 percent of the fresh milk is processed for export as dried milk powder, ultra-heat treated milk, cheese, and butter. In the 2010s, Australia was the world's eighth-largest exporter of dairy products. Imports of milk products into Australia are restricted.

Sugar

Sugarcane is grown in a series of small regions along the tropical coast of Queensland, extending across the border into northern New South Wales. A warm, wet climate is required for the successful cultivation of sugarcane, so it is confined to parts of the coastal plain with good deep soils and reliable rainfall. More than 80 percent of all sugar produced in Australia is exported as bulk raw sugar, making Australia the second-largest raw sugar exporter in the world. Australian sugar farmers are among the least subsidized in the world, and the country is one of the most efficient sugar producers, far more cost-effective than the heavily subsidized American sugar production. Nevertheless, the Australian sugarcane industry has been struggling to compete with the huge Brazilian industry (benefitting from the economy of scale) and India's industry, with its high subsidies at every level.

In Australia, sugarcane is grown on more than 4,000 small, individually owned farms. Until the 1960s, the cane was cut by hand. Now, it is harvested mechanically and taken by rail to a nearby mill. In 2020, there were twenty-four sugar mills in Queensland and New South Wales. The Australian sugar industry is an interesting case study in attitudes of Europeans. In the nineteenth century, men were brought from the islands of Melanesia to work in the sugar fields, as it was believed that no white person could survive physical work in a tropical climate. These workers were called "kanakas" (from the Hawaiian word for "person" or "man," but the word acquired a derogatory connotation in Australian English); they were virtually slaves of the plantation managers. The practice was stopped in 1901 when Queensland became part of the new Commonwealth of Australia. Italians then were encouraged to migrate to Queensland to do the hard work. Today, many sugarcane farms are owned by families of Italian descent.

Fruit

Fruit growing has a long history in Australia and is strongly influenced by climatic considerations. In Queensland, tropical fruits such as bananas, pineapples, mangoes, and papaya (called pawpaw in Australia) are cultivated. In the cooler south, apples, peaches, apricots, cherries, and grapes are grown. Fruits and nuts today account for 8 percent of the country's agricultural production by value.

Grapes are grown for both eating fresh and dried as raisins, although far more important is wine production, especially in the Barossa Valley and the Riverland (South Australia), Hunter Valley and Riverina (New South Wales), Margaret River area (Western Australia), and Yarra Valley (Victoria). John Macarthur, the aforementioned founder of the Australian wool industry, also established the first commercial vineyard, in New South Wales. In the 1960s, modern plantings and production methods were introduced. By 2020, Australia had almost 361,000 acres (146,000 hectares) of vineyards and almost 2,500 wineries. Wine is an important export for Australia, with China purchasing 40 percent of wine exported in 2019 and 2020, followed by the United States (14 percent), and the United Kingdom (12 percent).

Other Agricultural Products

Cotton is grown mainly in drier interior parts of New South Wales, in parts of central and southern Queensland, and in northern Victoria. It is usually grown on family farms in conjunction with sheep farming. Asia is the major customer for Australian cotton, with leading import destinations being China, Bangladesh, Vietnam, India, Indonesia, Turkey, and Thailand. In an average year, Australia produces enough cotton to clothe 500 million people! Thanks to biotechnology and advances in precision irrigation and timing, the Australian cotton industry has become the most water-efficient cotton industry in the world. Its Whole Farm Irrigation Efficiency (an index showing the percentage of water used by the crop instead of being lost on the farm) has increased from 57 percent in the late 1990s to 81 percent in 2020.

Rice has been grown commercially in Australia since 1924, using irrigation. New South Wales is the main producer, where the Murrumbidgee Irrigation Area dominates rice production. Australia ex-

A granny smith apple orchard in Australia. (Alfio Manciagli)

ports about 80 percent of its rice crop to more than seventy markets, mainly in the Middle East and the Asia-Pacific region. Oats are grown where the climate is too cool and too moist for wheat. In Australia, this means in the interior southeast of Western Australia. This state and New South Wales are the biggest producers of oats, which is used mainly for fodder.

Other agricultural products of Australia include grain sorghum; corn (maize); vegetables, including potatoes, peas, tomatoes, and beans; oil seeds such as canola and sunflower; soybeans; and tea and coffee in northern Queensland. Australia is a major producer of honey, with hundreds of commercial operations and thousands of registered apiarists. Blossoms of the eucalyptus tree produce distinctive-tasting honey, which is sold mainly to Southeast Asian countries.

Australian agriculture suffered when one of its then-largest importers of agricultural products, the United Kingdom, joined the European Economic Community (predecessor of the European Union) in 1972. Some of Australia's exports were displaced by European products, others became subjects to the European Community tariffs. However, Australia's export-oriented agriculture recovered and expanded as the country successfully negotiated bilateral free trade agreements with New Zealand and a number of Asian countries and regional trade blocs.

Ray Sumner

AGRICULTURE OF THE PACIFIC ISLANDS

Agricultural practices in the Pacific islands have an unsettled history, and they share some common problems. Many of these islands are separated from one another by hundreds to thousands of miles. Typhoons, storms, drought, and volcanic activity can wreak havoc on their agriculture. Low-lying islands are especially vulnerable to the rise of sea levels caused by climate change. Throughout history, attempts by foreigners to encourage plantation agriculture have threatened fragile island environments. Large-scale land clearing and the use of fertilizers and pesticides have caused erosion, contamination of soils and lagoons, loss of important trees and native crops, and depletion of precious soils.

The Pacific islands can be divided into two main types: the high islands, which are generally volcanic islands, and the low islands, which are primarily atolls. Volcanic lava weathers rapidly and provides reasonably fertile soil for cultivation on high islands, whereas atolls are low-lying coral reefs, which generally have an inadequate supply of freshwater and poor soils. Coconuts, which can ripen throughout the year and can grow on poor soil, are one of the few crops that thrive on the low islands.

Three farmers harvest coconuts in Vanuatu. (lkonya)

Pepper seed harvested from Pohnpei. (Supersmario)

The coconut is believed to have originated in the coastal areas of Southeast Asia and Melanesia. In prehistoric times, it spread naturally on ocean currents. Around 4,500 years ago, voyaging Polynesians and Indo-Malayans introduced their preferred varieties to various Pacific islands. Today, the coconut palm continues to be an important economic and subsistence crop in the Pacific region. Pacific islanders use almost every part of the coconut palm: it is a major source of food, oil, fiber (coir), wood, and fodder for livestock. Copra (the dried meat or kernel of the coconut) was for a time an important industrial product used in the production of detergents, synthetic rubber, and glycerin. It is still exported, but in smaller quantities, mainly for the extraction of coconut oil used in cooking and beauty products.

For making copra, the nuts of the coconut palm are harvested, husked (using sharp-edged or point-edged tools), split with a chopping knife, and, most commonly, left to dry in the Sun. (Smoke drying, kiln drying, and indirect hot-air drying in a heated tunnel are less commonly used methods.) After a few days, the dried copra can be easily scraped out and dried further on racks. The average yield of a coconut palm is eighty to 100 nuts a year, depending on the variety. About thirty nuts provide meat for 10 pounds (4.5 kilograms) of copra. Coconut oil is made by pressing fresh coconut meat (virgin oil) or copra (refined oil). A metric ton of copra yields about 1,345 pounds (610 kilograms) of oil. The rest—the copra cake—is mainly used as fodder.

Islands show great variations in the amounts of rainfall they receive, the steepness of their slopes, and the variety of plant life that can grow. Great differences in rainfall and vegetation can exist on different sides or parts of the same island. For example, one side of the Big Island of Hawaii has one of the driest deserts on Earth, whereas a few miles away on the other side of the island, a tropical rain forest exists.

Many traditional Pacific island crops are little known to Europeans or Americans. These include, among others, betel (or areca) nut, breadfruit, Gnetum, kava, noni, taro, and pandanus (screw palm).

Melanesian Islands

These fairly large islands are located to the northeast of Australia. They are quite damp, have a hot climate, and display a mountainous terrain that is covered with dense vegetation. Papua New Guinea is the largest island in Melanesia, slightly bigger in size than California. It has a mountainous interior with rolling foothills that are surrounded by lowlands along the coastal areas. Its highest point is Mount Wilhelm, which rises to 14,795 feet (4,509 meters). Arable lands make up only one-tenth of a percent of its land, and permanent crops occupy 1 percent of the land. Crops often need to be terraced in areas having steep slopes. Irrigation water is often brought to the crops through bamboo pipes. In the twenty-first century, subsistence agriculture still provides a livelihood for some 85 percent of the population. Agricultural products grown include coffee, cocoa, coconuts, palm kernels (edible seeds of the oil palm fruit), tea, natural rubber from rubber trees, fruit, sweet potatoes, vegetables, poultry, and pork. Palm oil, coffee, and cocoa are exported. Today, Papua New Guinea is the world's largest exporter of copra, with 35 percent of the global trade.

Vanuatu, which includes eighty islands in the South Pacific, covers a total area a bit larger than the state of Connecticut. Its mostly mountainous ter-

Leading Agricultural Products of Pacific Island Nations

Country	*Products*	*Percent of Arable Land*
Fiji	Sugarcane, copra, ginger, tropical fruits, vegetables; beef, pork, chicken, fish	9.0
Kiribati	Copra, breadfruit, fish	2.5
Micronesia, Federated States of	Taro, yams, coconuts, bananas, cassava (manioc, tapioca), sakau (kava), Kosraen citrus, betel nuts, black pepper, fish, pigs, chicken	2.3
New Zealand	Dairy products, sheep, beef, poultry, fruit, vegetables, wine, seafood, wheat and barley	1.8
Palau	Coconuts, cassava (manioc, tapioca), sweet potatoes; fish, pigs, chickens, eggs, bananas, papaya, breadfruit, calamansi, soursop, Polynesian chestnuts, Polynesian almonds, mangoes, taro, guava, beans, cucumbers, squash/pumpkins (various), eggplant, green onions, kangkong (watercress), cabbages (various), radishes, betel nuts, melons, peppers, noni, okra	2.2
Papua New Guinea	Coffee, cocoa, copra, palm kernels, tea, sugar, rubber, sweet potatoes, fruit, vegetables, vanilla; poultry, pork; shellfish	0.7
Samoa	Coconuts, nonu, bananas, taro, yams, coffee, cocoa	2.8
Solomon Islands	Cocoa, coconuts, palm kernels, rice, fruit; cattle, pigs; fish; timber	0.7
Vanuatu	Copra, coconuts, cocoa, coffee, taro, yams, fruits, vegetables; beef; fish	1.6

Source: The World Factbook, Central Intelligence Agency

rain provides only minimal arable land: approximately 2 percent, with another 2 percent used for pasture. About two-thirds of the population is involved in subsistence or small-scale agriculture. The main agricultural products are copra, coconuts, cocoa, yams, coffee, fruits, kava, fish, and beef. Copra, beef, cocoa, and kava are exported. In the 2010s, they accounted for more than 60 percent of Vanuatu's total exports. Vanuatu produced about 12 percent of the global supply of copra and was one of the two largest producers of kava (the other being Fiji). Kava, a root crop, has long been used throughout the Pacific region in traditional medicine and for making beverages and is also exported as herbal medicine to countries outside the region.

New Caledonia, located east of Australia in the South Pacific Ocean, is almost the size of New Jersey. It consists of coastal plains with interior mountains over 4,900 feet (1,500 meters) high. New Caledonia, which is known for its nickel resources, imports much of its food supply. A few vegetables are grown, but raising livestock is more common. Ten percent of New Caledonia's land is in permanent pasture used for raising beef cattle.

The island nation of Fiji includes 332 islands, only 110 of which are inhabited. These islands are volcanic in origin, and approximately 10 percent of the land is arable and 10 percent is in permanent pasture. Approximately 40 percent of the labor force is involved in subsistence agriculture. Sugarcane is an important crop in Fiji; cane sugar constitutes a major industry and 8 percent of Fiji's exports. Other agricultural products grown in Fiji are coconuts, cassava (tapioca), rice, sweet potatoes, taro, kava, cattle, pigs, and goats. Fried taro is a staple food. Taro is one of the oldest cultivated plants in Pacific island history and was once a staple food for many island peoples. This starchy, edible rhizome can be cultivated by clearing or partially clearing a patch in the tropical rain forest and planting the taro in the moist ground.

The Solomon Islands are a cluster of small islands that collectively cover an area almost the size of Maryland. They are located in the Solomon Sea

A macadamia orchard on the Big Island, Hawaii. (JTSorrell)

between Papua New Guinea and Vanuatu. Some of the islands have rugged mountainous terrain, others are low coral atolls. Only 1 percent of the land is arable, and 1 percent is devoted to pastures. Over 80 percent of the islanders live in rural areas and depend on subsistence agriculture. The main subsistence crops are sweet potato, cassava, banana, yam, beans, cabbage, watercress, watermelon, and breadfruit. Palm oil, cocoa beans, copra, and betel nuts are exported.

Micronesian Islands

Micronesia comprises thousands of relatively small islands located along the equator in the central Pacific Ocean and slightly north of the equator in the western Pacific Ocean. The region covers approximately 1.54 million square miles (4 million sq. km.) and is subdivided into four areas: the Kiribati group, the Marshall Islands, the Federated States of Micronesia, and Guam, a US territory located within the western cluster of islands of the Federated States of Micronesia.

The Kiribati group includes primarily low-lying atolls encircled by extensive living reefs. Twenty of the thirty-three islands in the group are inhabited. Agriculture is mainly subsistence, with copra being one of Kiribati's few exports. Grown on the island for local consumption are giant swamp taro, breadfruit, sweet potatoes, vegetables, and coconuts. Giant swamp taro is a root crop similar to taro but with bigger leaves and larger, coarser roots. The roots need to be cooked for hours to reduce toxicity but they are rich in nutrients. In the harsh atoll environments of the Central Pacific, especially Kiribati and Tuvalu, swamp taro is not only an important food source but also an essential part of cultural heritage. The crop is now under threat, being displaced by imported food products.

The Marshall Islands contain two island chains of 30 atolls and 1,152 islands. The economy of the Marshall Islands is dependent upon aid from the US government. Agriculture exists as small farms that provide commercial crops of tomatoes, melons, coconuts, and breadfruit. Coconuts, cacao, taro, breadfruit, pigs, and chickens are produced for local consumption.

The Federated States of Micronesia include four groups of islands divided into the states of Pohnpei, Chuuk, Yap, and Kosrae. There are a total of 607 individual islands, some high and mountainous, others low-lying atolls. Volcanic outcroppings are found in Kosrae, Pohnpei, and Chuuk. Agriculture on the islands is mainly subsistence farming. Products grown include peppers, coconuts, tropical fruits and vegetables, cassava, and sweet potatoes. Black and white peppers were introduced to the area in the 1930s, although commercial pepper growing only began in Pohnpei in the 1960s. Peppers have since become an important export crop, thanks to the island's heavy rainfall and rich volcanic soil.

Guam, the largest island in the Mariana archipelago, is of volcanic origin and surrounded by coral reefs. It has a flat coral limestone plateau that serves as a source of freshwater for the islands. In Guam, 15 percent of the land is used as permanent pasture and another 11 percent is arable. Although fruits, vegetables, and livestock are raised on the island, much of its food is imported because the economy relies heavily on US military spending and the tourist trade.

Polynesia

Polynesia comprises a diverse set of islands lying within a triangular area having corners at New Zealand, the Hawaiian Islands, and Easter Island. French Polynesia is at the center of the triangle and includes 118 islands and atolls.

The five archipelagoes of French Polynesia include four volcanic island chains (the Society Islands, the Marquesas, the Gambiers, and the Australs) and the low-lying atolls of the Tuamotus. The mountainous volcanic islands contain quite fertile soils along their narrow coastal strips. The atolls have little soil and lack permanent water supply. Permanent pasture covers 5 percent of French Polynesia, and another 6 percent is used for permanent crops. Agriculture, once of primary importance, now only provides a small portion of the gross domestic product of French Polynesia. Exports include vanilla and coffee, but the main ex-

A dairy farm in New Zealand. (JenS)

ports from French Polynesia are pearls and tourism-related services.

Samoa is a chain of seven islands lying several thousand miles to the west of Tahiti. Collectively, the islands are almost the size of Rhode Island, and they are covered in rugged mountains with a narrow coastal plain. Approximately 24 percent of the land sustains crops of bananas, taro, yams, and coconuts. Samoans practice both subsistence and crop agriculture; over two-thirds of Samoans are involved in agricultural production and are dependent on it. Samoan agricultural lands, however, have been frequently devastated by natural disasters. Coconuts grown on the islands are processed into creams and copra for export. Taro was a staple subsistence crop and the largest export commodity until the 1990s, when the crops were repeatedly destroyed by hurricanes, which were followed by the rapid spread of taro leaf disease. Taro currently accounts for less than 1 percent of export revenue. Samoa is no longer self-sufficient in food and must now rely on imported foodstuffs.

Tuvalu is a group of nine coral atolls located almost halfway between Hawaii and Australia. The soil is poor on these islands and there are no streams or rivers. Water for the islands is captured in catchment systems and put into storage. Islanders live by subsistence farming and fishing. Coconuts and giant swamp taro are among the few crops that can grow in the poor soils, and coconut farming allows the islanders to export copra.

The Cook Islands comprise a combined area almost the size of Washington, D.C. Located halfway between Hawaii and New Zealand, the northern islands are primarily low coral atolls, and the southern Cook Islands are hilly volcanic islands. The nation of the Cook Islands was a major exporter of fresh and processed citrus, pineapples, and bananas in the 1950s to 1970s. However, it has since lost its market share to more competitive producers. The population of the country has sharply declined since the 1990s owing to emigration to Australia and New Zealand. Many agricultural lands were lost to housing developments for tourism and private use. Current food production in the Cook Islands is insufficient to meet demand. The country is highly dependent on imported food products, especially given the rapidly growing number of tourists.

Tonga is a kingdom consisting of more than 170 mostly uninhabited islands. Tongans rely on both subsistence and plantation agriculture. Today, the majority of Tonga's agriculture is still based on traditional farming systems. All land in the kingdom belongs to the crown, and each Tongan male is entitled to lease an allotment, called *api*, measuring 8.25 acres (3.34 hectares) when he reaches the age of 16. Root crops such as taro, cassava, sweet potato, and kava are grown on two-thirds of the agricultural land. Other plants grown both as market cash crops and for home use include bananas, coconuts, coffee beans, vanilla beans, and squash. Kava and vanilla from larger plantations are now the main agricultural exports, together with squash. Agriculture and fishing contribute some 30 percent to the gross domestic product of Tonga.

The Hawaiian Islands are a volcanic chain of more than 130 islands. The US state of Hawaii, with more than 1.2 million people in 2018, has the largest population in the Polynesian island group. The Hawaiian Islands represent one of the most striking examples of the transformation of an agricultural economy into a tourism-based economy. For over a century, when Hawaii was still a kingdom and then later, an annexed US territory, it was among the leading producers and exporters of pineapples and cane sugar. Both industries had been introduced to the islands by Westerners and dominated by large American corporations. After Hawaii became a US state (in 1959), large commercial producers of pineapples and sugarcane started relocating their acreage elsewhere in the world, primarily due to high US labor and land costs. The last of the large food-production companies, Del Monte, harvested its final pineapple crop in Hawaii in 2008. Hawaii's last sugar mill produced its final shipment of sugar in 2016.

The areas of Hawaii once devoted to sugarcane and pineapple plantations are now occupied by hotels and resorts. Between the 1980s and the 2010s, Hawaii's total land use for agricultural production

A kiwi plantation in New Zealand. (Fyletto)

shrank by about 68 percent. Now the corn seed industry is the state's dominant agricultural land user, followed by commercial forestry and the production of macadamia nuts.

New Zealand

About the size of Colorado, New Zealand is divided into two large islands and numerous smaller islands. It is a mountainous country with large coastal plains. About 50 percent of its land is in permanent pasture, 9 percent of the land is arable, and 5 percent is planted in permanent crops. Since European settlement, New Zealand has been a farming country, and in the twenty-first century, agriculture remains a vital component of the national economy, forming its single largest tradable sector. Agriculture accounts, on a year-to-year average, for about 8 percent of the gross national product and employs approximately 5 percent of the labor force. Today, New Zealand produces enough food to feed more than 20 million people, around four times its population.

In New Zealand, sheep outnumber people six to one, and there are also more cows than people. Wool, lamb, mutton, beef, and dairy products account for more than one-third of New Zealand's exports. New Zealand now exports 95 percent of its dairy products and is the world's largest dairy exporter.

New Zealand is the third-largest producer of kiwifruit, after China and Italy. Kiwifruit, native to China, was originally known as Chinese gooseberry. In 1962, New Zealand fruit producers gave the fruit a new name, kiwifruit, after the country's iconic bird and the nickname by which New Zealanders refer to themselves—kiwis. In recent years, the New Zealand wine industry has grown rapidly, and the country now ranks among the world's top-ten wine exporters.

With the exception of New Zealand and Hawaii, subsistence agriculture and cash crop production have continued to coexist in Pacific island nations and territories in the twenty-first century, with semi-subsistence economies prevailing. Agriculture, in general, has been declining and giving way to the tourism sector.

Toby R. Stewart

Industries of Australia

The basis for much of Australia's industrial activity is its minerals. Few countries possess the large number of minerals found in Australia, nor are Australia's immense reserves matched anywhere else. Australia is self-sufficient in most minerals, and is a world-class producer of bauxite (the raw material for aluminum), copper, iron ore, lead, zinc, nickel, diamonds, gold, silver, salt, and uranium. Australia's mineral resources provide the raw materials for the country's processing and manufacturing industries. They also represent a valuable export commodity that generates foreign currency earnings. Australia has also emerged as an important exporter of natural gas. Taken together, in 2020, more than 40 percent of Australia's export earnings was derived from minerals and natural gas.

Historic Background

From the time the British established the first colonies in Australia, minerals have been extracted. Before 1800, coal was being exported from mines near Newcastle, New South Wales. Along with coal extraction, whaling and the export of such byproducts as whale oil became Australia's first primary industries. Whale oil was highly prized in Europe and North America as a transmission fluid and a lubricant. It was also used for candles and as a base for perfumes and soaps. Whalebone was used in the making of corsets, umbrellas, and hoop skirts. The whaling industry, started in the 1790s, helped the fledgling colonies survive and prosper. By 1806, an efficient method of bay whaling had been developed: a small ship would harpoon a whale off the

A bauxite mine in Queensland, Australia.

coast and tow it to a station on land. One whaling station could therefore serve many ships. The industry flourished until the eventual overharvesting of many species. The development of harpoon guns, explosive harpoons, and steam-driven whaling boats in the late nineteenth century made large-scale commercial whaling so efficient that species such as the humpback, blue, and southern right whale came very near to extinction. The last whaling station in Australia closed in 1978.

In the 1840s, copper and lead mines were established first in South Australia, then in New South Wales and Victoria. In the 1850s and 1860s, gold was discovered at diverse locations in Victoria and New South Wales. The ensuing gold rush led to a rapid increase in Australia's population. Through the years, other minerals were discovered and mined locally. By 1900, Australia was exporting large amounts of copper, lead, zinc, tin, gold, and silver.

As mineral exploration techniques were improved, enormous deposits of coal, iron ore, and bauxite were discovered in the 1950s. The following decade, large deposits of petroleum, natural gas, nickel, and manganese were found.

Australia has very large reserves of extremely high-quality thermal coal (also called steam coal as it is burned for making steam to run electricity-generating turbines) and coking coal (used for steelmaking). Most coal mines are located in New South Wales and Queensland. Coal production is the mineral industry's largest employer. Western Australia is Australia's leading mineral-producing state for most other minerals, although extensive resources are distributed widely across the country. Many of the mines, smelters (plants where minerals are extracted from ore), and refineries (plants for purifying minerals) are located in remote, arid regions of the country, factors that contribute to the high production costs for several minerals. These same factors have boosted the development of Australia's transportation infrastructure as roads and railroads had to be built to serve the mines, smelters, and refineries.

Australia's mineral industries, along with its farms and ranches, provide many of the raw materials for the country's manufacturing. Although some manufacturing took place in the nineteenth century, substantial industrial development did not occur until after 1900. During the first half of the twentieth century, manufacturing centered around the processing of minerals. This led to the development and expansion of such basic manufacturing industries as steel, aluminum, petroleum, shipbuilding, and fabricated metals. These industries, and many others, were expanded during World War II when domestic industrial growth was stimulated by a reduction in imports. Rapid expansion of manufacturing continued in the 1950s and 1960s as the demand for both consumer goods and products used by businesses accelerated. During those two decades, manufacturing employment rose 70 percent. By the early 1970s, more than 40 percent of Australia's gross domestic product was derived from manufacturing.

The last quarter of the twentieth century saw a rapid decline in manufacturing in both Australia and throughout the rest of the industrialized world, from the United States to the European Union. Manufacturers were moving production to other world regions where labor and production costs were lower. Australia also experienced a phenomenon known as "Dutch disease," a reference to the impact of the Netherlands' gas discovery in the

Australia's Diamonds

Diamonds in Australia were first recorded in 1851 in New South Wales, and were mined there from 1867 to 1922. Geologists discovered Northwest Australia's famous Argyle pipe in 1979, and the Argyle Diamond Mine opened in 1985, going on to supply a third of the world's diamond supply every year since the late 1980s. Only about 5 percent of the mine's diamonds were of gem quality, but these included the rare pink diamond, valued at fifty times that of white diamonds. With the mine expected to close in 2021, Australia's rough diamond output is expected to decline from 14.2 million carats to 134.7 thousand carats. Other notable mines include Merlin, which reached full production in 2018, and Ellendale, which the Western Australian government rehabilitated after suspending operations in 2015.

A natural gas pipeline in Western Australia. (Glen Dillon)

North Sea on the Dutch manufacturing sector. Australia followed the same pattern. As revenues in natural-resource exports grew and inflows of foreign capital increased, the Australian dollar became much stronger than the currencies of Australia's import partners. Australian exports therefore became more expensive and imports became cheaper.

Ultimately, many of Australia's manufacturing plants were forced to close. Employment in textiles, apparel, coal-mining, steel, aluminum, shipbuilding, and the metal-fabricating industries declined dramatically. The de-industrialization of Australia took place at the same time as the workforce in agriculture, mining, and other parts of the primary sector was shrinking as a result of automation and other technological advances. In the 2010s, the mining sector contributed around 8.5 percent to Australia's gross domestic product (GDP) while accounting for about 2 percent of the workforce.

Australia's Service Industries

As both the manufacturing industries and the primary sector workforce declined, Australia's service industries gained prominence. By 2020, nearly 80 percent of the labor force was employed in services and more than 60 percent of the GDP was derived from this sector. Major service industries included retail trade, education, finance, recreation and tourism, health care, and information services.

Australia has developed a strong educational service industry by providing schooling to overseas students. International full-tuition students constitute a major source of income for domestic colleges, universities, and other institutions of higher learning. A significant number of these foreign students ultimately choose to reside in Australia after completing their studies, providing a valuable addition to Australia's skilled labor force.

A group of tourists in Sydney, Australia. (Mlenny)

Since the late twentieth century, one of the most rapidly growing tertiary sectors of the Australian economy has been tourism. Domestic travel has been the most important component, providing almost 75 percent of tourism revenues in the 2010s. International tourism has been also increasing in importance. In 2018, 9.3 million foreign tourists visited Australia, including many Europeans and North Americans arriving on cruise ships during the Northern Hemisphere winter. Top international visitor destinations were Sydney, Melbourne, the Gold Coast, tropical North Queensland, and Brisbane. The tourism industry employed more than 5 percent of the labor force. The coronavirus (COVID-19) pandemic in 2020 had a catastrophic impact on Australia's flourishing tourist industry, though an eventual recovery was widely anticipated.

The Australian government provides financial incentives for producers of big-budget films to use Australian locations, cast, crew, and service providers. As a result, Australia has become popular as a film-making destination, which also enhances various film production-related services, from sound recording to graphic design to movie prop production.

The health care and social assistance sector employed over 14 percent of the workforce in 2020. Since 2011, it has been the country's largest (by employment) and fastest-growing economic sector.

Robert R. McKay

Industries of the Pacific Islands

With the exception of New Zealand and Fiji, the industries of the Pacific islands are less developed than those of many other world regions. The small population and limited market potential of the Pacific region have prevented the island countries from taking advantage of economies of scale that exist in regions with large populations. Another disadvantage is the long distances between the islands and their export markets. Transportation costs for both raw materials and finished products tend to be high, putting producers at a competitive disadvantage. Many of the countries in the region are further limited by an overall lack of natural resources.

Mineral, forestry, and agricultural resources are poor throughout the region. In some cases, the political-economic structure has not been conducive to industrial development. Migration, remittances from persons working abroad to their home countries, and foreign aid have long been important for Pacific island economies, along with a few agricultural exports such as copra and sugar. At the same time, the tourism industry and commercial fishing licenses have assumed important places in the economies of most Pacific island nations and territories.

Manufacturing

In the twenty-first century, New Zealand has remained the dominant Pacific island industrial power. From an industrial perspective, other countries of the region have neither had the strength nor the economic diversity of New Zealand. In 2017, manufacturing made up 21.5 percent of New Zealand's GDP and employed 20.7 percent of the total workforce. Food processing was New Zealand's largest industry in terms of employment and output, contributing one-third of manufacturing's share to the GDP. The processing of meat and dairy products was the most significant component of this sector. Most of these products were produced for international markets, and in 2019, the sector accounted for the second-largest amount of New Zealand's foreign exchange earnings, after the tourism sector.

New Zealand manufactures a wide range of aluminum, steel, and consumer goods. While manufacturing has decreased in recent decades, as part of the global trend affecting all developed economies, in New Zealand it has remained an important source of employment and exports.

In Fiji, sugar production has remained a significant industry, and sugar is a major export. Also important is the manufacturing and export of textiles and garments. Since the 2010s, however, the top export product of Fiji has been bottled water. The water is derived from a huge, 17-mile (27.4-km.)-long aquifer on a remote part of Fiji's largest island, Viti Levu. It is bottled at the source and shipped to the United States and other countries, under the *Fiji Water* brand, a privately owned US company.

In recent years, New Zealand has also been rapidly expanding its production and exportation of bottled water, most of which is destined for China. Local producers as well as foreign investors in New Zealand bottling plants have been encouraged by the success of *Fiji Water.*

Mining

A major area in the basin of the Pacific Ocean where many earthquakes and volcanic eruptions occur is known as the Ring of Fire. Substantial deposits of valuable minerals are associated with this volcanic activity. On a number of the Pacific islands, mining

Copra being loaded at a port in Kiribati. (Government of Kiribati)

has been an important industry since the nineteenth and early twentieth centuries. Papua New Guinea and New Caledonia are the region's leading producers of minerals, which are the mainstays of those economies.

Copper, gold, silver, natural gas, and petroleum are being extracted in Papua New Guinea. That country is extremely rich in natural resources but plagued by poverty, crime, and land-tenure issues; rugged terrain has also hampered the development of mining and the necessary associated infrastructure. Nevertheless, minerals extraction accounted for almost two-thirds of export earnings of Papua New Guinea in the 2010s. The first Papua New Guinea Liquefied Natural Gas (PNG LNG) project, a consortium led by ExxonMobil, began exporting gas to Asian markets in 2014.

New Caledonia is among the world's leading producers of nickel, which in the 2010s accounted for close to 10 percent of the GDP of that French collectivity. Large-scale mining there started in the 1870s and dozens of nickel mines have been established since then; two new ones opened very recently. New Caledonia also produces chromium, cobalt, manganese, and iron ore. Unprocessed nickel-containing ores have been exported to many countries. However, due to the remoteness of the New Caledonia archipelago from the world markets, part of the extracted volume has been smelted locally to produce ferronickel, which has been shipped to France. In the twenty-first century, the nickel mining and processing industry in New Caledonia has been dominated by companies based in France, Brazil, and Switzerland.

In Fiji, the Emperor Mine at Vatukoula has been producing gold since the 1930s. The Solomon Islands are also rich in mineral resources such as gold, lead, zinc, and nickel, but those have remained largely undeveloped due to political and economic instability. The country's only working mine, at Gold Ridge on Guadalcanal, was abandoned for several years as a result of ethnic violence but reopened in 2017. It is one of the country's biggest employers and largest contributors to the national GDP.

Phosphates had been profitably exported from Banaba Island of the Republic of Kiribati since the turn of the twentieth century, but the deposits were exhausted by the end of the 1970s. The only other Pacific island country where mining was ever of significance is Nauru, where the production of phosphate was a mainstay of the economy for decades. This mineral was exported to Australia and several Asian countries, where it was used as fertilizer. By 1998, Nauru's phosphate reserves were virtually exhausted. The mining had led to an environmental catastrophe on the island, with 80 percent of the land surface having been strip-mined.

Wood Industries

Many Pacific islands are deeply forested. In several of the island countries, tropical hardwood and softwood forests have provided the basis for the wood products industry. The lumber operations on islands export primarily roundwood (logs) but many sawmills and wood processing plants also produce lumber and wood products. In some areas, small furniture manufacturing plants have operated; elsewhere, wood is burned to produce charcoal, which is used for cooking by many rural residents.

The production of pulp, paper, and other wood products in New Zealand is based on the processing of timber from pine forests on North Island, which have been planted since 1970. In Papua New Guinea, where almost all land, including forested land, is owned by tribal groups, logging activities must be agreed upon between the government authorities and tribal chiefs. The country's most valu-

Bottles of Fiji Water awaiting distribution. (Jon Roig)

Stray logs from a logging operation in the Solomon Islands. (Velvetfish)

able (for wood export) tree species are eucalyptus, rosewood, and pine. Commercial logging in the Solomon Islands began only in the 1990s. Since then, however, the islands have lost over 20 percent of their forest cover—mainly because of logging and, to a much lesser extent, as a result of land clearing for oil palm, cocoa, and coconut plantations.

Wood is Fiji's third-largest export commodity. Fiji is noted for its roundwood, plywood, veneers, pulp, and wood chips. This island country has nearly 136,000 acres (55,000 hectares) of plantation of the high-value mahogany tree species, introduced to Fiji during the British colonial rule. Harvesting began in 2003. If well-managed and marketed, Fiji's substantial mahogany assets could emerge as a significant contributor to the growth of the national economy.

Processing of Agricultural Products

Locally grown agricultural products have been processed for centuries on many of the islands. The most widespread such industry is the processing of coconuts. Both copra (dried coconut meat) and coconut oil have been produced on many Pacific islands since the early nineteenth century when Western missionaries and colonists began to establish plantations for copra cultivation to meet the demand for coconut oil in Europe. Although copra and coconut oil were traditionally used by indigenous islanders, the copra trade had never been geared toward a local market. Copra was a valuable cash crop whose production was controlled entirely by foreigners for overseas export. This legacy continues today. Even as Oceania's copra production for industrial purposes has dwindled, the marketing of copra derivatives, such as coconut-oil products for cosmetic use, has endured across the Central and South Pacific. In Kiribati, Oceania's poorest country, copra accounted for about two-thirds of export revenue in the 2010s.

Sugarcane is believed to have originated in Melanesia and, at one time, a large number of traditional varieties of sugarcane was grown on the Pacific islands. In Fiji, among other islands, sugar-refining was the most important industry from the 1870s until the mid-twentieth century. The milling efficiency in the region has been in decline since the mid-1980s as sugarcane plantation productivities in the Pacific have not kept pace with the changing international sugar market.

Tourism and Other Service Industries

In the twenty-first century, several countries and territories in the Pacific have become centers of offshore financial services.

Samoa, the Marshall Islands, the Cook Islands, and Vanuatu are the largest centers, followed by Nauru and the Federated States of Micronesia. Samoa's International Finance Centre is one of the major conduits for foreign investment in China. The Marshall Islands and Vanuatu offer flags of convenience for ships and offshore oil rigs. These countries benefit from being the legal domiciles of many of the world's largest maritime companies which, in turn, benefit from reduced labor, tax, and regulatory costs. Vanuatu is an important center of offshore insurance, banking, and gambling.

The film industry, based on foreign movie and television production, has played an increasing role in the economies of New Zealand and Fiji, whose islands are famous for their scenic beauty. Like the Australian government, the governments of these two countries have provided major financial incentives to foreign-film-production companies. The incentives have led to a significant boost in productions. In New Zealand, the screen industry has contributed to the development of a wide range of creative industries such as audio-visual design and digital-content services. The screen industry in Fiji still faces challenges in attracting bigger-budget film productions due to the lack of adequately equipped film studios and skilled crew support.

A cruise ship docking in Hilo, Hawaii. (dani3315)

A single industry—tourism—dominated the economies of many Pacific island countries and territories in 2020. These destinations included New Zealand, Samoa, American Samoa, Hawaii, Fiji, Papua New Guinea, French Polynesia, the Cook Islands, Tonga, Guam, Palau, and Vanuatu. For countries such as New Zealand and Fiji, the export of tourism-related services has become the largest earner of foreign exchange.

In 2018, total visitor arrivals (by air and cruise ships) to the Pacific island region reached 3.16 million. Fiji was the Pacific's largest destination, with a 40.7-percent share of the total volume. It was followed by French Polynesia—with its top destinations of Tahiti and Bora Bora—at 10.1 percent. Tourism receipts contributed over 11 percent to the region's combined GDP. On an individual basis, tourism contributed the largest shares to the GDP of the Cook Islands (87 percent), French Polynesia (85 percent), Palau (64 percent), Vanuatu (45.1 percent), Fiji (38.5 percent), Samoa (30.4 percent), and Tonga (25.5 percent). On the supply side, the region offered more than 3,000 hotels, resorts, and other tourism-related accommodations, with some 75,000 beds. As in Australia, the coronavirus (COVID-19) pandemic in 2020 had a devastating impact on tourism throughout the Pacific islands.

Efforts were underway to promote tourism in several other island countries, including Tuvalu, Nauru, Kiribati, and Niue. However, a shortage of facilities to accommodate tourists, especially adequate ports and airports, has put constraints on the development of the tourism industry in these small and geographically remote countries.

In recent decades, local and international development agencies have focused on the potential of tourism as a vehicle for the economic development of the Pacific islands. The natural beauty, mild climate, diverse cultures, and even the isolation of the islands have been increasingly viewed—and marketed—as major economic assets. However, some researchers and economists have questioned whether tourism has contributed to the prosperity and welfare of the majority of the island populations. They have also pointed out that tourism-related misuse of land and the exploitation of island culture as a promotional and entertainment tool have driven a wedge between the human and nature relationship that has always been at the core of Pacific island life.

Fisheries

Subsistence fishing around Oceania's many islands and atolls has been practiced for thousands of years, and fish remain an extremely important food source for local populations. Much of the region's nutrition, employment, recreation, traditions, and culture are centered around the living resources in the coastal zone between the shoreline and the outer reefs. Local small-scale commercial fisheries supply domestic markets and produce some fresh and processed fish products as an export commodity.

Large-scale industrial offshore fisheries in Oceania are dominated by foreign fleets operating in the Exclusive Economic Zones (EEZs) of the various Pacific island countries and territories. EEZs extend to 200 nautical miles (230.2 miles, or 370.4 km.) from each nation's land boundary, making Oceania a unique region since its nations, scattered across multiple small islands, have huge—by comparison to their land—oceanic EEZs. For instance, Tuvalu's EEZ is almost 30,000 times larger than its land area. Altogether, Pacific island EEZs account for approximately 10 percent of the planet's ocean surface. Commercial fleets from countries such as China, Japan, Taiwan, the United States, Spain, and France harvest fish resources in the Pacific island EEZs using industrial-scale fishing vessels.

The economies of several Pacific island nations and territories have been largely dependent on fishing; these islands possess few other natural resources such as minerals or timber. For the nation of Tuvalu, consisting of nine coral atolls, fisheries have constituted the only significant natural resource, and earnings from substantial fishing license fees paid by foreign commercial fleets have been a major source of government revenue; in the 2010s, they totaled almost half of the national GDP. Kiribati and the Federated States of Micronesia also heavily rely on earnings from fishing licenses.

Fish-processing canneries have been built on American Samoa, Fiji, and the Solomon Islands. As Oceania is home to the largest tuna fishery on the

A catch of tuna being offloaded on the Pacific island of Niube. (ctbctb8)

planet, tuna fishing and onshore tuna processing have been especially significant to these island economies, generating up to 8 percent of regional wage employment in the 2010s. Tuna fishing alone contributed to 42 percent of government revenue in Kiribati, 17 percent in Nauru, and 11 percent in Tuvalu. Papua New Guinea has strong domestic tuna, prawn, and lobster resources and exports fresh chilled, canned, and frozen fishing products as well as fishmeal.

Diminished tuna populations in other regions have made the Pacific island region highly attractive to global tuna fleets. Without concerted regional effort at regulation and resource management, Oceania's offshore fisheries may quickly become overexploited. Several strong regional organizations are active in the fisheries sector, including the Pacific Community, the Forum Fisheries Agency, the Western and Central Pacific Fisheries Commission, and the Parties to the Nauru Agreement Concerning Cooperation in the Management of Fisheries of Common Interest. However, the presence of extensive areas of international waters among the region's EEZs greatly complicates the region's efforts at fishery management.

Robert R. McKay

Discussion Questions: Industries

Q1. What role do mineral resources play in Australia's economy? What are the country's leading industries? What is Australia's single most important economic sector?

Q2. How did tourism come to dominate the economies of many Pacific island nations? Which islands are the principal destinations of tourists? What challenges do these islands face in their quest to attract visitors?

Q3. What are Exclusive Economic Zones (EEZs)? How big are they? What local industries benefit from the establishment of EEZs? How do the individual countries benefit?

Transportation in Australia

Australia is located a great distance from Europe and the United States and is almost as large as the United States. Distance means isolation, both from other countries and from one place to another within the Australian continent. This isolation has shaped the human geography of Australia through its effects on immigration, trade, and even on the national character of Australians.

Early Shipping

Early European explorers of Australia and Oceania sailed on ships that had been originally designed for transporting cargo. The expedition of the Dutch seafarer Abel Tasman reached Tasmania and New Zealand on two ships—a war yacht and a type of cargo vessel called a fluyt. The ships commanded by the British explorer of Australia James Cook included merchant vessels called barques and colliers converted to carry coal and other bulk cargo. The so-called "convict ships" that carried convicted felons to the British penal colonies in Australia were ordinary British merchant ships randomly commandeered for convict transport.

Early in the nineteenth century, US shipbuilders developed what came to be regarded as the world's most beautiful sailing ships—the clippers. These graceful, slender wooden vessels with their tall masts and vast expanses of canvas were also the fastest sailing ships ever. Initially, they brought tea from China to America and Europe; later, they carried miners to California. Clipper design reached a peak of perfection in the 1850s, just as the ships were being replaced by steamships, especially for the Atlantic trade. On the long route between Europe and Australia, however, the clipper ships excelled, carrying grain, wool, and gold. A fast time for the voyage was around 100 days.

Racing in the Roaring Forties—the fierce westerly winds of the south Atlantic—clipper ships achieved recorded speeds of 400 miles (640 km.) per day. The sleek *Champion of the Seas*, designed by Donald McKay of Boston, claimed the record of 465 miles (744 km.) in twenty-four hours, on December 12, 1854, while sailing to Australia at 50° south latitude. The British shipper James Baines of Liverpool bought many McKay clippers for the Australia passage. His Black Ball Line held the record—from Liverpool, England, to Melbourne, Australia, in seventy-four days—but voyages through great stormy seas with huge areas of sail aloft were a frightening experience for many passengers. The rocky south coast of Australia was particularly dangerous, and many ships were wrecked there as they neared the end of the perilous voyage.

Steamships

The invention of the steam engine in England led to the building of steamships, which eventually meant the end of sailing ships to Australia. The opening of the Suez Canal in 1869 shortened the route considerably, so that by 1914 the voyage be-

"Transportation" in Australian History

Transportation means the carrying of goods or people from one place to another, but there is also an older meaning for the word. Transportation, in the earliest days of European settlement in Australia, meant the exile of British convicts to the penal colony of New South Wales, a harsh punishment in the eyes of British lawmakers. Prisoners who received this sentence were transported to one of the colonies and not permitted to return to Great Britain. During the nineteenth century some 165,000 convicts were transported to the Australian colonies, especially in the peak years 1820 to 1840. The last British convicts were received in the colony of Western Australia in 1868.

The Ghan train arriving at Alice Springs Station. (bennymarty)

tween Europe and Australia took only five weeks. Coastal shipping around Australia also rapidly made the change from sail to steam but with time, road and rail transport have largely overtaken coastal shipping in importance. Most Australian ships are now used for recreation or commercial fishing. Sea transport remains significant for the transport of minerals, which are carried long distances around the coast by vessels operating mostly under the flags of other countries.

In 2019, only fourteen large vessels, including four bulk carriers and seven tankers, were registered under the Australian flag. Over the past several decades, many Australian maritime companies have phased out Australian-registered and Australian-crewed ships and replaced them with those registered under flags of convenience and run by foreign crews. Flags of convenience—such as those of Liberia, Panama, and the Marshall Islands—allow ship owners to avoid Australia's stringent tax regime, environmental regulations, and laws regarding employment conditions and safety.

Modern Australian ports use bulk handling for agricultural exports of sugar and wheat, as well as for mineral exports such as iron ore, coal, bauxite, and oil. Container ships are commonly used for transport of manufactured goods, both imports and exports. Australia has few long rivers, so inland shipping was of only minor importance for internal transportation even in the nineteenth century, when some steamboats operated on the Murray River. In 2019, Melbourne and Sydney were Australia's largest container ports. Iron ore was exported from the neighboring ports of Dampier, Port Hedland, and Port Walcott; coal was exported from Darlymore Bay and Hay Point. Other major seaports included Brisbane, Fremantle, Cairns, Darwin, Geelong, Gladstone, and Hobart.

Railroads

The first railways—as they are called in Australia—were built by the government of each separate British colony in Australia, which caused problems in the twentieth century. Victoria used the broad, or Irish, gauge—5 feet, 3 inches—wanting large, safe trains for its passengers. The earliest lines from Melbourne to Bendigo and from Geelong to Ballarat, built in the 1860s, have massive stone bridges and other structures. Railway builders in New South Wales had difficulties crossing the Blue Mountains, but that colony chose the standard English gauge—4 feet, 8.5 inches—for its tracks. Many lines ran straight from a port to the hinterland. The colony of Queensland, which had chosen the economical, narrow gauge of 3 feet, 6 inches, was the best example of this—rail lines ran inland from Brisbane, Maryborough, Bundaberg, Rockhampton, Townsville, and Cooktown, but these towns were connected with each other only by steamship until the 1920s.

In the United States, railroad construction linked the continent, spread economic prosperity, and helped to create a powerful nation. By contrast, the colonial lines of Australia simply brought agricultural products or minerals to the ports of respective colonies for export overseas. By 1880, Australia had nearly 4,000 miles (6,400 km.) of railways, but there was no connectivity. Rail travelers going from Brisbane to Perth had to change trains five times because of the break of gauge, what the Victorian Institute of Engineers called "the most lamentable engineering disaster in Australia." Railway construction boomed through the 1880s, with lines being laid "from nowhere to nowhere." The trains ran on coal, which Australia still has in abundance. Modernization in the 1950s meant conversion to diesel trains and air-conditioned carriages.

After the colonies united to form the Commonwealth of Australia in 1901, the government of the Commonwealth was responsible for the construction of the rail line from Port Pirie to Kalgoorlie, the heart of Western Australia's Goldfields region, since this rail link was part of the bargain made to secure the entry of Western Australia into the federation. It forms the central part of the Trans-Australian Railway, opened in 1917. That railway is the only rail freight corridor between Western Australia and the eastern states and thus is strategically important because Western Australia is the country's largest producer of iron ore, gold, bauxite, and wheat. The Trans-Australian Railway includes the world's longest section of completely straight track.

The extended Perth-to-Adelaide line, now known as the Indian Pacific, is run as a private passenger rail service. Although the Australian National (previously known as the Australian National Railways Commission) was sold to private consortia in late 1997, the Australian rail systems are still largely operated by the state governments. In 2018, the Australian rail network had over 22,000 miles (35,400 km.) of track, most of it government- owned.

Freight is the most important part of railway operations, with 1.34 billion tons carried in 2016. Passenger services include numerous long-distance, regional, and suburban routes, with over 90 percent being suburban trips. Suburban commuter trains and urban rail are an important part of the infrastructure of Australian cities such as Melbourne, Sydney, Adelaide, and Brisbane.

Private freight railway lines have been built in remote parts of Australia to carry primary commodities such as sugarcane, timber, and coal. Australia has numerous privately owned heritage railways operated as living history excursions that re-create the railway experience of the past. Some of them,

Camels in Australia

Throughout the nineteenth century, places in the interior of Australia far from railroads were served by mules, packhorses, and donkeys, but the largest area of the interior was served by teams of camels that could cope well with the arid, hot conditions and the tough vegetation. Camels were imported from Pakistan and used in exploration parties. They also were used for the construction of the Overland Telegraph line through the arid heart of Australia from Darwin to Adelaide in 1872. Although many camel drivers came from Pakistan, they were called Afghans. The modern train from Adelaide to Alice Springs is named The Ghan in honor of the early camel teams and their masters. When camel teams were replaced in the 1930's by motorized transport, many of the camels were turned loose and became wild, roaming the center of Australia. By 2008, the feral camel population had grown to an estimated one million camels (although that figure would later be lowered), posing a danger to environmental and cultural sites. In response, the government introduced the Australian Feral Camel Management Project, devoting $19 million to culling the animals. The project ran from 2009-2013, reducing the population by 160,000 camels. A five-day cull also occurred in South Australia in early 2020, after widespread heat, drought, and bush fires prompted camels to inundate indigenous communities.

such as the Puffing Billy Railway in Victoria, run original rolling stock built in the late nineteenth century. The Kuranda Scenic Railway in north Queensland is a small, picturesque tourist enterprise that takes passengers from the humid coastal plains of Cairns up a steep track to the Atherton Tableland town of Kuranda, passing waterfalls and affording spectacular views of the coast, gorges, and rain forests.

Rail lines are still being built in Australia, which is uncommon around the world. Planned for some 120 years, the Adelaide-Darwin railway was completed in 2003. It is a south-north transcontinental railroad running for 1,849 miles (2,975 km.) and providing both freight and passenger services. The travel on *The Ghan*, the line's passenger train, takes 54 hours and is considered one of the world's great train journeys.

High-speed rail services between Australia's largest cities, most commonly between Sydney and Melbourne, have been proposed since the 1980s, but in 2020, all these projects were still in the planning stage.

Road Transportation

In the twentieth century, roads increased in importance, and Australians rapidly adopted automobiles as the preferred mode of travel. Australia has a relatively extensive road system for its small population and huge territory. In 2020, the country had about 542,820 miles (873,580 km.) of roadways—the ninth-largest road network in the world.

The continent of Australia is flat and has a generally mild climate, so road construction would at first glance seem to have been easier and less expensive than in many other regions. But Australia has a small population and the great distances to be covered added to the comparative cost per person and per mile. About 40 percent of Australia's roads are paved, compared to 65 percent in the United States. Major highways have an excellent-quality bitumen (asphalt) surface, and dirt roads are usually well-maintained, even in the Outback, Australia's vast, remote interior.

In 2020, Australians had more than one passenger vehicle for every two people, making this country one of the world's most mobile. Counting buses,

"Road trains" in Australia gassing up. (travellinglight)

trucks, and other commercial vehicles, there were almost 720 vehicles per 1,000 persons. Australian cities experience traffic congestion, especially Sydney, where there is only one bridge into the inner city. The Outback roads have little traffic, but travelers are cautioned about the danger of accidents that can be caused by fatigue or drowsiness when driving on the long, straight, and often empty roads with the little-changing landscape.

Most domestic freight is carried by road transport. Road trains—huge trucking vehicles carrying two or more trailers—move live cattle, fuel, mineral ores, milk, and all kinds of other freight. Australia has the longest and heaviest road-legal (that is, equipped and licensed for use on public roads) road trains in the world, consisting of up to four trailers weighing a maximum of 200 tons.

Australia's National Highway is a system of roads connecting the nation's large cities and major regional centers of all mainland states and territories. Highway 1 is a ring road around Australia joining all mainland state capitals. At a length of some 9,000 miles (14,500 km.), it is the longest national highway in the world, longer than the Trans-Siberian Highway and the Trans-Canada Highway.

Cobb and Company

The discovery of gold in 1851 in several places inland in New South Wales and Victoria led to a gold rush attracting miners from all over the world. Transportation was needed from the ports to the goldfields and the oxen teams laden with provisions were too slow. Horse-drawn coaches were the solution. In 1853, Freeman Cobb imported four thoroughbrace coaches from the United States—light coaches with the body suspended on two long leather straps instead of metal springs. This made the passengers' ride more comfortable over the rutted tracks and reduced breakdowns. Other specially designed coaches were built for Cobb by Australian factories. Cobb and Company employed experienced drivers, changed horses every 10 to 15 miles (16 to 24 km.), and achieved speeds of 9 miles (14.5 km.) per hour between Melbourne and Castlemaine. Within a decade, the company expanded its operations to New South Wales and Queensland, carrying passengers, gold, and mail. Soon robbers, known as "bushrangers," began to hold up the coaches and steal gold shipments or rob passengers.

In the 1870s, Cobb and Company was one of the world's largest coaching firms, with 6,000 horses working every day. When railways were built, Cobb continued to operate farther inland. The last route, in inland Queensland, closed in 1924. Many of the coaches are preserved in Australian museums today.

Air Transportation

The continent is well suited to air transport because of the great distances between cities and the generally fine weather. The absence of major mountains has contributed to Australian aviation's remarkable safety record. There were more than 400 airports in Australia in 2020, and thousands of registered aircraft, about half of which were privately owned small airplanes. Only about 100 small airfields lack paved runways. Passenger traffic in the country's busiest airports was close to 42 million passengers (Sydney Airport) and 36.7 million (Melbourne Airport).

Australia's first airmail was flown from Melbourne to Sydney in 1914. The first scheduled domestic air service in Australia, for transporting both

A Qantas jet. (Jetlinerimages)

airmail and passengers, began in the remote north of Western Australia in 1921. Sir Charles Kingsford Smith is Australia's best-known aviator and an Australian national hero. With Charles Ulm, Kingsford Smith was the first to cross the Pacific Ocean, flying in 1828 from Oakland, California, to Brisbane in the *Southern Cross*, a monoplane, with landings at Hawaii and Fiji. They covered 7,400 miles (11,900 km.) in eighty-three hours flying time. In 1934, Kingsford Smith made the first west-to-east Pacific crossing. Six weeks later, Ulm was lost near Hawaii; in 1935, Kingsford Smith vanished on a flight from England to Australia. The Sydney airport is named Kingsford Smith Airport, and the *Southern Cross* is preserved at the Brisbane Airport.

At the close of World War I, the flight from Britain to Australia took four days, and a flight from the United States to Australia took ten days. In 2020, one could fly nonstop from Los Angeles to Sydney in less than fifteen hours.

Qantas

In 1920, Qantas (the Queensland and Northern Territory Aerial Services) was begun in the Queensland Outback. In 1933, the company became Qantas Empire Airways, an international service to the United Kingdom. In the 1940s the airline began its first services outside of the British Empire, to Tokyo; it was later rebranded Qantas Airways. Now Qantas is Australia's international airline and its largest domestic carrier, with an admirable safety record. In 2019, Qantas completed the world's longest commercial flight; the flight between New York and Sydney using a Boeing 787-9 Dreamliner took 19 hours and 20 minutes.

Smaller airlines operate within Australia at the regional level. There are many private airplanes, since Outback pastoralists are able to have airstrips on their properties. Private airstrips also enable visits by the Royal Flying Doctor Service of Australia.

A Royal Flying Doctor Service plane. (moisseyev)

The Flying Doctor

A Presbyterian minister, the Reverend John Flynn, established the Australian Inland Mission in 1912 in the remote interior of Australia, to provide medical assistance to isolated settlers in the lonely Outback. He saw the possibilities of using aircraft to cover trackless expanses when there was a medical emergency, combined with having the new "pedal radio" at all remote cattle and sheep stations, missions, and other isolated settlements to keep in contact with a centralized medical base. In 1928, Flynn established the Aerial Medical Service at Cloncurry in western Queensland, with one small plane leased from Qantas. In 1942 this became the Flying Doctor Service, and the prefix "Royal" was granted in 1955.

This was the first comprehensive aerial medical organization in the world, and it continues to provide emergency medical care and routine health services (including telehealth sessions by radio, telephone, or video) to tens of thousands of people in remote and isolated parts of Australia each year. In 2020, the Royal Flying Doctor Service maintained more than twenty bases throughout Australia, served by a fleet of sixty-eight aircraft.

Ray Sumner

Transportation in the Pacific Islands

The Pacific island region, also known as Oceania, is divided into three subregions: Melanesia, Micronesia, and Polynesia. Approximately 30,000 islands make up the region. Thousands of years ago, the first inhabitants of the Pacific islands are believed to have traveled by boat to Melanesia from somewhere in Southeast Asia. The early Pacific islanders developed a twin-hulled sailing canoe, an extremely efficient seagoing vessel. In boats of this kind, they continued the process of migrating through Micronesia and Polynesia.

It was not until the sixteenth century that the first Europeans arrived in the South Pacific. European exploration of the Pacific began with the Spanish and the Portuguese. By the late 1500s, the Spanish traders had discovered several of the Caroline Islands in Micronesia, the Solomon Islands in Melanesia, and the Marquesas Islands in Polynesia. The Spanish sailed on galleons, large and heavy vessels that usually had three or more decks and masts and could be used both for trade and war. Not until the late eighteenth century and Europe's second Age of Exploration were most of the Pacific islands explored and mapped, and the cultures of their inhabitants described.

As modes of transportation evolved through the centuries, it became somewhat easier for the Pacific islands to be more fully integrated into the rest of the world. Between the 1830s and the 1950s, ocean liners were the mainstay of long-distance passenger transport. However, due to the growth of air travel, the passenger ships declined by the early 1960s. (Many liners continued their service as luxury cruisers.) The advent of air travel has made getting to Oceania far quicker. However, the cost of maintaining links with the Pacific islands is high for the airlines. Ironically, because of the cost of air travel, it is now more expensive to travel to the Pacific islands than it was in the 1950s when less-expensive passenger sea travel was common.

Within the Pacific's island groups, the predominant modes of transportation are now air travel services by local carriers, motorboats, ferries, water taxis, and freighters (some cargo ships have reserved spaces for travelers, who are mostly locals). An islander who wishes to travel to places such as the United States, Canada, or Japan must use an international airline. On the islands, automobiles are used primarily in urban centers, and shuttle buses are used for travel between ports and cities or resorts. Except for New Zealand and Hawaii, there are no well-developed road systems in the Pacific islands. For coastal and inter-island travel, the native population, which traditionally relied on the canoe for traveling short distances, now uses mostly motorboats.

Ports

Outside of private yachts and expensive cruise ship tours, few passengers traveled by boat or ship to the Pacific islands after the 1960s. Cargo was still transported, however. The island nations and territories export agricultural and mining products as well as seafood to places such as Australia, New Zealand, and East and Southeast Asia and import a wide variety of products, including foodstuffs and manufactured goods. Imports far outweigh exports and are carried mostly in containerized form, while major exports are shipped by bulk carriers. Major challenges that have persisted into the twenty-first century include the long distances between ports, low trade volumes, and widely varying port facilities with generally inadequate funding for operation and maintenance. Cargo shippers, often large multinational companies, charter vessels that sail at the times and to the ports determined by the cargo interests. Imports come mainly from Australia and Asia but also from as far as the United States and Europe.

All Pacific island countries and territories have ports, but most of those are very small, serving only motorboats and yachts. Their infrastructure is limited to basic wharves and paved docking areas. In 2020, French Polynesia had five small ports and a medium-sized cruise port in Papeete, its capital. The tiny countries of Vanuatu, Tonga, and the Marshall Islands had a few small ports. At the same time, Vanuatu, Tonga, and the Marshall Islands had numerous large merchant marine ships on their ship registries, including bulk cargo carriers and oil tankers. French Polynesia had seventeen cargo carriers and large yachts registered under its flag. These countries and the territory of French Polynesia have open registries, colloquially called flag-of-convenience registries, allowing foreign ship owners to take advantage of attractive fiscal regimes on these Pacific islands and their flexible rules governing maritime safety and working conditions for ship crews. Various fees and charges associated with registering foreign-owned ships under their flags are major sources of government revenues for Vanuatu, Tonga, and the Marshall Islands.

In 2020, New Zealand had over a dozen small ports, eight medium-sized ports, and a large seaport—Ports of Auckland—with container and cruise ship terminals and four additional freight hubs. New Zealand's largest container terminal, however, was at the recently upgraded nearby Port of Tauranga. The nation of the Solomon Islands had two deepwater ports, at Honiara and Noro, with anchorage facilities for international cargo and cruise ships.

In recent years, there has been some progress in port development in the rest of the Pacific island region. The ports of Suva and Lautoka on Fiji's main island have been significantly upgraded, with funds provided by the Asian Development Bank. The Fiji Ports Corporation plans to build a major new container and multipurpose port facility at Rokobili.

Traffic in Papeete, Tahiti. (Patrick Cooper)

The Japan International Cooperation Agency has developed a project for expanding the Pohnpei Port, the largest and the most commercially important port in the Federated States of Micronesia. The government of Papua New Guinea has approved two major port projects, including the relocation of Port Moresby's main port from the city's central business district to nearby Motukea Island.

The Pacific Forum Line is a regional shipping line run collectively by twelve Pacific island governments. Established in 1978, it has been held up as a model of how a regional approach to transport can succeed. In the 2010s, it operated eight vessels capable of carrying containerized and break bulk cargoes on five routes between various Pacific islands and also between these islands and Australia.

Air Travel

In the second half of the twentieth century, air travel became the most efficient mode of transportation to and from the Pacific islands. In 2020, most air traffic to the Pacific islands came from North America, Australia, New Zealand, Japan, South Korea, and Europe. Air New Zealand, Qantas, United Airlines, and Air France were among the major airlines with daily flights to the Pacific islands. Air Tahiti Nui, the main airline of French Polynesia, had several daily flights from Paris to Papeete via Los Angeles. Fiji Airways (Air Pacific) was bringing two-thirds of all visitors to Fiji. It operated flights to twenty-three cities in thirteen countries, including several Pacific island nations.

Profitability was a major problem for smaller Pacific island countries and territories that were funding regional airlines. Pacific island nations came to believe that regional cooperation initiatives were a viable solution to the profitability problem. In the mid-1990s, the Marshall Islands, Nauru, Tuvalu, and Kiribati agreed to share aircraft in order to cut the massive losses that they had been incurring. Agreements of such kind in aviation are called "codeshare." These and other lesser-traveled South Pacific destinations have made use of small commuter planes.

Most Pacific island countries and territories have more than one airport. The only territory that did not have an airport in 2020 was Tokelau: its three small atolls could be reached only by boat from Samoa or by private helicopter. In 2020, Papua New Guinea had almost 500, mostly very small airports, of which only about twenty had paved runways. New Zealand had more than 120 airports, a third of which had paved runways.

Roads and Motor Vehicles

The size of a particular island, and its level of economic advancement, have greatly influenced what kind of highway system has been built. Since New Zealand is one of the most developed countries in the world, let alone in the Pacific region, it has a far more elaborate highway system than any of its Pacific island neighbors. Although the terrain is primarily mountainous, which makes road construction both hazardous and expensive, New Zealand has invested large funds in order to have

A gate marking off a grass landing strip in Fiji. (Janos)

first-rate roads. It has spent more on the construction of roads than on any other mode of transport. By the late 2010s, New Zealand had more than 58,400 miles (94,000 km.) of roads, more than 60 percent of which were paved, including 224 miles (360 km.) of motorways and expressways. More than 100 additional miles (160 km.) of highways were expected to be built by 2030.

By comparison, Papua New Guinea—which is also mountainous—had only about 12,000 miles (20,000 km.) of roads—only about 1,860 miles (3,000 km.) of which were paved. The terrain and land ownership issues have worked against Papua New Guinea having a truly adequate highway system. Still, several major cities in Papua New Guinea were connected by the Highlands Highway, which ran 435 miles (700 km.).

Motor vehicles are a common mode of transportation in the urban areas of the Pacific islands. Papeete, the capital of French Polynesia located on the island of Tahiti had a four-lane freeway connecting it to the suburbs in 2020. The area around Papeete has become so urbanized that it takes commuters as much as two hours to go to and from work during the week. While other Polynesian islands remained less developed, Tahiti became the hub of the area and one of its most popular tourist locations. Because land is at such a premium, the introduction of roads and motor vehicles created major problems on other islands where the growth of tourism necessitated their use. In Honolulu, Hawaii, with a metropolitan population of 953,207 in 2019, traffic congestion was extremely heavy.

The Polynesian country of Tonga is a small, relatively poor island country, comprising an archipelago of 170 islands, of which only forty are inhabited. With a population of little more than 106,000 in 2020, Tonga had 420 miles (680 km.) of roads, 115 miles (184 km.) of which were paved and the rest surfaced with compacted coral.

Railroads

The small size of the islands and the hazardous terrain have made rail travel impractical on the majority of the Pacific islands. The most sophisticated rail system of the South Pacific exists in New Zealand. In 2020, the main rail services were provided by KiwiRail, a state-owned enterprise. The nationwide rail network had over 2,565 miles (4,128 km) of track linking major cities on both the country's large islands. The islands themselves were connected by inter-island rail and road ferries.

In Fiji, rail transport hauls cut sugarcane to crushing mills. In 2020, there were 370 miles (597 km.) of rail track, run by the government-owned Fiji Sugar Corporation.

The only other rail line of note in the Pacific region was in Nauru. Its 2.4-mile (3.9-km.)-long rail line that transported phosphates from the interior to the coast was built by a British phosphate company in 1907. By the beginning of the twenty-first century, Nauru's phosphate deposits were essentially exhausted, and the future of the railway remains uncertain.

Without financial backing from governments, international institutions, or overseas investors, it is unlikely that rail, road, or regional maritime transport will make significant further inroads into the smaller Pacific islands in the twenty-first century, where size or terrain, populations, and economies continue to make new transportation infrastructure a difficult proposition.

Jeffry Jensen

DISCUSSION QUESTIONS: TRANSPORTATION

Q1. What were the main challenges faced during the construction of Australia's railways? Where does the Trans-Australian Railway run? Which railroad crosses north-south through the heart of Australia?

Q2. What is the mission of the Royal Flying Doctor Service? How does it meet the medical needs of people living in remote areas? How has it grown and evolved since its inception in 1928?

Q3. What major challenges have impeded the development of efficient transportation networks in the Pacific islands? Which islands have well-built road systems? How has the emergence of aviation facilitated travel among the islands?

Trade in Australia

From the earliest days of European settlement in the 1780s, Australia was regarded as a place for the extraction and exploitation of natural resources. Throughout the nineteenth century, the British colonies on the Australian continent maintained this typical trade relationship with the home country, exporting raw materials and natural products—such as gold and whale oil—and importing manufactured goods.

Wool was the first commercially successful agricultural product produced in Australia specifically for export markets; it was soon followed by wheat and animal hides. The invention of the steamship and refrigeration expanded the exports to include frozen meat and fruits. One of the main arguments in favor of the unification (in 1901) of the six British colonies in Australia into a single dominion of the British Empire was the elimination of customs duties, or tariffs, that had been applied on goods traded among the six separate colonies. Even after gaining independence, Australia remained a member of the British Commonwealth of Nations (now known simply as the Commonwealth) and participated in trade agreements with other Commonwealth members, predominantly with the United Kingdom.

Wool, wheat, and gold continued to dominate Australian exports until the 1960s, when discoveries of rich ore and petroleum deposits occurred. Iron ore and coal quickly became valuable exports,

An Australian container ship. (ugurhan)

and new markets in Asia overtook traditional European markets. In the twenty-first century, Australian trade still maintained this pattern, with minerals and agricultural products dominating exports. In 2020, Australia remained the world's largest exporter of wool, coal, and iron ore; the fifth-largest exporter of alumina; and the sixth-largest exporter of gold.

The lobby of the Australian Securities Exchange. (Jason7825)

As a high-income economy, Australia has great demand for manufactured goods. It imports large volumes of consumer goods, such as automobiles and trucks, electronic devices, and pharmaceuticals. It also imports such capital goods as machinery, telecommunications equipment, aircraft, and oil. The manufacturing industries of Australia have the disadvantages of a small domestic market and a long distance to overseas markets, together with an Australian system of high wages and strong trade unionism. On the other hand, Australia has the advantages of being a stable and democratic country with a highly educated, urbanized workforce and a rich resource base.

Australian agriculture and mining are highly productive, using the most modern technology while employing only 3 percent of the workforce. Because of the strong export performance of its primary industries, the country has maintained a positive trade balance for the past several decades. Australia has encouraged the growth of tourism, which is regarded as an export industry. Foreign companies, especially the United States, the United Kingdom, and Japan, have invested strongly in Australian industry.

International Trade Agreements

The Australian government actively seeks new markets for Australian goods and services through bilateral and multilateral trade agreements at different levels. The necessity for these kinds of trade agreements became obvious in Australia in the 1960s when access to the traditional market, the United Kingdom, was restricted by that country's joining the European Economic Community (later the European Union). This forced Australia to look for new markets, especially in Asia, where several countries had rapidly developing economies. In 2020, Australia's top trading partners were China, Japan, South Korea, and the United States. At a regional level, Australia is a member of the Asia-Pacific Economic Cooperation (APEC), which was established in 1989. It has twenty-one member economies: nineteen countries that are members of the United Nations, plus Taiwan and Hong Kong. The APEC has been highly successful in establishing Asia-Pacific markets for agricultural products and raw materials as well as in reducing the costs of business transactions across that huge region.

Other important trade agreements are the Australia-New Zealand Closer Economic Relations (CER) trade agreement, in effect since 1983, and the China-Australia Free Trade Agreement (ChAFTA), which entered into force in 2015. Australia also has free trade agreements (FTAs) with Japan, South Korea, Chile, Malaysia, Singapore, Thailand, and the United States, and a regional FTA with the Association of Southeast Asian Nations (ASEAN). Australia is a strong proponent of the proposed Regional Comprehensive Economic Partnership (RCEP) among the ten ASEAN member states and five of ASEAN's FTA partners—Australia, China, Japan, New Zealand, and South Korea.

Ray Sumner

Trade in the Pacific Islands

The Pacific island region consists of more than 20,000 separate islands, ranging from lone specks of barely visible rock to landmasses covering thousands of square miles. They are traditionally divided into three subregions. Melanesia is located in the southwestern Pacific Ocean off the northern and eastern coasts of Australia and includes Papua New Guinea (the eastern half of the island of New Guinea), Fiji, Vanuatu, New Caledonia, and the Solomon Island chain. Micronesia is situated just south of Japan and contains strategic US military installations on Guam and Wake Island as well as the Federated States of Micronesia, the Republic of the Marshall Islands, and the small countries of Kiribati, Nauru, and Palau. Polynesia stretches from the Pacific islands of Midway and Hawaii to the southern tip of New Zealand. These latter islands historically relied on agricultural exports such as coffee, bananas, pineapples, and coconut oil and dominated the region's economic infrastructure. However, with the advent of jet air travel, a global tourist industry has transformed many Polynesian island nations and territories, including Fiji, Tahiti, Samoa, New Zealand, and the Hawaiian chain, into international vacation getaways.

Early Trade Contacts

Thousands of years ago, people began to migrate from Southeast Asia to the Pacific islands, facilitating the evolution of civilizations in the region. Be-

A row of duty-free stores in Tuom, Guam. (Hiroshi_H)

ginning with the early settlements in Australia and New Guinea, various tribes systematically moved outward from Melanesia to Polynesia—into New Caledonia, Fiji, Tonga, Samoa, and Hawaii; others sailed north and relocated to parts of Micronesia. The development of outrigger canoes and the sail helped spread agricultural commodities such as sweet potatoes, coconuts, and pigs throughout the region, and profitable trade relations developed throughout the island chains. Trade networks stretched from Fiji to Tahiti and connected New Caledonia and New Hebrides (now Vanuatu). In Hawaii, before European colonization, the islanders formed an interdependent trade system that furnished various communities with goods and services from each inhabited island.

Throughout the Pacific island region, different shells were used as money, from the common pearl shell broken into flakes to cowry shells to elaborate shell beads worked into strips of cloth. To this day, the shell currency is still used to some extent in the Solomon Islands and the Papua New Guinea island of East New Britain. The island of Yap (now part of the Federated States of Micronesia) is famous for its stone money—rai stones—carved disks with a hole in the middle. Rai stones vary in diameter from about 1.4 inches (3.5 centimeters) up to 13 feet (4 meters). The latter weigh thousands of pounds; those were placed in front of houses or along pathways and were rarely moved. Until the twentieth century, rai stones were the principal form of currency on Yap island. Even now, they are used as representative money for important social obligations such as dowry or inheritance transactions.

The arrival of European traders drastically altered the nature of trade relations in the Pacific islands. While earlier Spanish, Portuguese, and Dutch missions established the first western contact in the Pacific islands as far south as New Zealand, the explorations of English captain James Cook from 1768 to 1779 revolutionized trade in the region and expedited the integration of the region into international commerce. Cookand others discovered that the islands contained vast quantities of sandalwood, which could be exchanged in China for a handsome profit. Indigenous resources were quickly exploited and devoured as islanders swapped sandalwood for European cloth, guns, ironware, tobacco, and liquor.

This trade also produced profitable networks among the islands, since Europeans often paid for sandalwood with indigenous products. In Melanesia, traders obtained tortoise shells in the Solomons and exchanged them for pigs in Fiji. These pigs, along with stores of tortoise shells and whale teeth, were used to purchase sandalwood in the Vanuatu islands, advancing the growth of inter-island exchanges. These contacts also intensified both military conflict and political unification in the islands. Many island chiefs desired firearms to further their own ambitions, and the sandalwood trade enhanced the distribution of guns in the region.

Hawaii's King Kamehameha I was able to consolidate his hold over the Hawaiian Islands after securing control of hundreds of muskets, some small cannons, and sailing vessels from European merchants. In Tahiti, chief Tu (future King Pomare I) sold pigs to Australia for weapons, conquered his opponents, and united the different chiefdoms into the single Kingdom of Tahiti. This same process occurred in Fiji and the Marquesas, as pigs were swapped for firearms. Although the sandalwood trade had been depleted by the 1860s, it helped integrate the Pacific economies and led to a vast movement of people and products from New Zealand to Hawaii.

As a result of European and US domination during the nineteenth century, the Pacific islands were systematically incorporated into the global economy. However, for the most part, this process retarded the development of local manufacturing and processing industries and led to the emergence of one-crop export economies, tourism, and military bases in the region. As a result, many island economies became structurally dependent upon foreign trade for their survival.

Micronesia

The small islands of Micronesia are sparsely populated and, for the most part, play an inconsequential role in global trade. The Marshall Islands contain a plentiful supply of fish among their coral reefs, but with the exception of coconuts, few plants can survive in the sandy soil. Export earnings of the Marshall Islands come mostly from the service sector—lease payments by the United States for the use of Kwajalein Atoll as a US military base. Guam is the most vibrant island in Micronesia. As a location of several strategic US military operations, Guam benefits from American national defense spending. It is also a principal vacation spot for Japanese tourists, and since it has been established as a duty-free port, it generates significant revenue from service exports, mainly spending by foreign tourists visiting Guam's duty-free shopping outlets and malls. In 2018, service exports accounted for almost 18 percent of the island's gross domestic product (GDP).

The proximity to Guam also bolsters the tourism industry and service exports of the tiny nation of Palau. In Nauru, another tiny nation in Micronesia, the phosphates industry is all but dead, and most necessities—even water—must be imported. Imports exceed exports in all Micronesian economies, and the small population base has failed to attract or generate large-scale trade.

Melanesia

Located off the northern coast of Australia, the Melanesian islands are situated in some of the most strategic shipping lanes in the Pacific Ocean. As a result, these islands experienced fierce amphibious battles during World War II. However, most of the islanders continue to rely on subsistence farming for their livelihood, and this region remains highly underdeveloped. The Solomon Islands export some fish, timber, and coconut meat to East Asia, but they are dependent upon imports from China, Australia, Malaysia, Singapore, and Taiwan for most of their food supply, manufactured goods, and fuel.

Papua New Guinea benefits from profitable wood, fish, palm oil, and coffee exports, and with the discovery of copper, gold, oil, and natural gas, it also has developed lucrative mining industries. Mineral resources account for nearly two-thirds of the country's export earnings. Along with Australia, Papua New Guinea is a member of the Asia-Pacific Economic Cooperation (APEC) forum, a major Asia-Pa-

Tourists prepare to board a plane at the Daniel K. Inouye International Airport in Honolulu, Hawaii. (hapabapa)

cific trading bloc. In 2017, the country launched its first-ever national trade policy, setting the goal to reduce imports and significantly increase exports and foreign direct investment by 2032.

Fiji's major industry is tourism, and its airport at Nadi is an important hub for South Pacific travel. Fijians export bottled water (the popular *Fiji Water* brand) as well as fish, gold, sugar, and clothing. Since the 2010s, Fiji has become the world's second-largest exporter of water, after China. Along with this new source of export earnings, Fiji's main economic resources continue to be its exotic beaches, remote island locations, and comfortable tropical climate. In the first decades of the twenty-first century, the export of tourism-related services has remained Fiji's largest earner of foreign exchange.

Offshore financial services and tourism are foundations of the service trade in Vanuatu, but this nation, made up of some eighty small, low-lying islands, is highly vulnerable to natural disasters, even more so than most other Pacific island countries. In 2015, its leading tourist destination, the island of Efate, was severely impacted by Tropical Cyclone Pam. Another severe tropical cyclone, Harold, hit Vanuatu in 2020. It also caused widespread destruction in the Solomon Islands and Fiji.

Polynesia

Due to its long-standing, interactive, historic relationship with Asia, the United States, and Europe, Polynesia possesses the most modernized infrastructure and has prospered from a more diversified economic background than the two other island subregions of Oceania. It is the cornerstone of the Pacific tourist industry, and its import-export trade has created many profitable links throughout the global economy.

Following the construction of luxury hotels and international airports, Samoa and Tonga have gained economically thanks to the rise of tourism and related service trade. Tahiti and other parts of French Polynesia long considered tropical paradises, have reaped the same benefits. Jet air travel has allowed Tahiti to become one of the most exotic vacation spots in the world. French Polynesia's largest export, however, is pearls, especially the popular Tahitian black pearls. Pearls account for over half of annual exports from French Polynesia. Tourism-related services comprise the second-largest export category. In contrast, Tuvalu, one of the smallest and most isolated countries in the world, is almost entirely dependent on imports, particularly of food and fuel. The French collectivity of New Caledonia exports large quantities of nickel; food accounts for about 20 percent of New Caledonia's imports as only a small amount of the land is suitable for cultivation.

Hawaii and New Zealand are the two most developed and sophisticated Polynesian economies. Hawaii's economy has been directly tied to the United States since the mid-nineteenth century, leading to the formal annexation of the islands in 1898 and statehood in 1959. Initially colonized by US shippers and sugarcane plantation owners, Hawaii has evolved into the most important military base in the Pacific, and a large percentage of its people are employed in tourism-related sectors. Some islanders remain extremely critical of the US presence in Hawaii. They claim that the United States has destroyed indigenous rights in exchange for a system of dependency which, by 2017, had left approximately 15 percent of Native Hawaiians at the poverty level.

For about a century, sugar and pineapples were Hawaii's two principal agricultural exports. Pineapple production emerged as a profitable crop following the creation of the Hawaiian Pineapple Company by James Dole in 1901. As a result, Honolulu had one of the largest fruit-canning businesses in the world. By the mid-1970s, Dole and other fruit companies moved the production of canned fruits to countries such as the Philippines and Thailand, where labor costs are lower. As the Hawaii canneries closed, the local industry shifted to the production of fresh pineapples. Other products, such as Kona coffee, papayas, and macadamia nuts, have eclipsed pineapples as notable agricultural exports. Similarly, sugar production moved to India, South America, and the Caribbean. Former plantation lands were used by the "Big Five" sugarcane processing corporations to build hotels and develop Hawaii's modern tourism-based economy.

Imported cars in New Zealand being transported. (Nid Goloti)

New Zealand has achieved the most significant economic and trading success of any of the Pacific island nations. New Zealand farmers benefit from a mild climate and have efficiently utilized modern scientific farming techniques to create a profitable business in meat, dairy, and wool exports. Sheep provide a variety of products, including wool, mutton, sausage casings, and animal fat. Beef and dairy cattle also help fuel the nation's export-oriented primary sector. Commercial crops include barley, wheat, and potatoes, and the country is one of the world's most important suppliers of kiwifruit.

Products of fishing, forestry, and ore processing are also significant in New Zealand's export mix. Timber grows quickly in North Island's volcanic region, and the nation exports considerable amounts of pine and evergreen wood. Despite its limited supply of mineral resources, New Zealand has developed a processing base that includes the production of steel, aluminum, and iron. In 2020, New Zealand's top exports were dairy products (milk powder, butter, and cheese), meat, wood, fruits, wine, and aluminum. Export of services (mostly tourism-related) accounted for about 30 percent of all New Zealand's exports. Over the years, New Zealand has tended to import more than it exports, but the trade deficit has been relatively small. The country's top imports include cars and trucks, oil, and telecommunications equipment.

In the twenty-first century, China has become New Zealand's main trading partner. The New Zealand-China free-trade agreement (FTA), signed in 2008, was the first FTA that China has signed with any developed country. It was also New Zealand's largest trade deal since the Closer Economic Relations agreement with Australia, signed in 1983. In 2019, New Zealand and China agreed to upgrade their FTA. New provisions gave New Zealand preferential access to China's wood and paper market. New Zealand's other major trading partners have been Australia, the European Union, the United States, and Japan.

Robert D. Ubriaco, Jr.

Communications in Australia

Australia, the only country in the world that occupies a whole continent, is approximately the same size as the continental United States. It is the sixth-largest nation by area but is sparsely populated. Australia is a nation of extreme differences: more than 80 percent of the population live within 31 miles (50 km.) of the eastern shoreline and the southwestern coast of Western Australia; the rest of the people live in the remote interior area referred to as the Outback, which covers approximately 70 percent of Australia's landmass. Among all countries of the world with a population over 10 million, Australia is the least densely populated one. In 2020, the population density was about 9 people per square mile (3.3 per sq. km.). In the Northern Territory, defined by the Outback, it was 0.4 person per square mile (0.16 per sq. km.).

Communications have always been a major concern for the people of Australia because of the remoteness of the rural population. The federal government has traditionally played a prominent role in the development of the country's wire communications system. Since the completion of the Overland Telegraph Line in 1872, the government has maintained and strengthened the telecommunications infrastructure throughout the continent. By 2020, Australia had landing points for more than twenty submarine cables providing links between the Australian coastal population centers and between Australia and the Asia-Pacific region, the

Residents of Sydney, Australia mail letters at post office boxes on the street. (asiafoto)

United States, Europe, and the Middle East. It had a domestic satellite system and more than twenty international satellite Earth stations. Government funds have been offered on a competitive basis to service carriers to address communication gaps in rural areas.

Postal Services

For millennia, the widely dispersed Aboriginal Australian peoples communicated with each other by using ancestral routes that crisscrossed the continent. These routes led to meetings at particular locations of great cultural and mythical importance. There, the tribes traded goods and renewed and reinforced friendship ties, which involved ceremonial exchanges of objects and creation accounts known as Dreamtime songs. Dreamtime songs and stories were considered especially valuable for their spiritual, cultural, and artistic worth; they were passed from one tribal group to another. Between the meetings, messages from one group to another were transmitted on objects called message sticks: sticks carved or painted with symbols that could be understood by groups speaking the many different indigenous languages. The sticks were small and easy to carry over long distances.

When Europeans began arriving and settling in Australia in the 1780s, the early settlers had to rely on travelers, transporters of goods, or storekeepers to pass on letters or parcels. The colonial postal service was established in 1809. It later transformed into the Postmaster-General's Department, then the Australian Postal Corporation, and finally, the Australia Post— the government business enterprise that provides postal services in Australia today.

Australia was one of the first countries to introduce an airmail service (in 1914) and mechanical mail-processing centers (in 1967). Like all postal operators worldwide, the Australia Post has faced an accelerating decline in its letter-delivery service since the advent of digital communications. In 2015, it started reducing its workforce. In 2019, however, the Australia Post reported profit, largely due to record parcel-delivery numbers. In 2020, Australia had 7,950 postal routes, mostly serviced by motorcycle delivery. Cars and trucks are used in Australia only on the longest mail routes.

Key Dates in Australian Communications

1803: First newspaper, Sydney Gazette and New South Wales Advertiser, appears.

1842: First capital city newspaper, Sydney Morning Herald Daily is established.

1843: First rural newspaper, Maitland Mercury is established.

1872: Overland Telegraph is completed.

1923: Shortwave radio is introduced.

1927: Radio is introduced.

1932: Australian Broadcasting Commission Act establishes national broadcasting presence.

1950: Black and white television is introduced nationally.

1953: Television Act of 1953 establishes broadcast regulations.

1974: Color television is introduced.

1992: Federal government initiates Telecenter program.

1997: National telephone service is deregulated.

2000: One hundred Australian newspapers have Web sites on the Internet.

2001: Digital terrestrial television is introduced.

2007: The National Broadband Network initiative is announced, slated for completion in 2020.

2020: News Corporation announces the closure of 36 community and regional newspapers, and the transition of 76 to digital-only—a move reflective of the age of digital media and the COVID-19 pandemic.

Print and Online Media

In 2020, Australia had two national newspapers (the *Australian* and the *Australian Financial Review*), ten state- and territory-wide daily newspapers, thirty-seven regional dailies, and several hundred smaller local newspapers issued once or several times a week. By that time, the disruption of traditional print media by digital media that began late in the twentieth century was well underway. Year by

A newsstand in Sydney, Australia.

year, the circulation of print newspapers and magazines has been dropping, and a number of newspapers have closed or switched to an online-only format. The *Herald Sun* daily (published in Melbourne) remained the newspaper with the largest daily circulation, and the Sydney *Sunday Telegraph* had the largest weekend circulation. Twenty-eight percent of Australians still read print newspapers in 2020, a higher number than in many other developed countries.

Among the thirty dominant Australian magazines, the ones that had fared better than the others were magazines that served readership groups with specialized interests. In 2020, the largest-circulation magazine in Australia was *Club Marine*, devoted to boating and fishing.

The ownership of print and broadcast media in Australia is highly concentrated. Leading conglomerates include News Corp Australia (a spin-off of Rupert Murdoch's News Limited), Seven West Media, and Fairfax Media-Nine Entertainment. As the number and circulation of the print media decreased, the number of online content creators in Australia has more than doubled since the beginning of the twenty-first century. According to Nielsen's Digital Content Ratings, the most visited news and current-events sites in the late 2010s were *News.com.au*, *Nine.com.au*, *ABC News*, and *Daily Mail Australia*.

Broadcasting

There are two nationwide public broadcasting systems in Australia. Both systems are financially supported by the government but are free from government control of programming. The Special Broadcasting Service of Australia (SBS) broadcasts both radio and television programs in more than seventy languages. The service supplies limited news information to overseas news outlets.

The Australian Broadcasting Corporation (ABC), the larger of the two public broadcasting systems, runs multiple national and local radio networks and television stations as well as ABC Australia, a television channel that broadcasts throughout the Asia-Pacific region and is Australia's main international broadcaster. The ABC also operates another international service, Radio Australia, broadcasting online and via digital audio broadcasting (DAB). The ABC has a vast news-gathering capability and provides a major service for international communications. The federal government finances the operation and guarantees editorial and programming freedom.

Australia has three national commercial television networks that operate in a similar manner as those in the United States. They have affiliates throughout the country that also broadcast regional and local programming. Both the public and private television networks buy programs from local and international suppliers. Cable and satellite systems are widely available.

In the 2010s, besides the ABC and SBS radio networks, Australia had over 260 commercial radio stations and numerous community radio stations operating on the AM and FM bands, on the Internet, and (since 2009) also via DAB.

All television and radio broadcasters in Australia are required to carry a minimum percentage of Australian-made programs. The Australian Communications and Media Authority (ACMA) was formed in 2005 through the merger of the Australian government's previously separate broadcasting and communications authorities. ACMA regulates all forms of broadcasting and telecommunications, including television, radio, the Internet, and converged mass communication over digital platforms.

Telephone Systems

Australia's telephone system, and telecommunication services in general, underwent a major change

A classroom in Australia with the kids using computers. (SolStock)

in direction, from a monopoly to competition. As telephone communications became increasingly merged with other telecommunications, the state-owned service providers also merged into a single telecommunications corporation, which was privatized, in stages, between 1997 and 2011. Now known as Telstra, it is Australia's largest telecommunications company by market share. Telstra builds and operates telecommunications networks involving fixed-line and mobile telephony, Internet access, pay television, and other products and services. Other telecommunications companies operate on a more limited scale. The Australian mobile market is dominated by three major operators—Telstra, Optus, and TPG Telecom Limited.

In 2018, the country had thirty-three fixed-line telephone subscriptions per 100 people, placing Australia among the twenty countries with the highest per-capita number of landline telephone subscriptions. The use of mobile cellular phones was rapidly expanding; in 2018, there were 114 cell phone subscriptions per 100 Australians. In recent years, 90 percent of all mobile device sales in Australia were smartphones. Radiotelephone was still widely used in remote areas with low population densities.

Australia's Schools of the Air

The School of the Air is a unique Australian distance-learning institution for primary- and secondary-school students. In Australia, tens of thousands of school-aged children live in remote, isolated areas, mostly on cattle and sheep farms, called "stations" in Australia. These stations are scattered across huge areas, comparable to the size of large countries. From the early 1900s, the education of these children was handled by correspondence schools located in the capital cities of Australia's states and territories. Lessons and completed homework took days, and sometimes weeks, to reach their destinations.

This changed with the establishment of the first School of the Air at Alice Springs, a remote town in the Northern Territory, in 1951. The school used the radio network maintained by Australia's long-established Royal Flying Doctor Service, which provides health and medical treatment to remotely located families. The same radios were used by the school to make two-way, high-frequency radio broadcasts to the children of those families. By the 1980s, more than twenty Schools of the Air were in operation, covering almost the entire landmass of Australia. Distance learning in Schools of the Air is supplemented by three or four annual gatherings where the children travel to the school to spend a week with their teacher and classmates. Teachers also sometimes travel to visit individual students.

By 2010, traditional radio has been replaced by teleconferencing using satellite telephone set-ups, fax, email, and a special type of interactive online whiteboard. By the 2020s, some of the more remote areas of Australia did not yet have broadband Internet access. However, the federal government has been investing in broadband connectivity, and more and more locations in the Outback—and students of Australia's Schools of the Air—have been gaining access to live interactive video broadcasts over the Internet. The Australian Schools of the Air system has proved to be highly successful, and Australian educators have been consulting other countries on distance learning.

The Internet

When Australia joined the global Internet on June 23, 1989, via a connection made by the University of Melbourne, it was mostly used by computer scientists. Three decades later, in 2019, more than 86 percent of Australian households were connected to the Internet. According to a 2015 report by Oxford University's Reuters Institute, Australians are the most likely people in the world to use a digital device to obtain their news. In 2019, over 70 percent of Australians were active social media users, with Facebook being the top social platform.

Wireless Internet connectivity is well-suited to the remote rural areas of Australia because large distances and low population density make traditional lines costly. Several different long-term projects have been designed to bring the Internet to rural areas. The vast majority of Australia's Internet connectivity is achieved via undersea fiber-optic cables connecting Australia's large population centers, all of which

A group of Australian teens consult their smartphones as they walk down the street. (DisobeyArt)

are located on its coasts, and international cables connecting Australia with Asia, the United States, and Europe. They carry over 90 percent of Australia's total Internet traffic. Australian companies also finance the building of submarine Internet cable connections for its Pacific island neighbors.

The impact on the Australian society of new telecommunication technologies has been enormous. For much of its history, Australia and its cultural and creative institutions were isolated from other English-speaking markets, and the world in general, by distance. But distance is no longer a barrier to accessing or transmitting information, and Australians have promptly and eagerly adopted digital methods of communication and entertainment, growing much closer to—and becoming a bigger part of—the rest of the world.

Earl P. Andresen

Communications in the Pacific Islands

The communications systems found among the scattered islands of the Pacific Ocean are shaped by the unique geographic situation in which they exist. These far-flung islands, many of them tiny coral atolls with only a few hundred people, are separated by thousands of miles of open ocean. Communications among them have generally been difficult, and the low population densities prevented the development of economies of scale that reduced communications costs in larger societies.

Obstacles to Communication

Communications among the islands have been further complicated by the enormous variety of languages within the region. Linguists estimate that as many as 1,200 languages have been spoken in the Pacific islands. However, many of these languages have an extremely restricted range, often confined to a few hundred people on one or two coral islands. Almost none of them had a written form before the arrival of Western missionaries. Many of these smaller languages are considered to be severely or critically endangered, with few or no children learning them and their use restricted to an ever-shrinking elderly population. It is likely that as many as half of these languages will be extinct by 2050.

English, in both British and American forms, has become a trade language in many areas of the Pacific islands, enabling people from various language groups to do business. Japanese serves a similar function in many areas of the Western Pacific, particularly in those islands given to Japan by the League of Nations as mandated territories after World War I. French is used in French Polynesia and New Caledonia, the territories controlled by France. Although many people continue to use their native languages at home, the primary language in Oceania is English.

Two striking exceptions to the general patterns of economic development and communications systems in the Pacific islands are New Zealand and Hawaii. These archipelagos of volcanic islands were both originally settled in prehistoric times by Polynesian seafarers—ancestors of the Maori in New Zealand and Native Hawaiians in Hawaii. However, both of these island groups have been so thoroughly integrated into the Western world (New Zealand as a British colony and later a member of the Commonwealth of Nations, and Hawaii as a US-annexed territory and later a state of the United States) that they have developed Western-style economies and telecommunications systems. They are also larger than all other Pacific island nations and territories except Papua New Guinea, and thus have the resources to sustain such systems.

Postal Systems

Postal service to the various Pacific islands has often been spotty and unreliable. Until the second half of the twentieth century, the only way to get letters and parcels between the islands was by ship, which was often slow and always subject to the vagaries of wind and weather (particularly in the days of sailing ships). By the 1950s, airmail had become available, but was often expensive, particularly for the people of smaller islands. Because their small populations made it uneconomical to set up regular postal service, mail deliveries to those islands were generally irregular and dependent upon calls by unscheduled merchant ships. Mail delivery was further complicated by the fact that many rural Pacific islanders lived on unnamed roads without street numbers. In

2017, the kingdom of Tonga became the first Pacific island nation to adopt a new GPS-based geolocation technology for home mail delivery.

Many of the small island nations' postal systems are best known for their colorful stamps, which are prized by collectors. The sales of mint-quality stamps to collectors abroad have created a stream of foreign exchange revenues for these countries and territories.

New Zealand and Hawaii are notable exceptions to the pattern of slow and unreliable mail delivery typical of the Pacific island region. New Zealand has its own privately owned and operated postal system based upon a British model. Hawaii, as a state of the United States, is served by the United States Postal Service (USPS), which regularly airlifts mail to the mainland without additional charge. A USPS priority-mail shipment from New York to Maui costs the same price as when one is shipping the package from New York to New Jersey.

Telephone

Prior to the development of inexpensive telecommunications satellites and cellular telephones, the telephone service in many parts of the Pacific island region was often prohibitively costly, particularly on smaller islands. Thus radiotelephones were often the only form of two-way telecommunications among islands.

In the twenty-first century, high frequency (HF) and very high frequency (VHF) radiotelephones have remained essential means of communication with remote, outlying islands, although satellite-based mobile services have become the primary form of voice communication across the region. Larger islands have satellite Earth stations and microwave relay stations for radio signal relay.

In 2018, New Zealand had thirty-seven fixed-line telephone subscriptions and 134 cellular mobile phone subscriptions per 100 inhabitants. Fiji, the most advanced economy in the Pacific island region outside of New Zealand and Hawaii, had eight landline telephone subscriptions and 112 cell phone subscriptions per 100 people in 2018. Kiribati, among the poorest and smallest Pacific island countries, had no fixed-line phones. Its inhabitants continued to rely on HF/VHF radiotelephones and had fifty-four cell phones per 100 people.

A father and a son in Polynesia use a walkie talkie radio. (Bobby Coutu)

Submarine Cables

The ocean floor topography of the region has made connectivity via submarine cables a serious challenge for the Pacific islands as submarine fiber-optic networks are expensive to build and maintain. Nevertheless, more and more Pacific island nations and territories are being linked to undersea cable networks connecting Australia and New Zealand with the United States.

The Southern Cross Cable, completed in 2000, has landing points in Australia, New Zealand, and Fiji as well as the US states of Hawaii, Oregon, and California. Other cables interconnecting with Southern Cross include the Tonga Cable System, the Interchange Cable System to Vanuatu, the TUI-Samoa cable linking Samoa to Fiji, and the Gondwana-1 system linking Australia to New Caledonia. The Hantru-1 cable system, which became operational in 2010, has landing points in the Republic of the Marshall Islands, Guam, and the Federated States of Micronesia. Ten more cables were in various stages of development in 2020, including Southern Cross NEXT.

Newspapers

The largest daily newspapers of the Pacific islands are published—in print and online—in English in New Zealand and Hawaii. In 2020, New Zealand had about thirty daily papers, the largest of which was Auckland's *New Zealand Herald.* Many of New Zealand's newspapers are independents that provide thoughtful in-depth coverage of local events, but have few foreign correspondents and must rely heavily on wire services for international news. Hawaii had eleven daily and weekly newspapers. The largest was the *Honolulu Star-Advertiser,* formed in 2010 with the merger of the *Honolulu Advertiser* and the *Honolulu Star-Bulletin*. Although most of Hawaii's newspapers are in English, there are smaller print and online publications for the Native Hawaiian, Chinese, Filipino, Japanese, and Korean communities.

Other Pacific island nations and territories had their own newspapers, but few were dailies and almost none were known beyond their own shores. Most of them relied entirely upon wire services for off-island news. Fiji's largest newspaper, the *Fiji Times*, had a daily circulation of 2,000 copies in 2019.

The Rongorongo Script of Easter Island

Easter Island is one of most remote Polynesian Islands. Its aboriginal culture, long shrouded in mystery, has become the subject of much speculation. Along with the monumental statues erected along its coasts, the people of Easter Island produced a large number of puzzling wooden boards inscribed with rows of mysterious characters. By the time Western scholars took serious interest in Easter Island's culture, no one survived who could read this script, known as Rongorongo.

Because of the characters' notable similarities to those on seals found in the ruins of the Indus Valley civilization in India, a connection between the two cultures has been posited. Some writers have even suggested that these similarities are evidence of undiscovered ancient civilizations that were destroyed in some catastrophe before the accepted beginning of written history.

In the late 1990's a Russian linguist, Sergei V. Rjabchikov, claimed to have deciphered the Rongorongo. In his analysis, it is a hieroglyphic script containing a combination of ideographic characters—those that represent ideas, rather like Chinese characters, and phonetic characters—those that represent sounds, such as Western alphabets. Rjabchikov suggested that the Rongorongo script was used to record sacred legends and astronomical observations.

However, these theories are often discounted by serious scholars. Some linguists discount the idea that the Rongorongo is a native writing system for Rapa Nui, the language of Easter Island. Instead, they suggest that it is of relatively recent origin, probably imitative of European scripts, and was likely decorative, or at most a memory aid. Others suggest that Rongorongo is a form of proto-writing, now impossible to decipher.

Radio

Radio has long been an important medium of mass communication for the peoples of the Pacific islands. Because a radio requires little power and radio signals in certain frequencies can carry over long distances via reflection from the ionosphere, radio has generally been the most cost-efficient means to broadcast to those scattered islands. Radio listening also does not require any level of literacy, an important factor given the high rates of illiteracy on some islands. The smaller island nations have just one, state-run radio station. The larger ones have multiple stations, although their news programs generally consist of some local material combined with relays of news from Australia and New Zealand. Radio Australia is available throughout the Pacific island region via satellite feed; Radio New Zealand is available in the South Pacific via shortwave service.

New Zealand has a large number of radio stations, both public and commercial, including a Maori-owned, Maori-language station and a national government-funded station for the Pacific islander communities. Hawaii, as a territory and later a state of the United States, has a long history of radio broadcasting. Its first radio stations, KDYX and KGU of Honolulu, began operation in 1922. In 2020, it had more than 100 radio stations, most of which were headquartered in Honolulu.

Television

In New Zealand, multiple television networks are operated by the state-owned Television New Zealand. It also does some television broadcasting in the languages of the other Pacific islands for rebroadcast by local stations, although many of the smaller islands do not have their own television stations. Multi-channel pay television is available on these islands in hotels and resorts.

Three of Hawaii's television stations merged in 2009 under the banner *Hawaii News Now*. Since then, the *Hawaii News Now* group has controlled almost half of the state's television market. All Hawaii's television broadcast stations are located in Honolulu. Programming is transmitted throughout the islands of the Hawaiian archipelago via satellites and through Oceanic Time Warner Cable.

A TV satellite dish attached to a person's home in Port Moresby, Papua New Guinea. (seraficus)

Broadband Internet

Except for New Zealand and Hawaii, most of the Pacific islands have generally lagged in Internet usage. The populations of many of the islands are generally too poor to afford computers. In recent years, Internet-enabled cellular phones have allowed them to connect to the digital world for the first time. However, traditional dial-up Internet access in the region was generally slow and unreliable. The launch of Kacific-1 satellite has vastly improved Internet connectivity and the quality of Internet-related services.

The first satellite of the Kacific Broadband Satellites Group, providing high-speed Internet access for South East Asia and Pacific islands, was placed in orbit in December 2019. It was designed and manufactured by Boeing and launched into orbit from Cape Canaveral, Florida, on a SpaceX Falcon 9 launch vehicle. Kacific-1 uses spot beams to deliver high-speed broadband Internet service to more than twenty-five countries and territories of Southeast Asia and the Pacific, from American Samoa to Vanuatu.

Leigh Husband Kimmel

Engineering Projects in Australia

Early Australian engineers had been handicapped by the country's small, dispersed population, its great size, and the largely inhospitable physical environment. Prior to World War I, Australia was primarily an exporter of agricultural products and minerals, and its capital resources were limited. As a result, engineering activity was concentrated in mining, transportation, and water supply. The single largest engineering feat in nineteenth-century Australia was the construction of the Australian Overland Telegraph Line.

After World War I, Australian engineering structures were put on the map with the construction of the Sydney Harbour Bridge (1923-1932). The Snowy Mountains scheme was the next major engineering feat in Australia, followed by the Sydney Opera House. Although the engineering aspects of the Sydney Opera House have been overshadowed by its architectural and cultural fame, engineers agree that because of its design and unique construction techniques, the building stands as a model for engineering, and it remains, along with the Snowy Mountains scheme, one the most significant engineering monuments in Australia.

Although the Murray River Barrage and the Adelaide Aqueduct deserve mention, they have been subsumed in importance. The Dingo Fence, built in the 1880s, although perhaps the longest fence in the world, is merely a wire-link fence of no particular engineering significance. Its only claim to fame is that it is one of the world's longest such structures and that it was partially successful as a pest-exclusion barrier that helped reduce losses of sheep to dingoes (although wild dingoes can still be found in parts of Australia's southern states, thanks to holes in the fence through which many dingo offspring have passed).

Salt-interception works along the Murray River, which started in the 1980s, have offered an ingenious solution to the problem of soil salinization. In the twenty-first century, the construction of the Adelaide-Darwin railway was the second-largest civil engineering project in Australia since the Snowy Mountains scheme, and the Gorgon Gas Project has become the single-largest natural-resource development project ever undertaken in Australia.

The Overland Telegraph Line

Completed in 1872, the Australian Overland Telegraph Line runs coast to coast for 2,000 miles (3,200 km.) across the middle of the continent between Darwin, the capital of the Northern Territory, and Port Augusta, South Australia. Before the construction of the line was started, its projected route through the largely uninhabited interior of the continent had to be surveyed for waterholes and areas with sufficient supplies of timber for the telegraph poles. The line featured 36,000 telegraph poles and eleven repeater stations (necessary to boost the strength of the electrical signal); the latter were located about 125 miles (over 200 km.) apart.

Within months of completion, the Overland Telegraph Line was linked to the Java-to-Darwin submarine telegraph cable, reducing Australia's communication time with Europe from months to hours. The line also opened up the huge central part of the continent. Prospectors and cattle raisers used the repeater stations as starting points from which to explore and stake claims on the land. Established as a repeater station, the town of Alice Springs, which is situated roughly in Australia's geographic center, became the administrative hub for central Australia and, later, an important tourist hub.

The Snowy Mountains Scheme

The Snowy Mountains scheme is Australia's foremost engineering project of all time. This irrigation and hydroelectricity complex supplies 555 billion gallons (2.1 trillion liters) of water to 1.31 million acres (530,000 hectares) of farmland in the Murrumbidgee Irrigation Area on the Murrumbidgee River in New South Wales and in the Goulburn-Murray region in New South Wales and Victoria. As a result, additional primary agricultural production is estimated to generate about 60 million Australian dollars (A$) a year. While falling 2,640 feet (800 meters), the water generates 3.74 million kilowatts of electricity.

One system of tunnels and reservoirs diverts water from the Eucumbene River, a tributary of the east-flowing Snowy River, into the Tumut River, a tributary of the west-flowing Murrumbidgee River. The Upper Murrumbidgee is diverted into the Eucumbene and the Tooma River, a tributary of the Tumut. Another system diverts Snowy River water directly into the Murray River. The two systems connect through reversible flow tunnels, so that excess water can be diverted to storage.

Sixteen major dams, many smaller dams, seven large and three small power stations, and 140 miles (225 km.) of tunnels, pipelines, and aqueducts make up the project. When built, the Eucumbene main storage dam, with a capacity of 3,860,000 acre-feet (4.76 cubic kilometers), was the world's largest earthen-rock fill dam.

More than 100 miles (160 km.) of highway and 625 miles (1,000 km.) of secondary roads built for the project opened previously inaccessible wilderness for tourism and recreation. The Kosciuszko National Park, which includes over 2,660 square miles (690,000 hectares) of nine distinct wilderness zones surrounding Australia's highest mountain—Mount Kosciuszko (7,313 feet/2,229 meters)—is a heritage of the project.

The Snowy Mountains Hydroelectric Authority (later renamed Snowy Hydro), owned by the Australian federal government, began constructing the project in 1949. More than 100,000 people from thirty countries were involved in the construction. A major upgrade of Tumut 3, one of the underground power stations on the Tumut River, was undertaken between 2009 and 2011, adding additional capacity. The federal government, through Snowy Hydro, financed the construction cost of the Snowy Mountains scheme, which ran to approximately A$820 million. That would have been equivalent to almost A$6.5 billion, or $4.6 billion, in 2020.

The project's economics and impacts have not been without criticism. Providing water for intensive agriculture has been said to yield less return than an equivalent investment in grain-growing.

The Snowy Mountains Hydro-Electric Scheme

An innovative hydrological engineering project that began in 1949 was the Snowy Mountains Hydro-Electric Scheme. The Snowy River, which was fed by snowmelt from the Australian Alps, was diverted westward into the Murray River. The project involved construction of twelve long tunnels through the mountains, as well as sixteen dams and seven power-generating plants. Immigrants from more than thirty other countries, many of whom had lost everything during World War II, came to Australia to construct these facilities. The project was therefore not only a tremendous engineering challenge for Australia, but also a great social experiment for a country whose existing population consisted mainly of people of British origin. The workers' living conditions were spartan, and the environment was harsh, but the immigrants saw the project as a wonderful financial opportunity. The work, especially the tunneling, was sometimes dangerous, and some 120 people lost their lives during the construction. The Snowy Mountains Hydro-Electric Scheme has been honored twice by the American Society of Civil Engineers as one of the great engineering achievements of the twentieth century, and was added to the Australian National Heritage List in 2016. The following year, prime minister Malcolm Turnbull announced an expansion to the scheme, known as Snowy 2.0, which involves linking two existing dams (Tantangara and Talbingo) through new tunneling and the construction of a new underground power station. Snowy 2.0 is intended to provide an additional 2,000 megawatts of generating capacity and around 350,000 megawatt hours of storage to Australia's electricity system, with first power expected in 2025.

The Murray 1 Hydroelectric Power Station at the Snowy Mountains Scheme in New South Wales, Australia. (Martin Kraft)

The cost of the massive storage requirements needed to offset Australia's highly variable rainfall has been deemed by the critics as exorbitant. Environmentalists charge that the project has destroyed the Snowy River and violated the wilderness.

The expansion of the Snowy Mountains scheme, which started in 2020 and was dubbed Snowy 2.0, includes carving new tunnels through 17 miles (27 km.) of rock connecting two reservoirs and the building of a new underground power station capable of generating 10 percent of Australia's electricity. Snowy Hydro officials have disputed claims from environmentalists that up to 38.6 square miles (100 sq. km.) of the Kosciuszko National Park would be damaged by the dumping of excavated rock.

Sydney Opera House

Comprising a series of spherical, shell-like components, the Sydney Opera House is an iconic building whose original design concept was developed by the Danish architect Jørn Utzon. The building's unique construction techniques were devised by the Danish-English engineer Ove Arup, who helped make Utzon's vision a reality. Assembled from modular precast concrete elements, it is famous both as an architectural object of great beauty and for groundbreaking use of precast concrete.

The Sydney Opera House is an entertainment center housing a concert hall seating 2,700, an opera theater seating 1,500, a drama theater seating 540, a cinema-music room, rehearsal and recording halls, exhibition areas, and restaurants. As early as 1947, Eugene Goosens, then conductor of the Sydney Symphony Orchestra, called for construction of a musical center for symphony, opera, and chamber music. In 1954, J. J. Cahill, premier in the Labor government of New South Wales, initiated an international design competition for the center. In 1957, Jørn Utzon won the opera house competition, although his entry included only rough drawings of the proposed building. Utzon was appointed architect, and the Opera House Executive Committee selected the British firm Ove Arup and Partners as design engineers. Cahill, fearing a politically inspired cancellation of the project, insisted that construction begin before the 1959 elections.

Work began in March 1959 on the massive podium (the foundations and platform upon which the building was to rest). Because the concepts associated with the final form of the building and how it was to be built were still evolving, work was repeatedly interrupted and much of completed design work had to be discarded. In some cases, portions of the building were demolished and rebuilt. When the podium was completed in 1963, estimated costs had increased to more than A$24,000,000.

Utzon's original, free-hand sketches of elliptical paraboloid shells were not working plans for the walls and roof of the building. His first plan to cast the shells in place proved impractical. Arup designed shells composed of double skins supported by an internal steel frame, but Utzon rejected this design. Utzon and Arup then designed self-supporting shells composed of radiating ribs, each built of precast concrete sections—an utterly new concept. The original ellipsoidal form was discarded for spherical surfaces of a uniform radius, allowing mass-production of the sections.

Utzon's spectacular plans for the building's interior were never realized because of mounting conflicts among Arup, Utzon, the Executive Committee, and the New South Wales government. In 1965, after a Liberal Party election victory, a new Public Works Director, Davis Hughs, assumed personal control over the project. Questioning innovative designs and citing excessive costs, he vetoed Utzon's plans for the interior. Utzon was forced out in 1966, and succeeding architects and a new engineering firm redesigned the interior. (Utzon's removal triggered protests and marches through the streets of Sydney. In the late 1990s, the Sydney Opera House Trust initiated reconciliation with Utzon, but the architect never returned to Australia to see his masterpiece in its completed form.)

Queen Elizabeth II formally opened the new Sydney Opera House in October 1973. Largely financed by state lotteries, it was completed at a cost of A$102,000,000. The original cost estimate to build it had been A$7,000,000. Construction was expected to take four years; it took fourteen years.

The building is a world-famous engineering and architectural monument, but also has been criti-

The Murray-Darling Basin in Victoria, Australia. (Ingrid_Hendriksen)

cized. Some architects argue that the shells were designed wholly for aesthetics and thus cause overwhelming problems in seating, acoustics, and staging. (Original acoustics designs, commissioned by Utzon but never implemented, were later modeled and found to be very good.) Persistent structural shortcomings arose from the innovative techniques and materials. For example, the tiles on the outside of the building require continuous regrouting. The issue of poor acoustics in the concert hall was finally addressed in 2020 with the start of a two-year renovation, providing for a new acoustic ceiling, new automated drapes, and a 3D surround-sound system.

The building is 600 feet (183 meters) long and 394 feet (120 meters) wide at its widest point. The total roof surface of the structure, composed of "shells," is approximately 4 acres (1.62 hectares). Each shell is a section of a sphere with a radius of 246 feet (75.2 meters). While some believe that the shell form was inspired by boat sails, clouds, or seashells, the architect himself said that they were inspired by a peeled orange. Almost 11 million people visit the Sydney Opera House every year. It was added to UNESCO's World Heritage List in 2007.

Sydney Harbour Bridge

The Sydney Harbour Bridge spans 1,650 feet (503 meters) of open water from Dawes Point to Milsons Point. Five steel approach spans increase the bridge's total length to 3,770 feet (1,149 meters). The top of the main span is 440 feet (134 meters) above mean sea level, and the crest of the deck stands 160 feet (49 meters) above the water. The deck originally accommodated a central six-lane roadway, two tram lines, two rail lines, a footway on the eastern side, and a bikeway on the west. In 1959 the tram lines were removed, widening the roadway to its current eight lanes.

An English firm began construction in 1923. The two wings of the arch, based on two steel pins in each abutment, were cantilevered outward and built

The Sydney Harbour Bridge at sunset. (zorazhuang)

from large creeper cranes advancing on top of the growing arch. After the arch was closed in 1930, the cranes retreated, lifting deck panels from pontoons and hanging them from the arch. A total of 58,000 tons of steel (almost all imported from Middlesbrough, England, where the engineering firm was based) were used in construction. The completed bridge was opened on March 19, 1932. The bridge cost A$20,230,000, raised by loans and contributions from local governments. Although the loans were paid off in 1988, tolls remain in effect to offset maintenance costs.

When built, the bridge was the longest single-span bridge in the world. Until 2012, it was also the world's widest long-span bridge and, in 2020, it remained the tallest steel arch bridge in the world. The monumental stone pylons at the ends of the bridge have been criticized as entirely decorative and nonfunctional, and the bridge was called too costly and overdesigned. Nevertheless, the bridge, which perfectly fits its beautiful harbor location, has come to symbolize Australia and has even been commemorated on postage stamps. Although some people thought the bridge was not justified by the expected traffic, its designed capacity of 6,000 vehicles per hour was exceeded before the end of the twentieth century. In the 2010s, more than 20,000 cars crossed the Sydney Harbour Bridge every day. A combination land-and-immersed tunnel near the bridge was completed in 1992 to alleviate the congestion on the bridge.

Groundwater Interception Schemes

Since 1988, the New South Wales, Victoria, and South Australia governments, together with the Commonwealth government, have funded the construction and operation of salt-interception works along the Murray River in the Murray-Darling basin, one of the most significant agricultural areas in Australia. Clearance of native plants and drainage from long-term, wide-scale irrigation have added water to salty groundwater aquifers, leading to the elevation of groundwater levels, which deposited salt into the soil and the river. Another factor was the combination of low rainfall, which is common in parts of the basin, and high evaporation that leaves salt behind in the top layers of the soil. Soil salinity significantly limits the productivity of crop plants, and an excess amount of salt in the water can render it unusable for human or animal consumption. The annual cost to Australia in lost production alone is close to A$100 million; the environmental cost can only be conjectured.

Salt-interception schemes involve a system of bores and pumps that take out the highly saline groundwater from underground and pipe it for several miles into salt management basins. The water slowly evaporates, concentrating the salt or very gradually leaking it back into the groundwater systems; this way, it will not return to the river for thousands of years. In the 2010s, there were eighteen of these schemes in the Murray-Darling basin; they moved approximately half a million tons of salt away from the Murray each year. After decades of operation, irrigators in the area have started seeing the benefits of the lowered level of saline groundwater on their lands, and healthy flood-plain vegetation has begun reemerging.

Adelaide-Darwin Railway

The construction of a south-north transcontinental railway between South Australia and the country's sparsely populated northern region was envisaged even before the creation of the Northern Territory in 1911. The vision was not realized, however, until 2003. One of the largest civil engineering projects in Australia's history, the Adelaide-Darwin railway runs for 1,849 miles (2,975 km.) connecting the capital of South Australia, Adelaide, which is the state's most populous city, with Darwin, the capital of the Northern Territory. The Adelaide-Darwin line has both freight and passenger services.

Aboriginal people hold title to much of the land that the railway crosses, so the construction was preceded by extensive negotiations. Many additional difficulties were associated with the building of a railway across some of the most inhospitable terrain in the world, including the Tanami Desert, where summer temperatures reach 122° Fahrenheit (50° Celsius), and a three-month monsoon season. Teams worked from both final destinations toward each other, thereby minimizing the distances work trains had to travel. Continuous welded rail on concrete sleepers was used throughout, and up to 1.24 miles (2 km.) of track were laid per day.

With the completion of the railway, the population of the Northern Territory, including its many indigenous communities, gained important access to a major transportation artery. The economy of the Northern Territory heavily depends on visitors, and a major railway signified a new era of tourism development in the region. Darwin is also a large deepwater port serving as a gateway to Asian markets, and the line has facilitated bulk commodity exports originating from iron-ore and copper mines of central Australia. The railway network is integrated with Port Darwin's new East Arm wharf, which includes a railway embankment and an intermodal container terminal.

Gorgon Gas Project

Australia's largest natural-resource undertaking by capital expenditure is the Gorgon LNG (liquefied natural gas) project, a joint venture of Australian subsidiaries of the global energy companies Chevron, Shell, and Exxon Mobil. The project was started in 2009 and completed in 2017. It involved the development of Western Australia's Greater Gorgon gas fields and the building of an LNG plant on Barrow Island, 37 miles (60 km.) off the coast of Western Australia.

Gas is delivered to the island via an advanced subsea gas-gathering system. It consists of tree-like pipeline structures fixed to wellheads at each field that control the production of gas through groups of valves, spools, and monitors; gather the gas undersea; and deliver it to the LNG plant. The "trees" were designed by General Electric and manufactured in Scotland. The Barrow Island plant has three processing units designed to produce 15.6 million metric tons of LNG per year. The estimated production lifespan of the project is forty to fifty years.

M. Casey Diana

Engineering Projects in the Pacific Islands

The Pacific island region, or Oceania, is the site of one of the world's greatest feats of ancient monumental art and engineering—the huge stone sculptures of Easter Island (Rapa Nui). Carved from volcanic tuff, the statues, called moai, weigh an average of 14 tons each, with the largest weighing more than 80 tons. They were created by descendants of Polynesian seafarers who had settled the remote Polynesian island around the close of the first millennium C.E. The statues consist of massive heads, each on a sketchily carved top half of a body; probably representing ancestors, they are designed to stand facing inland on ceremonial platforms.

The moai must have been carved over a long period, for there are almost 900 of them on a 64-square-mile (166-sq.-km.) island. Not all of them were installed; some have remained in the quarry or in various places throughout the island, probably on their way to intended locations. It was those moai that had been moved and erected that have caused an intense and long-standing scholarly debate over possible means by which they had been transported. One theory involves a sledge with cross pieces, pulled with ropes by dozens of men. Another of the several theories is that the statues were "walked" to their destinations by attaching

A group of statues on Easter Island. (daboost)

ropes to the statue and rocking it, tugging it forward as it rocked. In any case, the Easter Islanders accomplished something that is still admired and talked about today.

Mining and Natural Gas Extraction in Papua New Guinea

Substantial deposits of valuable minerals are associated with the volcanic activity in the Pacific Ring of Fire, the active tectonic boundary between the Pacific and Indo-Australian tectonic plates. In Oceania, New Caledonia and Papua New Guinea have extensive mineral deposits; Papua New Guinea also has a rapidly developing natural gas industry, born out of several recent important discoveries.

Papua New Guinea's mining industry relies heavily on foreign technology. Australian, North American, and Asian mining companies have used core-sample drilling, satellite-based mapping, and other sophisticated prospecting technologies to identify new deposits of gold, aluminum, and other metals. In the late 1990s, significant deposits of gold and copper were discovered near the Frieda River in remote Sanduan and East Sepik provinces. The government of Papua New Guinea and a succession of foreign companies have been planning the construction of a A$1.5-billion mining, ore processing, and shipping complex at the Frieda River site, along with a hydroelectric scheme. The project has been billed as hugely beneficial for Papua New Guinea, geared to stimulate the growth of other sectors of the economy, including the chemicals, communications, and transportation industries. However, it has been proceeding very slowly and has remained controversial due to vehement opposition by environmental groups.

There are four liquefied natural gas (LNG) projects in Papua New Guinea. The first one, PNG LNG (Papua New Guinea Liquified Natural Gas), has been fully operational since 2014, exporting liquefied natural gas to Asian markets. PNG LNG is a consortium led by ExxonMobil, and the project has consistently operated above its intended capacity. Gas is sourced from seven fields, conditioned at the source, and then transported by gas pipeline to an LNG plant located 12.5 miles (20 km.) northwest of the country's capital, Port Moresby. The gas is liquefied at the plant prior to loading onto tankers to be shipped to Asian gas customers.

The success of the PNG LNG venture has encouraged other companies to look at similar LNG projects. An agreement was made between the Papua New Guinea government and a consortium of foreign companies to develop three other LNG projects, one of which was at an advanced planning stage in 2020. The benefits of LNG development for the country is a controversial issue. Government participation in the projects and environmental issues have emerged as a dominant theme in the politics of modern-day Papua New Guinea.

Papua New Guinea has almost 5 percent of the world's known reserves of nickel. A huge mining and refinery project known as Ramu NiCo began operations in 2012. New roads, bridges, and quarries were built to access nickel deposits in Madang province, in the southeastern part of the country. An 83-mile (133-km.)-long slurry pipeline carries unprocessed ore to a nickel refinery. The Chinese-owned mining and refining operation was temporarily shut down in 2019 after the plant spilled tens of thousands of liters of highly acidic slurry into the Basamuk Bay. It was

WASTE DISPOSAL IN THE PACIFIC ISLANDS

Long-term waste disposal options for the Pacific Islands are limited, due to a lack of planning, funding, and land, among other hindrances. To protect their fragile environments, island governments attempt to turn wastes into reusable resources, such as fertilizers for agriculture. Island governments are also considering limiting imports of nonbiodegradable and hazardous substances. The Moana Taka Partnership was launched in 2018 between the Secretariat of the Pacific Regional Environment Programme (SPREP) and China Navigation Company (CNCo). Under the agreement, CNCo vessels transport containers of recyclable waste from eligible Pacific island ports for treatment and recycling in suitable ports in the Asia-Pacific region—free of charge. As of 2020, approximately 686 tonnes of waste have been shipped under the agreement.

A row of pipes in Papua New Guinea which will be set up to transport natural gas. (dane-mo)

only one in a succession of incidents in the short history of Ramu NiCo. Local landowners have been unsuccessfully trying to stop the company from dumping waste into the sea by a process known as deep-sea tailings disposal. In 2020, they were joined by the provincial government in a lawsuit against the company, seeking the highest amount of restitution for environmental damages in the history of Papua New Guinea.

Telecommunications Engineering

The Pacific islands have faced a singular set of difficulties in the operation of effective communications systems. Huge distances separate the inhabited islands, and extreme weather conditions interfere with conventional communications. The first trans-Pacific submarine communications cables providing telegraph service were completed in 1902 and 1903, linking the US mainland to Hawaii and Guam to the Philippines. The SEACOM (South East Asia Commonwealth) submarine telephone cable system opened for traffic in 1967. It included Australia, Papua New Guinea, Guam, Hong Kong, Malaysia, and Singapore. Early undersea cables were coaxial; they suffered from attacks by sea life and had limited bandwidths (data transfer rates).

In the 1980s, fiber-optic cables were developed. Optical fibers permit the transmission of infrared or near-infrared light over long distances with much lower signal loss compared to electrical cables. This allows for higher bandwidth with fewer repeaters, or signal regeneration devices. By 2020, numerous fiber-optic submarine cables were crisscrossing the Pacific Ocean floor and linking the Pacific islands with Australia, Asia, and the United States, as well as some islands with each other.

These cables include the SEA-US (opened in 2017, with landing points in Indonesia, the Philippines, Guam, Hawaii, and the continental United States), the Honotua Cable System (laid in 2009-2010, it connected several islands of French Polynesia via Tahiti to Hawaii, and was later extended to the Cook Islands, Niue, and Samoa), the Belau Submarine Cable (opened in 2017 between

Palau and Guam), HANTRU-1 (since 2010, it connects Guam, the Federated States of Micronesia, and the Marshall Islands), and the Southern Cross Cable Network (commissioned in 2000, it has nine landing stations in Australia, Fiji, New Zealand, Hawaii, and the continental United States).

In 2020, more than a dozen new submarine cables and cable extensions were in various stages of construction, including the Samoa-American Samoa Cable and the Southern Cross NEXT, with branching units linking Fiji, Tokelau, and Kiribati. The benefits of submarine cables to the Pacific subregion are significant because they not only bring high bandwidth capacity but also considerably lower telecommunications costs.

In the late 1970s, several Pacific island countries such as the Cook Islands, Fiji, the Solomon Islands, Tonga, and Vanuatu engaged with a British telecommunications company for gateway access to international satellites. At the time, the satellite connection was the only practical way for them to telecommunicate with other parts of the world. However, satellite communications were very expensive, had high maintenance costs, and were not always reliable, especially during extreme weather events such as heavy rain. With submarine cables in place, satellite became the secondary connection in many Pacific island countries.

This changed in the 2010s with the introduction of O3b satellites designed specifically for telecommunications and data backhaul from remote locations. Each O3b satellite is equipped with twelve steerable Ka-band antennas (Ka is a portion of the microwave section of the electromagnetic spectrum). The footprint of each antenna beam measures 435 miles (700 km.) in diameter. These satellites provide high-bandwidth capacity and lower costs. The first O3b satellites were constructed by the French-Italian aerospace manufacturer Thales Alenia Space, and the next generation, by Boeing Satellite Systems.

In 2019, Boeing built and Elon Musk's aerospace company SpaceX launched the first Ka-band HTS (high-throughput satellite), Kacific1. HTS has many times the throughput of a traditional satellite for the same amount of allocated frequency. Since then, Kacific1 HTS has been used by American Samoa, the Cook Islands, Kiribati, Nauru, Norfolk Island, Palau, Papua New Guinea, Samoa, the Solomon Islands, and the Federated States of Micronesia for their international satellite gateway access.

The Nippon Causeway in Kiribati. To combat climate change, many experts have proposed replacing causeways like these in the Pacific Ocean with elevated bridge roads. (Flexmaen)

Sea-Level Rise Solutions

Climate change is causing sea levels to rise. Among the places expected to be most hard-hit by rising sea-levels in the twenty-first century are the islands of the tropical Pacific Ocean, ranging from sparsely developed low-lying archipelagos in Micronesia to heavily populated coastal areas on the Hawaiian Islands. But even long before the Pacific Ocean subsumes low-lying islands and coastal areas, waves will begin washing over them frequently enough to ruin groundwater supplies and damage crops and fragile infrastructure.

On an island-by-island level, projects have been identified as possible solutions for mitigating the impact of rising sea levels. Approaches include the building of resilient towns and cities as well as improving and constructing related infrastructure. Some of the islands can be protected, at least for a time, by building seawalls, dikes, embankments, and surge barriers. But the projected costs of these infrastructure projects are extremely high, and they will continue to climb unless climate change and rising sea levels are minimized as much as possible through humanity's collective effort.

Laura M. Calkins

Gazetteer

Places whose names are printed in SMALL CAPS *are subjects of their own entries in this gazetteer.*

Adelaide. Capital and largest city of the state of South Australia. Located on a plain 9 miles (14 km.) inland from the Gulf of St. Vincent and at the base of the Mount Lofty Ranges that rise to the east. Has a Mediterranean climate and a 2020 population of 1.3 million. The Torrens River flows through the city. Known for its beautiful parklands, Adelaide has evolved from an early agricultural center (fruit, wheat, wool, and Barossa Valley wines) into an industrial power producing automobile parts, machinery, textiles, and chemicals. A hub of rail, sea, air, and road transportation, it is also connected by pipeline with the Gidgealpa natural gas fields in Cooper Basin, northeastern South Australia. Port Adelaide, its main harbor, is located 7 miles (11 km.) northwest. Established in 1836 and named for Queen Adelaide, wife of Britain's King William IV.

Adélie Coast. See Terre Adélie.

Admiralty Islands. Island group in the western Pacific Ocean, part of Papua New Guinea. The eighteen islands are part of the Bismarck Archipelago. The two largest are Rambutyo Island and Manus Island. The League of Nations mandated the islands to Australia in 1920. Japan occupied the islands in 1942, then the Allies in 1944. They were administered by Australia, then became a United Nations Trust Territory before becoming part of Papua New Guinea in 1975.

Akaka Falls. Waterfall along the Hamakua coast of Hawaii Island. Tallest waterfall in the Hawaiian Islands, at 442 feet (135 meters). The stream that forms the waterfall cuts through a layer of ash to the lava below.

Alice Springs. Outback town located in the Red Centre of Australia's Northern Territory. The Red Centre is an area approximately 100,000 square miles (260,000 sq. km.) in size of desert and rocky outcrops, flanked by the MacDonnell Ranges. Had a 2018 population of fewer than 30,000 and is 954 miles (1,535 km.) south of Darwin and 1,028 miles (1,654 km.) north of Adelaide. It is a major shipping point for cattle and minerals and an important tourist destination as the gateway to Australia's Outback, to Ayers Rock/Uluru and the Olgas/Kata Tjuta, and to Aboriginal lands. It is the regional headquarters for the Royal Flying Doctor Service and the School of the Air. Originally established along the ephemeral Henley River in 1871 as a station for the Overland Telegraph Line.

Alps, Australian. Section of the Great Dividing Range, located in the south easternmost corner of Australia in southeastern New South

Wales and eastern Victoria. Forms as the divide between the Murray River system flowing west and the Snowy River and other streams that flow east to the Pacific Ocean. Massive and covered with snow nearly half the year, the Alps are the tallest mountains on the continent and include Australia's tallest peak, Mount Kosciuszko, at 7,310 feet (2,229 meters).

Ambrym. Active volcano on Ambrym Island on northern Vanuatu in the southwestern Pacific Ocean. Reaches a height of 4,377 feet (1,334 meters). It last erupted in 2018.

Lava lake in Marum crater, Ambrym. (Geophile71)

American Samoa. Unincorporated territory of the United States, comprising seven islands in the southern Pacific Ocean: Tutuila, the Manua group (Tau, Olosega, Ofu), Aunu'u, Rose Island, and Swains Island. Total area of 86 square miles (224 sq. km.); population was 49,437 in 2020. Tutuila, the Manua group, and Aunu'u are of volcanic origin; Rose and Swains islands are coral atolls. Pago Pago, a harbor town on Tutuila, is the capital and largest city.

Amery Ice Shelf. Ice shelf on the Indian Ocean in southeast Antarctica. Area is about 23,200 square miles (60,000 sq. km.); fed by the Lambert Glacier, one of Antarctica's largest. Gigantic iceberg broke off in 2019. Scientists believe that this was due to the ice shelf's normal calving cycle.

Amundsen-Scott South Pole Station. US research station located at 90 degrees south latitude, directly on the South Pole. Buildings are covered by a huge geodesic dome to protect them from the punishing winds that howl constantly across the pole. Named for Antarctic explorers Roald Amundsen and Robert Falcon Scott.

Andesite Line. Geological line forming a boundary between the high (continental) and the low (oceanic) islands of the Pacific Ocean. The line follows the island arcs from the Aleutians southward to the Yap and Patou arcs, eastward through the Bismarcks and Solomons, and southward through Samoa, Tonga, Chatham, and Macquaries to Antarctica.

Antarctic Circle. Imaginary line at 66°32′ south latitude, south of which lies all the Antarctic landmass but a small portion of the Antarctic Peninsula.

Antarctic Convergence. Point just north of the Antarctic Circle where cold polar waters meet the warmer waters of the Atlantic, Pacific, and Indian oceans. The cold water sinks, forcing the warmer currents to rise toward the surface. Also called the Antarctic Polar Front.

Antarctic Peninsula. Peninsula that juts north into the South Atlantic Ocean. With water on three sides, it has the most moderate climate on Antarctica, with summer temperatures sometimes reaching above 50 degrees Fahrenheit (10 degrees Celsius). Warm enough to have occasional light rainfall and to support the two flowering plants that grow south of 60 degrees south latitude, the pearlwort and Antarctic hairgrass, as well as several mosses and lichens.

Antarctic Polar Front. See Antarctic Convergence.

Antarctica. World's fifth-largest continent. Has 10 percent of Earth's landmass, covering 5.48 million square miles (14.2 million sq. km.). Twice the size of Europe or Australia, 1.5 times the size of the United States. Earth's highest continent, with an average altitude of about 6,000 feet (1,830 meters), excluding its covering of snow and ice. Its ice and snow cover, with an average thickness of about 7,086 feet (2,160 meters), increases its average altitude to 14,633 feet (4,460 meters), just a little less than Mount Rainier in Washington. The most isolated continent, located 600 miles (960 km.) from the southern tip of South America, 1,550 miles (2,500 km.) from Australia, and 2,500 miles (4,023 km.) from the southernmost point in Africa. Round in shape except for the Antarctic Peninsula, which juts northward into the South Atlantic Ocean, and the indentations of the Ross and Weddell seas. Has 11,165 miles (17,968 km.) of coastline.

Aoraki/Mount Cook. New Zealand's highest mountain peak, at 12,218 feet (3,724 meters). Located in the Southern Alps of South Island. Perpetual snowcap; popular skiing site. Known to the Maori as Aorangi, meaning "cloud piercer."

Arafura Sea. Shallow sea in the western Pacific Ocean. Covers 250,000 square miles (650,000 sq. km.) at depths of 165 to 260 feet (50 to 80 meters). Located between the north coast of Australia and the Gulf of Carpentaria and the south coast of New Guinea. Bordered by the Timor Sea to the west and connected to the Coral Sea to the east by the Torres Strait. Known as a navigational hazard because of its many uncharted shoals.

Arnhem Land. Historical region located at Australia's Top End in the eastern half of a large peninsula forming the northernmost portion of the Northern Territory. Covers 37,000 square miles (95,900 sq. km.), including islands off the coast. Set aside in 1931 as Aboriginal reserves, it has a tropical climate. Named for his ship the *Arheim* (*Aernem*) by Dutch explorer Willem van Colster in 1623.

Arthur's Pass. New Zealand's highest alpine pass, at 3,032 feet (924 meters). The road-railway pass crosses over the Southern Alps of South Island, linking the Canterbury Plains on the east with Westland on the west. Named after the surveyor Arthur Dobson, who rediscovered the old Maori alpine route in 1864.

Atherton Tableland. Highly fertile plain, part of the Great Dividing Range (Eastern Highlands) in northeastern Queensland, Australia. Covers 12,000 square miles (31,000 sq. km.), bordered by the Palmer River on the north and Burdekin River on the south. Originally settled in the 1870s as a mining region.

Atlantic Ocean. The South Atlantic Ocean lies north and east of Antarctica. The Antarctic Peninsula, the Weddell Sea, and Queen Maud Land in East (Greater) Antarctica all face toward the South Atlantic.

Auckland. Largest city in New Zealand, and largest Polynesian city in the world. Population was 1.47 million in 2020. Located on a narrow neck of land in the north of North Island, surrounded by two harbors, Waitemata to the east and Manukau to the west. Built on forty-eight extinct volcanoes, including Rangitoto, the last and largest of which became extinct 800 years ago.

Austral Islands. Groups of islands in French Polynesia, South Pacific Ocean. Small islands form a chain 850 miles (1,370 km.) long. Total area is 54 square miles (140 sq. km.); population was 6,509 in 1998. The inhabited islands are Rimatara, Rurutu, Tubuai, Raivavae, and Rapa. Also known as Îles Australes or Tubuai Islands.

Australasia. Loosely defined term for the region, which, at the least, includes Australia and New Zealand; at the most, it also includes other South Pacific islands in the region and takes on a meaning similar to that of Oceania.

Australia. Continent occupied by a single nation of the same name, covering 2.97 million square miles (7.7 million sq. km.), including its island state of Tasmania. Located between the Indian and Pacific oceans, Australia is the sixth-largest country and the smallest continent in the world. Population in 2020 was nearly 25.5 million people. Other states are New South Wales, Queensland, Victoria, South Australia, and Western Australia. It also has the Northern Territory and Australian Capital Territory, which contains the national capital, Canberra. Australia is the flattest and driest continent; two-thirds is either desert or semiarid. Geologically, it is the oldest and most isolated continent.

Australian Capital Territory. Site of Canberra and surrounding lands of Australia's national capital. Covers 910 square miles (2,350 sq. km.) within the Southern Tablelands of NEW SOUTH WALES. Established as a result of the British Parliament's passing the Commonwealth of Australia Constitution Act in 1901. The site was chosen in 1908; parliament moved from the temporary capital, MELBOURNE, into the official Parliament House in 1927. Population was 427,419 in 2019.

Ayers Rock. See ULURU.

Bagana. Active volcano on south central Bougainville Island, PAPUA NEW GUINEA. Reaches 6,086 feet (1,855 meters). Its most recent eruption occurred in 2020.

Balleny Islands. Island group straddles the ANTARCTIC CIRCLE. Unlike such coastal islands as Ross, Alexander, and Roosevelt, is in the sea off VICTORIA LAND. Connected to the mainland by pack ice during the winter.

Bass Strait. Channel separating VICTORIA, AUSTRALIA, from the state of TASMANIA. It is 150 miles (240 km.) wide and 180 to 240 feet (50 to 70 meters) deep and known for rough crossing. Bordered by King Island and the INDIAN OCEAN to the west and by the FURNEAUX GROUP to the west. Named in 1798 by the English navigator Matthew Flinders for the surgeon-explorer George Bass.

Bay of Islands. See ISLANDS, BAY OF.

Bay of Whales. See WHALES, BAY OF.

Beardmore Glacier. Largest glacier thus far discovered in the Antarctic. Originates at the SOUTH POLE and continues its flow to the ROSS ICE SHELF, which is roughly the size of France.

Belau. See PALAU, REPUBLIC OF.

Bellingshausen Sea. Sea in West ANTARCTICA. Extends from the eightieth south parallel to the ANTARCTIC PENINSULA. Named after Russian explorer Admiral Fabian von Bellingshausen.

Bentley Subglacial Trench. Lowest spot in ANTARCTICA and the lowest altitude on any continent. At 8,383 feet (2,555 meters) below sea level, it is 7,000 feet (2,134 meters) lower than Israel's Dead Sea. There is little snow, and the earth is often exposed.

Bikini Atoll. Small atoll in the MARSHALL ISLANDS group in the western PACIFIC OCEAN. In the 1940s and 1950s the United States tested nuclear bombs on Bikini and neighboring atolls, forcing their residents to leave the islands.

Bismarck Archipelago. Group of mostly volcanic islands in the west PACIFIC OCEAN, north of the east end of New Guinea; part of PAPUA NEW GUINEA. Total area is 19,173 square miles (49,658 sq. km.); population was 399,149, primarily Melanesian, in 1990. Contains New Britain, New Ireland, New Hanover, ADMIRALTY ISLANDS, and 200 other islands and islets. There are active volcanoes on New Britain.

Blue Mountains. Section of the GREAT DIVIDING RANGE, NEW SOUTH WALES, AUSTRALIA. Rises to 3,781 feet (1,180 meters). Steep and riddled with waterfalls and ravines, its bluish hue is caused by light rays diffusing through oils dispersed into the air by varieties of indigenous eucalyptus trees. The location of many ancient rock formations, including the Three Sisters and the Jenolan Caves, as well as the Blue Mountains National Park, a 662,130-acre (267,954-hectare) nature reserve. SYDNEY is 40 miles (60 km.) to the east.

Bora-Bora. Volcanic island in FRENCH POLYNESIA in the South Pacific. One of the Leeward Islands, SOCIETY ISLANDS group. Area is 15 square miles (39 sq. km.); population was 410,605 in 2017. Leading industries are tourism and copra production; was a US air base during World War II.

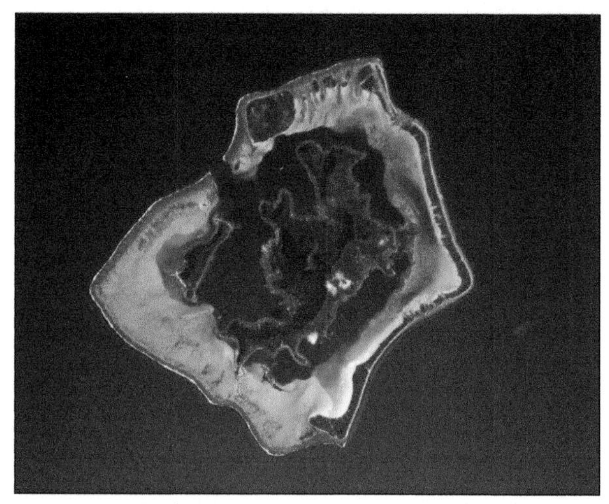

Bora Bora and its lagoon seen from the International Space Station. (NASA Johnson Space Center)

Botany Bay. Inlet of the South Pacific Ocean on the southeastern coast of Australia. Located 9 miles (15 km.) south of Port Jackson, in the city of Sydney, New South Wales. The site of Captain James Cook's first landing on Australian soil, in 1770.

Brisbane. State capital and major port of Queensland, and Australia's third-largest city. Built along the banks of the Brisbane River, 12 miles (19 km.) from where it empties into Moreton Bay. Balmy and tropical, it had a 2019 population of 2.5 million. A shipping and manufacturing hub, its proximity to the agricultural lands of the Eastern Highlands and Darling Downs, and its port, which can accommodate the largest cruise ships, have made Brisbane a major exporter of wool, meat, dairy products, sugar, and grains. Among its industrial facilities are food-processing plants, oil refineries, shipyards, sawmills, and factories that produce automobiles, rubber goods, cement, and fertilizer. Also the gateway to Australia's famous Gold Coast resorts and casinos. Originally established as a convict settlement in 1824; became the capital in 1859.

Brooks Islands. See Midway Islands.

Burdekin River. Largest river in Australia's Queensland state. Enters the sea in a big delta near Ayr, where groundwater makes the cultivation of sugarcane, a principal crop, possible. After flooding in the summer months, it dries up into waterholes for much of the year. Queensland's largest dam was constructed at the Burdekin Falls. Named by explorer Ludwig Leichhardt in 1845 in honor of the Burdekin family, which financed his expedition.

Byrd Glacier. Largest glacier feeding the Ross Sea and the Ross Ice Shelf in Antarctica. Wider than the English Channel and produces more ice annually than all the other six glaciers that feed the Ross Ice Shelf.

Byron, Cape. Easternmost point of Australia's mainland; located in northeastern New South Wales. Shelters Byron Bay. Originally the main port between the town of Newcastle and Brisbane, 90 miles (145 km.) to the north. Discovered by Captain James Cook in 1770 and named in honor of Commodore John Byron, the grandfather of British poet Lord Byron.

Cairns. City and port in North Queensland, Australia; gateway to the Great Barrier Reef. Located 860 miles (1,380 km.) north of Brisbane, with a 2018 population of 152,729. Served by international air, rail, and the Bruce Highway. An important tourist center for access to the Great Barrier Reef, coastal resort islands, and big-game fishing. Founded in the 1870s on Trinity Inlet of Trinity Bay; grew as a result of nearby gold and tin mining and sugarcane cultivation.

Canberra. Federal capital of Australia, located in the Australian Capital Territory. The country's first planned city; carved from the bush in southeastern Australia, about 150 miles (240 km.) southwest of Sydney. Site was chosen in 1909; an international contest was launched in 1911 to find a design for the new capital. Construction began in 1913; in 1927 ceremonies marked the official transfer of the federal parliament from its temporary home in Melbourne to the new Parliament House in Canberra. Low buildings, large expanses of park lands, and the artificial lake Burleigh Griffin form the city's core. Name is derived from an Aboriginal word for "meeting place." Population in 2019 was 426,704.

Canterbury Plains. Largest lowland area in New Zealand. About 200 miles (320 km.) long and 40 miles (64 km.) wide; located in the east-central region of South Island. The breadbasket of the country. Once home of the moa, a large flightless bird that was hunted into extinction by the Maori. Home of the nor'wester, a dry, warm wind that blows from the northwest across the plains, similar to the chinook of North America.

Cape York Peninsula. Large triangular area comprising the northernmost extension of Australia's Queensland state, with the Gulf of Carpentaria on the west and the Coral Sea to the east. A remote area with mostly dirt roads; travel in wet lowlands is made difficult by flooding streams. The twenty-five national parks on the peninsula represent the wide variety of

Queensland vegetation and habitats, ranging from tropical rain forest to woodland and tropical grasslands and swamps. Cattle raising and gold mining on the Palmer River were important during the nineteenth century; bauxite mining and tourism now support the local economy.

Caroline Islands. Archipelago in the western PACIFIC OCEAN, with more than 930 hundred islands, atolls, and islets. Area of 450 square miles (1,165 sq. km.) is divided into the Republic of PALAU (1994) and the FEDERATED STATES OF MICRONESIA (1979), consisting of Chuuk, Kosrae, Pohnpei, and Yap. Native inhabitants are Micronesians. The larger islands are volcanic in origin; the smaller are coral atolls. From 1947 to 1979, the islands, previously in Japanese hands, were the Trust Territory of the Pacific Islands, administered by the United States. Population was 131,200 in 2007.

Carpentaria, Gulf of. Shallow inlet of the ARAFURA SEA along the northern coast of AUSTRALIA. Covers 120,000 square miles (310,000 sq. km.) with a maximum depth of 230 feet (70 meters). Bordered on the east by the CAPE YORK PENINSULA and on the west by ARNHEM LAND. More than twenty meandering rivers drain into the gulf. The town of Karumba, located on the gulf, is a major center for the prawn-fishing industry; huge deposits of manganese and bauxite are mined in the area.

Chatham Islands. Remote group of islands belonging to NEW ZEALAND. Situated 528 miles (850 km.) east of CHRISTCHURCH. Population was 663 in 2018. Main population center is located at WAITANGI. Island economy is based on the export of sheep, wool, and fish to the mainland. Original inhabitants of the islands were the Moriori, of Polynesian origin. Tommy Soloman, the last full-blooded Moriori, died in 1933.

Christchurch. Largest city on SOUTH ISLAND, NEW ZEALAND. Situated on the coastal edge of the CANTERBURY PLAINS, at the foothills of Banks Peninsula, an extinct volcano. Population was 377,200 in 2019. Founded in 1850 as an Anglican settlement. Most of the city is built on flat terrain, making it ideal for bicycling and walking. The international airport is the primary gateway to ANTARCTICA. Earthquakes in 2010 and 2011 destroyed one-third of the city.

Christmas Island. See KIRITIMATI ISLAND.

Clutha River/Mata-Au. Largest river by volume and second-longest river in NEW ZEALAND. Flows southeast 200 miles (322 km.) across central Otago on SOUTH ISLAND, beginning at Lake Wanaka and reaching the PACIFIC OCEAN south of DUNEDIN. Major source of hydroelectric power for NEW ZEALAND. Center of the Otago gold rush in the eighteenth century.

Coober Pedy. Mining town in central SOUTH AUSTRALIA; most of the world's opals come from this area. Located 590 miles (950 km.) northwest of ADELAIDE in the Stuart Range on the edge of the GREAT VICTORIA DESERT. Population was 1,762 in 2016. Its name is derived from the Aboriginal words meaning "boy's waterhole," referring to miners' practice of building homes underground to escape temperatures as high as 125 degrees Fahrenheit (52 degrees Celsius) during the summer months. The first opals were found here in 1911.

Cook, Mount. NEW ZEALAND's highest mountain peak, at 12,317 feet (3,754 meters). Located in the SOUTHERN ALPS of SOUTH ISLAND. Perpetual snow cap; popular skiing site. Known to the Maori as Aorangi, meaning "the cloud piercer."

Cook Islands. Self-governing islands in association with NEW ZEALAND in the South PACIFIC OCEAN. About 2,000 miles (3,220 km.) northeast of New Zealand. Comprises fifteen islands with a total area of about 90 square miles (240 sq. km.); population was 8,574 in 2020. Northern islands are sparsely populated atolls; southern islands are volcanic in origin and include Rarotonga. Most of the islands' people live on Rarotonga and are of mixed Polynesian and European ancestry.

Cook Strait. Stretch of ocean separating NORTH and SOUTH Islands of NEW ZEALAND. Major inter island shipping link; 16 miles (25 km.) at its narrowest point. Subject to rough seas and site of maritime disasters; acts like a wind tunnel. In 1968 the inter-island roll-on, roll-off ferry sank, costing fifty-one lives, during one of the worst storms in the country's history. Named after Captain James Cook, who sailed through the strait in 1770.

Cooktown. Town and port located at the mouth of the Endeavour River on the Coral Sea in northeastern Queensland. Population was about 2,631 in 2016. Linked by the Mulligan Highway to Cairns, 120 miles (190 km.) to the south; faces the Great Barrier Reef. Founded in 1873 during the Palmer River gold rush; economy relies on agriculture mining, and ecotourism. Named after British explorer Captain James Cook, who beached his ship *Endeavour* there for repairs.

Coral Sea. Part of the southwestern Pacific Ocean; extends east from its meeting with the Arafura Sea, through the Torres Strait, east of the Australian continent and New Guinea, west of New Caledonia and the New Hebrides, and south of the Solomon Islands. Covers 1,849,800 square miles (4,791,000 sq. km.); named for its many coral formations, particularly the Great Barrier Reef, which extends 1,250 miles (2,010 km.) down Australia's northeast coast. Subject to typhoons from January to April; provides a 200-mile (320-km.) shipping channel between eastern Australia and the South Pacific Islands and China. Site of a naval battle in 1942 that prevented the Japanese invasion of Australia.

Darling Downs. Tableland located in southeastern Queensland, Australia. Agricultural and pastoral region covering 5,500 square miles (14,200 sq. km.); a major wheat and dairy belt.

Darwin. Capital city of Australia's Northern Territory. Located in the far north, or Top End, it was founded in 1869 on a low peninsula northeast of the entrance to its harbor, Port Darwin, an inlet of the Clarence Strait of the Timor Sea. A center for government administration and mining; also a tourist destination, providing access to Bathurst and Melville Islands. Population was 148,564 in 2018. Violence was part of the area's history, as the local Aborigines resisted European settlement. As a result of its proximity to Asia, Darwin was bombed by the Japanese in 1942 during World War II. The city was largely rebuilt after being devastated by Cyclone Tracy in 1974. Port Darwin was named in 1839 by John Lort Stokes aboard H.M.S. *Beagle*, after his former shipmate, Charles Darwin.

Dry valleys. Ice-free valleys in Antarctica that receive almost no precipitation and can sustain some rudimentary plant life. Dry valleys with oases are found in Victoria Land, Wilkes Land, and Princess Elizabeth Land, all in East (Greater) Antarctica. Protected by the Royal Society Mountain Range; consist of the Taylor, Wright, and Victoria valleys, which were once filled by glaciers. Collectively, they cover an area of 9 to 15 miles by 93 miles (15 to 25 km. by 150 km.).

Dunedin. Second-largest city on New Zealand's South Island. Located on the southeast coast of Otago; population was 131,700 in 2019. Founded as a Scottish settlement in 1848. New Zealand's first university was established there in 1869. The name is Gaelic for "Edinburgh."

East (Greater) Antarctica. Portion of Antarctica east of the Transantarctic Mountains. Includes Wilkes Land, Enderby Land, and Queen Maud Land. The geographic and magnetic South Poles lie in this area. Considerably larger than West (Lesser) Antarctica.

Easter Island. Chilean dependency in the South Pacific, about 2,400 miles (3,862 km.) off the coast of Chile. Area of 46 square miles (119 sq. km.); population was 7,750 in 2017. The island, discovered on Easter Sunday in 1722 by Dutch Admiral Jacob Roggeveen, is known for the hundreds of monolithic statues of people, averaging 15 feet (4.6 meters) tall, that dot its landscape. Also known as Isla de Pascua or Rapa Nui.

Ellice Islands. See Tuvalu.

Ellsworth Mountains. Range located south of the Antarctic Peninsula in Marie Byrd Land. Some of the world's highest peaks are located there: Vinson Massif soars to 16,050 feet (4,892 meters); Mount Tyree reaches 15,500 feet (4,723 meters). Named for Lincoln Ellsworth, the US aviator who discovered their northern portion, Sentinel Ridge.

Equator. Great circle of Earth that is everywhere equally distant from the two poles and divides Earth into the Northern and Southern hemispheres. The direction of the currents and winds around the equator have a profound effect on the Pacific Islands.

Erebus, Mount. Largest active volcano on Ross Island, Antarctica. Rises to 12,448 feet (3,794 meters) above sea level. Discovered in 1841 by Antarctic explorer Sir James Clark Ross, who named it after one of his ships. When discovered, it was spewing smoke and flame into the sky to a height that Ross estimated to be 2,000 feet (600 meters).

Aerial view of Mount Erebus craters in the foreground with Mount Terror in the background, Ross Island, Antarctica. (NSF/Josh Landis)

Eyre, Lake. Great salt lake located in the southeastern corner of the Great Artesian Basin in the state of South Australia. Covers 3,861 square miles (10,000 sq. km.). The surface of the lake bed is covered by a thin crust of salt, the result of deposits made by rapidly evaporating waters.

Federated States of Micronesia. See Micronesia, Federated States of.

Fiji. Archipelago in the South Pacific Ocean, comprising about 332 islands in a horseshoe configuration in Melanesia. Area is 7,056 square miles (18,274 sq. km.). Capital city is Suva. Population was 935,974 in 2020, 57 percent native Fijians and 38 percent Asian Indians. This has created continuing ethnic strife. Two main islands are Viti Levu and Vanua Levu. The islands are volcanic in origin, but coral reefs have built up along them. Also known as Fiji Islands.

Filchner-Ronne Ice Shelf. Enormous ice shelf on the northeastern side of the Weddell Sea in East (Greater) Antarctica. Became much smaller in 1986, when all the ice north of its great chasm calved, floating into the Weddell Sea. The area that broke away covered 5,000 square miles (13,000 sq. km.). Another large area of ice broke away in 2010. Some scientists predict that climate change may cause this shelf to disappear altogether by the end of the twenty-first century.

Fiordland National Park. Largest national park in New Zealand. Covers 4,853 square miles (12,570 sq. km.). Established in 1952 in the remote southwestern corner of South Island. Its geographical features include fjords, rugged mountain ranges, and vast beech forests. The park is a World Heritage Site.

Flinders Island. Largest and northernmost island in the Furneaux Group, Tasmania, Australia. Located in the eastern Bass Strait between Tasmania and the Australian mainland; covers approximately 800 square miles (2,080 sq. km.). Named for English navigator Matthew Flinders, who first surveyed it in 1798. Population was 833 in 2016.

Fraser Island. World Heritage-listed sand island, located off the southeastern coast of Queensland state, Australia; also called Great Sandy Island. It is 75 miles (120 km.) long and covers 625 square miles (1,620 sq. km.). Named for Captain James Fraser, who was killed there by indigenous Australians in 1836. The island is noted for its large population of dingoes, which have become a menace to residents and tourists.

Fremantle. One of Australia's largest ports and the principal port of Western Australia state. Located on the Indian Ocean at the mouth of the Swan River. Originally a major whaling center; later served the gold fields of Coolgardie-Kalgoorlie; a principal submarine base for the Allies in World War II; now a major industrial center. Population in 2016 was 28,893.

French Oceania. See French Polynesia.

French Polynesia. An overseas collectivity of France located in the South Pacific Ocean, comprising 118 islands and atolls. Area is 1,609 square miles (4,167 sq. km.); population was 295,121 in 2020. Contains Society, Marquesas, Gambier, and Austral island groups and Tuamotu Archipelago. Capital is Papeete, located on Tahiti, its largest island. Tahiti is also home to the largest peak on the islands, Mount Orohena, which reaches 7,352 feet (2,241 meters). About 78 percent of the population are Polynesian, 12 percent are Chinese, and 6 percent are of French origin. Also known as Polynésie Française. Formerly French Oceania.

Friendly Islands. See Tonga.

Furneaux Group. Cluster of mountainous, rocky islands situated in the Bass Strait that separates the Australian mainland from the island state of Tasmania. Primarily used for cattle and sheep raising. Sighted by Tobias Furneaux in 1773.

Gambier, Mount. Mountain located in southeastern South Australia, approximately 280 miles (450 km.) southeast of the capital city of Adelaide. An extinct volcano with four crater lakes; rises to a height of 623 feet (190 meters).

Gaua. One of the Banks Islands (also known as Santa Maria), in northern Vanuatu in the southwestern Pacific Ocean. Volcanic Mount Gharat, which reaches 2,615 feet (797 meters) in height, most recently erupted in 2013.

Gaussberg, Mount. Only volcano discovered in East (Greater) Antarctica.

Geographic South Pole. See South Pole.

George VI Ice Shelf. Antarctic ice shelf with a notable ice melt in 2020 of about 90 miles (140 km.). Located below the George VI Sound west of Palmer Land on the Antarctic Peninsula, it receives deep, warm waters that spill over the continental shelf. These waters are 3.6 Fahrenheit degrees (2 Celsius degrees) warmer than those beneath the Ross Ice Shelf, enough to drastically affect the speed at which the ice melts.

Gibson Desert. Vast dry area in the interior of Western Australia and the Northern Territory. Located between the Great Sandy Desert to the north and the Great Victoria Desert to the south. First crossed in 1876 by Ernest Giles; named for Alfred Gibson, a member of his party who perished there.

Gilbert Islands. Group of islands with sixteen atolls in Kiribati, west Pacific Ocean, across the equator. Area is 102 square miles (264 sq. km.); population was 83,382 in 2005. Part of the Gilbert and Ellice Islands (*see* Tuvalu) Colony until 1976. Major islands are Tarawa, Butaritari, Abaiang, Abemama, Tabiteuea, Nonouti, and Beru.

Gippsland. Fertile region of southeastern Victoria, Australia. Covers 13,600 square miles (35,200 sq. km.). Center of the state's dairy industry. Originally attracted settlers with its gold finds; farmers arrived with the completion of a rail line from Melbourne in 1887.

Graham Land. Southern half of the Antarctic Peninsula; the northern half is called Palmer Land. Named after a British nobleman.

Grampians. System of mountain ranges extending southwest from the Great Dividing Range in southwest central Victoria, Australia. Covers approximately 400 square miles (1,035 sq. km.). Mount William, the highest peak, rises to 3,827 feet (1,166 meters). Noted for sandstone rock formations, deep gorges, and colorful vegetation.

Great Antarctic Horst. See Transantarctic Mountains.

Great Artesian Basin. One of the largest areas of artesian water in the world, covering approximately 656,250 square miles (1.7 million sq. km.). Underlies approximately one-fifth of the Australian continent, including most of Queensland, with portions extending into New South Wales, the Northern Territory, and South Australia. Overexploitation of the Great Artesian Basin is a major environmental concern.

Great Australian Bight. Embayment of the Indian Ocean indenting the southern coast of the Australian continent. Its head abuts the arid Nullabor Plain. Has a continuous border of sea cliffs that rise 200 to 400 feet (60 to 120 meters) high.

Great Barrier Reef. Limestone formation created by the largest group of coral reefs in the world; extends 1,250 miles (2,000 km.) along the northeast coast of Australia. Parts of the reef are more than 100 miles (160 km.) from the coast; the closest point is 10 miles (16 km.) distant. Covers approximately 80,000 square miles (207,000 sq. km.). Thought to have begun growing millions of years ago, the reef is formed from the skeletal waste of tiny marine organisms called coral polyps and hydrocorals. Supports about 400 species of these organisms, more than 1,500 species of fish, and myriad other sea life and birds. Current biggest threats are thought to be climate change and overfishing. A 2012 study estimated that over half the reef has disappeared since the 1980s. Australia proposed a 35-year plan in 2014 to save the reef. In addition to its scientific interest, the reef is a major tourist attraction. In 1981 it was named a UNESCO World Heritage Site. The Great Barrier Reef was first explored by Europeans in 1770, when British explorer Captain James Cook's ship ran aground on it.

Great Dividing Range. Series of mountain ranges and plateaus that form the main watershed of eastern Australia. Also known as the Great Divide or Eastern Highlands, the range begins to the north on the Cape York Peninsula and parallels the coasts of Queensland, New South Wales, and Victoria, ending in the Grampians. One section, the Australian Alps, includes Australia's highest peak, Mount Kosciuszko. Several of the country's major rivers form their headwaters in the range, including the Snowy, Darling, and Murrumbidgee rivers. An important agricultural, lumbering, and mining region.

Great Sandy Desert. Vast, flat expanse of sand hills and salt marshes located in northern Western Australia, between the Pilbara and Kimberley ranges. Sparsely populated, with no significant settlements. Daytime temperatures in the summer are among the hottest in Australia, frequently rising above 100 degrees Fahrenheit (38 degrees Celsius) at times.

Great Sandy Island. See Fraser Island.

Great Victoria Desert. Vast, arid expanse of land in Western and South Australia, situated between the Nullarbor Plain to the south and the Gibson Desert to the north, and stretching eastward from Kalgoorlie nearly to the Stuart Range. Composed of salt marshes and hills covered with *Spinifex* grass. First explored by Ernest Giles, who named it.

Greater Antarctica. See East (Greater) Antarctica.

Guadalcanal. Largest of the Solomon Islands in the South Pacific Ocean. Area is 2,180 square miles (5,646 sq. km.). Population was 109,382 in 1999. Extending the length of the island (92 miles/148 km.) are the Kavo Mountains. During World War II, the site of fighting between US and Japanese forces, beginning in August 1942, and ending in February 1943, with a US victory.

Guahan. See Guam.

Guam. Largest and southernmost of the Mariana Islands in the western North Pacific Ocean; unincorporated US territory. Area is 212 square miles (549 sq. km.); population was 168,485 in 2020. Capital is Agana. Its northern part, largely a plateau, was formed by coral; the southern part, hilly with streams, is volcanic in origin. Occupied by the Japanese in 1941 and taken by the United States in 1944. Many US Navy, Army, and Air Force installations are located on Guam, which was an important air base in the Vietnam War. Also known as Guahan.

Halycon Island. See Wake Island.

Hamersley Range. Mountains in the Pilbara region of Western Australia. Includes the state's highest peak, Mount Meharry (4,111 feet/1,253 meters). Important for mining, particularly iron ore.

Haupapa/Tasman Glacier. Largest glacier in New Zealand. Flows down the eastern slopes of Aoraki/Mount Cook in South Island. First explored by geologist Julius Von Haast and surveyor Arthur Dobson in 1862. Melting has occurred since the 1990s.

Hawaii. State of the United States consisting of a chain of volcanic and coral islands in the North

Pacific Ocean. The Hawaiian archipelago contains eight major islands—Hawaii (also known as the Big Island), Kauai, Oahu, Maui, Lanai, Molokai, Niihau, and Kahoolawe—and 124 islets, reefs, and shoals. The islands' total area is 10,931 square miles (28,311 sq. km.). Hawaii's population was 1.41 million in 2018. Largest city is Honolulu, on Oahu Island. The islands are almost all volcanic in origin, with some areas formed from coral reefs. The islands contain mountains rising to more than 13,000 feet (4,000 meters) above sea level, numerous sandy beaches, dense rain forests, arid scrublands, fields, forests, barren lava beds, and canyons. European explorer Captain James Cook arrived in 1778 and named the islands the Sandwich Islands. Originally an independent kingdom created by Kamehameha I in the late eighteenth century, the islands lost their independence in the late nineteenth century when American planters overthrew the monarchy and declared a republic. Hawaii was annexed to the United States in 1898 and became the nation's fiftieth state in 1959.

Hawaii Island. Largest and southernmost of the Hawaiian Islands, United States, in the North Pacific. Area is 4,021 square miles (10,414 sq. km.); population was 186,738 in 2011. Has four volcanic mountains: Mauna Kea, Hualalai, Mauna Loa, and Kilauea.

Hawke's Bay. Largest production region of apples, pears, and peaches in New Zealand. A wide bay located on the east central coast of North Island. Boasts a Mediterranean climate and fertile soils. Contains Cape Kidnappers, the largest known mainland gannet-nesting colony in the world. Named by Captain James Cook after Sir Edward Hawke, then Lord of the British Admiralty, in 1769.

Hobart. Largest city, main port, and capital of Australia's island state, Tasmania. Australia's smallest capital city, with an estimated population of 240,342 in 2019. Established as a convict settlement at the mouth of the Derwent River in the southeastern corner of the state in 1804. Existed in conflict with the Aboriginal people until the last of them was exterminated in 1876. Situated at the foot of lofty Mount Wellington (4,165 feet/1,270 meters) and with an excellent deep-water harbor, Hobart has prospered, particularly in the whaling trade, shipbuilding, and the export of agricultural products such as corn and Merino wool.

Honolulu. Seaport city on Oahu Island, and Hawaii state capital. In the Pacific Ocean. Metropolitan population was 953,207 in 2019. Its convenient location in the North Pacific Ocean and its protected harbor make it a valuable trade port. Located at the mouth of the Nuuanu Valley, with the Punchbowl and Mount Tantalus behind it. A superior beach is located in its southeastern suburb of Waikiki. Main industries are tourism, sugar processing, and pineapple canning.

Howe, Cape. Southeastern point of mainland Australia, located at the New South Wales-Victoria border. Named by Captain James Cook in 1770 for Richard, Lord Howe, of the Royal Navy.

Hunter Valley. Fertile agricultural region in east central New South Wales, Australia; watered by the Hunter River and its tributaries. Agriculture includes fruit orchards and vineyards; cattle, sheep, and poultry are also raised.

Îles Australes. See Austral Islands.

Îles Marquises. See Marquesas Islands.

Indian Ocean. One of the three oceans on which the coast of Antarctica fronts. The Indian Ocean is off East (Greater) Antarctica, from about the fiftieth to the hundred-twentieth parallel.

Invercargill. Southernmost city in New Zealand. Located on the Waihopai River at the bottom of South Island; population was 56,200 in 2019. Has some of the widest streets of any city in New Zealand.

Isla de Pascua. See Easter Island.

Islands, Bay of. Site of New Zealand's first permanent European settlement. Located on the east coast in the Northland region of North Island. Contains 150 islands, most uninhabited, within an irregular coastline that stretches 497 miles (800 km.).

Jervis Bay. Inlet of the TASMAN SEA, in southeastern NEW SOUTH WALES, AUSTRALIA. One of the country's finest natural harbors, 10 miles (16 km.) wide and 6 miles (10 km.) long. Discovered and named in 1770 by Captain James Cook.

Joseph Bonaparte Gulf. Inlet of the TIMOR SEA indenting the northern coast of AUSTRALIA for nearly 100 miles (160 km.). Site of Wyndham, the area's main port; termination point for the Ord, Victoria, and other rivers. Aboriginal reserves are located here.

Kalgoorlie. Historical Gold Capital of the World, located in the OUTBACK northern goldfields of WESTERN AUSTRALIA, 372 miles (600 km.) from PERTH. Population was 29,849 in 2018. Mining began with (Paddy) "Hannan's Find"—the discovery of gold in 1893; production peaked in 1903. Nickel is now the region's chief mineral product. An arid region, its name is a corruption of the Aboriginal word "galgurlie," an indigenous plant.

Kalgoorlie Court House, Western Australia. (Bonga)

Kangaroo Island. AUSTRALIA's third-largest offshore island. Located at the entrance to the Gulf of Saint Vincent in SOUTH AUSTRALIA, 80 miles (130 km.) southwest of the state's capital city, ADELAIDE. Kingscote is the main town; population was 4,702 in 2016. A low, cliffed plateau, 90 miles (145 km.) long and 27 miles (44 km.) wide. Separated from the mainland by Investigator Strait. Named by English explorer Matthew Flinders in 1802 for the abundance of kangaroos.

Karkar. Active volcano on a small island in the PACIFIC OCEAN, off the east coast of New Guinea Island, in the BISMARCK ARCHIPELAGO, PAPUA NEW GUINEA. Reaches a height of 6,033 feet (1,839 meters). One of its most recent eruptions occurred in 2013.

Kauai. Island in the northwest part of the HAWAIIAN ISLANDS, North PACIFIC OCEAN. Area is 555 square miles (1,437 sq. km.). Mountainous, with two major peaks, Kawaikini and WAIALEALE; three major bays are Nawiliwilli, Hanalei, and Hanapepe. Largest city, Lihue, had a population of 6,455 in 2010.

Kavachi. Active submarine volcano in the SOLOMON ISLANDS, south of Vangunu Island, in the PACIFIC OCEAN. Located 18.6 miles (30 km.) from the boundary of the Indian and Australian tectonic plates. Rises from a depth of about 3,600 feet (1,100 meters); base measures about 4.9 miles (8 km.) in diameter. Since the first recorded eruption in 1939, it has created at least eight islands up to about 480 feet (150 meters) in length. Last eruption occurred in 2014.

Kerr Point. Northernmost tip of NEW ZEALAND. Situated on North Cape at 34°24′ south latitude, longitude 172°59′ east. Extends northward into the PACIFIC OCEAN.

Kilauea. World's most active volcano. In Hawaii Volcanoes National Park, on the east slope of volcanic mountain of MAUNA LOA, south central HAWAII ISLAND, Hawaii. Crater is at an elevation of 4,091 feet (1,247 meters), more than 10,000 feet (3,000 meters) below the summit. It began erupting in 1983, spewing 525,000 cubic yards (400,000 cubic meters) of lava per day on average.

Kimberley Plateau. Sandstone plateau with deep river gorges, located in northern WESTERN AUSTRALIA. Covers 140,000 square miles (360,000 sq. km.). Mining and cattle grazing are primary industries. Many of the inhabitants are indigenous Australians.

Kiribati. Island nation, part of the Commonwealth of Nations, in the central PACIFIC OCEAN. Area is

313 square miles (811 sq. km.); population was 111,796 in 2020. This Micronesian republic comprises thirty-three coral atolls in three groups: the GILBERT, PHOENIX, and Line Islands. Seventeen of the twenty-one permanently inhabited islands are in the Gilbert Islands, where its capital, Tarawa, is located. Except for Banaba, the islands are atolls, few rising more than 13 feet (4 meters) above sea level. Kiribati, among other Pacific Island nations, has expressed concern over climate change, which could cause sea levels to rise and submerge its islands.

Kiritimati Island. One of the Line Islands of KIRIBATI, in central PACIFIC OCEAN. Largest atoll in the Pacific, at 150 square miles (388 sq. km.); population was 6,456 in 2015. Part of KIRIBATI since 1979. Also known as Christmas Island.

Kosciuszko, Mount. AUSTRALIA's highest mountain peak, located in NEW SOUTH WALES. Rises to 7,310 feet (2,229 meters). Snow covers the mountain through the winter. In 1840 explorer Paul Edmund de Strzelecki named the mountain after Thaddeus Kosciuszko, a Polish army officer who fought with the Americans in the Revolutionary War.

Lagoon Islands. See TUVALU.

Lambert Glacier. Sizable Antarctic glacier; 25 miles (40 km.) wide and more than 248 miles (400 km.) long. Flowing through the PRINCE CHARLES MOUNTAINS, it drains the interior reaches of the continent of ice and snow, discharging 8.4 cubic miles (35 cubic km.) of ice every year into the AMERY ICE SHELF.

Lesser Antarctica. See WEST (LESSER) ANTARCTICA.

Little America. Name of several bases in ANTARCTICA. The first was established by explorer Richard Byrd near the Bay of WHALES in 1928. Four subsequent bases called Little America were built after this one. Little America V was disbanded in 1979 after the International Geophysical Year program ended.

Lopevi. Active volcano on the island of Lopevi, in south central VANUATU in the southwestern PACIFIC OCEAN. Reaches a height of 4,755 feet (1,449 meters). One of its latest eruptions occurred in 2007.

McMurdo Station. One of six US stations built in ANTARCTICA during the International Geophysical Year (1957-1958). Situated on the east side of the ROSS SEA off VICTORIA LAND in the McMurdo Sound. Staffed by over 1,200 researchers and explorers during the Antarctic summer; about 200 people remain during the long winter.

Magnetic South Pole. See SOUTH MAGNETIC POLE.

Mariana Islands. PACIFIC OCEAN island group east of the Philippines that comprises the sixteen coral and volcanic islands of the Commonwealth of the NORTHERN MARIANA ISLANDS and GUAM.

Mariana Trench. Deepest seafloor depression in the world, 36,210 feet (11,034 meters) deep. Located just east of the MARIANA ISLANDS in the PACIFIC OCEAN at 11°22′ north latitude, longitude 142°36′ east. The arc-shaped trench, which averages 44 miles (70 km.) in width, stretches from northeast to southwest for about 1,554 miles (2,500 km.). The Challenger Deep, at the southwestern end of the trench, is the deepest point on Earth at 35,856 feet (10,929 meters).

The bathyscaphe Trieste *(designed by Auguste Piccard), the first manned vehicle to reach the bottom of the Mariana Trench. (U.S. Navy Electronics Laboratory)*

Marie Byrd Land. Large area in WEST (LESSER) ANTARCTICA. Established near the one hundredth parallel; named for the wife of Antarctic explorer Richard Byrd.

Marquesas Islands. Fourteen-island group in French Polynesia, in the South Pacific Ocean. Area is 405 square miles (1,049 sq. km.); population was 9,346 in 2017. Major islands fall in three groups: Nuku Hiva, Ua Pu, and Ua Huka (center); Hiva Oa, Tahuata, and Fau Hiva (southeast); and Eiao and Hatutu (northwest). Islands are of volcanic origin. Also known as Îles Marquises.

Marshall Islands. Archipelago and republic in central North Pacific Ocean. Contains 30 atolls and 1,152 islands. Area is 70 square miles (181 sq. km.); population was 77,917 in 2020. Majuro is the capital island of these Micronesian islands. The February 1944, capture of Majuro by the United States in World War II was the first seizure by the United States of a Japanese possession. The United States used Bikini Atoll in the Marshall Islands as a nuclear testing ground in 1946. A trusteeship of the United States after 1947; became self-governing in 1979. Officially, the Republic of the Marshall Islands.

Maui. Second-largest of the Hawaiian Islands. Area is 728 square miles (1,886 sq. km.). East end of the island has Haleakala National Park, with a peak, Red Hill, reaching 10,023 feet (3,055 meters); the west end has Puu Kukui, at 5,787 feet (1,764 meters). On its south coast is Maalaea Bay. Population was 154,834 in 2015.

Mauna Kea. Extinct volcano in north central Hawaii Island, Hawaii. At 13,796 feet (4,205 meters), the highest island mountain in the world from base to peak.

Mauna Loa. Active volcano in Hawaii Volcanoes National Park on south central Hawaii Island, Hawaii. Reaches 13,680 feet (4,170 meters) in height; is the largest mountain in the world in cubic content. Large lava flows occurred in 1919, 1950, and 1984. Last eruption occurred in 1984.

Melanesia. One of three divisions of the Pacific Islands, the other two being Micronesia and Polynesia. Located in the western Pacific Ocean. Papua New Guinea, the eastern half of the island of New Guinea, forms the westernmost part of Melanesia (the western half of New Guinea, Irian Jaya, is part of Indonesia, and thus considered part of Asia.) The other islands include the Solomon Islands, Vanuatu, New Caledonia, Fiji, Admiralty Islands, and the Bismarck Archipelago. Population in 2020 was 9.47 million.

Melbourne. State capital of Victoria, Australia; country's second-largest city. Situated 3 miles (4.5 km.) inland at the head of Port Phillip Bay on the southeastern coast. Population was 5 million in 2019. Established in 1835 on the banks of the Yarra River, the cosmopolitan city is the country's financial center. The densely populated city center and outlying suburbs cover approximately 2,359 square miles (6,109 sq. km.). Hosted the Olympic Games in 1956.

Micronesia. One of three divisions of the Pacific Islands, the other two being Melanesia and Polynesia. The name Micronesia means "little islands." The islands are mostly atolls and coral islands; however, many of the islands in the Carolines are of volcanic origin. The more than 2,000 islands of Micronesia are located in the Pacific Ocean east of the Philippines, mostly above the equator. Micronesia contains the Federated States of Micronesia, Nauru, Kiribati, the Marshall Islands, the Republic of Palau, Guam, and the Commonwealth of the Northern Mariana Islands. Population in 2020 was approximately 545,000.

Micronesia, Federated States of (FSM). Islands of the Carolines except Palau, in free association with the United States. Area is 271 square miles (702 sq. km.); population was 1102,436 in 2020. Comprises four states—Kosrae, Pohnpei (formerly Ponape), Chuuk (formerly Truk), and Yap—and is located north of the equator. Contains more than 600 islands, many of which are low-lying coral atolls. Pohnpei, a volcanic island, is FSM's largest and is home to the capital, Palikir. The islands of Yap state, unlike the others of FSM, are continental (formed by an uplifting of the continental shelf).

Midway Islands. Two islands, Eastern and Sand islands, parts of a coral atoll in central Pacific Ocean. Area is 2 square miles (5 sq. km.). Unpopulated before occupied by the United

States since 1867; administered by the US Navy and has been used extensively by the US Navy and Air Force. In June 1942, during World War II, U.S. forces defeated a Japanese fleet nearby. Formerly Brooks Islands.

Milford Sound. One of the wettest places in New Zealand, with an annual average rainfall of 256 inches (650 centimeter). A 10-mile-long (16-km.)-long fjord located in the remote corner of South Island. Part of Fiordland National Park. Contains a number of rock-sheer cliffs including Mitre Peak, which rises 5,560 feet (1,695 meters) out of the sea. Ending point of the Milford hiking track.

Molokai. One of the Hawaiian Islands. Area is 259 square miles (671 sq. km.). Largest city is Kaunakakai; the island has Mauna Loa at its west end and Kamakou at its east end. A leprosy-treatment center operated on its north coast until 1969.

Eastern Moloka'i with a portion of Kamakou and Moloka'i Forest Reserve. (Travis.Thurston)

Murray River. Australia's principal river. Flows 1,560 miles (2,500 km.) across southeastern Australia, from the Snowy Mountains to the Great Australian Bight on the Indian Ocean. Main tributaries include the Darling and Murrumbidgee rivers. The fertile Murray Valley, the most irrigated land on the continent, is of great economic importance.

Nauru. Raised coral island and republic in central Pacific Ocean. Area is 8.2 square miles (21.2 sq. km.); population was 11,000 in 2020 most living in a narrow coastal strip around the edges of the island. Naurans are of Polynesian, Micronesian, and Melanesian origin. The island contained large amounts of phosphate, depleted by 2006, produced royalties that made the republic, per capita, one of the richest in the world, but caused severe damage to the environment. Today, Nauru relies heavily on financial aid from Australia. Formerly Pleasant Island.

New Caledonia. French collectivity in the southwestern Pacific Ocean. Contains New Caledonia Island, Loyalty Islands, Isle of Pines, Chesterfield Islands, and Huon Islands. Total area is 7,358 square miles (19,058 sq. km.); population was 290,009 in 2020, with 39 percent Melanesian; 34 percent Vietnamese, Polynesian, and Indonesian; and 27 percent European (mainly French). Economic activities include mining (nickel, iron, and manganese), agriculture, livestock production, fishing and forestry, and tourism.

New Hebrides. See Vanuatu.

New Hebrides Basin. Part of the Coral Sea, located east of Australia and west of the New Hebrides island chain. The basin contains volcanic islands, both old and recent.

New Hebrides Trench. Ocean trench in the southwestern Pacific Ocean, about 750 miles (1,200 km.) long and 25,000 feet (7,600 meters) deep.

New South Wales. Australia's most populous state. Located in the southeastern section of the continent; bounded by the Pacific Ocean to the east, the state of South Australia to the west, Queensland state to the north, and the state of Victoria to the south. Population was more than 8 million in 2020. Environments range from tropical rain forests to snow-clad mountains. Geographically, it is a narrow coastal plain bordered in the east by the Great Dividing Range, which runs the length of the state; slopes and plains lie to the west. In 1788 it became the site of the first British settlement on the continent; during the nineteenth century the colonies of Victoria, South Australia, Tasmania, and Queensland were carved out of its vast original territory. Highly industrialized and economically stable, the state is famous for its beautiful harbor and capital city, Sydney.

New Zealand. Large South Pacific island nation, located more than 1,200 miles (1,900 km.) southeast of Australia, its nearest neighbor. It was a British colony until 1947. The country comprises two main islands, North and South islands, and a number of smaller islands, which collectively cover an area of about 103,500 square miles (268,000 sq. km.). New Zealand administers the South Pacific island group of Tokelau and claims a section of Antarctica. Niue and the Cook Islands are self-governing states in free association with New Zealand. Its economy is based on the export of agricultural products. About 16.5 percent of its 4.92 million people (2020) are native Maoris. Capital city is Wellington.

Ninety Mile Beach. Famous beach in New Zealand, located along the west coast of the top of North Island. Despite its name, it is about 55 miles (88 km.) long. The fringe of the beach is lined by spectacular white sand dunes.

Niue. Raised coral island in the south central Pacific Ocean, east of Tonga, in free association with New Zealand. Area is 100 square miles (260 sq. km.); population was 2,000 in 2019. Consists of a plateau (200 feet/61 meters) surrounded by a lower level (90 feet/27 meters) and rimmed by uneven, steep cliffs. Most of its residents are Polynesian; about half live in or around the capital, Alofi (population 597 in 2017). Also known as Savage Island.

North Island. Smaller of the two principal islands of New Zealand, separated from South Island by Cook Strait. North Island has an area of 44,010 square miles (113,985 sq. km.), as well as a majority of the national population. Most of its people are concentrated in the vicinity of the major urban areas, Wellington and Auckland. Its population was 3,760,900 in 2019.

Northern Mariana Islands, Commonwealth of the. Island group in the Pacific Ocean, east of the Philippines, a commonwealth of the United States. Area is 184 square miles (477 sq. km.); population was 51,433 in 2020. Consists of sixteen coral and volcanic islands that are the Mariana Islands minus Guam. Main islands are Saipan (where the seat of government is), Tinia, and Rota. The islands' economy is based on agriculture, light manufacturing, and tourism. Also known as Northern Marianas.

Northern Territory. Self-governing territory of Australia. Located in the central section of the northern part of the continent; bounded by the Timor and Arafura seas to the north, the state of Western Australia to the west, Queensland and the Gulf of Carpentaria to the east, and South Australia to the south. Population was 244,000 in 2020. Occupies nearly one-sixth of the Australian landmass. Climate is tropical in the north and semiarid in the south. Central section is crossed east-to-west by the Macdonnell Ranges; main rivers include the Victoria and the Adelaide. Uluru monolith is located near the southwest corner of the territory. Major towns are Darwin (the capital) and Alice Springs.

Nullarbor Plain. Flat, treeless plateau located in the states of South Australia and Western Australia. An ancient limestone seabed rising about 150 to 500 feet (50 to 200 meters) above sea level; southern edge is bordered by the Great Australian Bight and features 124 miles (200 km.) of spectacular eroded sea cliffs. Covers 100,000 square miles (259,000 sq. km.) and is crossed by the world's longest stretch (330 miles/530 km.) of straight railroad track; it is also accessed nearer the coast by the Eyre Highway. Has an average rainfall of less than 10 inches (254 millimeters) and is nearly featureless with no surface streams. Name is Latin for "no tree."

Oahu. Third-largest of the Hawaiian Islands. Area is 600 square miles (1,555 sq. km.). Contains two mountain ranges: the Koolau along the northeast coast and the Waianae along the southwest coast. Honolulu, capital of the state of Hawaii, and Pearl Harbor are located on its south coast.

Oceania. Collective name for the more than 10,000 islands scattered throughout the Pacific

Ocean, excluding the Japanese archipelago, Indonesia, Taiwan, the Philippines, and Alaska's Aleutian Islands. The term also generally excludes Australia, while including Papua New Guinea and New Zealand. Oceania has traditionally been divided into four parts: Australasia, Melanesia, Micronesia, and Polynesia. Its total land area, excluding Australia, is approximately 317,000 square miles (821,000 sq. km.); its population (again excluding Australia) was about 41.5 million people in 2018.

Ord River. River in the Kimberley plateau region of northeastern Western Australia. Flows east then north nearly 400 miles (650 km.) to the Cambridge Gulf. Site of one of Australia's most expensive and controversial irrigation plans, the Ord River Project, designed to prevent seasonal flooding and store water for irrigating areas subject to drought.

Outback. Name by which Australians refer to any place away from their cities. More specifically, the semiarid region west of the Great Dividing Range (Eastern Highlands), which covers about 80 percent of the continent, including most of Queensland, all of the Northern Territory, and all of Western Australia except the southwest corner. The Macdonnell, Musgrave, and Petermann mountain ranges and the country's four major deserts (Victoria, Great Sandy, Gibson, and Tanami) are situated in the Outback. Economy is mainly supported by sheep raising, opal mining, and oil production. Home of the Royal Flying Doctor Service, which provides medical assistance to the residents of this area.

View across sand plains and salt pans to Mount Conner, Central Australia. (Gabriele Delhey)

Pacific Ocean. Largest body of water on Earth, covering more than one-third of the world's surface. Its area of about 70 million square miles (181 million sq. km.) is greater than the entire land area of the world. It is framed by North and South America in the northeast, Asia in the west, and Australia in the southwest. The equator separates the North Pacific Ocean from the South Pacific Ocean. The ocean's average depth is about 12,900 feet (3,900 meters). Its bottom is more geologically varied than those of the earth's other great oceans—the Indian and Atlantic Oceans. The Pacific has more volcanoes, ridges, trenches, seamounts, and islands. Historically, the vast size of the Pacific was a formidable barrier to travel, communications, and trade well into the nineteenth century.

Pagan. Active volcano on Pagan Island in the Northern Mariana Islands. Reaches 1,870 feet (570 meters) in height. Its most recent eruptions occurred in 2012.

Palau, Republic of. Republic and group of about 200 islands and islets in the west Pacific Ocean, in free association with the United States. Area is 188 square miles (488 sq. km.). Its population was 21,685 in 2020, with about 70 percent of its people living on Koror. Part of Micronesia and considered western part of Carolines. The republic includes islands of Babeldaob (the largest), Urukthapel, Peleliu, Angaur, Eil Malk, and Koror. Palau's islands, the larger of which are volcanic, are encompassed by a fringing reef. Also known as Belau.

Palau Trench. Seafloor depression in the western Pacific Ocean; about 400 miles (650 km.) long and 27,976 feet (8,527 meters) deep. Located near Palau at 7°52′ north latitude, longitude 134°56′ east.

Palmer Land. Northern half of the Antarctic Peninsula; the southern half is called Graham Land. Named after Nathaniel Palmer, an American who discovered the South Orkney Islands.

Papua New Guinea. Independent country in the southwestern Pacific Ocean. Area is 178,704 square miles (462,840 sq. km.); population was 7.25 million in 2020. Capital is Port Moresby. Located on the eastern half of the island of New Guinea; includes Bougainville Island, Buka Island, Louisiade Archipelago, Trobriand Islands, D'Entrecasteaux Islands, the Bismarck Archipelago (New Britain, New Ireland, Manus), and hundreds of smaller islands. (The western part of New Guinea is Irian Jaya, part of Indonesia and thus considered part of Asia.) Mountains on the mainland stretch from east to west and rise to 14,795 feet (4,509 meters) at Mount Wilhelm in the Bismarck Range. Between this range and the Owen Stanley range in the southeast lie broad, high valleys. North of the valleys is a swampy plain; on its north are coastal mountain ranges, running from west to east. Major rivers include the Fly, Purari, Kikori, Sepik, and Ramu. Located along the Ring of Fire, a belt of seismic activity circling the Pacific Ocean, it has forty active volcanoes and frequent earthquakes. Formerly known as the Territory of Papua and New Guinea.

Perth. Capital of Western Australia state, Australia. Located on the banks of the Swan River, 12 miles (19 km.) from the river's mouth, which forms the harbor of Fremantle. A city since 1856, it is now the fourth-largest in the country; more than 2 million (2018) people—nearly three-quarters of the state's population—live there. An important industrial center; has a mild climate nearly eight months of the year. Nicknamed the "city of lights" by astronauts orbiting Earth in the 1960s. Closer in distance to Asia's Singapore than to Sydney.

Petermann Range. Low mountain range extending for 200 miles (320 km.) from east-central Western Australia to the southwest corner of the Northern Territory. Rises to 3,800 feet (1,158 meters). First explored by Ernest Giles in 1874, and named for August Petermann, a German geographer.

Phoenix Islands. Island group consisting of eight small coral atolls, in the central Pacific Ocean; part of Kiribati. Area is 11 square miles (28 sq. km.). The Phoenix Islands contain Canton, Rawaki, Enderbury, Birnie, Manra, Orona, Nikumaroro, and McKean. Formerly a stop on trans-Pacific flights.

Pitcairn Island. Group of islands forming a British dependency in the south Pacific Ocean. Consists of Pitcairn Island and the uninhabited islands of Henderson, Ducie, and Oeno. Area is 18 square miles (47 sq. km.); population was 50 in 2020. Pitcairn, of volcanic origin, contains a single village, Adamstown. It was uninhabited when mutineers of the HMS *Bounty* and a group of Tahitians occupied it in 1790; the remains of their settlement were discovered in 1808. Overpopulation resulted in 200 islanders moving to Norfolk Island in 1856; some later returned to Pitcairn.

Polynesia. One of three divisions of the Pacific Islands, the other two being Melanesia and Micronesia. Group of islands in the central and southern Pacific Ocean. Polynesia means "many islands." Includes American Samoa, the Cook Islands, Easter Island, French Polynesia, Hawaiian Islands, Line Islands, New Zealand, Niue, Pitcairn Island, Samoa, Tokelau, Tonga, Tuvalu, and the Wallis and Futuna Islands. The islands, mostly small, are predominantly coral atolls, but some are of volcanic origin. Population was 7.05 million in 2020.

Port Arthur. Inlet of the Tasman Sea on the south coast of the Tasman Peninsula, Tasmania, Australia. Located 63 miles (101 km.) northwest of the capital city of Hobart. Site of one of the most notorious British penal colonies, established in 1830, the ruins of which have become a popular tourist attraction.

Port Jackson. Inlet of the Pacific Ocean and principal port of New South Wales, Australia; also known as Sydney Harbour. Its 1.5-mile (2.4 km.)-long entrance lies between North and South Heads; it is 12 miles (19 km.) long, with a total area of 21 square miles (55 sq. km.), and is spanned by the Sydney Harbour Bridge at its northern end. Comprises three smaller harbors—North Harbour, Middle Harbour, and Sydney Harbour—and the deepwater mouths of two rivers, the Parramatta and the Lane Cove.

Port Moresby. Seaport and capital of Papua New Guinea in the Pacific Ocean. Located on the southeast coast of the Gulf of Papua. Population was 364,000 in 2019. Has a large, sheltered harbor; was an Allied base in World War II.

Prince Charles Mountains. Area through which the Lambert Glacier in East (Greater) Antarctica flows, annually carrying 8.4 cubic miles (13.4 cubic km.) of ice to the Amery Ice Shelf.

Queen Maud Land. Section of Antarctica located west of Enderby Land, between 20 degrees west and 45 degrees east latitude.

Queensland. Australia's second-largest state. Covers 668,200 square miles (1,730,648 sq. km.) including almost 2,700 square miles (7,000 sq. km.) of islands. The state constitutes 22.5 percent of Australia's total area. In 2020 its population was 5.1 million; the population is heavily concentrated in the southeast. Capital is Brisbane, with 2.5 million people in 2019. The state has a tropical climate and is known for its unusual flora and fauna, and for its proximity to the Great Barrier Reef. Originally a British penal colony named Moreton Bay.

Rabaul. Active volcano on New Britain Island, in the Bismarck Archipelago, part of Papua New Guinea in the Pacific Ocean. Reaches a height of 2,257 feet (687 meters). Most recent eruption occurred in 2014.

Rapa Nui. See Easter Island.

Red Centre. See Alice Springs.

Rockhampton. City and commercial center in central Queensland, Australia. Located 38 miles (60 km.) inland at the head of the Fitzroy River, which empties into Keppel Bay. Population was 78,592 in 2018. Regional commercial center served by the deepwater Port Alma; important tourist base for ocean access to the Great Barrier Reef.

Ronne Ice Shelf. Large Antarctic ice shelf in the Weddell Sea. Named after Edith Ronne, who spent nearly a year in 1947 on Stonington Island with her husband, explorer Finn Ronne. Originally called Edith Ronne Land, the name was later changed to the Ronne Ice Shelf.

Ross Ice Shelf. Largest ice shelf in Antarctica. About the size of France, it is located along the Ross Sea. Presents an impenetrable surface along its entire face, high cliffs of ice totally without openings, making it impossible for early explorers to move from the Ross Sea into the interior. Scientists have observed melting of the ice shelf due to warming Antarctic temperatures.

Ross Ice Shelf, Ross Sea, Antarctica. (NOAA Corps Collection)

Ross Island. Large island in the Ross Sea of Antarctica. Has four volcanoes, Mount Erebus being the only active one. Covered most of the year by ice that connects it to the mainland. The cold temperatures there prevent rain from falling, but it regularly receives snow. Named after British explorer Sir James Clark Ross.

Ross Sea. Large sea south of the Pacific Ocean on Antarctica's west side; ice-free part of the year. The Transantarctic Mountains run 3,000 miles (4,838 km.) from the Ross Sea across the South Pole to the Weddell Sea. Named after Sir James Clark Ross, who explored the area 1839-1843.

Rotorua. Active thermal region of New Zealand, situated in central North Island. Site of sulfur fumes, bubbling mud, and hot springs. One of the few areas in the world where geysers are found. An important center of Maori culture.

Saipan. One of the NORTHERN MARIANA ISLANDS in the west PACIFIC OCEAN. Area is 47 square miles (120 sq. km.); population was 52,263 in 2017. High cliffs mark the north end of the hilly island. Was included in the Japanese mandate under the League of Nations in 1920; Japan developed it as a naval base. Captured by US troops in 1944 and converted to a US air base for the duration of World War II.

Samoa. Independent island nation in the southern PACIFIC OCEAN. Located on the western part of the Samoan archipelago, west of longitude 171 degrees west. (The part of the archipelago east of this line is AMERICAN SAMOA.) Consists of seven volcanic islands. Area is 1,093 square miles (2,831 sq. km.); 99 percent of this area is Savavi'i and Upolu, the two largest islands, which contain rain forests. Samoa's population was 203,774 in 2016, most of whom live on Savai'I and Upolu. Samoans make up more than 90 percent of the population. Also known as Samoa Islands; formerly Western Samoa.

Sandwich Islands. See HAWAII.

Savage Island. See NIUE.

Sentinel Ridge. Northern ridge of the ELLSWORTH MOUNTAINS in the Antarctic. VINSON MASSIF, ANTARCTICA's highest mountain, is found there.

Shark Bay. Inlet of the INDIAN OCEAN, WESTERN AUSTRALIA. Bisected by Peron Peninsula and sheltered to the west by Dorre, Benier, and Dirk Hartog islands. Named in 1699 by the pirate William Dampier after numerous sharks were sighted. Notable for preserving the last remaining species of dugong, and for rich beds of sea grass.

Shirase. ANTARCTICA's fastest-moving glacier. Located in QUEEN MAUD LAND on the Atlantic side of the continent.

Simpson Desert. Vast arid region of central AUSTRALIA, situated mainly in the southeastern corner of the NORTHERN TERRITORY. Covers approximately 56,000 square miles (145,000 sq. km.), mostly uninhabited. The last refuge for some of Australia's rarest animal species. Characterized by miles of sand dunes and ridges, some rising 70 to 120 feet (20 to 37 meters) in height, and interspersed with growths of *Spinifex* grass. First noted by explorer Charles Sturt in 1845.

Snowy River. River in NEW SOUTH WALES, Australia. Rises from the Snowy Mountains (a section of the Australian ALPS). Fed by melting snow, it flows about 270 miles (435 km.) southeast, then west, and south to the BASS STRAIT. Waters are diverted by the Snowy Mountains scheme, which includes mountain tunnels, sixteen dams, and several reservoirs; this is one of the world's largest power and irrigation projects.

Society Islands. Group of fourteen islands in the west part of FRENCH POLYNESIA in the South PACIFIC OCEAN. Area is 621 square miles (1,608 sq. km.); population was 275,918 in 2018. Volcanic in origin. Made up of two island groups, the Windward Islands (TAHITI, Mooréa, and several islets) and the Leeward Islands.

Solomon Islands. Independent nation, member of the Commonwealth of Nations, in the west PACIFIC OCEAN, east of New Guinea. Area is 11,157 square miles (28,896 sq. km.); population was 685,097 in 2020, mainly Melanesians. Capital is Honiara, on GUADALCANAL. Comprises thirty islands and many atolls. Includes most of the Solomon Islands group: Guadalcanal, New Georgia, Santa Isabel, Malaita, Choiseul, San Cristóbal, Vella Lavella, Ontong Java Atoll, Rennell, and the Santa Cruz Islands. Mountainous islands, mostly of volcanic origin. Major exports include timber, fish, and palm oil.

South Australia. AUSTRALIA's driest state, occupying about one-eighth of the continent's total landmass. Bounded on the west, north, and east by other mainland states; flanked on the south by the GREAT AUSTRALIAN BIGHT of the INDIAN OCEAN. Covers 379,900 square miles (984,377 sq. km.). Capital is ADELAIDE; population was more than 1.75 million in 2020. More than 50 percent of the state is pastoral lands; the southeastern section is the economic center and where most of the population resides. The Murray is the only major river; Mount Lofty/Flinders Ranges are the most important mountains.

South Georgia. Island in the South Atlantic Ocean. The Andes Mountains of South America are part of a mountain chain that extends beneath the ocean and continues to the Antarctic Peninsula. South Georgia was formed when the peaks of some of these submarine mountains emerged above the water.

South Island. Larger and more southerly of the two principal islands of New Zealand in the southwest Pacific Ocean. Cook Strait separates South Island from North Island. Its population has grown much less rapidly than that of North Island. Its area is 58,084 square miles (150,437 sq. km.). Its population was 1,155,400 in 2019.

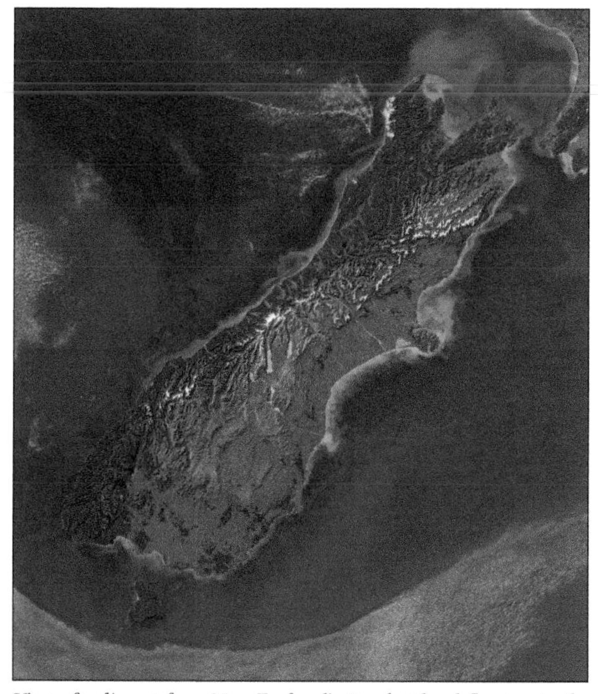

View of sediment from New Zealand's South Island flowing in the Pacific Ocean. (NASA Earth Observatory)

South Magnetic Pole. Southern pole of Earth's magnetic field, to which compass needles point. Located about 170 miles (272 km.) from the geographic South Pole, which is the southernmost point on Earth.

South Orkney Islands. Uninhabited islands in the South Atlantic Ocean off the Antarctic Peninsula. They are the tops of a mountain range that includes the Andes in South America and continues to the mountains on the Antarctic Peninsula. In 1904, Argentina set up a weather station at Laurie Island in the Orkneys.

South Pacific Gyre. Large mass of water, located in the Pacific Ocean in the Southern Hemisphere, that rotates counterclockwise. Warm water moves toward the pole and cold water moves toward the equator.

South Pole. Point which by definition is the southernmost point on Earth—a point from which the only possible direction is north. The geographical South Pole is at an elevation of 9,200 feet (2,804 meters). See also South Magnetic Pole.

Southern Alps. Highest mountain range in Australasia. The backbone of South Island of New Zealand, running almost the entire length of the island in a north-to-south direction. The highest peak is Aoraki/Mount Cook, at 12,317 feet (3,754 meters). The Southern Alps contains 360 glaciers. Permanent snowfields occur throughout most of the mountain range.

Southern Ocean. World's stormiest waters. Located between 60 and 65 degrees south latitude. The cold continent and the warmer ocean waters meet, creating a low-pressure situation that results in cyclones. The winds blow violently almost constantly from east to west until they spiral toward the coast of Antarctica, where some of their fury dissipates. This is nature's way of distributing tropical heat to the polar regions.

Spencer Gulf. Inlet of the Indian Ocean, indenting the southeastern coast of South Australia. Runs 200 miles (320 km.) with a maximum width of 80 miles (130 km.). Several islands are situated in its 50-mile,(80 km.)-wide mouth; major ports include Port Pirie and Port Augusta, which is located near the head of the inlet.

Stewart Island. Smallest of the three main islands of New Zealand. Located 15 miles (24 km.) south of Bluff, across Foveaux Strait. Population was 408 in 2018. Almost triangular in shape; covers 670 square miles (1,746 sq. km.). Most of the terrain is mountainous and covered by bushland.

Stonington Island. Island near the Antarctic Peninsula, where Antarctic explorer Finn

Ronne and his wife, Edith, spent nearly a year in 1947.

Sutherland Falls. Highest waterfall in NEW ZEALAND, fifth-highest in the world, at 1,903 feet (580 meters). Situated in FIORDLAND NATIONAL PARK, at 44°48′ south latitude, longitude 167°44′ east. Fed by waters from Lake Quill, the falls drop down a rockface in three stages. Named after Donald Sutherland, the Hermit of MILFORD SOUND, who discovered the falls in 1880.

Suva. Seaport town, capital of FIJI, on southeast coast of Viti Levu Island in the PACIFIC OCEAN. Metro population was 185,913 in 2017. An excellent harbor; major industries are coconut processing and tourism.

Sydney. AUSTRALIA's oldest and largest city and the capital of NEW SOUTH WALES. Established in 1788 as a British penal colony and located on PORT JACKSON, a magnificent deep harbor along the southeastern coast of the continent. Australia's leading industrial city and its major seaport. The city and suburbs cover about 4,700 square miles (12,000 sq. km.) with a population of 5.3 million in 2018. A major tourist destination, famous for its mild climate and white sand beaches, the Sydney Harbour Bridge, and the Opera House. In 2000 the city hosted the Olympic Games.

Tahiti. Mountainous island in the Windward Islands of the SOCIETY ISLANDS in FRENCH POLYNESIA, in the South PACIFIC OCEAN. Area is 408 square miles (1,057 sq. km.); population was 189,517 in 2017. Largest city is Papeete. Largest of the French islands in the Pacific and the commercial and political center of French Polynesia. Tourism is a major industry. Formerly Otaheite; also known as Taïte.

Tasman Sea. Area of the PACIFIC OCEAN off the southeast coast of AUSTRALIA, between Australia and TASMANIA on the west and NEW ZEALAND on the east. Also called the Tasmanian Sea.

Tasmania. AUSTRALIA's only island state, located about 150 miles (240 km.) south of VICTORIA state across the BASS STRAIT. Comprises the heart-shaped main island (Tasmania) and numerous smaller islands, including Bruny, King, and Flinders in the Bass Strait, and the subarctic Macquarie Island, located 900 miles (1,440 km.) to the southeast. The main island has a maximum width and length of about 200 miles (320 km.) with a cool, Mediterranean-like climate moderated by the SOUTHERN OCEAN. Mount Ossa (part of the GREAT DIVIDING RANGE) is the highest peak at 5,305 feet (1,617 meters). Capital is HOBART; population was just over 500,000 in 2020. Agriculture and mining form Tasmania's economic base. Originally called Van Diemen's Land by the Dutch navigator, Abel Tasman, who discovered it in 1642.

Taupo, Lake. NEW ZEALAND's largest natural lake. Located within the Volcanic Plateau in central NORTH ISLAND; covers 235 square miles (606 sq. km.). Considered one of the best trout-fishing spots in the world. Site of one of the largest volcanic eruptions in the world, around 135 C.E.

Taylor Valley. One of the DRY VALLEYS in EAST (GREATER) ANTARCTICA. Dry valleys receive almost no precipitation, so they do not have the ice and snowcover that characterizes 98 percent of the continent. They can support rudimentary plant life.

Terre Adélie. Area that has the thickest covering of ice in ANTARCTICA. Located 248 miles (400 km.) from the coast, in EAST (GREATER) ANTARCTICA between Wilkes Land and George V Land. Also called Adélie Coast.

Terror, Mount. Inactive volcano in ANTARCTICA. Located close to the active volcano Mount EREBUS. Sir James Clark Ross discovered it; nearby, he encountered a perpendicular wall of ice 200 feet (60 meters) high that extended as far as the eye could see and blocked his passage.

Timor Sea. Section of the INDIAN OCEAN, located northwest of AUSTRALIA, and southeast of Indonesia and the island of Timor. Covers approximately 235,000 square miles (615,000 sq. km.) and is about 300 miles (480 km.) wide. Opens east into the ARAFURA SEA and west into the Indian Ocean.

Tinian. Island in the NORTHERN MARIANA ISLANDS in the west PACIFIC OCEAN. Area is 20 square miles (52 sq. km.). Part of the Japanese mandate in 1919; occupied by US troops in 1944 during World War II. In 1945, planes sent from its airfield dropped atomic bombs on Hiroshima and Nagasaki.

Tokelau. Group of islands in central Pacific Ocean; part of New Zealand. Area is 4 square miles (10 sq. km.); population was 1,647 in 2019. Part of the Gilbert and Ellice islands (*see* Tuvalu). Includes islands of Atafu, Fakaofo, and Nukunonu. Formerly Union Islands.

Tonga. Independent island nation in the southern Pacific Ocean. Area is 290 square miles (750 sq. km.); population was 106,095 in 2020, two-thirds living on Tongatapu. Capital, chief port, and largest town is Nukualofa. Polynesian monarchy, comprising 170 islands (40 inhabited) divided into three main groups: Tongatapu, Ha'apai, and Vava'u. Largest island is Tongatapu. Western islands are of volcanic origin; eastern are coral islands. Active volcanoes dot several islands. Most Tonga residents (97 percent) are Polynesians, and Tonga has retained much of its traditional culture, although the island is Christian. Also known as Tonga Islands or Friendly Islands.

Tonga Trench. Seafloor depression in the Pacific Ocean, 35,433 feet (10,800 meters) deep. Located near Tonga, at 23°16′ south latitude, longitude 174°44′ west.

Tongariro National Park. First national park in New Zealand. Gifted by chiefs from the Maori Tuwharetoa tribe to the people of Aotearoa (New Zealand) in 1887. Located in central North Island, covering 196,602 acres (79,598 hectares). Site of three active volcanoes including Mount Ruapehu, highest peak in the North Island at 9,177 feet (2,797 meters).

Torres Strait. Passage through the Coral Sea to the east and the Arafura Sea to the west in the western Pacific Ocean; to the north is New Guinea and to the south lies Australia's Cape York Peninsula. Dangerous to navigate because of reefs and rocky shoals. Nearly 95 miles (150 km.) wide; some of its larger islands are inhabited. Discovered in 1606 by Spanish navigator Luis Vaez de Torres.

Townesville. City, commercial center, and major port of eastern Queensland, Australia. Located close to the Flinders Passage through the Great Barrier Reef and to the Atherton Plateau inland, making it an important gateway to both areas. Population was 180,820 in 2018.

Transantarctic Mountains. Range of mountains running across Antarctic from the Ross Sea to the Weddell Sea, passing over the South Pole. Divides Antarctica into East (Greater) Antarctica south of the Indian Ocean and West (Lesser) Antarctica south of the Pacific Ocean. Has a trough running alongside it with an accumulation of ice 2 miles (3.2 km.) deep, half of it below sea level. Also called the Great Antarctic Horst.

Tuamotu Archipelago. About eighty islands in French Polynesia in the South Pacific Ocean. Area is 331 square miles (857 sq. km.); population was 15,346 in 2017. Mostly coral atolls; its main islands are Makatéa, Fakarava, Rangiroa, Anaa, Hao, and Reao. France has used it for atomic testing since in the mid-1960s. Also known as Low Archipelago, Paumotu, or Dangerous Archipelago.

Tubuai Islands. See Austral Islands.

Tuvalu. Island country in the western Pacific Ocean; member of the Commonwealth of Nations. Area is 10 square miles (26 sq. km.); population was 11,342 in 2020. Consists of a chain of nine coral islands, none of which rises more than 16 feet (5 meters) above sea level. Funafuti, Nanumea, Nui, Nukufetau, and Hukulaelae are atolls; Namumanga, Niutao, Vaitupu, and Niulakita are single islands with lagoons. About one-third of Tuvalu's people, who are mostly ethnic Polynesians, live in Funafuti. Also known as Lagoon Islands; formerly Ellice Islands.

Ulawun. Active volcano in the Whiteman range, on the island of New Britain, part of Papua New Guinea in the Pacific Ocean. Located on the north coast and east end of the island; reaches 7,657 feet (2,334 meters) in height. Most recent eruption occurred in 2019. The volcano is also known as The Father.

Uluru. Giant red sandstone monolith located about 280 miles (450 km.) northeast of Alice Springs in Australia's Northern Territory. The largest in a group of tors (isolated and weathered rock masses) that includes the Olgas/Kata Tjuta; the Uluru-Kata Tjuta National Park is listed as a World Heritage Site. Rises more than 1,132 feet (345 meters) from a flat, sandy plateau; it is oval in shape, 1.5 miles (2.4 km.) long, 0.9 mile (2 km.) wide, and 6 miles (9.45 km.) around its base. Its coarse sandstone composition causes it to glow a brilliant red at sunrise and sunset. In 1985, ownership of the rock was returned to the Aboriginal Australians, who leased it back to the government for ninety-nine years. The rock takes its name from an Aboriginal word for giant pebble; the first European to sight the monolith, Ernest Giles, named it Ayers Rock after South Australian premier Sir Henry Ayers.

Union Islands. See Tokelau.

Vanuatu. Independent republic in the southwestern Pacific Ocean. Area is 4,707 square miles (12,190 sq. km.); population was 298,333 in 2020. Capital is Port-Vila. Consists of more than eighty islands—70 percent of which are inhabited—in a Y-shaped configuration. The largest island is Espiritu Santo; others are Malekula, Éfaté, Erromango, and Ambrim. Most of the islands are volcanic in origin; there are several active volcanoes. About 99 percent of the population are ethnic Melanesians (ni-Vanuatu); about 70 percent live on Anatom, Éfaté, Espiritu Santo, Futuna, Malekula, and Tanna. Formerly known as New Hebrides.

Victoria. Australia's smallest mainland state, located in the southeast corner of the continent. Bordered to the north by New South Wales (with the Murray River forming a natural boundary), to the east by the Australia Alps section of the Great Dividing Range, to the west by the deserts of South Australia, and to the south by Bass Strait. Covers 87,884 square miles (227,600 sq. km.), with a range of climates. Population was 6.6 million in 2020, with most living in the capital city of Melbourne and its suburbs. Victoria's major harbor is located in Port Phillip Bay.

Victoria Land. Region of Antarctica that presses against the Transantarctic Mountains east of the Ross Sea. Named after Britain's Queen Victoria.

Vinson Massif. Highest point in Antarctica. Soars to 16,050 feet (4,892 meters), higher than any mountain in the United States except for Denali.

Vostok. Russian research station near the South Magnetic Pole in Antarctica. Site of the lowest temperature ever recorded: -1286 degrees Fahrenheit (-89.2 degrees Celsius) in 1983.

Waialeale. Mountain in central Kauai Island, Hawaii, that is the rainiest place in the world, according to the National Geographic Society. Average annual rainfall is 460 inches (1,168 centimeters). Reaches 5,200 feet (1,585 meters).

Waikato River. Longest river in New Zealand, stretching 264 miles (425 km.). Flows northward across central North Island, from the headwaters at Mount Ruapehu, through Lake Taupo, reaching the Tasman Sea 30 miles (50 km.) southwest of Auckland. A major source of hydroelectric power for North Island.

Waitangi. Modern birthplace of New Zealand. Located on North Island in the Bay of Islands. Site of the signing of the Treaty of Waitangi on February 6, 1840, between representatives of the British Crown and a number of Maori chiefs. Under the terms of the treaty, the Maori chiefs ceded sovereignty to the Crown and were guaranteed full rights of tribal lands and personal protection as British subjects. Maori word for "weeping waters."

Wake Island. Coral atoll in the central Pacific Ocean, territory of the United States. Area is 3 square miles (7.7 sq. km.); there is no indigenous population. Consists of three islets (Wake, Peale, and Wilkes) around a shallow lagoon. The United States formally occupied Wake Island in 1898 and built a naval air base and submarine base on the island in 1941. Attacked by the Japanese on December 7, 1941, immediately after Pearl Harbor. It reverted to the United States after World War II and today is administered by the US Air Force. Formerly known as Halycon Island.

Wallis and Futuna Islands. Overseas collectivity of France, in the southwestern Pacific Ocean. Consists of the Wallis Archipelago (Wallis and twenty smaller islands and islets) and the Futuna Archipelago (two mountainous islands, Futuna and Alofi). Total area is 106 square miles (274 sq. km.); population was 15,854 in 2020. Capital is Matâ-Uti on Wallis.

Weddell Sea. Body of water formed by an indentation in the continent of Antarctica. Located southeast of South America; bordered by the Antarctic Peninsula, the Leopold Coast, the Cairn Coast, and Coats Land. Named for James Weddell, a British navigator who laid claim to it in 1823.

Wellington. Capital and third-largest city of New Zealand. Population was 215,400 in 2019. Located on the shores of a natural harbor, at the southern end of North Island, on a major earthquake fault. Built on steep, hilly terrain, similar to San Francisco.

West (Lesser) Antarctica. Area that lies west of the Transantarctic Mountains. Includes Marie Byrd Land, Ellsworth Land, the Antarctic Peninsula, and the Ross Sea.

Western Australia. Australia's largest state; occupies nearly one-third of the continent. Bounded to the north by the Timor Sea, to the south by the Antarctic Ocean, to the west by the Indian Ocean, and to the east by the deserts of South Australia and the Northern Territory. Covers 975,100 square miles (2,525,500 sq. km.), most of which is a semiarid plateau. Has no winter, only "wet" and "dry" seasons. Population exceeded 2.6 million in 2020, more than 70 percent residing in and around Perth, the capital city.

Western Samoa. See Samoa.

Whakaari/White Island. Active volcano in New Zealand. Located 31 miles (50 km.) off the Bay of Plenty on North Island. Characterized by intense thermal activity as steam, noxious gases, and toxic fumes that are released into the atmosphere from numerous vents in the crater. In 2019, a violent eruption killed twenty-one people, most of them tourists.

Whales, Bay of. Bay in the Ross Sea on Antarctica's west side. Has much less wind than other parts of Antarctica. Over a four-year period, the average wind speed was 11 miles (17.6 km.) per hour, about equivalent to that in Peoria, Illinois, or Kansas City, Missouri.

White Desert. Another name for the polar ice cap. Antarctica is Earth's driest continent. The annual snowfall on the polar ice cap is only 1 to 2 inches (2.5 to 5 centimeters); the coastal regions receive 10 to 20 inches (25 to 51 centimeters) of snow a year.

Wilhelm, Mount. Mountain in Bismarck Range, New Guinea Island, Papua New Guinea, in the Pacific Ocean. Highest peak in New Guinea at 14,795 feet (4,509 meters).

Wilsons Promontory. Granite peninsula projecting into the Bass Strait, forming the southernmost point of the Australian mainland. Located in the state of Victoria, 110 miles (177 km.) southeast of Melbourne. It is 22 miles (35 km.) long and a maximum of 14 miles (22.5 km.) wide; the 80-mile (130-km.) coastline rises to a mountainous interior, with Mount Latrobe (2,475 feet/754 meters) its highest peak.

Yasur. Active volcano on Tanna Island, south Vanuatu. Reaches 1,184 feet (361 meters) in height. Its most recent eruption occurred in 2020.

York, Cape. Northernmost tip of Australia, located at 10°41′ south latitude, longitude 142°32′ east. Both it and Cape York Peninsula were named by Captain James Cook in 1770, in honor of Great Britain's Duke of York. Reaches into Torres Strait, which separates Australia from New Guinea.

Cynthia Beres; Huia Richard Hutton; Dana P. McDermott; R. Baird Shuman; Rowena Wildin

Appendix

The Earth in Space

The Solar System

Earth's solar system comprises the Sun and its planets, as well as all the natural satellites, asteroids, meteors, and comets that are captive around it. The solar system formed from an interstellar cloud of dust and gas, or nebula, about 4.6 billion years ago. Gravity drew most of the dust and gas together to make the Sun, a medium-size star with an estimated life span of 10 billion years. Its system is located in the Orion arm of the Milky Way galaxy, about two-thirds of the way out from the center.

During the Sun's first 100 million years, the remaining rock and ice smashed together into increasingly larger chunks, or planetesimals, until the planets, moons, asteroids, and comets reached their present state. The resulting disk-shaped solar system can be divided into four regions—terrestrial planets, giant planets, the Kuiper Belt, and the Oort Cloud—each containing its own types of bodies.

Terrestrial Planets

In the first region are the terrestrial (Earth-like) planets Mercury, Venus, Earth, and Mars. Mercury, the nearest to the Sun, orbits at an average distance of 36 million miles (58 million km.) and Mars, the farthest, at 142 million miles (228 million km.). Astronomers call the distance from the Sun to Earth (93 million miles/150 million km.) an astronomical unit (AU) and use it to measure planetary distances.

Terrestrial planets are rocky and warm and have cores of dense metal. All four planets have volcanoes, which long ago spewed out gases that created atmospheres on all but Mercury, which is too close to the Sun to hold onto an atmosphere. Mercury is heavily cratered, like the earth's moon. Venus has a permanent thick cloud cover and a surface temperature hot enough to melt lead. The air on Mars is very thin and usually cold, made mostly of carbon dioxide. Its dry, rock-strewn surface has many craters. It also has the largest known volcano in the solar system, Olympus Mons, which is 16 miles (25 km.) high.

Average temperatures and air pressures on Earth allow liquid water to collect on the surface, a unique feature among planets within the solar system. Meanwhile, Earth's atmosphere—mostly nitrogen and oxygen—and a strong magnetic field protect the surface from harmful solar radiation. These are the conditions that nurture life, according to scientists. It is widely accepted that Mars had abundant water very early in its history, but all large areas of liquid water have since disappeared. A fraction of this water is retained on modern Mars as both ice and locked into the structure of abundant water-rich materials, including clay minerals (phyllosilicates) and sulfates. Studies of hydrogen isotopic ratios indicate that asteroids and comets from beyond 2.5 AU provide the source of Mars' water. Like Earth, Mars has polar ice caps, although those on Mars are made up mostly of carbon dioxide ice (dry ice), while those on Earth are made up of water ice.

A single natural satellite, the Moon, orbits Earth, probably created by a collision with a huge planetesimal more than 4 billion years ago. Mars has two tiny moons that may have drifted to it from the asteroid belt. A broad ring from 2 to 3.3 AU from the Sun, this belt is composed of space rocks as small as dust grains and as large as 600 miles (1,000 km.) in diameter. Asteroids are made of mineral compounds, especially those containing iron, carbon, and silicon. Although the asteroid belt contains

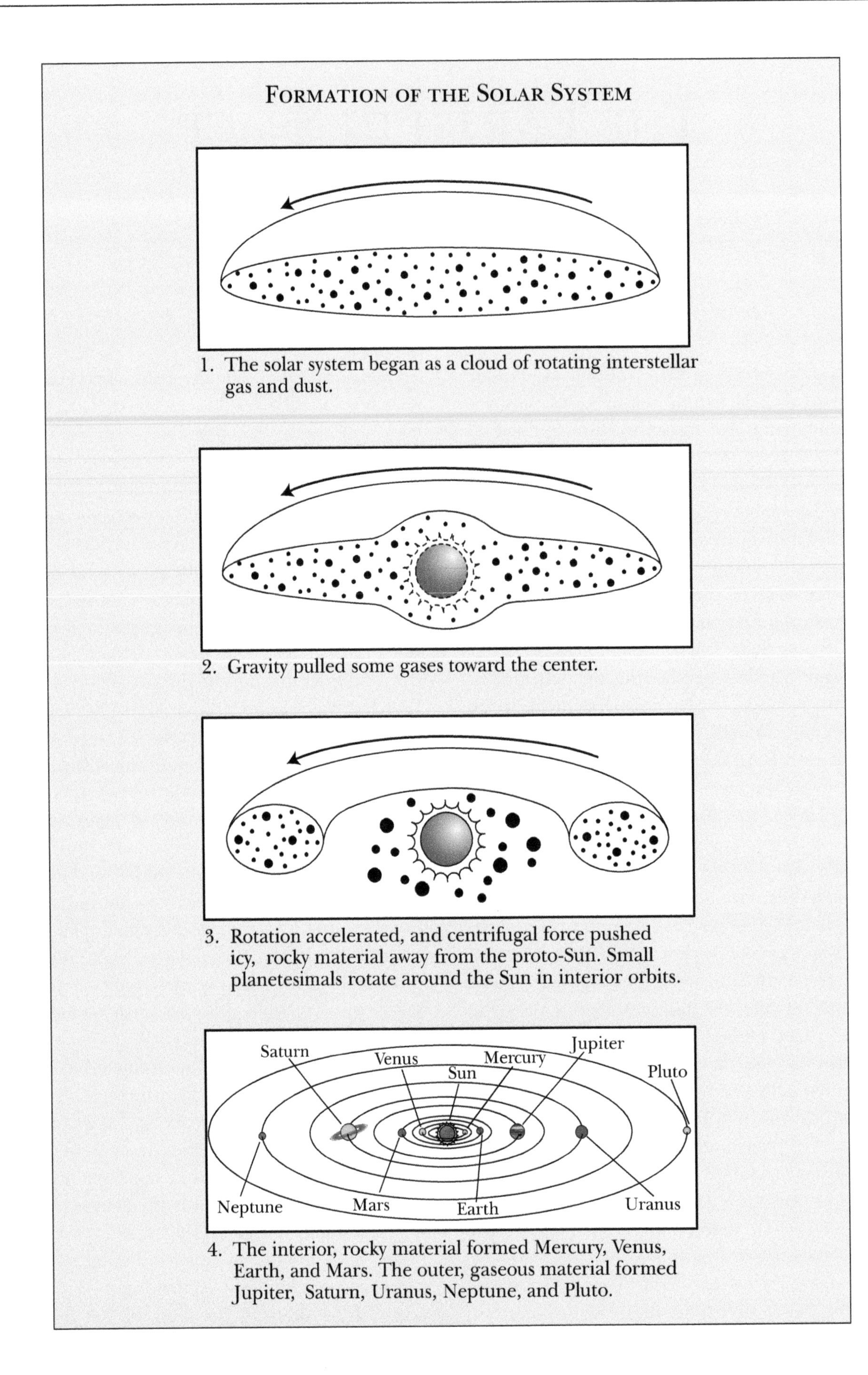

1. The solar system began as a cloud of rotating interstellar gas and dust.

2. Gravity pulled some gases toward the center.

3. Rotation accelerated, and centrifugal force pushed icy, rocky material away from the proto-Sun. Small planetesimals rotate around the Sun in interior orbits.

4. The interior, rocky material formed Mercury, Venus, Earth, and Mars. The outer, gaseous material formed Jupiter, Saturn, Uranus, Neptune, and Pluto.

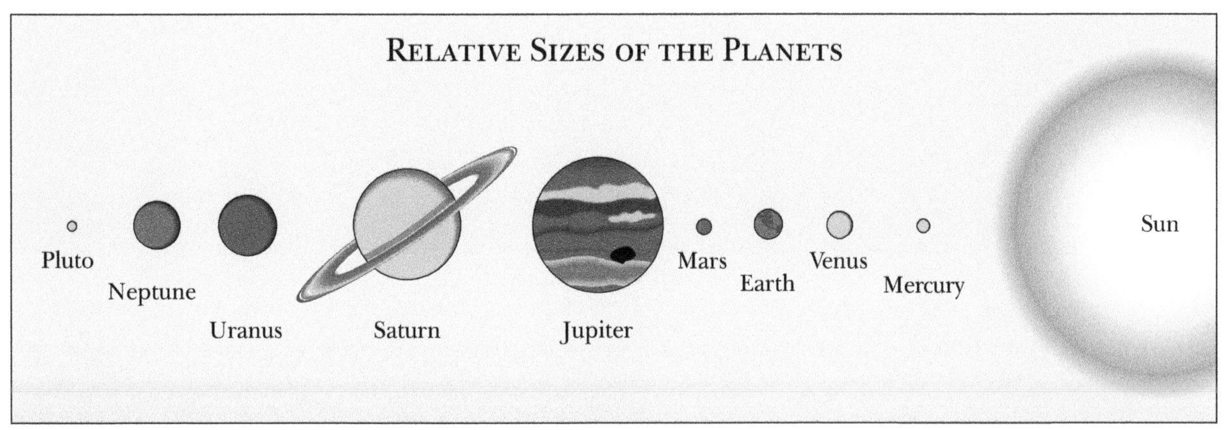

Note: The size of the Sun and distances between the planets are not to scale. Source: Data are from Jet Propulsion Laboratory, California Institute of Technology. The Deep Space Network. Pasadena, Calif.:JPL, 1988, p. 17.

enough material for a planet, one did not form there because Jupiter's gravity prevented the asteroids from crashing together. The belt separates the first region of the solar system from the second.

The Giant Planets

The second region belongs to the gas giants Jupiter, Saturn, Uranus, and Neptune. The closest, Jupiter, is 5.2 AU from the Sun, and the most distant, Neptune, is 30.11 AU. Jupiter is the largest planet in the solar system, its diameter 109 times larger than Earth's. The giant planets have solid cores, but most of their immense size is taken up by hydrogen, helium, and methane gases that grow thicker and thicker until they are like sludge near the core. On Jupiter, Saturn, and Uranus, the gases form wide bands over the surface. The bands sometimes have immense circular storms like hurricanes, but hundreds of times larger. The Great Red Spot of Jupiter is an example. It has winds of up to 250 miles (400 km.) per hour, and is at least a century old.

These planets have such strong gravity that each has attracted many moons to orbit it. In fact, they are like miniature solar systems. Jupiter has the most moons—eighteen—and Neptune has the fewest—eight—but Neptune's moon Triton is the largest of all. Most moons are balls of ice and rock, but Jupiter's Europa and Saturn's Titan may have liquid water below ice-bound surfaces. Several moons appear to have volcanoes, and a wispy atmosphere covers Titan. Additionally, the giant planets have rings of broken rock and ice around them, no more than 330 feet (100 meters) thick. Saturn's hundreds of rings are the brightest and most famous.

The Kuiper Belt

The third region of the solar system, the Kuiper Belt, contains the dwarf planet, Pluto. Pluto has a single moon, Charon. It does not orbit on the same plane, called the ecliptic, as the rest of the planets do. Instead, its orbit diverges more than seventeen degrees above and below the ecliptic. Its orbit's oval shape brings Pluto within the orbit of Neptune for a large percentage of its long year, which is equal to 248 Earth years. Two-thirds the size of the earth's moon, Pluto has a thin, frigid methane atmosphere. Charon is half Pluto's size and orbits less than 32,000 miles (20,000 km.) from Pluto's surface. Some astronomers consider Pluto and Charon to be a double planet.

The Kuiper Belt holds asteroids and the "short-period" comets that pass by Earth in orbits of 20 to 200 years. These bodies are the remains of

Other Earths

By the year 2000 astronomers had detected twenty-eight planets circling stars in the Sun's neighborhood of the galaxy. Planets, they think, are common. Those found were all gas giants the size of Saturn or larger. Earth-size planets are much too small to spot at such great distances. Where there are gas giants, there also may be terrestrial dwarfs, as in Earth's solar system. Where there are terrestrial planets, there may be liquid water and, possibly, life.

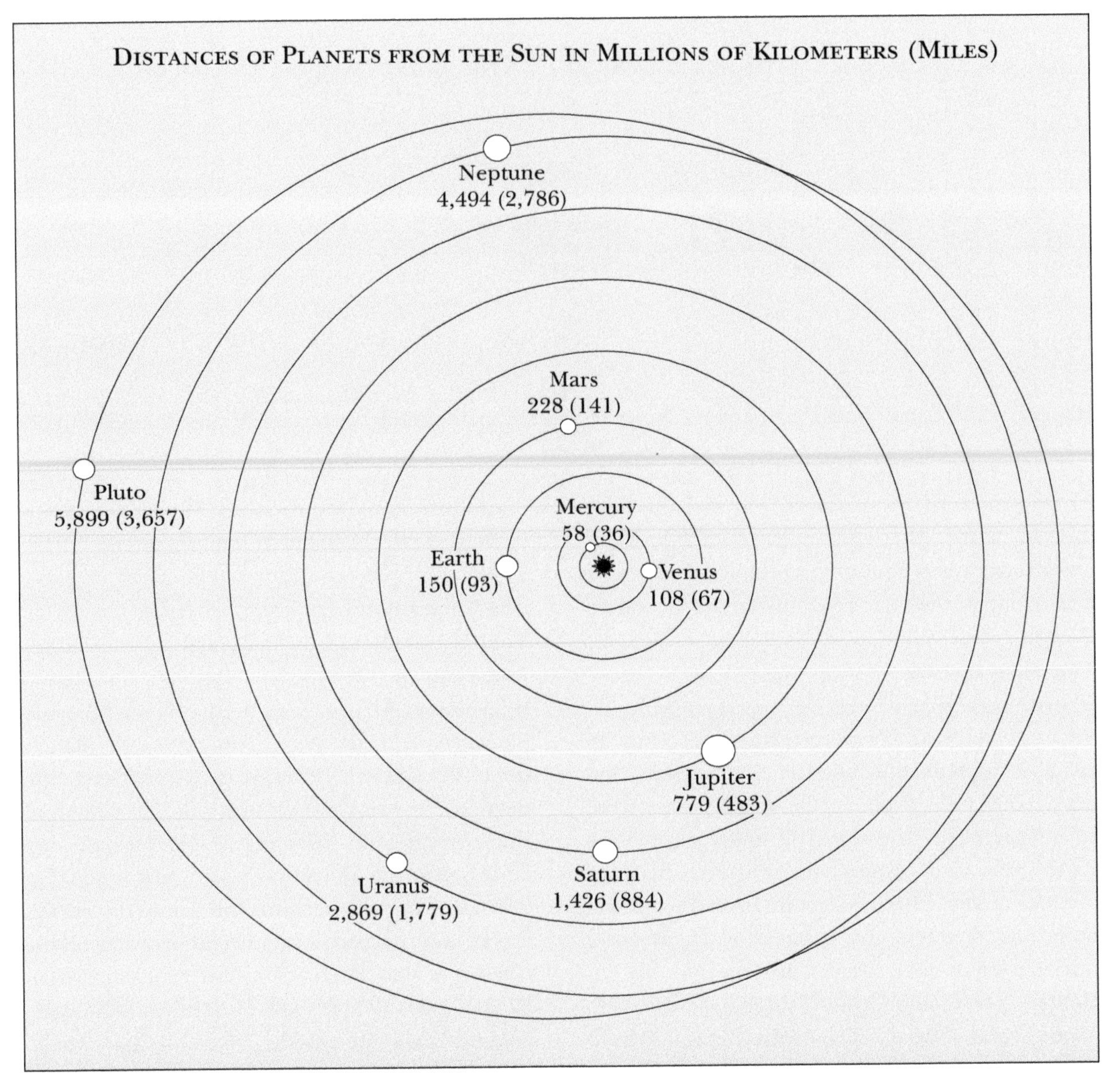

planet formation and did not collect into planets because distances between them are too great for many collisions to occur. Most of them are loosely compacted bodies of ice and mineral—"dirty snowballs," as they were termed by the famous astronomer Fred Lawrence Whipple (November 5, 1906–August 30, 2004). An estimated 200 million Kuiper Belt objects orbit within a band of space from 30 to 50 AU from the Sun.

The Oort Cloud

In contrast to the other regions of the solar system, the Oort Cloud is a spherical shell surrounding the entire solar system. It is also a collection of comets—as many as two trillion, scientists calculate. The inner edge of the cloud forms at a distance of about 20,000 AU from the Sun and extends as far out as 100,000 AU. The Oort Cloud thus gives the solar system a theoretical diameter of 200,000 AU—a distance so vast that light needs more than three years to cross it. No astronomer has yet detected an Oort Cloud object, because the cloud is so far away. Occasionally, however, gravity from a nearby star dislodges an object in the cloud, causing it to fall toward the Sun. When observers on Earth see such an object sweep by in a long, cigar-shaped orbit, they call it a long-period comet.

The outer edge of the Oort Cloud marks the farthest reach of the Sun's gravitational power to bind bodies to it. In one respect, the Oort Cloud is part of interstellar space.

In addition to light, the Sun sends out a constant stream of charged particles—atoms and subatomic particles—called the solar wind. The solar wind shields the solar system from the interstellar medium, but it only does so out to about 100 AU, a boundary called the heliopause. That is a small fraction of the distance to the Oort Cloud.

Roger Smith

Earth's Moon

The fourth-largest natural satellite in the solar system, Earth's moon has a diameter of 2,159.2 miles (3,475 km.)—less than one-third the diameter of Earth. The Moon's mass is less than one-eightieth that of Earth.

The Moon orbits Earth in an elliptical path. When it is at perigee (when it is closest to Earth), it is 221,473 miles (356,410 km.) distant. When it is at apogee (farthest from Earth), it is 252,722 miles (406,697 km.) distant.

The Moon completes one orbit around Earth every 27.3 Earth days. Because it rotates at about the same rate that it orbits the earth, observers on Earth only see one side of the Moon. The changing angles between Earth, the Sun, and the Moon determine how much of the Moon's illuminated surface can be seen from Earth and cause the Moon's changing phases.

Volcanism

Naked-eye observations of the Moon from Earth reveal dark areas called *maria*, the plural form of the Latin word *mare* for sea. The maria are the remains of ancient lava flows from inside gigantic impact craters; the last eruptions were more than 3 billion years ago. The lava consists of basalt, similar in composition to Earth's oceanic crust and many volcanoes. The maria have names such as Mare Serenitatis (15° to 40°N, 5° to 20°E) and Mare Tranquillitatis (0° to 20°N, 15° to 45°E). Some of the smaller dark areas on the Moon also have names that are water-related: lacus (lake), sinus (bay), and palus (marsh).

Impact Craters

Observing the Moon with an optical aid, such as a telescope or a pair of binoculars, provides a closer view of impact craters. Impact craters of various sizes cover 83 percent of the Moon's surface. More than 33,000 craters have been counted on the Moon.

One of the easiest craters to observe from the Earth is Tycho. Located at 43.3°S, longitude 11.2 degrees west, it is about 50 miles (85 km.) wide. Surrounding Tycho are rays of dusty material, known as

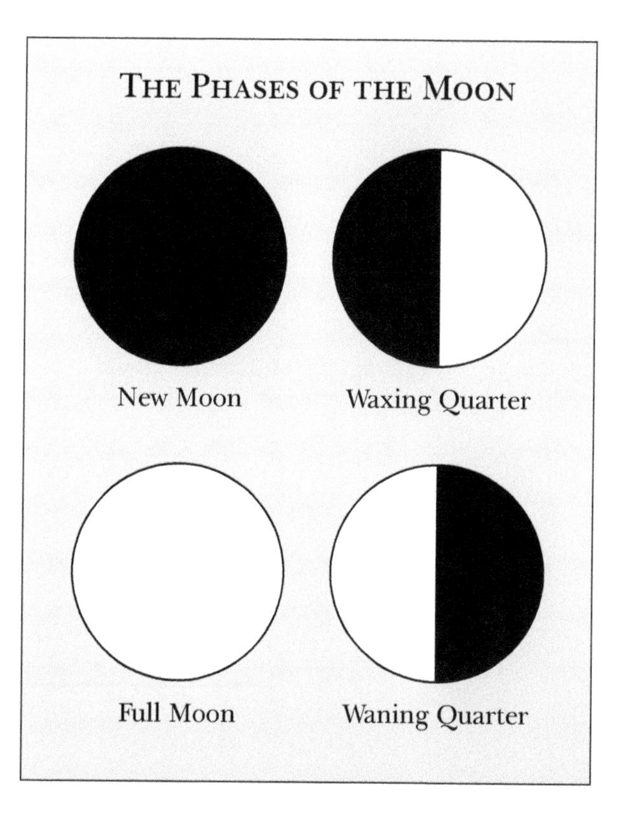

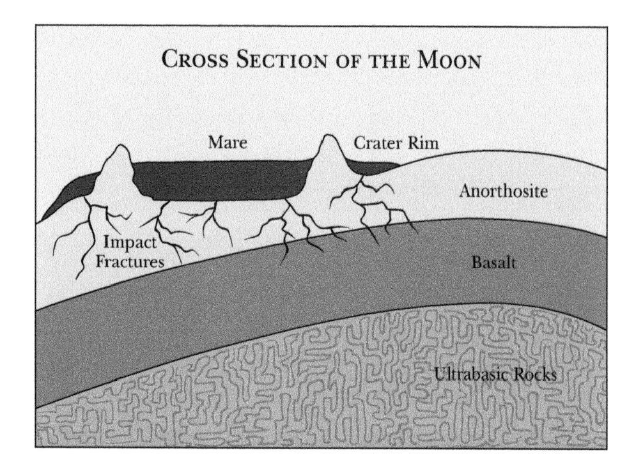

ejecta, that appear to radiate from the crater. When an object from space, such as a meteoroid, slams into the Moon's surface, it is vaporized upon impact. The dust and debris from the interior of the crater fall back onto the lunar surface in a pattern of rays. Because the ejecta is disrupted by subsequent impacts, only the youngest craters still have rays. Sometimes, pieces of the ejecta fall back and create smaller craters called secondary craters. The ejecta rays of Tycho extend to almost 1,865 miles (3,000 km.) beyond the crater's edge.

Other Lunar Features

Near the crater called Archimedes is the Apennines mountain range, which has peaks nearly 20,000 feet (60,000 meters) high—altitudes comparable to South America's Andes.

The Moon also has valleys. Two of the most well known are the Alpine Valley, which is about 115 miles (185 km.) long; and the Rheita Valley, located about 155 miles (250 km.) from the Stevinus crater, which is 238 miles (383 km.) long, 15.5 miles (25 km.) wide, and 2,000 feet (609 meters) deep.

Smaller than valleys and resembling cracks in the lunar surface are features called rilles, which are thought to be places of ancient lava flow. Many rilles can be seen near the Aristarchus crater. Rilles are often up to 3 miles (5 km.) wide and can stretch for more than 104 miles (167 km.).

A wrinkle in the lunar surface is called a ridge. Many ridges are found around the boundaries of the maria. The Serpentine Ridge cuts through Mare Serenitatis.

Exploration of the Moon

Robotic spacecraft were the first visitors to explore the Moon. The Russian spacecraft Luna 1 made the first flyby of the Moon in January, 1959. Eight months later, Luna 2 made the first impact on the Moon's surface. In October, 1959, Luna 3 was the first spacecraft to photograph the side of the Moon not visible from Earth. In 1994 the United States' *Clementine* spacecraft was the first probe to map the Moon's composition and topography globally.

The first humans to land on the Moon were the U.S. astronauts Neil Armstrong and Edwin "Buzz" Aldrin. On July 20, 1969, they landed in the *Eagle* lunar module, during the Apollo 11 mission. Armstrong's famous statement, "That's one small step for man, one giant leap for mankind," was heard around the world by millions of people who watched the first humans set foot on the lunar surface, at the Sea of Tranquillity. The last twentieth century human mission to reach the lunar surface, Apollo 17, landed there in December, 1972. Astronauts Gene Cernan and geologist Jack Schmitt landed in the Taurus-Littrow Valley (20°N, 31°E).

Noreen A. Grice

The Sun and the Earth

Of all the astronomical phenomena that one can consider, few are more important to the survival of life on Earth than the relationship between Earth and the Sun. With the exception of small amounts of residual (endogenic) energy that have remained inside the earth from the time of its formation some 4.5 billion years ago and which sustain some specialized forms of life along some oceanic rift systems, almost all other forms of life, including human, depend on the exogenic light and energy that the earth receives directly from the Sun.

The enormous variety of ecosystems on Earth are highly dependent on the angles at which the Sun's rays strike Earth's spherical surface. These angles, which vary greatly with latitude and time of year, determine many commonly observed phenomena, such as the height of the Sun above the horizon, the changing lengths of day and night throughout the year, and the rhythm of the seasons. Daily and seasonal changes have profound effects on the many climatic regions and life cycles found on earth.

The Sun

The center of Earth's solar system, the Sun is but one ordinary star among some 100 billion stars in an ordinary cluster of stars called the Milky Way galaxy. There are at least 10 billion galaxies in the universe, each with billions of stars. Statistically, the chances are good that many of these stars have their own solar systems. Late twentieth century astronomical observations discovered the presence of what appear to be planets, large ones similar in size to Jupiter, orbiting other stars.

Earth's Sun is an average star in terms of its physical characteristics. It is a large sphere of incandescent gas that has a diameter more than 100 times that of Earth, a mass more than 300,000 times that of Earth, and a volume 1.3 million times that of Earth. The Sun's surface gravity is thirty-four times that of Earth.

The conversion of hydrogen into helium in the Sun's interior, a process known as nuclear fusion, is the source of the Sun's energy. The amount of mass that is lost in the fusion process is minuscule, as evidenced by the fact that it will take perhaps 15 million years for the Sun to lose one-millionth of its total mass. The Sun is expected to continue shining through another several billion years.

Earth Revolution

The earth moves about the Sun in a slightly elliptical orbit called a revolution. It takes one year for the earth

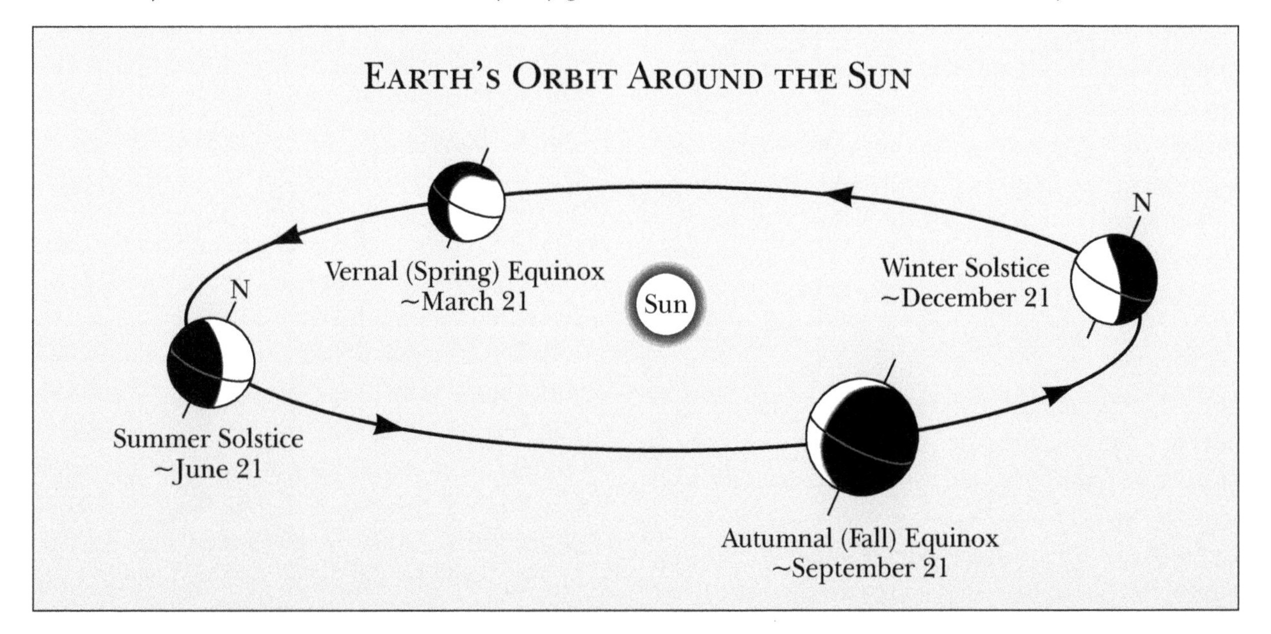

to make one revolution at an average orbital velocity of about 29.6 kilometers per second (18.5 miles per second). Earth-sun relationships are described by a tropical year, which is defined as the period of time (365.25 average solar days) from one vernal equinox to another. To balance the tropical year with the calendar year, a whole day (February 29) is added every fourth year (leap year). Other minor adjustments are necessary so as to balance the system.

Perihelion and Aphelion

The average distance between Earth and the Sun is approximately 93 million miles (150 million km.). At that distance, sunlight, which travels at the speed of light (186,000 miles/300,000 kilometers per second), takes about 8.3 minutes to reach the earth. Since the earth's orbit is an ellipse rather than a circle, the earth is closest to the Sun on about January 3—a distance of 91.5 million miles (147 million km.). This position in space is called perihelion, which comes from the Greek *peri*, meaning "around" or "near," and *helios*, meaning the Sun. Earth is farthest from the Sun on about July 4 at aphelion (Greek *ap*, "away from," and *helios*), with a distance of 152 million kilometers (94.5 million miles).

Axial Inclination

Astronomers call the imaginary surface on which Earth orbits around the Sun the plane of the ecliptic. The earth's axis is inclined 66.5 degrees to the plane of the ecliptic (or 23.5 degrees from the perpendicular to the plane of the ecliptic), and it maintains this orientation with respect to the stars. Thus, the North Pole points in the same direction to Polaris, the North Star, as it revolves about the Sun. Consequently, the Northern Hemisphere tilts away from the Sun during one-half of Earth's orbit and toward the Sun through the other half.

Winter solstice occurs on December 21 or 22, when the tilt of the Northern Hemisphere away from the Sun is at its maximum. The opposite condition occurs during summer solstice on June 21 or 22, when the Northern Hemisphere reaches its maximum tilt toward the Sun. The equinoxes occur midway between the solstices when neither the Southern nor the Northern Hemisphere is tilted toward the Sun. The vernal and autumnal equinoxes occur on March 20 or 21 and September 22 or 23, respectively.

The axial inclination of 66.6 degrees (or 23.5 degrees from the perpendicular) explains the significance of certain parallels on the earth. The noon sun shines directly overhead on the earth at varying latitudes on different days—between 23.5°S and 23.5°N. The parallels at 23.5°S and 23.5°N are called the Tropics of Capricorn and Cancer, respectively.

During the winter and summer solstices, the area on the earth between the Arctic Circle (at 66.5°N) and the North Pole has twenty-four hours of darkness and daylight, respectively. The same phenomena occurs for the area between the Antarctic Circle (at 66.5°S) and the South Pole, except that the seasons are reversed in the Southern Hemisphere. At the poles, the Sun is below the horizon for six months of the year.

For those living outside the tropics (poleward of 23.5 degrees north and south latitude), the noon sun will never shine directly overhead. Hours of daylight will also vary greatly during the year. For example, daylight will range from approximately

ECLIPSES

The Sun's diameter is 400 times larger than the moon's; however, the moon is 400 times closer to Earth than the Sun, making the two objects appear nearly the same size in the sky to observers on Earth. As the moon orbits Earth, it crosses the plane of the Earth-Sun orbit twice each month. If one of the orbit-crossing points (called nodes) occurs during a new or full moon phase, a solar or lunar eclipse can occur.

A solar eclipse occurs when the moon and the Sun appear to be in the exact same place in the sky during a new moon phase. When that happens, the moon blocks the light of the Sun for up to seven minutes. Because solar eclipses can be seen only from certain places on Earth, some people travel around the world—sometimes to remote places—to view them.

A lunar eclipse occurs when Earth is positioned between the Sun and the moon and casts its shadow on the moon. In contrast to solar eclipses, lunar eclipses are visible from every place on Earth from which the moon can be seen.

nine hours during the winter solstice to fifteen hours during the summer solstice for persons living near 40°N, such as in Philadelphia, Denver, Madrid, and Beijing.

Solar Radiation

Given the size of the earth and its distance from the Sun, it is estimated that this planet receives only about one two-billionth part of the total energy released by the Sun. However, this seemingly small amount is enough to drive the massive oceanic and atmospheric circulation systems and to support all life processes on Earth.

Solar energy is not evenly distributed on Earth. The higher the angle of the Sun in the sky, the greater the duration and intensity of the insolation. To illustrate this, note how easy it is look at the Sun when it is very low on the horizon—near dawn and sunset. At those times, the Sun's rays have to penetrate much more of the atmosphere, so more of the sunlight is absorbed. When the Sun's rays are coming in at a low angle, the same solar energy is spread over a larger area, thereby leading to less insolation per unit of area. Thus, the equatorial region receives much more solar energy than the polar region. This radiation imbalance would make the earth decidedly less habitable were it not for the atmospheric and oceanic circulation systems (such as the warm Gulf Stream) that move the excess heat from the Tropics to the middle and high latitudes.

Robert M. Hordon

The Seasons

Earth's 365-day year is divided into seasons. In most parts of the world, there are four seasons—winter, spring, summer, and fall (also called autumn). In some tropical regions—those close to the equator—there are only two seasons. In areas close to the equator, temperatures change little throughout the year; however, amounts of rainfall vary greatly, resulting in distinct wet and dry seasons. The polar regions of the Arctic and Antarctic also have little variation in temperature, remaining cold throughout the year. Their seasons are light and dark, because the Sun shines almost constantly in the summer and hardly at all in the winter.

The four seasons that occur throughout the northern and southern temperate zones—between the tropics and the polar regions—are climatic seasons, based on temperature and weather changes. Winter is the coldest season; it is the time when days are short and few crops can be grown. It is followed by spring, when the days lengthen and the earth warms; this is the time when planting typically begins, and animals that hibernate (from the French word for winter) during the winter leave their dens.

Summer is the hottest time of the year. In many areas, summer is marked by drought, but other regions experience frequent thunderstorms and humid air. In the fall, the days again become shorter and cooler. This is the time when many crops are harvested. In ancient cultures, the turning of the seasons was marked by festivals, acknowledging the importance of seasonal changes to the community's survival.

Each season is defined as lasting three months. Winter begins at the winter solstice, which is the time when the Sun is farthest from the equator. In the Northern Hemisphere, this occurs on December 21 or 22, when the Sun is directly over the tropic of Capricorn. Summer begins at the other solstice, June 20 or 21 in the Northern Hemisphere, when the Sun is directly over the tropic of Cancer. The winter solstice is the shortest day of the year; the summer solstice is the longest.

Spring and fall begin on the two equinoxes. At an equinox, the Sun is directly above the earth's equator and the lengths of day and night are approximately equal everywhere on Earth. In the Northern Hemisphere, the vernal (spring) equinox occurs on March 21 or 22; in the Southern Hemisphere, it is the autumnal (fall) equinox. The Northern Hemisphere's autumnal equinox (and the Southern Hemisphere's vernal equinox) occurs September 22 or 23.

Seasons and the Hemispheres

The relationship of the seasons to the calendar is opposite in the Northern and Southern Hemispheres. On the day that a summer solstice occurs in the Northern Hemisphere, the winter solstice occurs in the Southern Hemisphere. Thus, when it is summer in the Southern Hemisphere, it is winter in the Northern Hemisphere, and vice versa.

The Sun and the Seasons

The reason why summers and winters differ in the temperate zones is often misunderstood. Many people think that winter happens when the Sun is more distant from Earth than it is in summer. What causes Earth's seasons is not the changing distances between the earth and the Sun, but the tilt of the earth's axis. A line drawn from the North Pole to the South Pole through the center of the earth (the earth's axis) is not perpendicular to the plane of the earth's orbit (the ecliptic). The earth's axis and the perpendicular to the ecliptic make an angle of 23.5 degrees. This tilts the Northern Hemisphere toward the Sun when the earth is on one side of its orbit around the Sun, and tilts the Southern Hemisphere toward the Sun when the earth moves around to the Sun's opposite side. When the Sun appears to be at its highest in the sky, and its rays are most direct, summer occurs. When the Sun appears to be at its lowest, and its rays are indirect, there is winter.

Local Phenomena

Local conditions can have important effects on seasonal weather. At locations near oceans, sea breezes develop during the day, and evenings are characterized by land breezes. Sea breezes bring cooler ocean air in toward land. This results in temperatures at the shore often being 5 to 11 degrees Fahrenheit (3 to 6 degrees Celsius) lower than temperatures a few miles inland.

At night, when land temperatures are lower than ocean temperatures, land breezes move air from the land toward the water. As a result, coastal regions have less seasonal temperature variations than inland areas do. For example, coastal areas seldom become cold enough to have snow in the winter, even though inland areas at the same latitude do.

Hailstorms

Hail usually occurs during the summer, and is associated with towering thunderstorm clouds, called cumulonimbus. Hail is occasionally confused with sleet. Sleet is a wintertime event, and occurs when warmer layers of air sit above freezing layers near the ground. Rain that forms in the warmer, upper layer solidifies into tiny ice pellets in the lower, subfreezing layer before hitting the ground.

Hail is an entirely different phenomenon. When cold air plows into warmer, moist air—called a cold front boundary—powerful updrafts of rising air can be created. The warm, moist air propelled upward by the heavier cold air can reach velocities approaching 100 miles (160 kilometers) per hour. Ice crystals form above the freezing level in the cumulonimbus clouds and fall into lower, warmer parts of the clouds, where they become coated with water. Picked up by an updraft, the coated ice crystals are carried back to a higher, colder levels where their water coatings freeze. This cycle can repeat many times, producing hailstones that have multiple, concentric layers of ice.

Hailstorms can be very damaging. Hail can ruin crops, dent car bodies, crack windshields, and injure people. The Midwest of the United States is particularly susceptible to hailstorms. There, warm, moist air from the Gulf of Mexico often meets much colder, drier air originating in Canada. This combination produces the extreme atmospheric instability necessary for that kind of weather.

Alvin S. Konigsberg

EARTH'S INTERIOR

EARTH'S INTERNAL STRUCTURE

Earth is one of the nine known planets in the Sun's solar system that formed from a giant cloud of cosmic dust called a nebula. This event is thought to have happened between 4.44 billion years ago (based on the age of the oldest-known Moon rock) and 4.56 billion years ago (the age of meteorite bombardment). After Earth's formation, heat released by colliding particles combined with the heat energy released by the decay of radioactive elements to cause some or all of Earth's interior to melt. This melting began the process of differentiation, which allowed the heavier elements, mainly iron and nickel, to sink toward Earth's center while the lighter, rocky components moved upward, as a result of the contrast in density of the earth's forming elements.

This process of differentiation was probably the most important event of Earth's early history. It changed the planet from a homogeneous mixture with neither continents nor oceans to a planet with three layers: a dense core beginning at 1,800 miles (2,900 km.) deep and ending at Earth's center, 3,977 miles (6,400 km.) below the surface; a mantle beginning between 3 and 44 miles (5-70 km.) deep and ending at Earth's core; and a crust going from Earth's surface to about 3-6 miles (5-10 km.) deep for oceanic crust and 22-44 miles (35-70 km.) deep for continental crust.

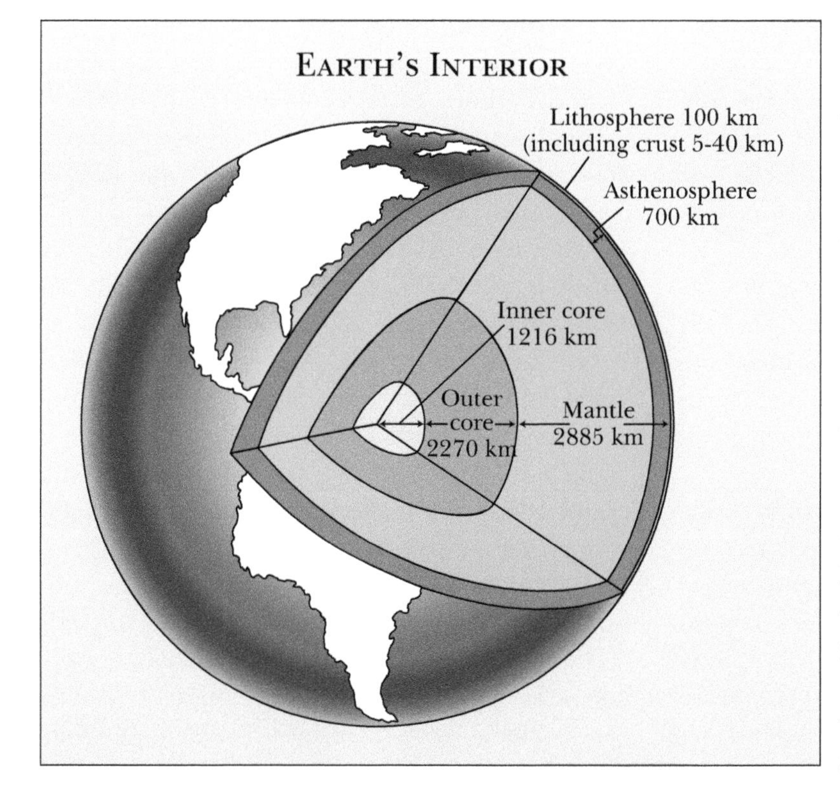

Layering of the Earth

Earth's layers can be classified either by their composition (the traditional method) or by their mechanical behavior (strength). Compositional classification identifies several distinct concentric layers, each with its own properties. The outermost layer of Earth is the crust or skin. This is divided into continental and oceanic crusts. The continental crust varies in thickness between 22 and 25 miles (35 and 40 km.) under flat continental regions and up to 44 miles (70 km.) under high mountains. The oceanic crust is made up of igneous rocks rich in iron and magnesium, such as basalt and peridotite. The upper continental crust is composed mainly of alumino-silicates. The old-

PROPERTIES OF SEISMIC WAVES

Seismologists use two types of body waves—primary (P-waves) and secondary (S-waves) waves—to estimate seismic velocities of the different layers within the earth. In most rock types P-waves travel between 1.7 and 1.8 times more quickly than S-waves; therefore, P-waves always arrive first at seismographic stations. P-waves travel by a series of compressions and expansions of the material through which they travel. P-waves can travel through solids, liquids, or gases. When P-waves travel in air, they are called sound waves.

The slower S-waves, also called shear waves, move like a wave in a rope. This movement makes the S-wave more destructive to structures like buildings and highway overpasses during earthquakes. Because S-waves can travel only through solids and cannot travel through Earth's outer core, seismologists concluded that Earth's outer core must be liquid or at least must have the properties of a fluid.

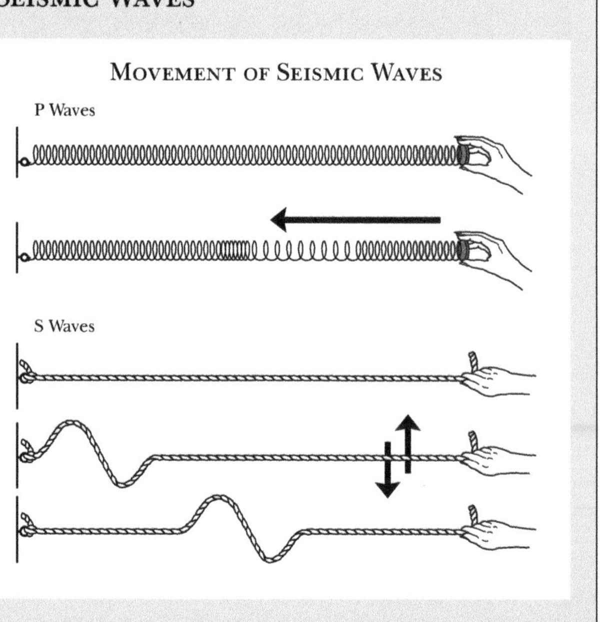

est continental crustal rock exceeds 3.8 billion years, while oceanic crustal rocks are not older than 180 million years. The oceanic crust is heavier than the continental crust.

Earth's next layer is the mantle, which is made up mostly of ferro-magnesium silicates. It is about 1,800 miles (2,900 km.) thick and is separated into the upper and lower mantle. Most of Earth's internal heat is contained within the mantle. Large convective cells in the mantle circulate heat and may drive plate-tectonic processes.

The last layer is the core, which is separated into the liquid outer core and the solid inner core. The outer core is 1,429 miles (2,300 km.) thick, twice as thick as the inner core. The outer core is mainly composed of a nickel-iron alloy, while the inner core is almost entirely composed of iron. Earth's magnetic field is believed to be controlled by the liquid outer core.

In the mechanical layering classification of the earth's interior, the layers are separated based on mechanical properties or strength (resistance to flowing or deformation) in addition to composition. The uppermost layer is the lithosphere (sphere of rock), which comprises the crust and a solid portion of the upper mantle. The lithosphere is divided into many plates that move in relation to each other due to tectonic forces. The solid lithosphere floats atop a semiliquid layer known as the asthenosphere (weak sphere), which enables the lithosphere to move around.

Exploring Earth's Interior

Volcanic activity provides natural samples of the outer 124 miles (200 km.) of Earth's interior. Meteorites—samples of the solar system that have collided with Earth—also provide clues about Earth's composition and early history. The most ambitious human effort to penetrate Earth's interior was made by the former Soviet Union, which drilled a super-deep research well, named the Kola Well, near Murmansk, Russia. This was an attempt to penetrate the crust and reach the upper mantle. The reported depth of the Kola Well is a little more than 7.5 miles (12 km.). Although impressive, the drilled depth represents less than 0.2 percent of the distance from the earth's surface to its center.

A great deal of knowledge about Earth's composition and structure has been obtained through computer modeling, high-pressure laboratory experiments, and meteorites, but most of what is known about Earth's interior has been acquired by studying seismic waves generated by earthquakes and nuclear explosions. As seismic waves are transmitted, reflected, and refracted through the earth, they carry information to the surface about the materials through which they have traveled. Seismic waves are recorded at receiver stations (seismographic stations) and processed to provide a picturelike image of Earth's interior.

Changes in P- and S-wave velocities within Earth reveal the sequence of layers that make up Earth's interior. P-wave velocity depends on the elasticity, rigidity, and density of the material. By contrast, S-wave velocity depends only on the rigidity and density of the material. There are sharp variations in velocity at different depths, which correspond to boundaries between the different layers of Earth. P-wave velocity within crustal rocks ranges from 3.6-4.2 miles (6-7 km.) per second.

The boundary between the crust and the mantle is called the Mohorovičić discontinuity or Moho. At Moho, P-wave velocity increases from 4.2-4.8 miles (7-8 km.) per second. Beyond the crust-mantle boundary, P-wave velocity increases gradually up to about 8.1 miles (13.5 km.) per second at the core-mantle boundary. At this depth, S-waves are not transmitted and P-wave velocity, decreases from 8.1 to 4.8 miles (13.5 to 8 km.) per second, which strongly supports the concept that the outer core is liquid, since S-waves cannot travel through liquids. As P-waves enter the inner core, their velocity again increases, to about 6.8 miles (11.3 km.) per second.

Earth's interior seems to be characterized by a gradual increase with depth in temperature, pressure, and density. Extensive experimental and modeling work indicates that the temperature at 62 miles (100 km.) is between 1,200 and 1,400 degrees Celsius (2,192 to 2,552 degrees Fahrenheit). The temperature at the core-mantle boundary—about 1,802 miles (2,900 km.) deep—is calculated to be about 8,130 degrees Fahrenheit (4,500 degrees Celsius). At Earth's center the temperature may exceed 12,092 degrees Fahrenheit (6,700 degrees Celsius). Although at Earth's surface, heat energy is slowly but continuously lost as a result of outgassing, such as from volcanic eruptions, its interior remains hot.

Seismic Tomography and Future Exploration

Seismic tomography is one of the newest tools that earth scientists are using to develop three-dimensional velocity images of Earth's interior. In seismic tomography, several crossing seismic waves from different sources (earthquakes and nuclear explosions) are analyzed in much the same way that computerized axial tomography (CAT) scanners are used in medicine to obtain images of human organs. Seismic tomography is providing two- and three-dimensional images from the crust to the core-mantle boundary. Fast P-wave velocities have been correlated to cool material—for example, a piece of sinking lithosphere (cool rigid layer) such as in regions underneath the Andes Mountains (subduction zone); slow P-wave velocities have been correlated with hot materials—for example, rising mantle plumes of hot spots such as the one responsible for volcanic activity in the Hawaiian Islands.

Rubén A. Mazariegos-Alfaro

Plate Tectonics

The theory of plate tectonics provides an explanation for the present-day structure of the large landforms that constitute the outer part of the earth. The theory accounts for the global distribution of continents, mountains, hills, valleys, plains, earthquake activity, and volcanism, as well as various associations of igneous, metamorphic, and sedimentary rocks, the formation and location of min-

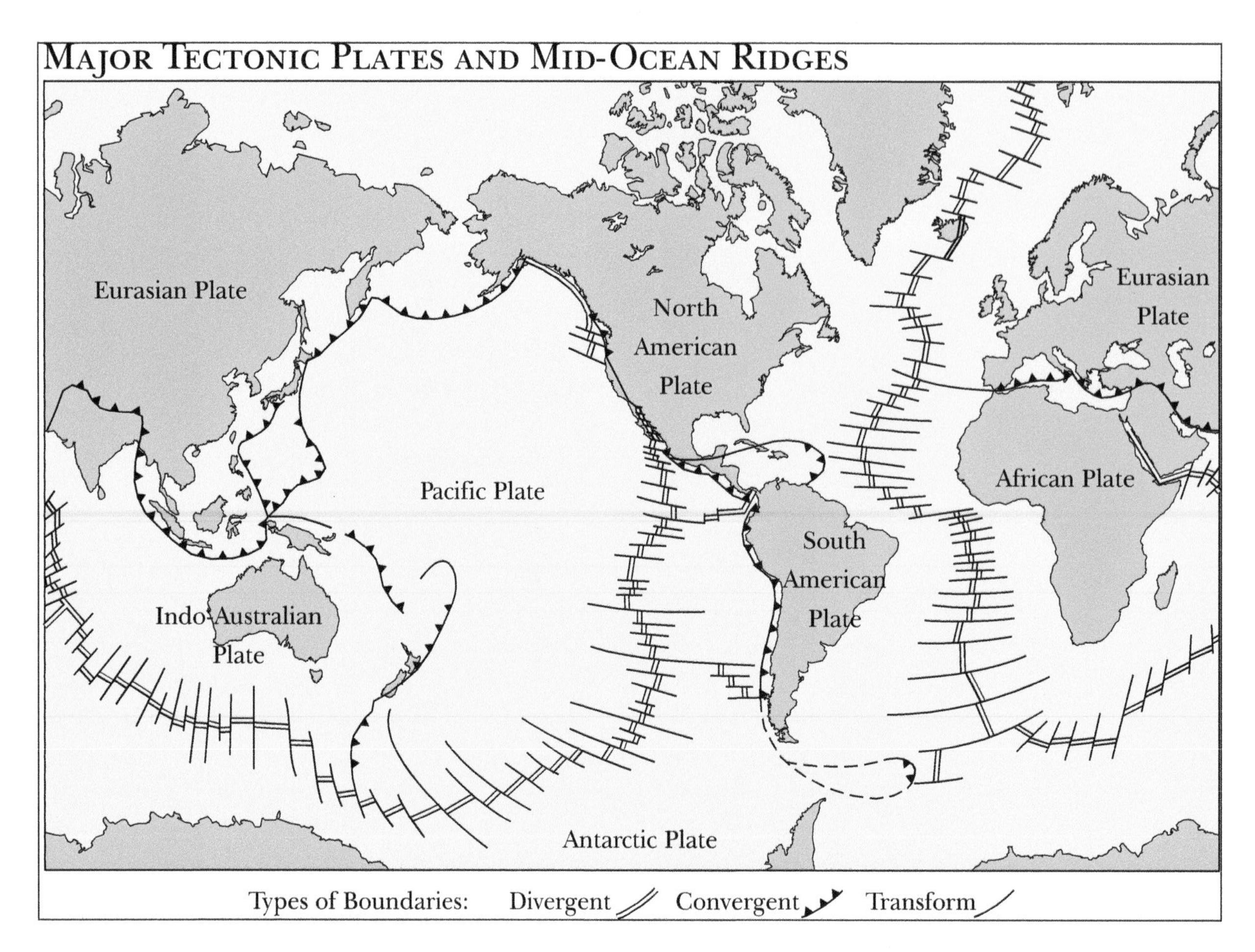

eral resources, and the geology of ocean basins. Everything about the earth is related either directly or indirectly to plate tectonics.

Basic Theory

Plate-tectonic theory is based on an Earth model in which a rigid, outer shell—the lithosphere—lies above a hotter, weaker, partially molten part of the mantle called the asthenosphere. The lithosphere varies in thickness between 6 and 90 miles (10 and 150 km.), and comprises the crust and the underlying, upper mantle. The asthenosphere extends from the base of the lithosphere to a depth of about 420 miles (700 km.). The brittle lithosphere is broken into a pattern of internally rigid plates that move horizontally relative to each other across the earth's surface.

More than a dozen plates have been distinguished, some extending more than 2,500 miles (4,000 km.) across. Exhibiting independent motion, the plates grind and scrape against each other, similar to chunks of ice in water, or like giant rafts cruising slowly on the asthenosphere. Most of the earth's dynamic activity, including earthquakes and volcanism, occurs along plate boundaries. The global distribution of these tectonic phenomena delineates the boundaries of the plates.

Geological observations, geophysical data, and theoretical models support the existence of three types of plate boundaries. Divergent boundaries occur where adjacent plates move away from each other. Convergent boundaries occur where adjacent plates move toward each other. Transform boundaries occur where plates slip past one another in directions parallel to their common boundaries.

The continents were formed by the movement at plate boundaries, and continental landforms were generated by volcanic eruptions and continental plates colliding with each other. The velocity of plate movement varies from plate to plate and even within portions of the same plate, ranging from 0.8 to 8 inches (2 to 20 centimeters) per year. The rates

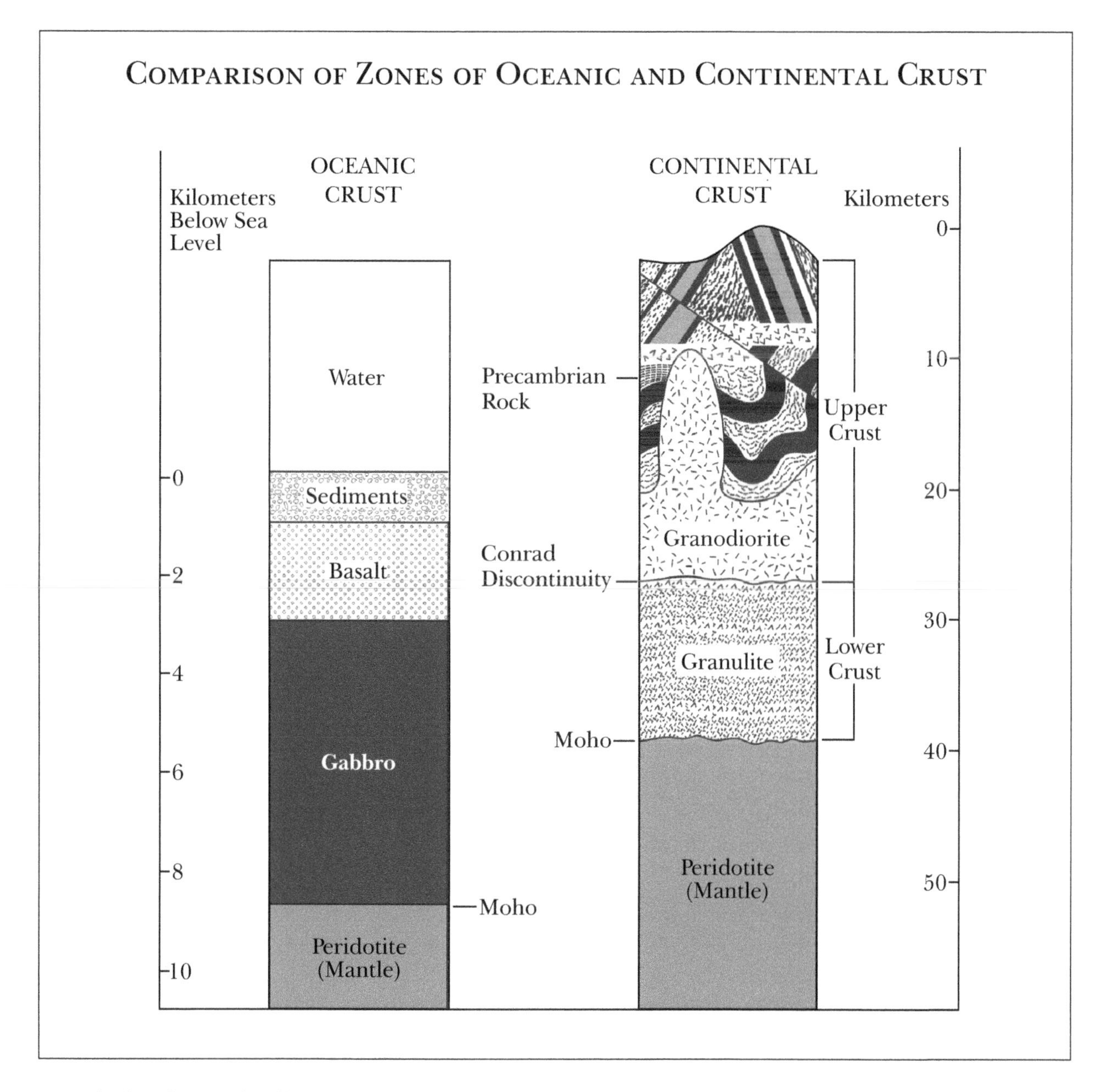

Comparison of Zones of Oceanic and Continental Crust

are calculated from the distance to the midoceanic ridge crests, along with the age of the sea floor as determined by radioactive dating methods.

Convection currents that are driven by heat from radioactive decay in the mantle are important mechanisms involved in moving the huge plates. Convection currents in the earth's mantle carry magma (molten rock) up from the asthenosphere. Some of this magma escapes to form new lithosphere, but the rest spreads out sideways beneath the lithosphere, slowly cooling in the process. Assisted by gravity, the magma flows outward, dragging the overlying lithosphere with it, thus continuing to open the ridges. When the flowing hot rock cools, it becomes dense enough to sink back into the mantle at convergent boundaries.

A second plate-driving mechanism is the pull of dense, cold, down-flowing lithosphere in a subduction zone on the rest of the trailing plate, further opening up the spreading centers so magma can move upward.

Divergent Plate Boundaries

During the 1950s and 1960s, oceanographic studies revealed that Earth's seafloors were marked by a nearly continuous system of submarine ridges,

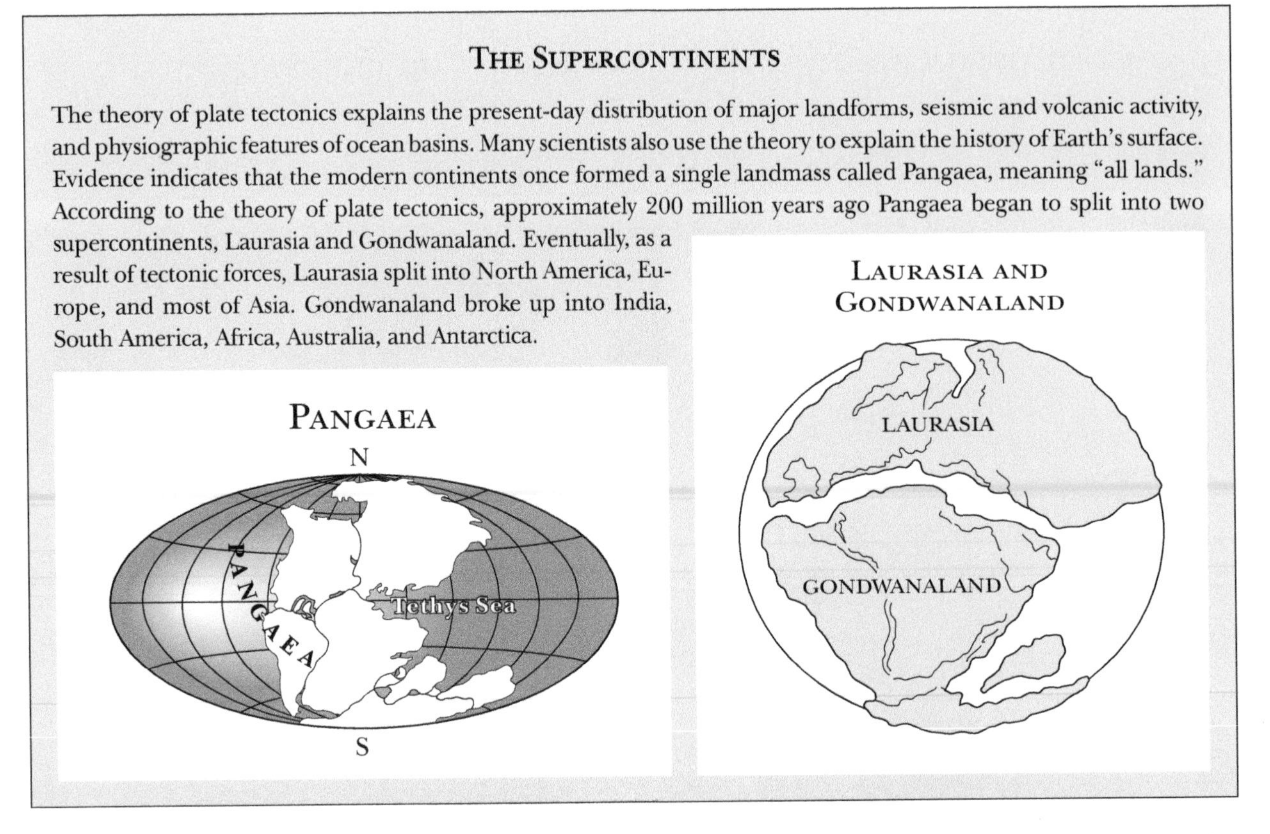

THE SUPERCONTINENTS

The theory of plate tectonics explains the present-day distribution of major landforms, seismic and volcanic activity, and physiographic features of ocean basins. Many scientists also use the theory to explain the history of Earth's surface. Evidence indicates that the modern continents once formed a single landmass called Pangaea, meaning "all lands." According to the theory of plate tectonics, approximately 200 million years ago Pangaea began to split into two supercontinents, Laurasia and Gondwanaland. Eventually, as a result of tectonic forces, Laurasia split into North America, Europe, and most of Asia. Gondwanaland broke up into India, South America, Africa, Australia, and Antarctica.

more than 40,000 miles (64,000 km.) in length. Detailed investigations revealed that the midoceanic ridge system has a central rift valley that runs along its length and that the ridge system is associated with volcanic and earthquake activity. The earthquakes are frequent, shallow, and mild.

Magnetic studies of the seafloor indicate that the oceanic lithosphere has been segmented into a series of long magnetic strips that run parallel to the axis of the midoceanic ridges. On either side of the ridge, the ocean floor consists of alternating bands of rock, magnetized either parallel to or exactly opposite of the present-day direction of the earth's magnetic field.

Midoceanic ridges, or divergent plate boundaries, are tensional features representing zones of weakness within the earth's crust, where new seafloor is created by the welling up of mantle material from the asthenosphere into cracks along the ridges. As rifting proceeds, magma ascends to fill in the fissures, creating new oceanic crust. Iron minerals within the magma become aligned to the existing Earth polarity as the rock cools and crystallizes. The oceanic floor slowly moves away from the oceanic ridge toward deep ocean trenches, where it descends into the mantle to be melted and recycled to the earth's surface to generate new rocks and landforms.

As the seafloor spreads outward from the rift center, about half of the material is carried to either side of the rift, which is later filled by another influx of molten basalt. When the polarity of the earth changes, the subsequent molten basalt is magnetized in the opposite polarity. The continuation of this process over geologic time leads to the young geologic age of the seafloor and the magnetic symmetry around the midoceanic ridges.

Not all spreading centers are underneath the oceans. An example of continental rifting in its embryonic stage can be observed in the Red Sea, where the Arabian plate has separated from the African plate, creating a new oceanic ridge. Another modern-day example of continental divergent activity is East Africa's Great Rift Valley system. If this rifting continues, it will eventually fragment Africa, producing an ocean that will separate the resulting pieces. Through divergence, large plates are made into smaller ones.

Convergent Plate Boundaries

Because Earth's volume is not changing, the increase in lithosphere created along divergent boundaries must be compensated for by the destruction of lithosphere elsewhere. Otherwise, the radius of Earth would change. The compensation occurs at convergent plate boundaries, where plates are moving together. Three scenarios are possible along convergent boundaries, depending on whether the crust involved is oceanic or continental.

If both converging plates are made of oceanic crust, one will inevitably be older, cooler, and denser than the other. The denser plate eventually subducts beneath the less-dense plate and descends into the asthenosphere. The boundary along the two interacting plates, called a subduction zone, forms a trench. Some trenches are more than 620 miles (1,000 km.) long, 62 miles (100 km.) wide, and 6.8 miles (11 km.) deep. Heated by the hot asthenosphere beneath, the subducted plate becomes hot enough to melt.

Because of buoyancy, some of the melted material rises through fissures and cracks to generate volcanoes along the overlying plate. Over time, other parts of the melted material eventually migrate to a divergent boundary and rise again in cyclic fashion to generate new seafloor. The volcanoes generated along the overriding plate often form a string of islands called island arcs. Japan, the Philippines, the Aleutians, and the Mariannas are good examples of island arcs resulting from subduction of two plates consisting of oceanic lithosphere. Intense earthquakes often occur along subduction zones.

If the leading edge of one of the two convergent plates is oceanic crust and the other is continental crust, the oceanic plate is always the one subducted, because it is always denser. A classic example of this case is the western boundary of South America. On the oceanic side of the boundary, a trench was formed where the oceanic plate plunged underneath the continental plate. On the continental side, a fold mountain belt—the Andes—was formed as the oceanic lithosphere pushed against the continental lithosphere.

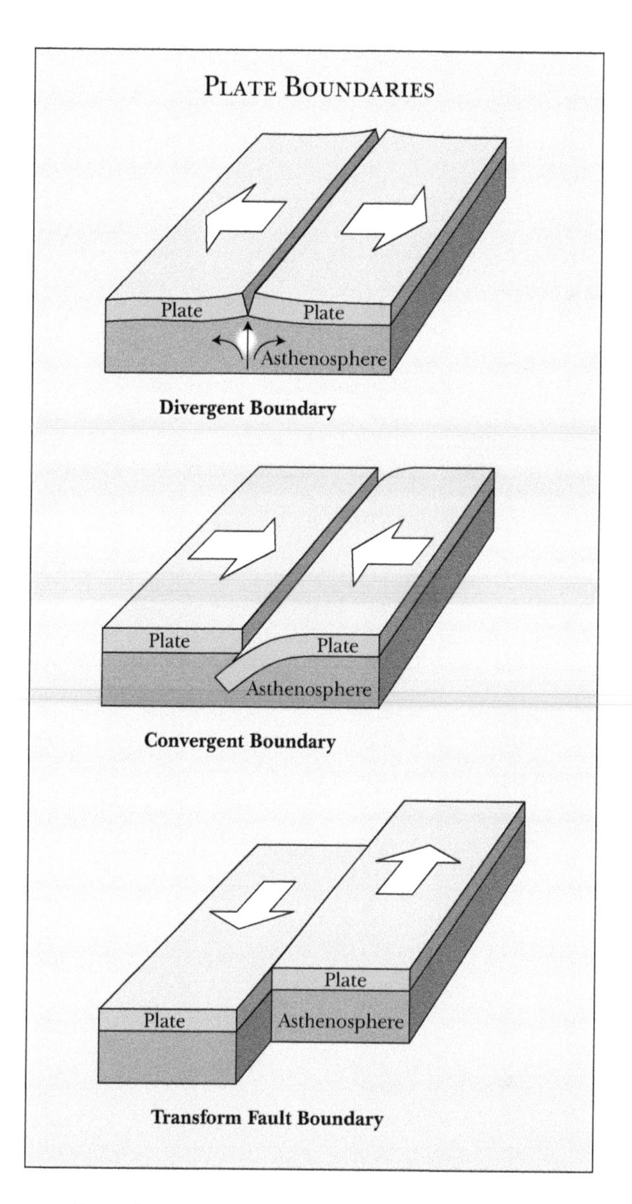

When the oceanic plate descends into the mantle, some of the material melts and works its way up through the mountain belt to produce rather violent volcanoes. The boundary between the plates is a region of earthquake activity. The earthquakes range from shallow to relatively deep, and some are quite severe.

The last type of convergent plate boundary involves the collision of two continental masses of lithosphere, which can result in folding, faulting, metamorphism, and volcanic activity. When the plates collide, neither is dense enough to be forced into the asthenosphere. The collision compresses and thickens the continental edges, twisting and deforming the rocks and uplifting the land to form unusually high fold mountain belts. The prototype example is the collision of India with Asia, resulting in the formation of the Himalayas. In this case, the earthquakes are typically shallow, but frequent and severe.

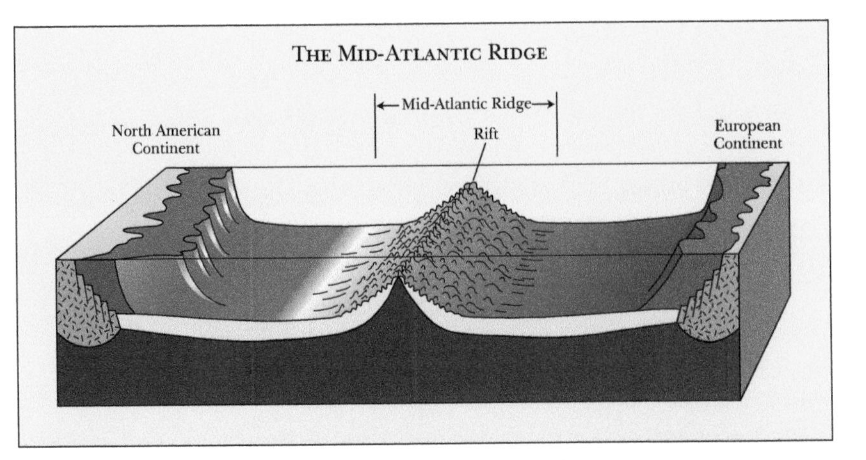

Transform Plate Boundaries

The actual structure of a seafloor spreading ridge is more complex than a single, straight crack. Instead, ridges comprise many short segments slightly offset from one another. The offsets are a special kind of fault, or break in the lithosphere, known as a transform fault, and their function is to connect segments of a spreading ridge. The opposite sides of a transform fault belong to two different plates that are grinding against each other in opposite directions.

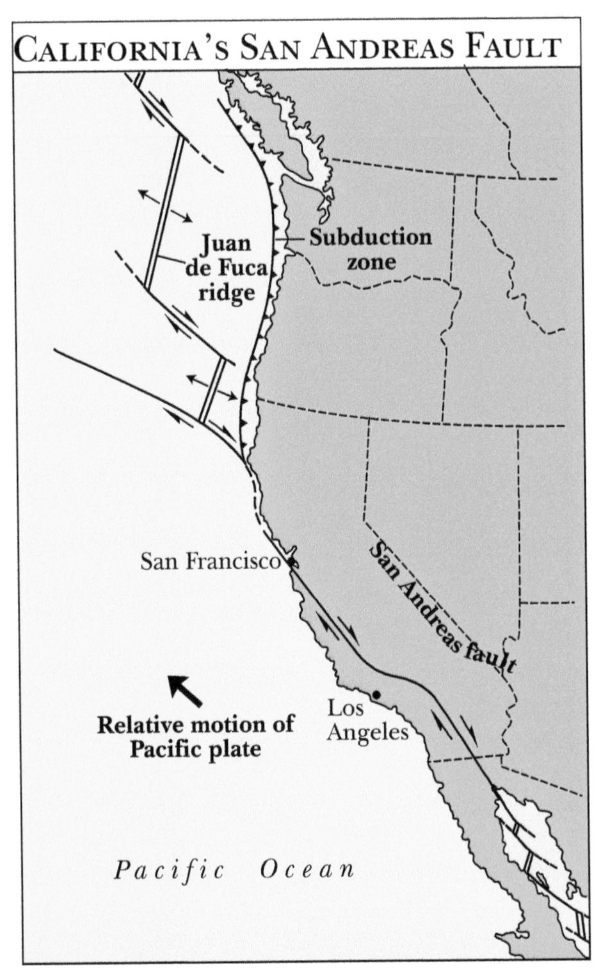

Transform faults form the boundaries that allow the plates to move relative to each another. The classic case of a transform boundary is the San Andreas Fault. It slices off a small piece of western California, which rides on the Pacific plate, from the rest of the state, which resides on the North American plate. As the two plates scrape past each other, stress builds up, eventually being released in earthquakes that can be quite violent.

Mantle Plumes and Hot Spots

Most plate tectonic features are near plate boundaries, but the Hawaiian Islands are not. In the late twentieth century, the only active volcanoes in the Hawaiian Islands were on the island of Hawaii, at the southeast end of the chain. Radiometric dating and examination of states of erosion show that, when proceeding along the chain to the northwest, successive islands are progressively older.

Evidently, the same heat source produced all the volcanoes in the Hawaiian chain. Known as a mantle plume, it has remained stationary while the Pacific plate rides over it, producing a volcanic trail from which absolute motion of the plate can be determined. Since mantle plumes do not move with the plates, the plumes must originate beneath the lithosphere, probably far below it. Resulting volcanoes are called hot spots to distinguish them from subduction-zone volcanoes. Iceland is a good example of a hot spot, as is Yellowstone. At least 100 hot spots are distributed around Earth.

Alvin K. Benson

Volcanoes

Volcanoes form mountains both on land and in the sea and either do it on a grand scale or merely create minute bumps on the seafloor. Volcanoes do not occur in a random pattern, but are found in distinct zones that are related to plate dynamics. Each of the three types of volcanism on Earth is characterized by specific types of eruptions and magma compositions. Molten magma is the rock material below the earth's crust that forms igneous rock as it cools.

Types of Volcanoes

Geologists generally group volcanoes into four main kinds—cinder cones, composite volcanoes, shield volcanoes, and lava domes.

Cinder cones are built from congealed lava ejected from a single vent. As the gas-charged lava is blown into the air, it breaks into small fragments that solidify and fall as *cinders* around the vent to form a circular or oval cone. Most cinder cones have a bowl-shaped *crater* at the summit and rarely rise more than a thousand feet or so above their surroundings. Cinder cones are numerous in western North America and in other volcanic terrains of the world.

Composite volcanoes —sometimes called stratovolcanoes—include some of the Earth's grandest mountains, including Mount Fuji in Japan, Mount Cotopaxi in Ecuador, Mount Shasta in California, Mount Hood in Oregon, and Mount St. Helens and Mount Rainier in Washington. The essential feature of a composite volcano is a conduit system through which magma deep in the Earth's crust rises to the surface. They are typically steep-sided, symmetrical cones of large dimension built of alternating layers of lava flows, volcanic ash, cinders, blocks, and bombs. They may rise as much as 8,000 feet above their bases. Most have a crater at the summit that contains a central vent or a clustered group of vents. Lavas either flow through breaks in the crater wall or fissures on the flanks of the cone.

Shield volcanoes, the third type of volcano, are built almost entirely of fluid lava flows that pour out in all directions from a central vent, or group of vents, building a broad, gently sloping cone. They are built up slowly as thousands of highly fluid lava flows—basalt lava—spread over great distances, and then cool into thin sheets. Some of the largest volcanoes in the world are shield volcanoes. The Hawaiian Islands are composed of linear chains of these volcanoes including Kilauea and Mauna Loa on the island of Hawaii—two of the world's most active volcanoes. The floor of the ocean is more than 15,000 feet deep at the bases of the islands. As Mauna Loa, the largest of the shield volcanoes (and also the world's largest active volcano), projects 13,679 feet above sea level, its top is over 28,000 feet above the deep ocean floor.

Volcanic Composition

Volcanoes in the midocean ridges and plume environments draw most of their magmas from the earth's mantle and produce mainly dark, magnesium-rich basaltic magmas. When basaltic magmas accumulate in the continental crust (for example, at Yellowstone), the large-scale crustal melting leads to rhyolitic volcanism, the volcanic equivalent of granites. Arc magmas cover a wider range of magmatic compositions, ranging from arc basalt to light-colored, silica-rich rhyolites; the latter are commonly erupted in the form of the silica-rich volcanic rock known as pumice, or the black volcanic glass known as obsidian. Andesites, named after the Andes Mountains, are a common volcanic rock in stratovolcanoes, intermediate in composition between basalt and rhyolite.

Magmas form from several processes that lead to partial melting of a solid rock. The simplest is adding heat—for example, plumes carrying heat from deep levels in the mantle to shallower levels, where melting occurs. Decompressional (lowering the pressure) melting of the mantle occurs where the

SOME VOLCANIC HOT SPOTS AROUND THE WORLD

ocean floor is thinned or carried away by seafloor spreading in midocean ridge environments.

Genesis of Magma

Adding a "flux" to a solid mineral mixture may lower the substance's melting point. The most common theory about arc magma genesis invokes the addition of a low-melting-point substance to the arc mantle, a layer of mantle material at about 60 to 90 miles (100 to 150 km.) below the volcanic arc. The relatively dry arc mantle would usually start to melt at about 2,100 to 2,300 degrees Fahrenheit (1,200 to 1,300 degrees Celsius). However, the addition of water and other gases can lower the melting point of the mixture. The water and its dissolved chemicals are supposedly derived from the subducted slab, the former ocean floor that is pushed back into the earth.

The sequence of events is as follows: New basaltic ocean floor forms at midocean ridge volcanoes. The new hot magma interacts with seawater, leading to vents at the seafloor with their mineralized deposits. The seafloor becomes hydrated, and sulfur and chlorine from seawater are locked up in newly formed minerals. During subduction, this altered seafloor with slivers of sediment, including limestone, is gradually warmed up and starts to decompose, adding a flux to the surrounding mantle rocks. The mantle rocks then start to melt, and these magmas with minor inherited oceanic materials start to rise and pond at the bottom of the crust. There the magmas sit and wait for an opportunity to erupt, while cooling and crystallizing. Thus, arc magmas bear a chemical signature of subducted oceanic components while their chemical compositions range from basalt to rhyolite.

Volcanic Eruptions

Volcanic eruptions occur as a result of the rise of magma into the volcano (from depths as great as several miles) and then into the throat of the volcano. In basaltic volcanoes, the magmas have relatively little gas, and the magma simply overflows and forms large lava flows, sometimes associated

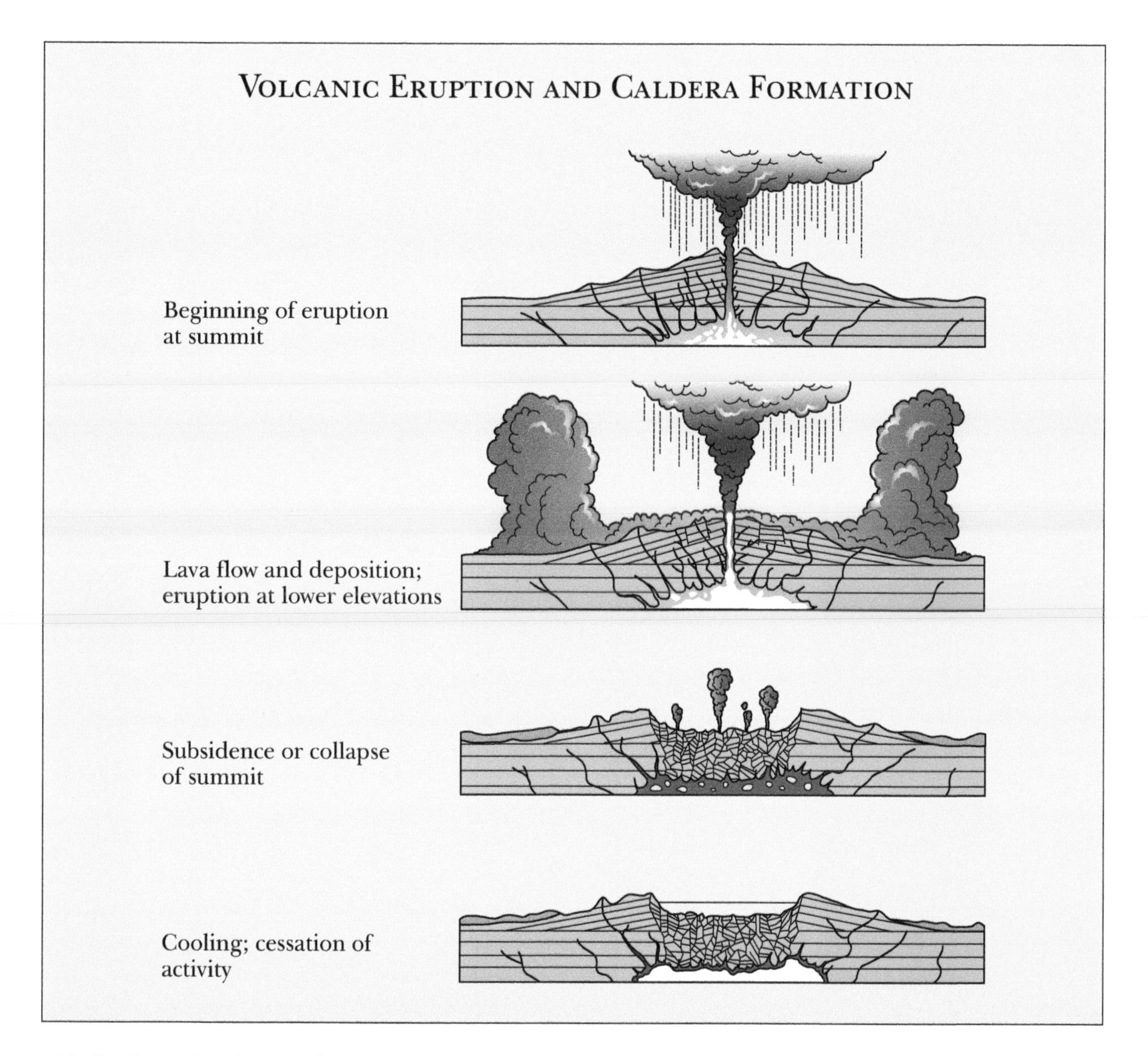

with fire fountains. Stratovolcanoes can erupt regularly with small explosions or catastrophically after long periods of dormancy. Mount Stromboli, a volcano in Italy, erupts every twenty minutes, with an explosion that creates a column 650 to 980 feet (200 to 300 meters) high. Mount St. Helens in the U.S. state of Washington had a catastrophic eruption in 1980 after about 200 years of dormancy. It emitted an ash plume that reached more than 12 miles (20 km.) into the atmosphere.

After long magma storage periods in the crust, crystallization and melting of crustal material can lead to silica-rich magmas. These are viscous and can have high dissolved water contents—up to 4 to 6 percent by weight. When these magmas break out, the eruption can be violent and form an eruption column 12 to 35 miles (20 to 55 km.) high. Many cubic miles of magma can be ejected. This leads to so-called plinian ash falls, with showers of pumice and ash over thousands of square miles, with the ash commonly carried around the globe by the high-level winds known as jet streams.

If the volume of ejected magma is large, the volcano empties itself and collapses into the hole, leading to a caldera—a volcanic collapse structure. The caldera at Crater Lake in Oregon is related to a large pumice eruption about 76,000 years ago. Basaltic volcanoes can also form collapse calderas when large volumes of lava have been extruded in a short time. Examples of famous basaltic calderas can be found in Hawaii's Mount Kilauea and the Galapagos Islands.

Volcanic Plumes

The dynamics of volcanic plumes has been studied from eruption photographs, experiments, and theoretical work. The rapidly expanding hot gases force the viscous magma out of the throat of the volcano, where it freezes into pumice. The kinetic energy of the ejected mass carries it 2 to 2.5 miles (3-4 km.) above the volcano. During this phase, air is entrained in the column, diluting the concentration of ash and pumice particles. The hot particles heat the entrained air, the mixture of hot air and solids becomes less dense than the surrounding atmosphere, and a buoyant column rises high into the sky.

The height of an eruption column is not directly proportional to the force of the eruption but is strongly dependent on the rate of heat release of the volcano. If little of the entrained air is heated up, the column will collapse back to the ground and an ash flow forms, which may deposit ash around the volcano. These types of eruptions are among the most devastating, creating glowing ash clouds traveling at speeds up to 60 miles (100 km.) per hour, burning everything in their path. The 1902 eruption of Mount Pelée on Martinique in the Caribbean was such an eruption and killed nearly 30,000 people in a few minutes.

Many volcanoes that are high in elevation are glaciated, and their eruptions lead to large-scale ice melting and possibly mixing of water, magma, and volcanic debris. Massive hot mudflows can race down from the volcano, following river valleys and filling up low areas. The 1980 Mount St. Helens eruption created many mudflows, some of which reached the Pacific Ocean, ninety miles to the west. A catastrophic mudflow event occurred in 1984 at Nevado del Ruiz, a volcano in Colombia, where 20,000 people were buried in mud and perished. When magma intrudes under the ice, meltwater can accumulate and then escape catastrophically, but such meltwater bursts are rare outside Iceland.

Minerals and Gases in Eruptions

The gas-rich character of arc magmas leads to fluid escape at various levels in the volcanoes, and these fluids tend to be rich in chlorine. They can transport metals such as copper, lead, zinc, and gold at high concentrations, and lead to the enrichment of these metals in the fractured volcanic rocks. Many of the world's largest copper ore deposits are associated with older arc volcanism, where erosion has removed most of the volcanic structure and laid the volcano innards bare. Many active volcanoes have modern hydrothermal (hot-water) systems, leading to acid hot springs and crater lakes and the potential to harness geothermal energy. Some areas in Japan, New Zealand, and Central America have an abundance of geothermal energy resources, which are gradually being developed.

Apart from the dangers of eruptions, continuous emissions of large amounts of sulfur dioxide, hydrochloric acid, and hydrofluoric acid present a danger of air pollution and acid rain. Incidences of emphysema and other irritations of the respiratory system are common in people living on the slopes of active volcanoes. The large lava emissions in Iceland in the eighteenth century led to acid fogs all over Europe. Many cattle died in Iceland during this period from the hydrofluoric acid vapors. High levels of fluorine in drinking water can lead to fluorosis, a disease that attacks the bone structure. The discharge of highly acidic fluids from hot springs and crater lakes can cause widespread environmental contamination, which can present a danger for crops gathered from fields irrigated with these waters and for local ecosystems in general.

Johan C. Varekamp

Geologic Time Scale

A major difference between the geosciences (earth sciences) and other sciences is the great enormity of their time scale. One might compare the magnitude of geologic time for geoscientists to the vastness of space for astronomers. Every geological process, such as the movement of crustal plates (plate tectonics), the formation of mountains, and the advance and retreat of glaciers, must be considered within the context of time.

Although certain geologic events, such as floods and earthquakes, seem to occur over short periods of time, the vast majority of observed geological features formed over a great span of time. Consequently, modern geoscientists consider Earth to be exceedingly old. Using radiometric age-dating techniques, they calculate the age of Earth as 4.6 billion years old.

Early miners were probably the first to recognize the need for a scale by which rock and mineral units could be compared over large geographic areas. However, before a time scale—and even geology as a science—could develop, certain principles had to be established. This did not occur until the late eighteenth century when James Hutton, a Scottish naturalist, began his extensive examinations of rock relationships and natural processes at work on the earth. His work was amplified by Charles Lyell in his textbook *Principles of Geology* (1830-1833). After careful observation, Hutton concluded that the natural processes and functions he observed had operated in the same basic manner in the past, and that, in general, natural laws were invariable. That idea became known as the principle of uniformitarianism.

The Birth of Stratigraphy

In 1669 Nicholas Steno, a Danish physician working in Italy, recognized that horizontal rock layers contained a chronological record of Earth history and formulated three important principles for interpreting that history. The principle of superposition states that in a succession of undeformed strata, the oldest stratum lies at the bottom, with successively younger ones above. The principle of original horizontality states that because sedimentary particles settle from fluids under gravitational influence, sedimentary rock layers must be horizon-

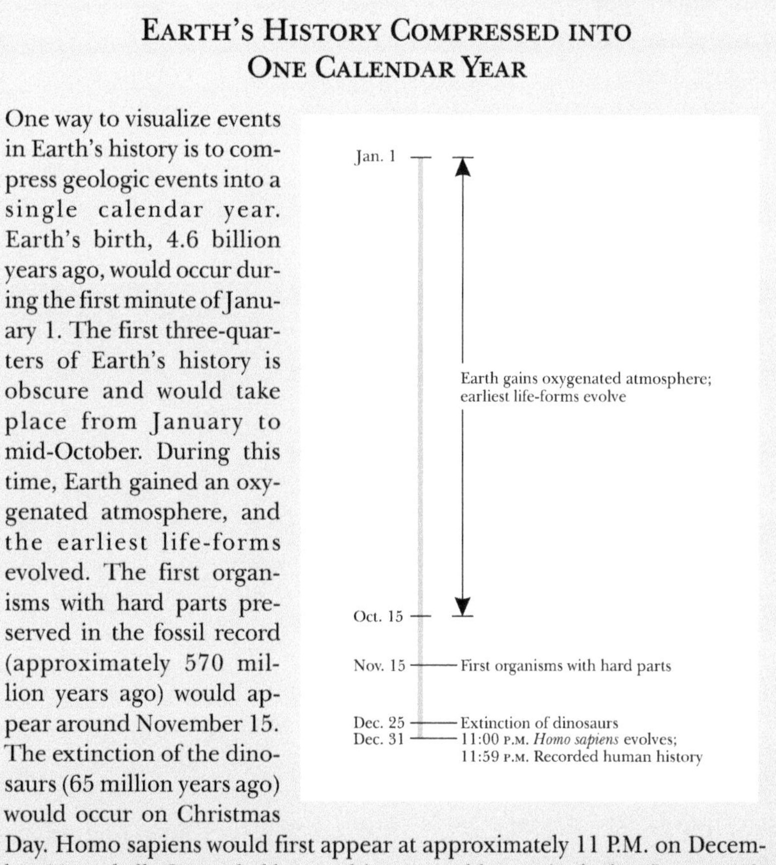

Earth's History Compressed into One Calendar Year

One way to visualize events in Earth's history is to compress geologic events into a single calendar year. Earth's birth, 4.6 billion years ago, would occur during the first minute of January 1. The first three-quarters of Earth's history is obscure and would take place from January to mid-October. During this time, Earth gained an oxygenated atmosphere, and the earliest life-forms evolved. The first organisms with hard parts preserved in the fossil record (approximately 570 million years ago) would appear around November 15. The extinction of the dinosaurs (65 million years ago) would occur on Christmas Day. Homo sapiens would first appear at approximately 11 P.M. on December 31, and all of recorded human history would occur in the last few seconds of New Year's Eve.

tal; if not, they have suffered from subsequent disturbance. The principle of original lateral continuity states that strata originally extended in all directions until they thinned to zero or terminated against the edges of the original area of deposition.

In the late eighteenth century, the English surveyor William Smith recognized the wide geographic uniformity of rock layers and discovered the utility of fossils in correlating these layers. By 1815, Smith had completed a geologic map of England and was able to correlate English rock layers with layers exposed across the English Channel in France.

From the need to classify and organize rock layers into an orderly form arose a subdiscipline of modern geology—stratigraphy, the study of rock layers and their age relationships. In 1835 two British geologists, Adam Sedgwick and Roderick Murchison, began organizing rock units into a formal stratigraphic classification. Large divisions, called eras, were based upon well known and characteristic fossils, and included a number of smaller subdivisions, called periods.

The periods are often subdivided into smaller units called epochs. Each period is defined by a representative sequence of rock strata and fossils. For instance, the Devonian period is named for exposures of rock in Devonshire in southern England, while the Jurassic period is defined by strata exposed in the Jura Mountains in northern Switzerland.

Approximately 80 percent of Earth's history is included in the Cryptozoic eon (meaning obscure life). Fossils from the Cryptozoic eon are rare, and the rock record is very incomplete. After the Cryptozoic eon came the Paleozoic (ancient life), Mesozoic (middle life), and Cenozoic (recent life) eras. Most of the life forms that evolved during the Paleozoic and Mesozoic eras are now extinct, whereas 90 percent of the life-forms that evolved up to the middle Cenozoic era still exist.

The Geologic Time Scale

The geologic time scale is continually in revision as new rock formations are discovered and dated. The ages shown in the table below are in millions of years ago (MYA) before the present and represent the beginning of that particular period. It would be impossible to list all the significant events in Earth's history, but one or two are provided for each period. Note that in the United States, the Carboniferous period has been subdivided into the Mississippian period (older) and the Pennsylvanian period (younger).

The Fossil Record

The word "fossil" comes from the Latin *fossilium*, meaning "dug from beneath the surface of the ground." Fossils are defined as any physical evidence of past life. Fossils can include not only shells, bones, and teeth, but also tracks, trails, and bur-

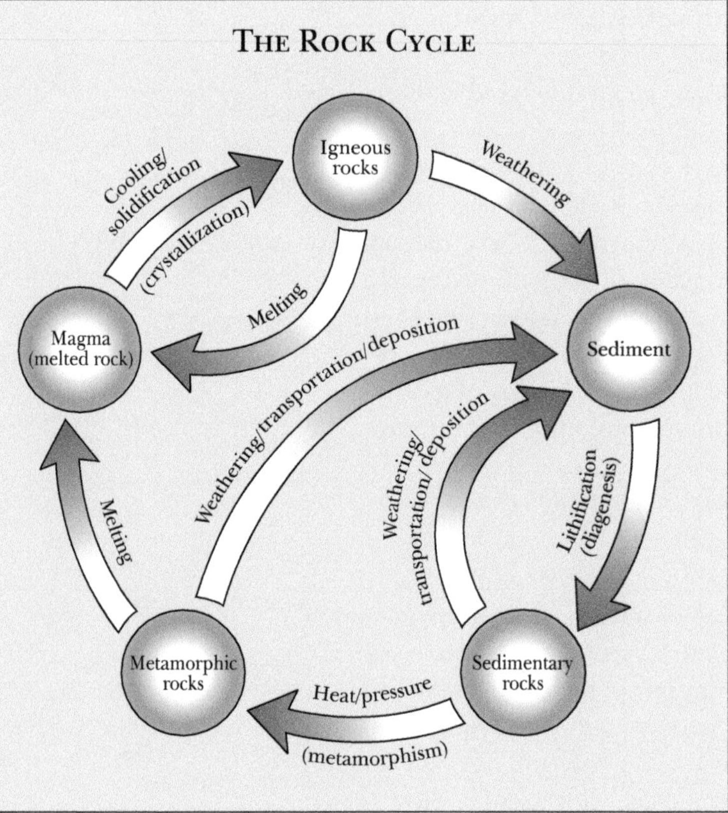

rows. The latter group are referred to as trace fossils. Fossils demonstrate two important truths about life on Earth: First, thousands of species of plants and animals have existed and later became extinct. Second, plants and animals have evolved through time, and the communities of life that have existed on Earth have changed.

Some organisms are slow to evolve and may exist in several geologic time periods, while others evolve quickly and are restricted to small intervals of time within a particular period. The latter, referred to as index fossils, are the most useful to geoscientists for correlating rock layers over wide geographic areas and for recognizing geologic time.

The fossil record is incomplete, because the process of preservation favors organisms with hard parts that are rapidly buried by sediments soon after death. For this reason, the vast majority of fossils are represented by marine invertebrates with exoskeletons, such as clams and snails. Under special circumstances, soft-bodied organism can be preserved, for instance the preservation of insects in amber, made famous by the feature film *Jurassic Park* (1993).

The Rock Cycle

A rock is a naturally formed aggregate of one or more minerals. Three types of rocks exist in the earth's crust, each reflecting a different origin. Igneous rocks have cooled and solidified from molten material either at or beneath Earth's surface. Sedimentary rocks form when preexisting rocks are weathered and broken down into fragments that accumulate and become compacted or cemented together. Fossils are most commonly found in sedimentary rocks. Metamorphic rocks form when heat, pressure, or chemical reactions in Earth's interior change the mineral or chemical composition and structure of any type of preexisting rock.

Over the huge span of geologic time, rocks of any one of these basic types can change into either of the other types or into a different form of the same type. For this reason, older rocks become increasingly more rare. The processes by which the various rock types change over time are illustrated in the rock cycle.

Larry E. Davis

Earth's Surface

Internal Geological Processes

The earth is layered into a core, a mantle, and a crust. The topmost mantle and the crust make up the lithosphere. Beneath this is a layer called the asthenosphere, which is composed of moldable and partly liquid materials. Heat transference within the asthenosphere sets up convection cells that diverge from hot regions and converge to cold regions. Consequently, the overlying lithosphere is segmented into ridged plates that are moved by the convection process. The hot asthenosphere does not rise along a line. This causes the development of a structure called a transform plate boundary, which is perpendicular to and offsetting the divergent boundary.

The topographic features at Earth's surface, such as mountains, rift valleys, oceans, islands, and ocean trenches, are produced by extension or compression forces that act along divergent, convergent, or transform plate boundaries. The extension and compression forces at Earth's surface are powered by convection within the asthenosphere.

Mountains and Depressions in Zones of Compression

Compression along convergent plate boundaries yields three types of mountain: island arcs that are partly under water; mountains along a continental edge, such as the Andes; and mountains at continental interiors, such as the Alps. At convergent plate boundaries, the denser of the two colliding plates slides down into the asthenosphere and causes volcanic activity to form on the leading edge of the upper plate. Island arcs such as the Aleutians and the Caribbean are formed when an oceanic plate descends beneath another oceanic plate.

Volcanic mountain chains such as the Andes of South America are formed when an oceanic plate descends beneath a continental plate. In both the island arc type and Andean type collisions, a deep depression in the oceans, called a trench, marks the place where neighboring plates are colliding and where the denser plates are pulled downward into the asthensophere. If the colliding plates are of similar density, neither plate will go into the asthenosphere. Instead, the edges of the neighboring plates will be folded and faulted and excess material will be pushed upward to form a block mountain, such as the mountain chain that stretches from the Alps through to the Himalayas. This type of mountain chain is not associated with a trench.

The Appalachians of the eastern United States are an example of the alpine type of mountain belt. When the Appalachians were forming 300 million years ago, rock layers were deformed. The deformation included folding to form ridges and valleys; fracturing along joint sets, with one joint set being parallel to ridges, while the other set is perpendicular; and thrust faulting, in which rock blocks were detached and shoved upward and northwestward.

Millions of years of erosion have reduced the height of the mountains and have produced topographic inversion in the foothills. Topographic inversion occurs because joints create wider fractures at upfolded ridges and narrower fractures at downfolded valleys. Erosion is then accelerated at upfolded ridges, converting ancient ridges into valleys, while ancient valleys stand as ridges. The Valley and Ridge Province of the Appalachians is noted for such topographic inversion.

West of the Valley and Ridge Province of the Appalachians is the Allegheny Plateau, which is bounded by a cliff on its eastern side. In general, plateaus are flat topped because the rock layer that covers the surface is resistant to weathering. The cliff side is formed by erosion along joint or fault surfaces.

The Sierra Nevada range, which formed 70 million years ago, is an example of an Andean type of mountain belt. Millions of years of erosion there has exposed igneous rocks that formed at depth. Over the years, the force of compression that formed the Sierras has evolved to form a zone of extension between the Sierras and the Colorado Plateau.

Mountains and Depressions in Zones of Extension

Extension is a strain that involves an increase in length and causes crustal thinning and faulting. Extension is associated with convergent boundaries, divergent boundaries, and transform boundaries.

Extension Associated with a Convergent Boundary

During the formation of the Sierra Nevada, an oceanic plate that was subducted beneath California declined at a shallow angle eastward toward the Colorado Plateau. Later, the subducted plate peeled off and molten asthenosphere took its place. From the asthenosphere, lava ascended through fractures to form volcanic mountains in Arizona and Utah, and lava flowed and volcanic ash fell as far west as California. The lithosphere has been heated up and has become buoyant, so the Colorado Plateau rises to higher elevations, and rock layers slide westward from it in a zone of extension that characterizes the Basin and Range Province.

In the extension zone, the top rock layers move westward on curved displacement planes that are steep at the surface and nearly horizontal at depth. When rock layers move westward over a curved detachment surface, the trailing edge of the rock layers roll over and are tilted toward the east so they do not leave space in buried rocks. On the other hand, a west-facing slope is left behind on a mountain from which the rock layers were detached. Therefore, movement along one curved detachment surface creates a valley, and movement along several such detachment surfaces forms a series of valleys separated by ridges, as in the Basin and Range Province. The amount of the displacement along the curved surfaces is not uniform. For example, more displacement has created wide zones of valleys such as the Las Vegas valley in Nevada, and Death Valley in California.

Extension Associated with a Divergent Boundary

The longest mountain chain on Earth lies under the Pacific Ocean. It is about 37,500 miles (60,000 km.) long, 31.3 miles (50 km.) wide, and 2 miles (3 km.) high. The central part of this midoceanic ridge is marked by a depression, about 3,000 feet (1,000 meters) deep, and is called a rift valley. A part of the submarine ridge, called the East Pacific Rise, forms the seafloor sector in the Gulf of California and reappears off the coast of northern California, Oregon, and Washington as the Juan de Fuca Ridge. Another part forms the seafloor sector in the Gulf of Aden and Red Sea seafloor, part of which is exposed in the Afar of Ethiopia. From the Afar southward to the southern part of Mozambique is the longest exposed rift valley on land, the East African Great Rift Valley.

A rift valley is the place where old rocks are pushed aside and new rocks are created. Blocks of rock that are detached from the rift walls slide down by a series of normal fault displacements. The ridge adjacent to the central rift is present because hot rocks are less dense and buoyant. If the process of divergences continues from the rifting stage to a drifting stage, as the rocks move farther away from the central rift, the rocks become older, colder, and denser, and push on the underlying asthenosphere to create basins. These basins will be flooded by oceanic water as neighboring continents drift away. However, not all processes of divergence advance from the rifting to the drifting stage.

Extension Associated with Transform Boundary

The best-known example of a transform boundary is the San Andreas Fault that offsets the East Pacific

Rise from the Juan de Fuca Ridge, and is exposed on land from the Gulf of California to San Francisco. Along transform boundaries, there are pull-apart basins that may be filled to form lakes, such as the Salton Sea in Southern California. Another example is the Aqaba transform of the Middle East, along which the Sea of Galilee and the Dead Sea are located.

H. G. Churnet

External Processes

Continuous processes are at work shaping the earth's surface. These include breaking down rocks, moving the pieces, and depositing the pieces in new locations. Weathering breaks down rocks through atmospheric agents. The process of moving weathered pieces of rock by wind, water, ice, or gravity is called erosion. The materials that are deposited by erosion are called sediment.

Mechanical weathering occurs when a rock is broken into smaller pieces but its chemical makeup is not changed. If the rock is broken down by a change in its chemical composition, the process is called chemical weathering.

Mechanical Weathering

Different types of mechanical weathering occur, depending on climatic conditions. In areas with moist climates and fluctuating temperatures, rocks can be broken apart by frost wedging. Water fills in cracks in rocks, then freezes during cold nights. As the ice expands and pushes out on the crack walls, the crack enlarges. During the warm days, the water thaws and flows deeper into the enlarged crack. Over time, the crack grows until the rock is broken apart. This process is active in mountains, producing a pile of rock pieces at the mountain base called talus.

Salt weathering occurs in areas where much salt is available or there is a high evaporation rate, such as along the seashore. Salt crystals form when salty moisture enters rock cracks. Growing crystals settle in the bottom of the crack and apply pressure on the crack walls, enlarging the crack.

Thermal expansion and contraction occur in climates with fluctuating temperatures, such as deserts. All minerals expand during hot days and contract during cold nights, and some minerals expand and contract more than others. This process continues until the rock loosens up and breaks into pieces.

Mechanical exfoliation can happen to a rock body overlain by a thick rock or sediment layer. If the heavy overlying layer over a portion of the rock body is removed, pressure is relieved and the exposed rock surface will expand in response. This expanding surface will break off into sheets parallel to the surface, but the remaining rock body remains under pressure and unchanged.

When plant roots grow into cracks in rocks, they enlarge the cracks and break up the rocks. Finally, abrasion can occur to rock fragments during transport. Either the fragments collide, breaking apart, or fragments are scraped against rocks, breaking off pieces.

Chemical Weathering

Water and oxygen create two common causes of chemical weathering. For example, dissolution occurs when water or another solution dissolves minerals within a rock and carries them away. Hydrolysis can occur when water flows through earth materials. The hydrogen ions or the hydroxide ions of the water may react with minerals in the

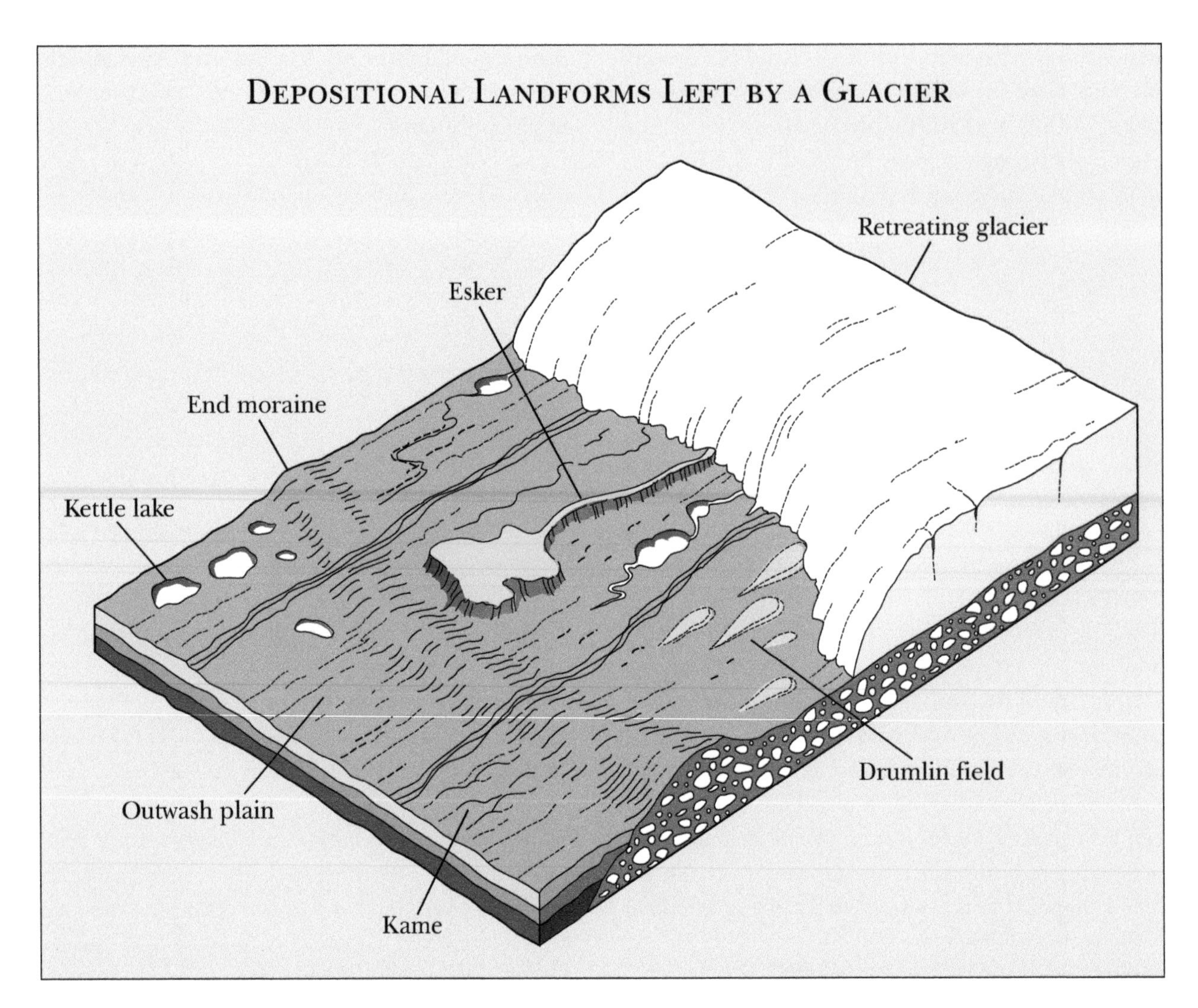

rocks. When this occurs, the chemical composition of the mineral is changed, and a new mineral is formed. Hydrolysis often produces clay minerals.

Some elements in minerals combine with oxygen from the atmosphere, creating a new mineral. This process is called oxidation. Some of these oxidation minerals are commonly referred to as rust.

Mass Movement

Weathered rock pieces (sediments) are transported (eroded) by one or more of four transport processes: water (streams and oceans), wind, ice (glaciers), or gravity. Mass movement transports earth materials down slopes by the pull of gravity. Gravity, constantly working to pull surface materials down, parallel to the slope, is the most important factor affecting mass movement. There is also a force involved perpendicular to the slope that contributes to the effects of friction.

Friction, the second factor, is determined by the earth material type involved. For example, weathering may create cracks in rocks, which form planes of weakness on which the mass movement can occur. Loose sediments always tend to roll downhill.

The third factor is the slope angle. Each earth material has its own angle of repose, which is the steepest slope angle on which the materials remain stable. Beyond this slope angle, earth materials will move downslope.

Water, the fourth factor, affects the stability of the earth material in the slope. Friction is weakened by water between the mineral grains in the rock. For example, water can make clay quite slippery, causing the mass movement.

HOW HYDROLOGY SHAPES GEOGRAPHY

Water and ice sculpt the landscape over time. Fast-flowing rivers erode the soil and rock through which they flow. When rivers slow down in flatter areas, they deposit eroded sediments, creating areas of rich soils and deltas at the mouths of the rivers. Over time this process wears down mountain ranges. The Appalachian Mountain range on the eastern side of the North American continent is hundreds of millions of years older than the Rocky Mountain range on the continent's western side. Although the Appalachians once rivaled the Rockies in size, they have been made smaller by time and erosion.

Canyons are carved by rivers, as the Grand Canyon was carved by the Colorado River, which exposed rocks billions of years old. Ice also changes the landscape. Large ice sheets from past ice ages could have been well over 1 mile (1,600 meters) thick, and they scoured enormous amounts of soil and rock as they slowly moved over the land surface. Terminal moraines are the enormous mounds of soil pushed directly in front of the ice sheets. Long Island, New York, and Cape Cod, Massachusetts, are two examples of enormous terminal moraines that were left behind when the ice sheets retreated.

The rooting system of vegetation, the fifth factor, helps make the surficial materials of the slope stable by binding the loose materials together.

Mass movements can be classified by their speed of movement. Creep and solifluction are the two types of slow mass movement, which are measured in fractions of inches per year. Creep is the slowest mass movement process, where unconsolidated materials at the surface of a slope move slowly downslope. The materials move slightly faster at the surface than below, so evidence of creep appears in the form of slanted telephone poles. During solifluction, the warm sun of the brief summer season in cold regions thaws the upper few feet of the earth. This waterlogged soil flows downslope over the underlying permafrost.

Rapid mass movement processes occur at feet per second or miles per hour. Falls occur when loose rock or sediment is dislodged and drops from a steep slope, such as along sea cliffs where waves erode the cliff base. Topples occur when there is an overturning movement of the mass. A topple can turn into a fall or a slide. A slide is a mass of rock or sediment that becomes dislodged and moves along a plane of weakness, such as a fracture. A slump is a slide that separates along a concave surface. Lateral spreads occur when a fractured earth mass spreads out at the sides.

A flow occurs when a mass of wet or dry rock fragments or sediment moves downslope as a highly viscous fluid. There are several different flow types. A debris flow is a mass of relatively dry, broken pieces of earth material that suddenly has water added. The debris flow occurs on steeper slopes and moves at speeds of 1-25 miles (2-40 km.) per hour. A debris avalanche occurs when an entire area of soil and underlying weathered bedrock becomes detached from the underlying bedrock and moves quickly down the slope. This flow type is often triggered by heavy rains in areas where vegetation has been removed. An earthflow is a dry mass of clayey or silty material that moves relatively slowly down the slope. A mudflow is a mass of earth material mixed with water that moves quickly down the slope.

A quick clay can occur when partially saturated, solid, clayey sediments are subjected to an earthquake, explosion, or loud noise and become liquid instantly.

Sherry L. Eaton

FLUVIAL AND KARST PROCESSES

Earth's landscape has been sculptured into an almost infinite variety of forms. The earth's surface has been modified by various processes for thousands, even hundreds of millions, of years to arrive at the modern configuration of landscapes.

Each process that transforms the surface is classified as either endogenic or exogenic. Endogenic processes are driven by the earth's internal heat and energy and are responsible for major crustal deformation. Endogenic processes are considered constructional, because they build up the earth's surface and create new landforms, such as mountain systems. Conversely, exogenic processes are considered destructional because they result in the wearing away of landforms created by endogenic processes. Exogenic processes are driven by solar energy putting into motion the earth's atmosphere and water, resulting in the lowering of features originally created by endogenic processes.

The most effective exogenic processes for wearing away the landscape are those that involve the action of flowing water, commonly referred to as fluvial processes. Water flows over the surface as runoff, after it evaporates into the atmosphere and infiltrates into the soil. The water that is left over flows down under the influence of gravity and has tremendous energy for sculpturing the earth's surface. Although flowing water is the most effective agent for modifying the landscape, it represents less than 0.01 percent of all the water on Earth's surface. By comparison, nearly 75 percent of the earth's surface water is stored within glaciers.

Drainage Basins

Fluvial processes can be considered from a variety of spatial scales. The largest scale is the drainage basin. A drainage basin is the area defined by topographic divides that diverts all water and material within the basin to a single outlet. Every stream of any size has its own drainage basin, and every portion of the earth's land surfaces are located within a drainage basin. Drainage basins vary tremendously in size, depending on the size of the river considered. For example, the largest drainage basin on earth is the Amazon, which drains about 2.25 million square miles (5.83 million sq. km.) of South America.

The Amazon Basin is so large that it could contain nearly the entire continent of Australia. By comparison, the Mississippi River drainage basin, the largest in North America, drains an area of about 1,235,000 square miles (3,200,000 sq. km.). Smaller rivers have much smaller basins, with many draining only an area roughly the size of a football field. While basins vary tremendously in size, they are spatially organized, with larger basins receiving the drainage from smaller basins, and eventually draining into the ocean. Because drainage basins receive water and material from the landscape within the basin, they are sensitive to environmental change that occurs within the basin. For example, during the twentieth century, the Mississippi River was influenced by many human-imposed changes that occurred either within the basin or directly within the channel, such as agriculture, dams and reservoirs, and levees.

Drainage Networks and Surface Erosion

Drainage basins can be subdivided into drainage networks by the arrangement of their valleys and interfluves. Interfluves are the ridges of higher elevation that separate adjacent valleys. Where an interfluve represents a natural boundary between two or more basins, it is referred to as a drainage divide. Valleys contain the larger rivers and are easily distinguished from interfluves by their relatively low, flat surfaces. Interfluves have relatively steep slopes and, for this reason, are eroded by runoff. The term erosion refers to the transport of material, in this case sediment that is dislodged from the surface.

Runoff starts as a broad sheet of slow-moving water that is not very erosive. As it continues to flow downslope, it speeds up and concentrates into rills, which are narrow, fast-moving lines of water. Because the runoff is concentrated within rills, the water travels faster and has more energy for erosion. Thus, rills are responsible for transporting sediment from higher points of elevation within the basin to the valleys, which are at a lower elevation. Rills can become powerful enough to scour deeply into the surface, developing into permanent channels called gullies.

The presence of many gullies indicates significant erosion on the landscape and represents an expensive and long-lasting problem if it is not remedied after initial development. The formations of gullies is often associated with human manipulation of the earth. For example, gullies can develop after improper land management, particularly intensive agricultural and grazing practices. A change in land use from natural vegetation, such as forests or prairie, can result in a type of land cover that is not suited for preventing erosion. Such land surfaces become susceptible to the formation of gullies during heavy, prolonged rains.

At a smaller scale, fluvial processes can be considered from the perspective of the river channel. River channels are located within the valleys of basins, offering a permanent conduit for drainage. Higher in the basin, river channels and valleys are relatively narrow, but grow larger toward the mouth of the basin as they receive drainage from smaller rivers within the basin. River channels may be categorized by their planform pattern, which refers to their overhead appearance, such as would be viewed from the window of an airplane.

The two major types of rivers are meandering and braided. Meandering rivers have a single channel that is sinuous and winding. These rivers are characterized as having orderly and symmetrical bends, causing the river to alternate directions as it flows across its valley. In contrast, braided rivers contain numerous channels divided by small islands, which results in a disorganized pattern. The islands within a braided river channel are not permanent. Instead, they erode and form over the course of a few years, or even during large flood events. Meandering channels usually have narrow and deep channels, but braided river channels are shallow and wide.

Sediment and Floodplains

Another distinction between braided and meandering river channels is the types of sediment they transport. Braided rivers transport a great amount of sediment that is deposited into midchannel islands within the river. Also, because braided rivers are frequently located higher in the drainage basin, they may have larger sediments from the erosion of adjacent slopes. In contrast, meandering river channels are located closer to the mouth of the basin and transport fine-grained sediment that is easily stored within point bars, which results in symmetrical bends within the river.

The sediments of both meandering and braided rivers are deposited within the valleys onto floodplains. Floodplains are wide, flat surfaces formed from the accumulation of alluvium, which is a term for sediment that is deposited by water. Floodplain sediments are deposited with seasonal flooding. When a river floods, it transports a large amount of sediment from the channel to the adjacent floodplain. After the water escapes the channel, it loses energy and can no longer transport the sediment. As a result, the sediment falls out of suspension and is deposited onto the floodplain. Because flooding occurs seasonally, floodplain deposits are layered and may accumulate into very thick alluvial deposits over thousands of years.

Karst Processes and Landforms

A specialized type of exogenic process that is also related to the presence of water is karst. Karst processes and topography are characterized by the solution of limestone by acidic groundwater into a number of distinctive landforms. While fluvial processes lower the landscape from the surface, karst processes lower the landscape from beneath the surface. Because limestone is a very permeable sedimentary rock, it allows for a large amount of groundwater flow. The primary areas for solution of the limestone occur along bedding planes and joints. This creates a positive feedback by increasing the amount of water flowing through the rock, thereby further increasing solution of the limestone. The result is a complex maze of underground conduits and caverns, and a surface with few rivers because of the high degree of infiltration.

The surface topography of karst regions often is characterized as undulating. A closer inspection reveals numerous depressions that lack surface outlets. Where this is best developed, it is referred to as cockpit karst. It occurs in areas underlain by extensive limestone and receiving high amounts of precipitation, for example, southern Illinois and Indiana in the midwestern United States, and in Puerto Rico and Jamaica.

Sinkholes are also common to karstic regions. Sinkholes are circular depressions having steep-sided vertical walls. Sinkholes can form either from the sudden collapse of the ceiling of an underground cavern or as a result of the gradual solution and lowering of the surface. Sinkholes can fill with sediments washed in from surface runoff. This reduces infiltration and results in the development of small circular lakes, particularly common in central Florida. Over time, erosion causes the vertical walls to retreat, resulting in uvalas, which are much larger flat-floored depressions.

Where there are numerous adjacent sinkholes, the retreat and expansion of the depressions causes them to coalesce, resulting in the formation of poljes. Unlike uvalas, poljes have an irregular shape, and the floor of the basin is not flat because of differences between the coalescing sinkholes.

Caves are among the most characteristic features of karst regions, but can only be seen beneath the surface. Caves can traverse the subsurface for miles, developing into a complex network of interconnected passages. Some caves develop spectacular formations as a result of the high amount of dissolved limestone transported by the groundwater. The evaporation of water results in the accumulation of carbonate deposits, which may grow for thousands of years. Some of the most common deposits are stalactites, which grow downward from the ceiling of the cave, and stalagmites, which grow upward and occasionally connect with stalactites to form large vertical columns.

Paul F. Hudson

GLACIATION

In areas where more snow accumulates each winter than can thaw in summer, glaciers form. Glacier ice, called firn, looks like rock but is not as strong as most rocks and is subject to intermittent thawing and freezing. Glacier ice can be brittle and fracture readily into crevasses, while other ice behaves as a plastic substance. A glacier is thickest in the area receiving the most snow, called the zone of accumulation. As the thickness piles up, it settles down and squeezes the limit of the ice outward in all directions. Eventually, the ice reaches a climate where the ice begins to melt and evaporate. This is called the zone of ablation.

Alpine Glaciation

Varied topographic evidence throughout the alpine environment attests to the sculpturing ability of glacial ice. The world's most spectacular mountain scenery has been produced by alpine glaciation, including the Matterhorn, Yosemite Valley, Glacier National Park, Mount Blanc, the Tetons, and Rocky

A FUTURE ICE AGE

If past history is an indicator, some time in the future conditions again will become favorable for the growth of glaciers. As recently as 1300 to 1600 CE, a cold period known at the Little Ice Age settled over Northern Europe and Eastern North America. Viking colonies perished as agriculture became unfeasible, and previously ice-free rivers in Europe froze over.

Another ice age would probably develop rapidly and be impossible to stop. Active mountain glaciers would bury living forests. Great ice caps would again cover Europe and North America, and move at a rate of 100 feet (30 meters) per day. Major cities and populations would shift to the subtropics and the topics.

Mountain National Park, all of which are visited by large numbers of people annually. Although alpine glaciation is still an active process of land sculpture in the high mountain ranges of the world, it is much less active than it was in the Ice Age of the Pleistocene epoch.

The prerequisites for alpine, or mountain, glaciation to become active are a mountainous terrain with Arctic climatic conditions in the higher elevations, and sufficient moisture to help snow and ice develop into glacial ice. As glaciers move out from their points of origin, they erode into the sides of mountains and increase the local relief in the higher elevations. The erosional features produced by alpine glaciation dominate mountain topography and usually are the most visible features on topographic maps. The eroded material is transported downvalley and deposited in a variety of landforms.

One kind of an erosional feature is a cirque, a hollow bowl-shaped depression. The bowl of the cirque commonly contains a small round lake or tarn. A steep-walled mountain ridge called an arête forms between two cirques. A high pyramidal peak, called horn, is formed by the intersecting walls of three or more cirques.

Erosion is particularly rapid at the head of a glacier. In valleys, moving glaciers press rock fragments against the sides, widening and deepening them by abrasion and forming broad U-shaped valleys. When glaciers recede, tributary streams become higher than the floor of the U-shaped valley and waterfalls occur over these hanging valleys. As the ice continues to melt, residual sediments called moraines may be deposited. Moraines are made up of glacier till, a collection of sediment of all sizes. Bands of sediment along the side of a valley glacier are lateral moraines; those crossing the valley are end or recessional moraines; where two glaciers join, a medial moraine is formed. Meltwater may also sort out the finer materials, transport them downvalley, and deposit them in beds as outwash.

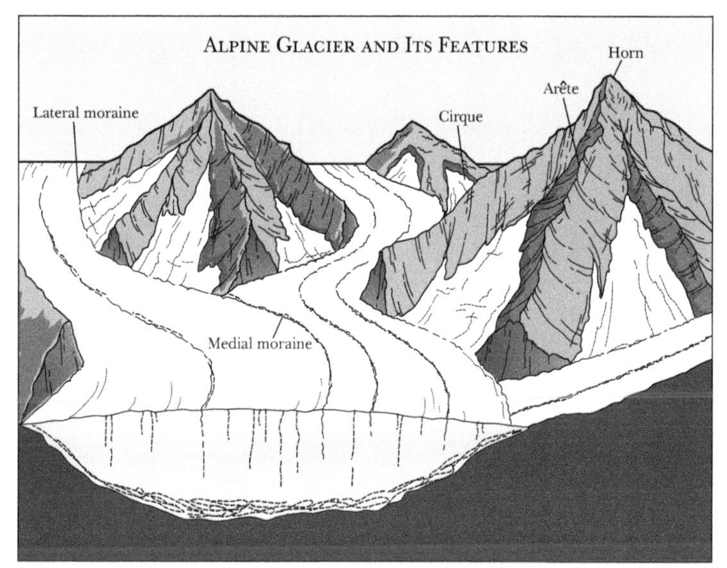

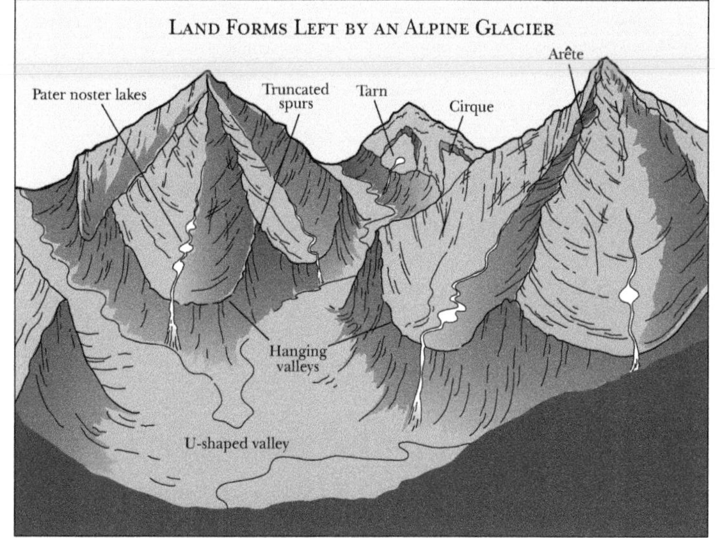

Continental Glaciation

In the modern world, continental glaciation operates on a large scale only in Greenland and Antarctica. However, its existence in previous geologic ages is evidenced by strata of tillite (a compacted rock formed of glacial deposits) or, more frequently, by surficial deposits of glacial materials.

Much of the geomorphology of the northeastern quadrant of North America and the northwestern portion of Europe was formed during the Ice Age. During that time, great masses of ice accumulated on the continents and moved out from centers near the Hudson Bay and the Fennoscandian Shield, extending over the continents in great advancing and retreating lobes. In North America, the four major

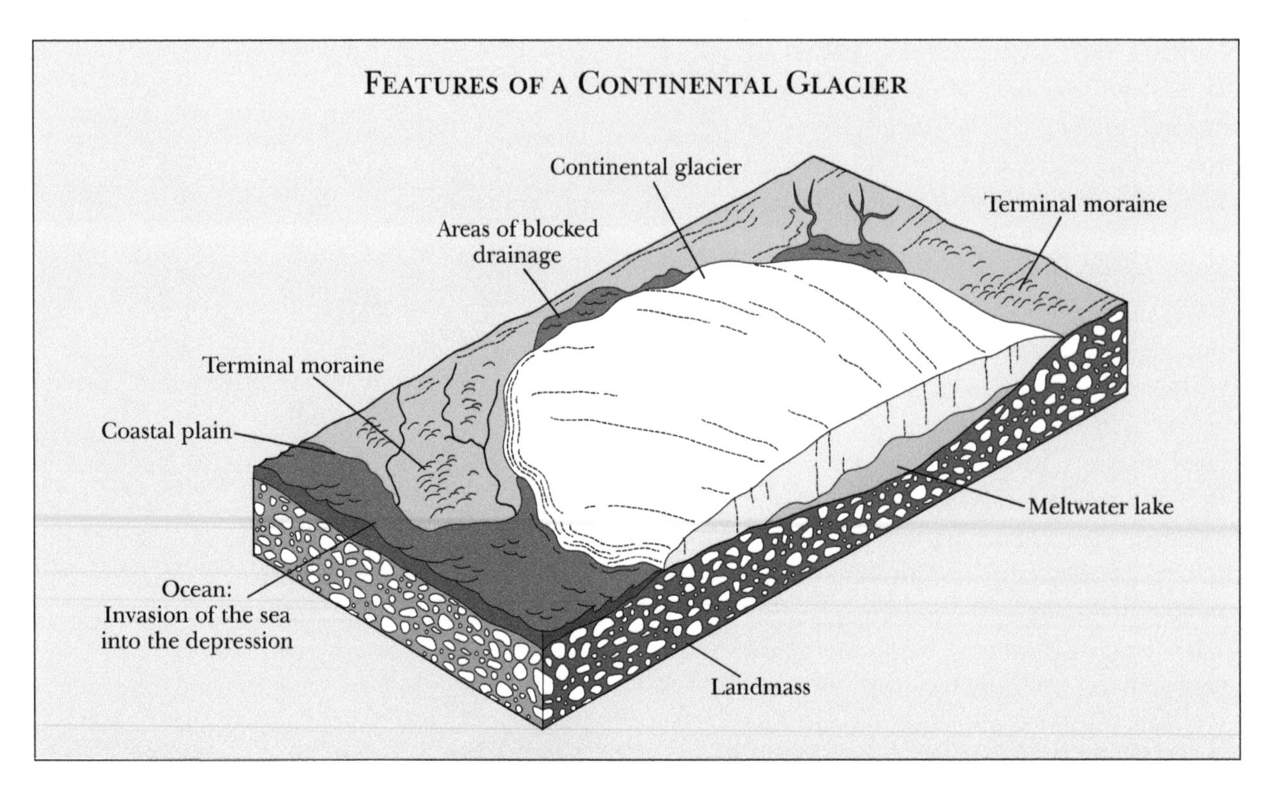

stages of lobe advance were the Wisconsin (the most recent), the Illinoian, the Kansan, and the Nebraskan (the oldest). Between each of these major advances were pluvial periods in which the ice melted and great quantities of water rushed over or stood on the continents, creating distinctive features which can still be detected today.

The two major functions of gradation are accomplished by the processes of scour (degradation) in the areas close to the centers and deposition (aggradation) adjacent to the terminal or peripheral areas of the lobes. Thus, the overall effect of continental glaciation is to reduce relief—to scour high areas and fill in lower regions—unlike the changes caused by alpine glaciation.

Although continental glaciation usually does not result in the spectacular scenery of alpine glaciation, it was responsible for creating most of the Great Lakes and the lakes of Wisconsin, Michigan, Minnesota, Finland, and Canada; for gravel deposits; and for the rich agricultural lands of the Midwest, to mention just a few of its effects.

While glaciers were leveling hilly sections of North America and Europe by scraping them bare of soil and cutting into the ice itself, they acquired a tremendous load of material. As a glacier warms and melts, there is a tremendous outflow of water, and the streams thus formed carry with them the debris of the glacier. The material deposited by glaciers is called drift or outwash. Glaciofluvial drift can be recognized by its separation into layers of finer sands and coarser gravels.

Kettles and kames are the most common features of the end moraines found at the outermost edges of a glacier. A kettle is a depression left when a block of ice, partially or completely buried in deposits of drift, melts away. Most of the lakes in the upper Great Lakes of the United States are kettle lakes. A kame is a round, cone-shaped hill. Kames are produced by deposition from glacial meltwater. Sometimes, the outwash material poured into a long and deep crevasse, rather than a hole. These tunnels have had their courses choked by debris, revealed today by long, narrow ridges, generally referred to as eskers.

Ron Janke

Desert Landforms

Deserts are often striking in color, form, or both. The underlying lack of water in deserts produces unique features not found in humid regions. Arid lands cover approximately 30 percent of the earth's land surface, an area of about 15.4 million square miles (40 million sq. km.). Arid lands include deserts and surrounding steppes, semiarid regions that act as transition zones between arid and humid lands.

Many of the world's largest and driest deserts are found between 20 and 40 degrees north and south latitude. These include the Mojave and Sonoran Deserts of the United States, the Sahara in northern Africa, and the Great Sandy Desert in Australia. In these deserts, the subtropical high prevents cloud formation and precipitation while increasing rates of surface evaporation.

Some arid lands, like China's Gobi Desert, form because they are far from oceans that are the dominant source for atmospheric water vapor and precipitation. Others, like California's Death Valley, are arid because mountain ranges block moisture from coming from the sea. The combination of mountain barriers and very low elevations makes Death Valley the hottest, driest desert in North America.

Sand Dunes

Many people envision deserts as vast expanses of blowing sand. Although wind plays a more important role in deserts than it does elsewhere, only about 25 percent of arid lands are covered by sand. Broad regions that are covered entirely in sand (such as portions of northwestern Africa, Arabia, and Australia) are referred to as sand seas. Why is wind more effective here than elsewhere?

The lack of soil water and vegetation, both of which act to bind grains together, allows enhanced eolian (wind) erosion. Very small particles are picked up and suspended within the moving air mass, while sand grains bounce along the surface. Removal of material often leaves behind depressions called blowouts or deflation hollows. Moving grains abrade cobbles and boulders at the surface, creating uniquely sculpted and smoothed rocks known as ventifacts. Bedrock outcrops can be streamlined as they are blasted by wind-borne grains to form features called yardangs. As these rocks are ground away, they contribute additional sediment to the wind.

Desert sand dunes are not stationary features—instead, they represent accumulations of moving sand. Wind blows sand along the desert floor. Where it collects, it forms dunes. Typically, dunes have relatively shallow windward faces and steeper slip faces. Sand grains bounce up the windward face and then eventually cascade down the slip face, the movement of individual grains driving movement of the entire dune in a downwind direction.

Four major dune types are found within arid regions. Barchan dunes are crescent-shaped features, with arms that point downwind. They may occur as isolated structures or within fields. They form where winds blow in a single direction and where the supply of sand is limited. With a larger supply of sand, barchan dunes can join with one another to form a transverse dune field.

There, ridges are perpendicular to the predominant wind direction. With quartering winds (that is, winds that vary in direction throughout a range of about 45 degrees) dune ridges form that are parallel to the average wind direction. These so-called longitudinal dunes have no clearly defined windward and slip faces. Where winds blow sand from all directions, star dunes form. Sand collects in the middle of the feature to form a peaked center with arms that spiral outward.

Badlands, Mesas, and Buttes

As scarce as it may be, water is still the dominant force in shaping desert landscapes. Annual precipitation may be low, but the amount of precipitation in a single storm may be a large fraction of the yearly total. An arid landscape that is underlain by poorly cemented rock or sediment, such as that found in western South Dakota, may form badlands as a result of the erosive ability of storm-water run-off. Overall aridity prevents vegetation from establishing the interconnected root system that holds soil particles together in more humid regions.

Cloudbursts cause rapid erosion that forms numerous gullies, deeply incised washes, and hoodoos, which are created when rock or sediment that is more resistant protects underlying material from erosion. Over time, protected sections stand as prominent spires while surrounding material is removed. Landscapes like those found in Badlands National Park in South Dakota are devoid of vegetation and erode rapidly during storms.

Arid regions that are underlain by flat-lying rock units can form mesas and buttes. Water follows fractures and other lines of weakness, forming ever-widening canyons. Over time, these grow into broad valleys. In northern Arizona's Monument Valley, remnants of original bedrock stand as isolated, flat-topped structures. Broad mesas are marked by their flat tops (made of a resistant rock like sandstone or basalt) and steep sides. Buttes are much narrower, with a small resistant cap, but are often as tall and steep as neighboring mesas.

Death Valley Playa

California's Death Valley is the driest desert in the United States, with an average rainfall of only 1.5 inches (38 millimeters) per year at the town of Furnace Creek. It is also consistently one of the hottest places on Earth, with a record high of 134 degrees Fahrenheit (57 degrees Celsius). In the distant past, however, Death Valley held lakes that formed in response to global cooling. Over 120,000 years ago, Death Valley hosted a 295-foot-deep (90 meters) body of water called Lake Manley. Evidence of this lake remains in evaporite deposits that make up the playa in the valley's center, in wave-cut shorelines, and in beach bars.

Desert Pavement and Desert Varnish

Much of the desert floor is covered by desert pavement, an accumulation of gravel and cobbles that forms a surface fabric that can interconnect tightly. Fine material has been removed by wind and water, leaving behind larger fragments that inhibit further erosion. In many areas, desert pavements have been stable for long periods of time, as evidenced by their surface patina of desert varnish. Desert varnish is a thin outer coating of wind-deposited clay mixed with iron and manganese oxides. Varying in color from light brown to black, these coatings are thought to adhere to rocks by the action of single-celled microorganisms. Under a microscope, desert varnish can be seen to be made up of very fine layers. A thick, dark patina means that a rock has been exposed for a long time.

Playas

Where neither dunes nor rocky pavements cover the desert floor, one may find an accumulation of saline minerals. A playa is a flat surface that is often blindingly white in color. Playas are usually found in the centers of desert valleys and contain material that mineralized during the evaporation of a lake. Dry lake beds are a common feature of the Great Basin in the western United States. During glacial stages, the last of which occurred about 20,000 years ago, lakes grew in what are now arid, closed valleys. As the climate warmed, these lakes shrank, and many dried completely. As a lake evaporates, minerals that were held in solution crystallize, forming salts, including halite (table salt). These salt deposits frequently are mined for useful household and industrial chemicals.

Richard L. Orndorff

OCEAN MARGINS

Ocean margins are the areas where land borders the sea. Although often referred to as coastlines or beaches, ocean margins cover far greater territory than beaches. An ocean margin extends from the coastal plain—the fertile farming belt of land along the seacoast—to the edge of the gently sloping land submerged in water, called the continental shelf.

Ocean margin constitutes 8 percent of the world's surface. It is rich in minerals, both above and below water, and is home to 25 percent of Earth's people, along with 90 percent of the marine life. This fringe of land at the border of the ocean is ever changing. Tides wash sediment in and leave it behind, just below sea level. This process, called deposition, builds up land in some areas of the coastline. At the same time, ocean waves, winds, and storms wear away or erode parts of the shoreline. As land is worn away or built up, the amount of land above sea level changes. Factors such as climate, erosion, deposition, changes in sea level, and the effects of humans constantly change the shape of the ocean margin on Earth.

Beach Dynamics

The two types of coasts or land formations at the ocean margin are primary coasts and secondary coasts. Primary coasts are formed by systems on land, such as the melting of glaciers, wind or water erosion, and sediment deposited by rivers. Deltas and fjords are examples of primary coasts. Secondary coasts are formed by ocean patterns, such as erosion by waves or currents, sediment deposition by waves or currents, or changes by marine plants or animals. Beaches, coral reefs, salt marshes, and mangrove swamps are examples of secondary coasts.

Sediment carried by rivers to the sea is deposited to form deltas at the mouths of the rivers. Some of the sediment can wash out to sea, causing formations to build up at a distance from the shore. These formations eventually become barrier islands, which are often little more than 10 feet (3 km.) above sea level. As a consequence, heavy storms, such as hurricanes, can cause great damage to barrier islands. Barrier islands naturally protect the coastline from erosion, however, especially during heavy coastal storms.

Sea level changes also affect the shape of the coastline. As oceans slowly rise, land is slowly consumed by the ocean. Barrier islands, having low sea levels, may slowly be covered with water. The melting of continental glaciers increased the sea level 0.06 inch (0.15 centimeter) per year during the twentieth century. As ocean waters warm, they expand, eating away at sea levels. Global warming caused by carbon dioxide levels in the atmosphere could cause sea levels to rise as much as 0.24 inch (0.6 centimeter) per year as a result of the warming of the water and glacial melting.

Human Influence

The shape of the ocean margin also changes radically as a result of human influence. According to the United States Geological Survey, 39 percent of the people living in the United States live directly on the coasts. According to UN Atlas of Oceans, about 44 percent of the world's population lives within 93 miles (150 kilometers) of the coast. Pollution from toxins, dredging, recreational boating, and waste disposal kills plants and animals along the ocean margin. This changes the coastal shape, as mangrove forests, coral reefs, and other coastal lifeforms die.

A greater concern along the coastal fringe, however, is human development. Not only are people drawn to the fertile soil along the coastal zone of the continent, but they also develop islands and coves into resort communities. To protect homes and hotels along the coastal zone from coastal erosion, people build breakwalls, jetties, and sand and stone bars called groins.

These human-made barriers disrupt the natural method by which the ocean carries material along the coast. Longshore drift, a zigzag movement, deposits sediment from one area of the beach farther along the shoreline. Breakwalls, jetties, and groins disrupt this flow. As the ocean smashes against a breakwall, the property behind it may be safe for the present, but the coastline neighboring the breakwall takes a greater beating. The silt and sediment from upshore, which would replace that carried downshore, never arrives. Eventually, the breakwall will break down under the impact of the ocean force. Areas with breakwalls and jetties often suffer greater damage in coastal storms than areas that remain naturally open to the changing forces of the ocean.

To compensate for the destructive nature of human-made barriers, many recreational beaches replace lost sand with dredgings or deposit truckloads of sand from inland sources. For example, Virginia Beach in the United States spends between US$2 million and US$3 million annually to restore beaches for the tourist season in this way.

Despite the changes in the shape of the ocean margin, it continues to provide a stable supply of resources—fish, seafood, minerals, sponges, and other marine plants and animals. Offshore drilling of oil and natural gas often takes place within 200 miles (322 km.) of shorelines.

Lisa A. Wroble

Earth's Climates

The Atmosphere

The thin layer of gases that envelops the earth is the atmosphere. This layer is so thin that if the earth were the size of a desktop globe, more than 99 percent of its atmosphere would be contained within the thickness of an ordinary sheet of paper. Despite its thinness, the atmosphere sustains life on Earth, protecting it from the Sun's searing radiation and regulating the earth's temperature. Storms of the atmosphere carry water to the continents, and weathering by its wind and rain helps shape them.

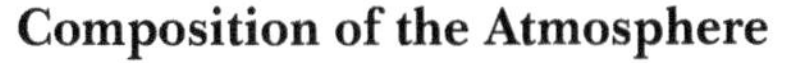

Composition of the Atmosphere

The earth's atmosphere consists of gases, microscopic particles called aerosol, and clouds consisting of water droplets and ice particles. Its two principal gases are nitrogen and oxygen. In dry air, nitrogen occupies 78 percent, and oxygen 21 percent, of the atmosphere's volume. Argon, neon, xenon, helium, hydrogen, and other trace gases together equal less than 1 percent of the remaining volume.

These gases are distributed homogeneously in a layer called the homosphere, which occurs between the earth's surface and about 50 miles (80 km.) altitude. Above 50 miles altitude, in the heterosphere, the concentration of heavier gases decreases more rapidly than lighter gases.

The atmosphere has no firm top. It simply thins out until the concentration of its gas molecules approaches that of the gases in outer space. The concentration of nitrogen and oxygen remains essentially constant in the atmosphere because a balance exists between the production and removal of these gases at the earth's surface. Decaying organic matter adds nitrogen to the atmosphere, while soil bacteria remove nitrogen. Oxygen enters the atmosphere primarily through photosynthesis and is removed through animal respiration, combustion, and decay of organic material, and by chemical reactions involving the creation of oxides.

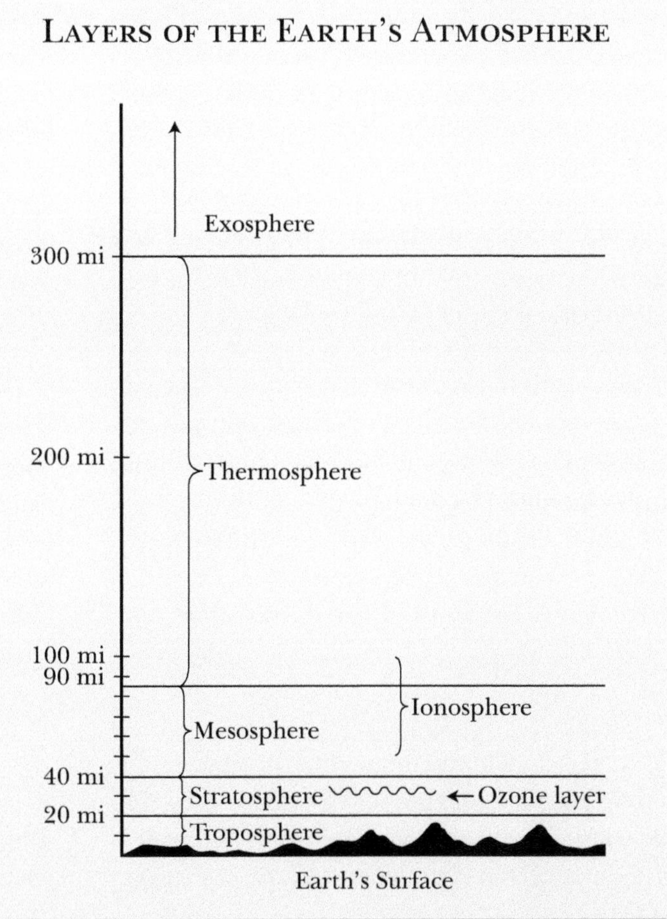

THE GREENHOUSE EFFECT

Clouds and atmospheric gases such as water vapor, carbon dioxide, methane, and nitrous oxide absorb part of the infrared radiation emitted by the earth's surface and reradiate part of it back to the earth. This process effectively reduces the amount of energy escaping to space and is popularly called the "greenhouse effect" because of its role in warming the lower atmosphere. The greenhouse effect has drawn worldwide attention because increasing concentrations of carbon dioxide from the burning of fossil fuels result in a global warming of the atmosphere.

Scientists know that the greenhouse analogy is incorrect. A greenhouse traps warm air within a glass building where it cannot mix with cooler air outside. In a real greenhouse, the trapping of air is more important in maintaining the temperature than is the trapping of infrared energy. In the atmosphere, air is free to mix and move about.

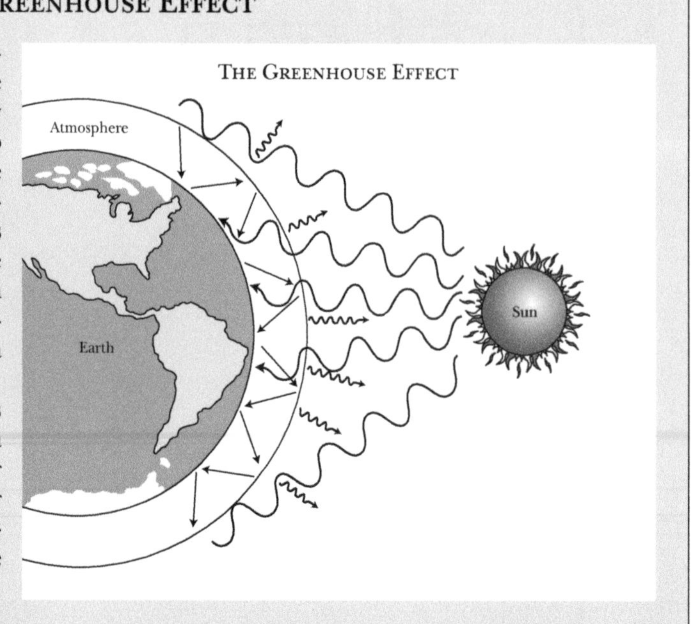

The atmosphere contains many gases that are present in small, variable concentrations. Three gases—water vapor, carbon dioxide and ozone—are vital to life on Earth. Water vapor enters the atmosphere through evaporation, primarily from the oceans, and through transpiration by plants. It condenses to form clouds, which provide the rain and snow that sustain life outside the oceans. The concentration of water vapor varies from about 4 percent by volume in tropical humid climates to a small fraction of a percent in polar dry climates. Water vapor plays an important role in regulating the temperature of the earth's surface and the atmosphere. Clouds reflect some of the incoming solar radiation, while water vapor and clouds both absorb earth's infrared radiation.

Carbon dioxide also absorbs the earth's infrared radiation. The global average atmospheric carbon dioxide in 2018 was 407.4 parts per million (ppm for short), with a range of uncertainty of plus or minus 0.1 ppm. Carbon dioxide levels today are higher than at any point in at least the past 800,000 years. The annual rate of increase in atmospheric carbon dioxide over the past 60 years is about 100 times faster than previous natural increases, such as those that occurred at the end of the last ice age 11,000–17,000 years ago.

Carbon dioxide enters the atmosphere as the result of decay of organic material, through respiration, during volcanic eruptions, and from the burning of fossil fuels. It is removed during photosynthesis and by dissolving in ocean water, where it is used by organisms and converted to carbonates. The increase in atmospheric carbon dioxide associated with the burning of fossil fuels has raised concerns that the earth's atmosphere may be warming through enhancement of the greenhouse effect.

Ozone, a gas consisting of molecules containing three oxygen atoms, forms in the upper atmosphere when oxygen atoms and oxygen molecules combine. Most ozone exists in the upper atmosphere between 6.2 and 31 miles (10 and 50 km.) in altitude, in concentrations of no more than 0.0015 percent by volume. This small amount of ozone sustains life outside the oceans by absorbing most of the Sun's ultraviolet radiation, thereby shielding the earth's surface from the radiation's harmful effects on living organisms. Paradoxically, ozone is an irritant near the earth's surface and is the major component of photochemical smog. Other gases that contribute to pollution include methane, nitrous oxide, hydrocarbons, and chlorofluorocarbons.

Aerosols represent another component of atmospheric pollution. Aerosols form in the atmosphere during chemical reactions between gases, through mechanical or chemical interactions between the earth, ocean surface, and atmosphere, and during evaporation of droplets containing dissolved or solid material. These microscopic particles are always present in air, with concentrations of about a few hundred per cubic centimeter in clean air to as many as a million per cubic centimeter in polluted air. Aerosols are essential to the formation of rain and snow, because they serve as centers upon which cloud droplets and ice particles form.

The Ozone Hole

Since the 1970s, balloon-borne and satellite measurements of stratospheric ozone have shown rapidly declining stratospheric ozone concentrations over the continent of Antarctica, termed the "ozone hole." The lowest concentrations occur during the Antarctic spring, in September and October. The decrease in ozone has been associated with an increase in the concentration of chlorine, a gas introduced into the stratosphere through chemical reactions involving sunlight and chlorofluorocarbons—synthetic chemicals used primarily as refrigerants. The ozone hole over Antarctica has raised concern about possible worldwide reduction in the concentration of upper atmospheric ozone.

Energy Exchange in the Atmosphere

The Sun is the ultimate source of the energy in Earth's atmosphere. Its radiation, called electromagnetic radiation because it propagates as waves with electric and magnetic properties, travels to the surface of the earth's atmosphere at the speed of light. This energy spans many wavelengths, some of which the human eye perceives as colors. Visible wavelengths make up about 44 percent of the Sun's energy. The remainder of the Sun's radiant energy cannot be seen by human eyes. About 7 percent arrives as ultraviolet radiation, and most of the remaining energy is infrared radiation.

The Sun is not the only source of radiation. All objects emit and absorb radiation to some degree. Cooler objects such as the earth emit nearly all their energy at infrared wavelengths. Objects heat when they absorb radiation and cool when they emit radiation. The radiation emitted by the earth and atmosphere is called terrestrial radiation.

The balance between absorption of solar radiation and emission of terrestrial radiation ultimately determines the average temperature of the earth-atmosphere system. The vertical temperature distribution within the atmosphere also depends on the absorption and emission of radiation within the atmosphere, and the transfer of energy by the processes of conduction, convection, and latent heat exchange. Conduction is the direct transfer of heat from molecule to molecule. This process is most important in transferring heat from the earth's surface to the first few centimeters of the atmosphere. Convection, the transfer of heat by rising or sinking air, transports heat energy vertically through the atmosphere.

Latent heat is the energy required to change the state of a substance, for example, from a liquid to a gas. Energy is transferred from the earth's surface to the atmosphere through latent heat exchange when water evaporates from the oceans and condenses to form rain in the atmosphere.

Only 48 percent of the solar energy reaching the top of the earth's atmosphere is absorbed by the earth's surface. The atmosphere absorbs another 23 percent. The remaining 30 percent is scattered back to space by atmospheric gases, clouds and the earth's surface. To understand the importance of terrestrial radiation and the greenhouse effect in the atmosphere's energy balance, consider the solar radiation arriving at the top of the earth to be 100 energy units, with 48 energy units absorbed by the earth's surface and 23 units by the atmosphere.

The earth's surface actually emits 117 units of energy upward as terrestrial radiation, more than twice as much energy as it receives from the Sun. Only 6 of these units are radiated to space—the atmosphere absorbs the remaining energy. Latent heat exchange, conduction, and convection account for another 30 units of energy transferred from the surface to the atmosphere. The atmosphere, in turn, radiates 96 units of energy back to the earth's surface (the greenhouse effect), and 64 units to space. The earth's and atmosphere's energy budget remains in balance, the atmosphere gaining and losing 160 units of energy, and the earth gaining and losing 147 units of energy.

Vertical Structure of the Atmosphere

Temperature decreases rapidly upward away from the earth's surface, to about –60 degrees Fahrenheit (–51 degrees Celsius) at an altitude of about 7.5 miles (12 km.). Above this altitude, temperature increases with height to about 32 degrees Fahrenheit (0 degrees Celsius) at an altitude of 31 miles (50 km.). The layer of air in the lower atmosphere where temperature decreases with height is called the troposphere. It contains about 75 percent of the atmosphere's mass. The layer of air above the troposphere, where temperature increases with height, is called the stratosphere. All but 0.1 percent of the remaining mass of the atmosphere resides in the stratosphere.

The stratosphere exists because ozone in the stratosphere absorbs ultraviolet light and converts it to heat. The boundary between the troposphere and stratosphere is called the tropopause. The tropopause is extremely important because it acts as a lid on the earth's weather. Storms can grow vertically in the troposphere, but cannot rise far, if at all, beyond the tropopause. In the polar regions, the tropopause can be as low as 5 miles (8 km.) above the surface, while in the tropics, the tropopause can be as high as 11 miles (18 km.). For this reason, tropical storms can extend to much higher altitudes than storms in cold regions.

The mesosphere extends from the top of the stratosphere, the stratopause, to an altitude of about 56 miles (90 km.). Temperature decreases with height within the mesosphere. The lowest average temperatures in the atmosphere occur at the mesopause, the top of the mesosphere, where the temperature is about –130 degrees Fahrenheit (–90 degrees Celsius). Only 0.0005 percent of the atmosphere's mass remains above the mesopause. In this uppermost layer, the thermosphere, there are few atoms and molecules. Oxygen molecules in the thermosphere absorb high-energy solar radiation. In this near vacuum, absorption of even small amounts of energy causes a large increase in temperature. As a result, temperature increases rapidly with height in the lower thermosphere, reaching about 1,300 degrees Fahrenheit (700 degrees Celsius) above 155 miles (250 km.) altitude.

The upper mesosphere and thermosphere also contain ions, electrically charged atoms or molecules. Ions are created in the atmosphere when air molecules collide with high-energy particles arriving from space or absorb high-energy solar radiation. Ions cannot exist very long in the lower atmosphere, because collisions between newly formed ions quickly restore ions to their uncharged state. However, above about 37 miles (60 km.) collisions are less frequent and ions can exist for longer times. This region of the atmosphere, called the ionosphere, is particularly important for amplitude-modulated (AM) radio communication because it reflects standard AM radio waves. At night, the lower ionosphere disappears as ions recombine, allowing AM radio waves to travel longer distances when reflected. For this reason, AM radio station signals can sometimes travel great distances at night.

The top of the atmosphere occurs at about 310 miles (500 km.). At this altitude, the distance between individual molecules is so great that energetic molecules can move into free space without colliding with neighbor molecules. In this uppermost layer, called the exosphere, the earth's atmosphere merges into space.

Robert M. Rauber

Global Climates

A region's climate is the sum of its long-term weather conditions. Most descriptions of climate emphasize temperature and precipitation characteristics, because these two climatic elements usually exert more impact on environmental conditions and human activities than do other elements, such as wind, humidity, and cloud cover. Climatic descriptions of a region generally cover both mean conditions and extremes. Climatic means are important because they represent average conditions that are frequently experienced; extreme conditions, such as severe storms, excessive heat and cold, and droughts, are important because of their adverse impact.

Important Climate Controls

A region's climate is largely determined by the interaction of six important natural controls: sun angle, elevation, ocean currents, land and water heating and cooling characteristics, air pressure and wind belts, and orographic influence.

Sun angle—the height of the Sun in degrees above the nearest horizon—largely controls the amount of solar heating that a site on Earth receives. It strongly influences the mean temperatures of most of the earth's surface, because the Sun is the ultimate energy source for nearly all the atmosphere's heat. The higher the angle of the Sun in the sky, the greater the concentration of energy, per unit area, on the earth's surface (assuming clear skies). From a global perspective, the Sun's mean angle is highest, on average, at the equator, and becomes progressively lower poleward. This causes a gradual decrease in mean temperatures with increasing latitude.

Sun angles also vary seasonally and daily. Each hemisphere is inclined toward the Sun during spring and summer, and away from the Sun during fall and winter. This changing inclination causes mean sun angles to be higher, and the length of daylight longer, during the spring and summer. Therefore, most locations, especially those outside the tropics, have warmer temperatures during these two seasons. The earth's rotation causes sun angles to be higher during midday than in the early morning and late afternoon, resulting in warmer temperatures at midday. Heating and cooling lags cause both seasonal and daily maximum and minimum temperatures typically to occur somewhat after the periods of maximum and minimum solar energy receipt.

Variations in elevation—the distance above sea level—can cause locations at similar latitudes to vary greatly in temperature. Temperatures decrease an average of about 3.5 degrees Fahrenheit per thousand feet (6.4 degrees Celsius per thousand meters). Therefore, high mountain and plateau stations are much colder than low-elevation stations at the same latitude.

Surface ocean currents can transport masses of warm or cold water great distances from their source regions, affecting both temperature and moisture conditions. Warm currents facilitate the evaporation of copious amounts of water into the atmosphere and add buoyancy to the air by heating it from below. This results in a general increase in precipitation totals. Cold currents evaporate water relatively slowly and chill the overlying air, thus stabilizing it and reducing its potential for precipitation.

The influence of ocean currents on land areas is greatest in coastal regions and decreases inland. The west coasts of continents (except for Europe) generally are paralleled by relatively cold currents, and the east coasts by relatively warm currents. For example, the warm Gulf Stream flows northward off the eastern United States, while the West Coast is cooled by the southward-flowing California Current.

Land can change temperature much more readily than water. As a result, the air over continents typically experiences larger annual temperature ranges (that is, larger temperature differences between summer and winter) and shorter heating and cooling lags than does the air over oceans. This same effect causes continental interiors and the leeward (downwind) coasts of continents typically to have larger temperature ranges than do windward (upwind) coasts. Climates that are dominated by air from landmasses are often described as continental climates. Conversely, climates dominated by air from oceans are described as maritime climates.

The seasonal heating and cooling of continents can also produce a monsoon influence, which has to do with annual shifts of wind patterns. Areas influenced by a monsoon, such as Southeast Asia, tend to have a predominantly onshore flow of moist maritime air during the summer. This often produces heavy rains. An offshore flow of dry air predominates in winter, producing fair weather.

Earth's atmosphere displays a banded, or beltlike, pattern of air pressure and wind systems. High pressure is associated with descending air and dry weather; low pressure is associated with rising air, which produces cloudiness and often precipitation. Wind is produced by differences in air pressure. The air blows outward from high-pressure systems and into low-pressure systems in a constant attempt to equalize air pressures.

The direction and speed of movement of weather systems, such as weather fronts and storms, are controlled by wind patterns, especially those several kilometers above the surface. The seasonal shift of global temperatures caused by the movement of the Sun's vertical rays between the Tropics of Cancer and Capricorn produces a latitudinal migration of both air pressure and wind belts. This shift affects the annual temperature and precipitation patterns of many regions.

Four air-pressure belts exist in each hemisphere. The intertropical convergence zone (ITCZ) is a broad belt of low pressure centered within a few degrees of latitude of the equator. The subtropical highs are high-pressure belts centered between 20 and 40 degrees north and south latitude, which are responsible for many of the world's deserts. The subpolar lows are low-pressure belts centered about 50 or 70 degrees north and south latitude. Finally, the polar highs are high-pressure centers located near the North and South Poles.

The air pressure gradient between these belts produces the earth's major wind belts. The regions between the ITCZ and the subtropical highs are dominated by the trade winds, a broad belt in each hemisphere of easterly (that is, moving east to west) winds. The middle latitudes are mostly situated between the subtropical highs and the subpolar lows and are within the westerly wind belt. This wind belt causes winds, and weather systems, to travel generally from west to east in the United States and Canada. Finally, the high-latitude zones between the subpolar lows and polar highs are situated within the polar easterlies.

The final factor affecting climate— orographic influence—is the lifting effect of mountain peaks or ranges on winds that pass over them. As air approaches a mountain barrier, it rises, typically producing clouds and precipitation on the windward (upwind) side of the mountains. After it crosses the crest, it descends the leeward (downwind) side of the mountains, generally producing dry weather. Most of the world's wettest locations are found on the windward sides of high mountain ranges; some deserts, such as those of the western interior United States, owe their aridity to their location on the leeward sides of orographic barriers.

World Climate Types

The global distribution of the world climate controls is responsible for the development of fourteen widely recognized climate types. In this section, the major characteristics of each of these climates will be briefly described. The climates are discussed in a rough poleward sequence.

Tropical Wet Climate

Sometimes called the tropical rainforest climate, the tropical wet climate exists chiefly in areas lying within 10 degrees of the equator. It is an almost seasonless climate, characterized by year-round warm, humid, rainy conditions that allow land areas to support a dense broadleaf forest cover. The warm temperatures, which for most locations average near 80 degrees Fahrenheit (27 degrees Celsius) throughout the year, result from the constantly high midday sun angles experienced at this low latitude.

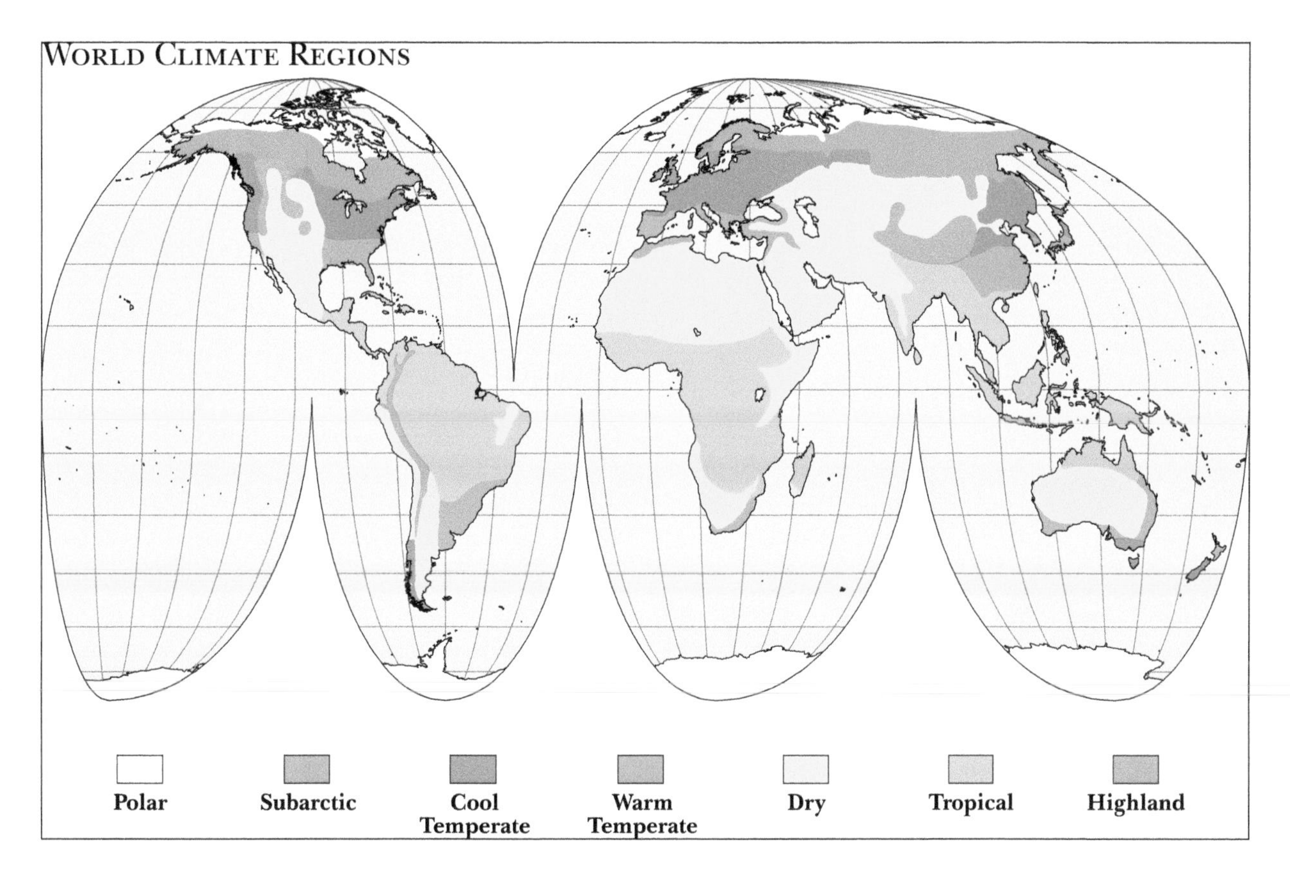

The heavy precipitation totals result from the heating and subsequent rising of the warm moist air to form frequent showers and thunderstorms, especially during the afternoon hours. The dominance of the ITCZ enhances precipitation totals, helping make this climate type one of the world's rainiest.

Tropical Monsoonal Climate

The tropical monsoonal climate occurs in low-latitude areas, such as Southeast Asia, that have a warm, rainy climate with a short dry season. Temperatures are similar to those of the tropical wet climate, with the warmest weather often occurring during the drier period, when sunshine is more abundant. The heavy rainfalls result from the nearness of the ITCZ for much of the year, as well as the dominance of warm, moist air masses derived from tropical oceans. During the brief dry season, however, the ITCZ has usually shifted into the opposite hemisphere, and windflow patterns often have changed so as to bring in somewhat drier air derived from continental sources.

Tropical Savanna Climate

The tropical savanna climate, also referred to as the tropical wet-dry climate, occupies a large portion of the tropics between 5 and 20 degrees latitude in both hemispheres. It experiences a distinctive alternation of wet and dry seasons, caused chiefly by the seasonal shift in latitude of the subtropical highs and ITCZ. Summer is typically the rainy season because of the domination of the ITCZ. In many areas, an onshore windflow associated with the summer monsoon increases rainfalls at this time. In winter, however, the ITCZ shifts into the opposite hemisphere and is replaced by drier and more stable air associated with the subtropical high. In addition, the winter monsoon tendency often produces an outflow of continental air. The long dry season inhibits forest growth, so vegetation usually consists of a cover of drought-resistant shrubs or the tall savanna grasses after which the climate is named.

Subtropical Desert Climate

The subtropical desert climate has hot, arid conditions as a result of the year-round dominance of the subtropical highs. Summertime temperatures in this climate soar to the highest readings found anywhere on earth. The world's record high temperature was 134 degrees Fahrenheit (56.7 degrees Celsius), recorded in Furnace Creek Ranch, California (formerly Greenland Ranch) on July 10, 1913. Rainfall totals in this type of climate are generally less than 10 inches (25 centimeters) per year. What rainfall does occur often arrives as brief, sometimes violent, afternoon thunderstorms. Although summer temperatures are extremely hot, the dry air enables rapid cooling during the winter, so that temperatures are cool to mild at this time of year.

Subtropical Steppe Climate

The subtropical steppe climate is a semiarid climate, found mostly on the margins of the subtropical deserts. Precipitation usually ranges from 10 to 30 inches (25 to 75 centimeters), sufficient for a ground cover of shrubs or short steppe grasses. Areas on the equatorward margins of subtropical deserts typically receive their precipitation during a brief showery period in midsummer, associated with the poleward shift of the ITCZ. Areas on the poleward margins of the subtropical highs receive most of their rainfall during the winter, due to the penetration of cyclonic storms associated with the equatorward shift of the westerly wind belt.

Mediterranean Climate

The Mediterranean climate, also sometimes referred to as the dry summer subtropics, has a distinctive pattern of dry summers and more humid, moderately wet winters. This pattern is caused by the seasonal shift in latitude of the subtropical high and the westerlies. During the summer, the subtropical high shifts poleward into the Mediterranean climate regions, blanketing them with dry, warm, stable air. As winter approaches, this pressure center retreats equatorward, allowing the westerlies, with their eastward-traveling weather fronts and cyclonic storms, to overspread this region. The Mediterranean climate is found on the windward sides of continents, particularly the area surrounding the Mediterranean Sea and much of California. This results in the predominance of maritime air and relatively mild temperatures throughout the year.

Humid Subtropical Climate

The humid subtropical climate is found on the eastern, or leeward, sides of continents in the lower middle latitudes. The most extensive land area with this climate is the southeastern United States, but it is also seen in large areas in South America, Asia, and Australia. Temperature ranges are moderately large, with warm to hot summers and cool to mild winters. Mean temperatures for a given location are dictated largely by latitude, elevation, and proximity to the coast. Precipitation is moderate. Winter precipitation is usually associated with weather fronts and cyclonic storms that travel eastward within the westerly wind belt. During summer, most precipitation is in the form of brief, heavy afternoon and evening thunderstorms. Some coastal areas are subject to destructive hurricanes during the late summer and autumn.

Midlatitude Desert Climate

This type of climate consists of areas within the western United States, southern South America, and Central Asia that have arid conditions resulting from the moisture-blocking influence of mountain barriers. This climate is highly continental, with warm summers and cold winters. When precipitations occurs, it frequently comes in the form of winter snowfalls associated with weather fronts and cyclonic storms. Rainfall in summer typically occurs as afternoon thunderstorms.

Midlatitude Steppe

The midlatitude steppe climate is located in interior portions of continents in the middle latitudes, particularly in Asia and North America. This climate has semiarid conditions caused by a combination of continentality resulting from the large distance from oceanic moisture sources and the presence of mountain barriers. Like the midlatitude desert climate, this climate has large annual temperature ranges, with cold winters and warm summers. It also receives winter rains and snows chiefly from weather fronts and cyclonic storms; summer rains occur largely from afternoon convectional storms. In the Great Plains of the United States, spring can bring very turbulent conditions, with blizzards in early spring and hailstorms and tornadoes in mid to late spring.

Marine West Coast

This type of climate is typically located on the west coasts of continents just poleward of the Mediterranean climate. Its location in the heart of the westerly wind belt on the windward sides of continents produces highly maritime conditions. As a result, cloudy and humid weather is common, along with frequent periods of rainfall from passing weather fronts and cyclonic storms. These storms are often well developed in winter, resulting in extended periods of wet and windy weather. Precipitation amounts are largely controlled by the presence and strength of the orographic effect; mountainous coasts like the northwestern United States and the west coast of Canada are much wetter than are flatter areas like northern Europe. Temperatures are held at moderate levels by the onshore flow of maritime air. As a consequence, winters are relatively mild and summers relatively cool for the latitude.

Humid Continental Climate

The humid continental climate is found in the northern interiors of Eurasia (Europe and Asia) and North America. It does not occur in the Southern Hemisphere because of the absence of large land masses in the upper midlatitudes of that hemisphere. This climate type is characterized by low to moderate precipitation that is largely frontal and cyclonic in nature. Most precipitation occurs in summer, but cold winter temperatures typically cause the surface to be frozen and snow-covered for much of the late fall, winter, and early spring. Temperature ranges in this climate are the largest in the world. A town in Siberia, Verkhoyansk, holds the Guinness world record for the greatest temperature range at a single location is 221 degrees Fahrenheit (105 degrees Celsius), from -90 degrees Fahrenheit (-68 degrees Celsius) to 99 degrees Fahrenheit (37 degrees Celsius). Winter temperatures in parts of both North America and Siberia can fall well below -49 degrees Fahrenheit (-45 degrees Celsius), making these the coldest permanently settled sites in the world.

Tundra Climate

The tundra climate is a severely cold climate that exists mostly on the coastal margins of the Arctic Ocean in extreme northern North America and Eurasia, and along the coast of Greenland. The high-latitude location and proximity to icy water cause every month to have average temperatures below 50 degrees Fahrenheit (10 degrees Celsius), although a few months in summer have means above freezing. As a result of the cold temperatures, tundra areas are not forested, but instead typically have a sparse ground cover of grasses, sedges, flowers, and lichens. Even this vegetation is buried by a layer of snow during most of the year. Cold temperatures lower the water vapor holding capacity of the air, causing precipitation totals to be generally light. Most precipitation is associated with weather fronts and cyclonic storms and occurs during the summer half of the year.

Ice Cap Climate

The most poleward and coldest of the world's climates is called the ice cap climate. It is found on the continent of Antarctica, interior Greenland, and some high mountain peaks and plateaus. Because monthly mean temperatures are subfreezing throughout the year, areas with this climate are glaciated and have no permanent human inhabitants.

The coldest temperatures of all occur in interior Antarctica, where a Russian research station named Vostok recorded the world's coldest temperature of -128.6 degrees Fahrenheit (-89.2 degrees Celsius) on July 21, 1983. This climate receives little precipitation because the atmosphere can hold very little water vapor. A major moisture surplus exists, however, because of the lack of snowmelt and evaporation. This causes the build up of a surface snow cover that eventually compacts to form the icecaps that bury the surface. Snowstorms are often accompanied by high winds, producing blizzard conditions.

Global Warming

Though warming has not been uniform across the planet, the upward trend in the globally averaged temperature shows that more areas are warming than cooling. According to the National Oceanic and Atmospheric Administration (NOAA) 2018 Global Climate Summary, the combined land and ocean temperature has increased at an average rate of 0.13°F (0.07°C) per decade since 1880; however, the average rate of increase since 1981 (0.31°F/ 0.17°C) is more than twice as great. It is strongly suspected that human activities that increase the accumulation of greenhouse gases (heat-trapping gases) in the atmosphere may play a key role in the temperature rise.

Levels of carbon dioxide (CO_2) in the atmosphere are higher now than they have been at any time in the past 400,000 years. This gas is responsible for nearly two-thirds of the global-warming potential of all human-released gases. Levels surpassed 407 ppm in 2018 for the first time in recorded history. By comparison, during ice ages, CO_2 levels were around 200 parts per million (ppm), and during the warmer interglacial periods, they hovered around 280 ppm. The recent rise in CO_2 shows a remarkably constant relationship with fossil-fuel burning, which is understandable when one considers that about 60 percent of fossil-fuel emissions stay in the air. Atmospheric carbon dioxide concentrations are also increased by deforestation, which is occurring at a rapid rate in several tropical countries. Deforestation causes carbon dioxide levels to rise because trees remove large quantities of this gas from the atmosphere during the process of photosynthesis.

Research indicates that if atmospheric concentrations of greenhouse gases continue to increase at the 1990s pace, global temperatures could rise an additional 1.8 to 6.3 degrees Fahrenheit (1 to 3.5 degrees Celsius) during the twenty-first century. That level of temperature increase would produce major changes in global climates and plant and animal habitats and would cause sea levels to rise substantially.

Ralph C. Scott

Cloud Formation

Clouds are visible manifestations of water in the air. Cloud patterns can provide even a casual observer with much information about air movements and the processes occurring in the atmosphere. The shapes and heights of the clouds and the directions from which they have come are valuable clues in understanding weather.

Importance of Cooling

Clouds are formed when water vapor in the air is transformed into either water droplets or ice crystals. Sometimes large amounts of moisture are added to the air, producing clouds, but clouds generally are formed when a large amount of air is cooled. The amount of water vapor that air can hold varies with temperature: Cold air can hold less water vapor than warmer air. If air is cooled to the point at which it can hold no more water vapor, the water vapor will condense into water droplets. The temperature at which condensation begins is called the dew point. At below freezing temperatures, the water vapor will turn or deposit into ice crystals.

Cloud droplets do not necessarily form even if the air is fully saturated, that is, holding as much water vapor as possible at a given temperature. Once formed, cloud droplets can evaporate again very easily. Two factors hasten the production and growth of cloud droplets. One is the presence of

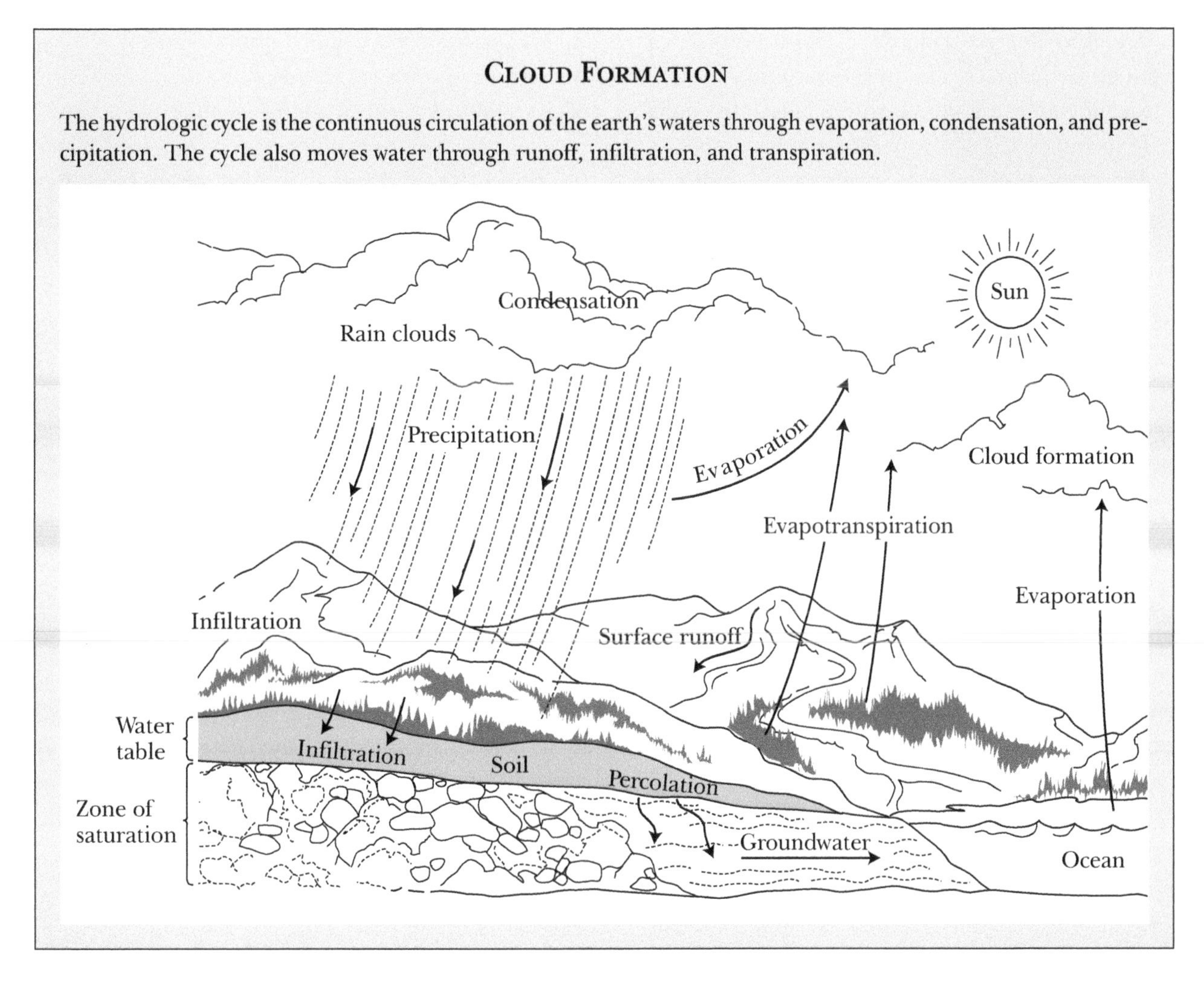

CLOUD FORMATION

The hydrologic cycle is the continuous circulation of the earth's waters through evaporation, condensation, and precipitation. The cycle also moves water through runoff, infiltration, and transpiration.

particles in the atmosphere that attract water. These are called hygroscopic particles or condensation nuclei. They include salt, dust, and pollen. Once water vapor condenses on these particles, more condensation can occur. Then the droplets can grow larger and bump into other droplets, growing even larger through this process, called coalescence.

Condensation and cloud droplet growth also is hastened when the air is very cold, at about -40 degrees Fahrenheit (which is also -40 degrees Celsius). At this temperature ice crystals form, but some water droplets can exist as liquid water. These water droplets are said to be supercooled. The water vapor is more likely to deposit on the ice crystals than on the supercooled water. Thus the ice crystals grow larger and the supercooled water droplets evaporate, resulting in more water vapor to deposit on ice crystals. Whether the cloud droplets start as hygroscopic particles or ice crystals, they eventually can grow in size to become a raindrop; around 1 million cloud droplets make one raindrop.

How and Why Rising Air Cools

In order for air to be cooled, it must rise or be lifted. When a volume of air, or an air parcel, is forced to rise through the surrounding air, the parcel expands in size as the pressure of the air around it declines with altitude. Close to the surface, the atmospheric pressure is relatively high because the density of the atmosphere is high. As altitude increases, the atmosphere declines in density, and the still air exerts less pressure. Thus, as an air parcel rises through the atmosphere, the pressure of the surrounding air declines, and the parcel takes up more space as it expands. Since work is done by the parcel as it expands, the parcel cools and its temperature declines.

An alternative explanation of the cooling is that the number of molecules in the air parcel remains the same, but when the volume is larger, the molecules produce less frictional heat because they do not bang into each other as much. The temperature of the air parcel declines, but no heat left the parcel—the change in temperature resulted from internal processes. The process of an air parcel rising, expanding, and cooling is called adiabatic cooling. Adiabatic means that no heat leaves the parcel. If the parcel rises far enough, it will cool sufficiently to reach its dewpoint temperature. With continued cooling, condensation will result—a cloud will be formed. At this height, which is called the lifting condensation level, an invisible parcel of air will turn into a cloud.

Uplift Mechanisms

An initial force is necessary to cause the air parcel to rise and then cool adiabatically. The three major processes are convection, orographic, and frontal or cyclonic.

With certain conditions, convection or vertical movement can cause clouds to form. On a sunny day, usually in the summer, the ground is heated unevenly. Some areas of the ground become warmer and heat the air above, making it warmer and less dense. A stream of air, called a thermal, may rise. As it rises, it cools adiabatically through expansion and may reach its dewpoint temperature. With continued cooling and rising, condensation will occur, forming a cloud. Since the cloud is formed by predominantly vertical motions, the cloud will be cumulus. With continued warming of the surface, the thermals may rise even higher, perhaps producing thunderstorm, or cumulonimbus, clouds. Thus, a sunny summer day can start off without a cloud in the sky, but can be stormy with many thunderstorms by afternoon.

Clouds also can form when air is forced to rise when it meets a mountain or other large vertical barrier. This type of lifting—orographic—is especially prevalent where air moves over the ocean and then is forced to rise up a mountain, as occurs on the west coast of North and South America. As the

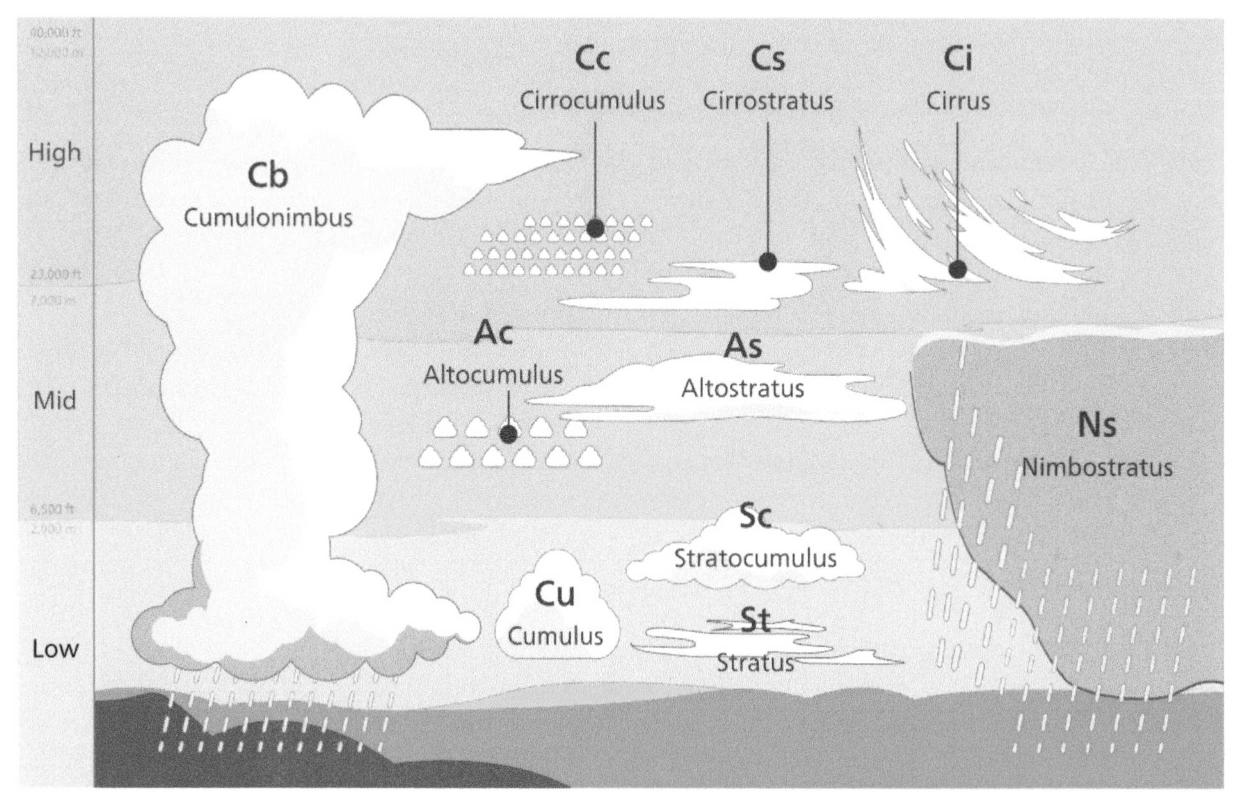

Cloud classification by altitude of occurrence. Multi-level and vertical genus-types not limited to a single altitude level include nimbostratus, cumulonimbus, and some of the larger cumulus species. (Illustration by Valentin de Bruyn)

air rises, it cools adiabatically and eventually becomes so cool that it cannot hold the water vapor. Condensation occurs and clouds form. The air continues to move up the mountain, producing clouds and precipitation on the side of the mountain from which the wind came, the windward side. However, the air eventually must fall down the other side of the mountain, the leeward side. That air is warmed and moisture evaporates, resulting in no clouds.

A third lifting mechanism is frontal, or cyclonic, action. This occurs when a large mass of cold air and a large mass of warm air—often hundreds of miles in area—meet. The warm air mass and the cold air mass will not mix freely, resulting in a border or front between the two air masses. The warm, less dense, air will always rise above the cold, denser, air mass. As the warm air rises, it cools, and when it reaches its dew point, clouds will form. If the warm air displaces the cold air, or a warm front occurs, the warm air will rise gradually, resulting in layered or stratiform clouds. The cloud types will change on an upward diagonal path, with the lowest being stratus, and nimbostratus if rain occurs, followed by altostratus, then cirrostratus, and cirrus.

On the other hand, if the cold air displaces the warm air, the warm air will be forced to rise much more quickly. The clouds formed will be puffy or cumuliform—cumulus at the lowest levels, altocumulus and cirrocumulus at the highest altitudes. Sometimes cumulonimbus clouds will also form.

Sometimes when a cold front meets a warm front, the whole warm air mass is forced off the ground. This forms a cyclone—an area of low pressure—as the warm air rises. As this air rises, it cools. If it reaches its dew point, condensation and clouds will result. In oceanic tropical areas, a cyclone can form within warm, moist air. This air also will cool and, if it reaches its dew point, will condense and form clouds. Sometimes, these tropical cyclones are the precursors of hurricanes. The clouds associated with cyclones are usually cumulus, including cumulonimbus, as they are formed by rapidly rising air.

Margaret F. Boorstein

STORMS

A storm is an atmospheric disturbance that produces wind, is accompanied by some form of precipitation, and sometimes involves thunder and lightning. Storms that meet certain criteria are given specific names, such as hurricanes, blizzards, and tornadoes.

Stormy weather is associated with low atmospheric pressure, while clear, calm, dry weather is associated with high atmospheric pressure. Because of the way atmospheric pressure and wind direction are related, low-pressure areas are characterized by winds moving cyclonically (in a counterclockwise direction in the Northern Hemisphere; clockwise in the Southern Hemisphere) around the center of the low pressure. Storms of all kinds are associated with cyclones, but two classes of cyclones—tropical and extratropical—produce most storms.

Tropical Cyclones

These storms develop during the summer and autumn in every tropical ocean except the South Atlantic and eastern South Pacific Oceans. Tropical cyclones that occur in the North Atlantic and eastern North Pacific Oceans are known as hurricanes; in the western North Pacific Ocean, as typhoons; and in the Indian and South Pacific Oceans, as cyclones.

All tropical cyclones develop in three stages. Arising from the formation of the initial atmospheric disturbance that is characterized by a cluster of thunderstorms, the first stage—tropical depres-

ANATOMY OF A HURRICANE

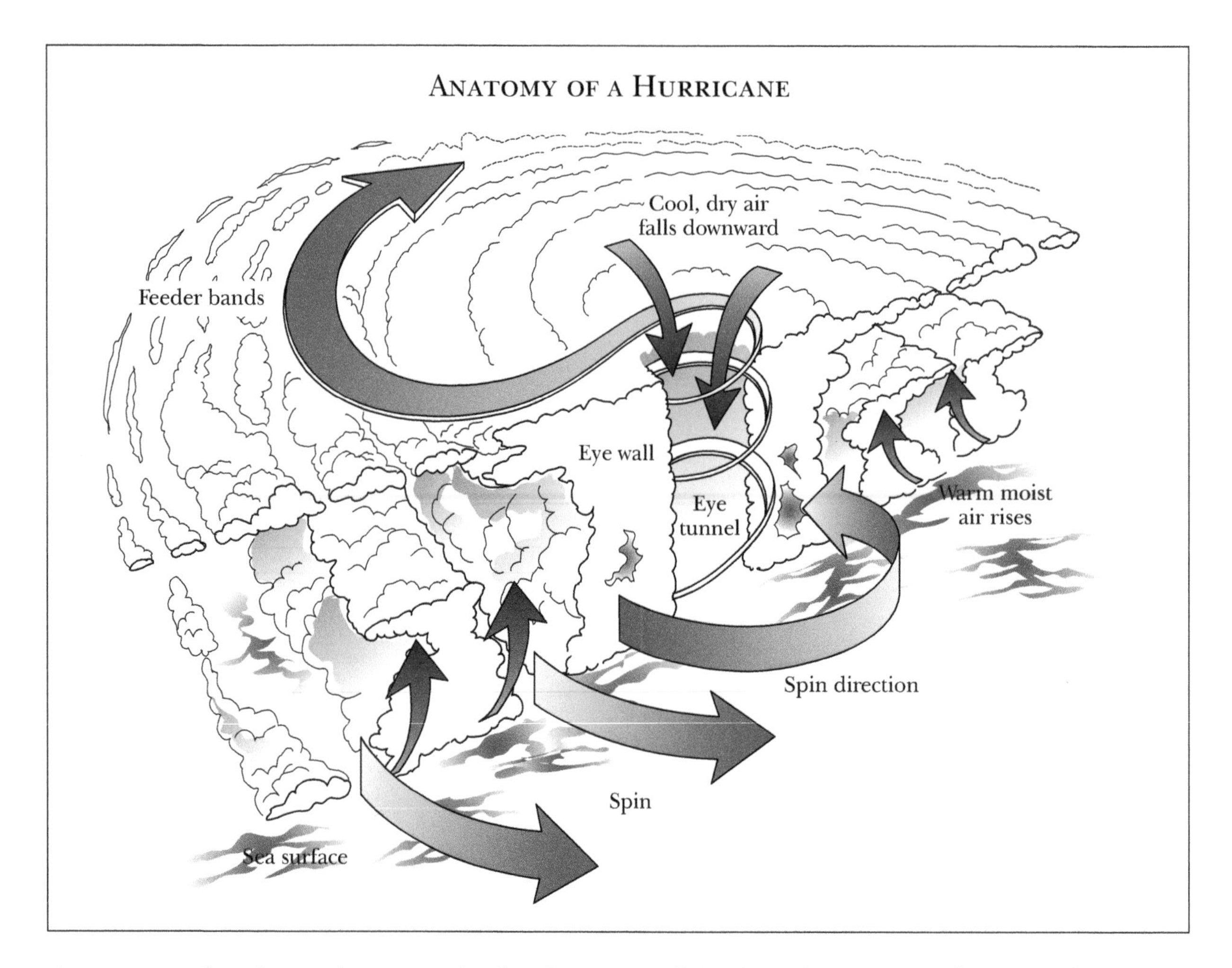

sion—occurs when the maximum sustained surface wind speeds (the average speed over one minute) range from 23–39 miles (37–61 km.) per hour. The second stage—tropical storm—occurs when sustained winds range from 40–73 miles (62–119 km.) per hour. At this stage, the storm is given a name. From 80 to 100 tropical storms develop each year across the world, with about half continuing to the final stage—hurricane—at which sustained wind speeds are 74 miles (120 km.) per hour or greater. Moving over land or into colder oceans initiates the end of the hurricane after a week or so by eliminating the hurricane's fuel—warm water.

A mature hurricane is a symmetrical storm, with the "eye" at the center; the eye develops as winds increase and become circular around the central core of low pressure. Within the eye, it is relatively warm, and there are light winds, no precipitation, and few clouds. This is caused by air descending in the center of the storm. Surrounding the eye is the "eye wall," a ring of intense thunderstorms that can extend high into the atmosphere. Within the eye wall, the strongest winds and heaviest rainfall are found; this is also where warm, moist air, the hurricane's "fuel," flows into the storm. Spiraling bands of clouds, called "rain bands," surround the eye wall. Precipitation and wind speeds decrease from the eye wall out toward the edge of the rain bands, while atmospheric pressure is lowest in the eye and increases outward.

Hurricanes can be the most damaging storms because of their intensity and size. Damage is caused by high winds and the flying debris they carry, flooding from the tremendous amounts of rain a hurricane can produce, and storm surge. A storm surge, which accounts for most of the coastal property loss and 90 percent of hurricane deaths, is a dome of water that is pushed forward as the storm moves. This wall of water is lifted up onto the coast as the eye wall comes in contact with land. For exam-

Naming Hurricanes

Hurricanes once were identified by their latitudes and longitudes, but this method of naming became confusing when two or more hurricanes developed at the same time in the same ocean. During World War II hurricanes were identified by radio code letters, such as Able and Baker. In 1953 the National Weather Service began using English female names in an alphabetized list. Male names and French and Spanish names were added in 1978. By 2000 six lists of names were used on a rotating basis. When a hurricane causes much death or destruction, as Hurricane Andrew did in August of 1992—its name is retired for at least ten years.

ple, a 25-foot (8-meter) storm surge created by Hurricane Camille in 1969 destroyed the Richelieu Apartments next to the ocean in Pass Christian, Mississippi. Ignoring advice to evacuate, twenty-five people had gathered there for a hurricane party; all but one was killed.

To help predict the damage that an approaching hurricane can cause, the Saffir-Simpson Scale was developed. A hurricane is rated from 1 (weak) to 5 (devastating), according to its central pressure, sustained wind speed, and storm surge height. Michael was a category 5 storm at the time of landfall on October 10, 2018, near Mexico Beach and Tyndall Air Force Base, Florida. Michael was the first hurricane to make landfall in the United States as a category 5 since Hurricane Andrew in 1992, and only the fourth on record. The others are the Labor Day Hurricane in 1935 and Hurricane Camille in 1969.

Extratropical Cyclones

Also known as midlatitude cyclones, these storms are traveling low-pressure systems that are seen on newspaper and television daily weather maps. They are created when a mass of moist, warm air from the south contacts a mass of drier, cool air from the north, causing a front to develop. At the front, the warmer air rides up over the colder air. This causes water vapor to condense and produces clouds and rain during most of the year, and snow in the winter.

Thunderstorms

Thunderstorms also develop in stages. During the cumulus stage, strong updrafts of warm air build the storm clouds. The storm moves into the mature stage when updrafts continue to feed the storm, but cool downdrafts are also occurring in a portion of the cloud where precipitation is falling. When the warm updrafts disappear, the storm's fuel is gone and the dissipating stage begins. Eventually, the cloud rains itself out and evaporates.

Thunderstorms can also form away from a frontal system, usually during summer. This formation is related to a relatively small area of warm, moist air

Storm Classifications

Tropical Classification	*Wind speed*
Gale-force winds	>15 meters/second
Tropical depression	20-34 knots and a closed circulation
Tropical storm (named)	35-64 knots
Hurricane	65+ knots (74+ mph)
Saffir-Simpson Scale for Hurricanes	
Category 1	63-83 knots (74-95 mph)
Category 2	83-95 knots (96-110 mph)
Category 3	96-113 knots (111-130 mph)
Category 4	114-135 knots (131-155 mph)
Category 5	>135 knots (>155 mph)

Notes: 1 knot = 1 nautical mile/hour = 1.152 miles/hour = 1.85 kilometers/hour.
Source: National Aeronautics and Space Administration, Office of Space Science, Planetary Data System. http:/atmos.nmsu.edu/jsdap/encyclopediawork.html

rising and creating a thunderstorm that is usually localized and short lived.

Wind, lightning, hail, and flooding from heavy rain are the main destructive forces of a thunderstorm. Lightning occurs in all mature thunderstorms as the positive and negative electrical charges in a cloud attempt to equal out, creating a giant spark. Most lightning stays within the clouds, but some finds its way to the surface. The lightning heats the air around it to incredible temperatures (54,000 degrees Fahrenheit/30,000 degrees Celsius), which causes the air to expand explosively, creating the shock wave called thunder. Since lightning travels at the speed of light and thunder at the speed of sound, one can estimate how many miles away the lightning is by counting the seconds between the lightning and thunder and dividing by five. People have been killed by lightning while boating, swimming, biking, golfing, standing under a tree, talking on the telephone, and riding on a lawnmower.

Hail is formed in towering cumulonimbus clouds with strong updrafts. It begins as small ice pellets that grow by collecting water droplets that freeze on contact as the pellets fall through the cloud. The strong updrafts push the pellets back into the cloud, where they continue collecting water droplets until they are too heavy to stay aloft and fall as hailstones. The more an ice pellet is pushed back into the cloud, the larger the hailstone becomes. The largest authenticated hailstone in the United States fell near Vivian, South Dakota, on July 23, 2010. It measured 8.0 inches (20 cm) in diameter, 18 ½ inches (47 cm.) in circumference, and weighed in at 1.9375 pounds (879 grams).

Tornadoes

For reasons not well understood, less than 1 percent of all thunderstorms spawn tornadoes. Called funnel clouds until they touch earth, tornadoes contain the highest wind speeds known.

Although tornadoes can occur anywhere in the world, the United States has the most, with an average of 1000 per year. Tornadoes have occurred in every state, but the greatest number hit a portion of the Great Plains from central Texas to Nebraska, known as "Tornado Alley." There cold Canadian air and warm Gulf Coast air often collide over the flat land, creating the wall cloud from which most tornadoes are spawned. May is the peak month for tornado activity, but they have been spotted in every month.

Because tornado winds cannot be measured directly, the tornado is ranked according to its damage, using the Fujita Intensity Scale. The scale ranges from an F0, with wind speeds less than 72 miles (116 km.) per hour, causing light damage, to an F5, with winds greater than 260 miles (419 km.) per hour, causing incredible damage. Most tornadoes are small, but the larger ones cause much damage and death.

Kay R. S. Williams

Earth's Biological Systems

Biomes

The major recognizable life zones of the continents, biomes are characterized by their plant communities. Temperature, precipitation, soil, and length of day affect the survival and distribution of biome species. Species diversity within a biome may increase its stability and capability to deliver natural services, including enhancing the quality of the atmosphere, forming and protecting the soil, controlling pests, and providing clean water, fuel, food, and drugs. Land biomes are the temperate, tropical, and boreal forests; tundra; desert; grasslands; and chaparral.

Temperate Forest

The temperate forest biome occupies the so-called temperate zones in the midlatitudes (from about 30 to 60 degrees north and south of the equator). Temperate forests are found mainly in Europe, eastern North America, and eastern China, and in narrow zones on the coasts of Australia, New Zealand, Tasmania, and the Pacific coasts of North and South America. Their climates are characterized by high rainfall and temperatures that vary from cold to mild.

Temperate forests contain primarily deciduous trees—including maple, oak, hickory, and beechwood—and, secondarily, evergreen trees—including pine, spruce, fir, and hemlock. Evergreen forests in some parts of the Southern Hemisphere contain eucalyptus trees.

The root systems of forest trees help keep the soil rich. The soil quality and color is due to the action of earthworms. Where these forests are frequently cut, soil runoff pollutes streams, which reduces fisheries because of the loss of spawning habitat. Racoons, opposums, bats, and squirrels are found in the trees. Deer and black bear roam forest floors. During winter, small animals such as groundhogs and squirrels burrow in the ground.

Tropical Forest

Tropical forests are in frost-free areas between the Tropic of Cancer and the Tropic of Capricorn. Temperatures range from warm to hot year-round, because the Sun's rays shine nearly straight down around midday. These forests are found in northern Australia, the East Indies, southeastern Asia, equatorial Africa, and parts of Central America and northern South America.

Tropical forests have high biological diversity and contain about 15 percent of the world's plant species. Animal life lives at different layers of tropical forests. Nuts and fruits on the trees provide food for birds, monkeys, squirrels, and bats. Monkeys and sloths feed on tree leaves. Roots, seeds, leaves, and fruit on the forest floor feed deer, hogs, tapirs, antelopes, and rodents. The tropical forests produce rubber trees, mahogany, and rosewood. Large animals in these forests include the Asian tiger, the African bongo, the South American tapir, the Central and South American jaguar, the Asian and African leopard, and the Asian axis deer. Deforestation for agriculture and pastures has caused reduction in plant and animal diversity.

Boreal Forest

The boreal forest is a circumpolar Northern Hemisphere biome spread across Russia, Scandinavia, Canada, and Alaska. The region is very cold. Evergreen trees such as white spruce and black spruce

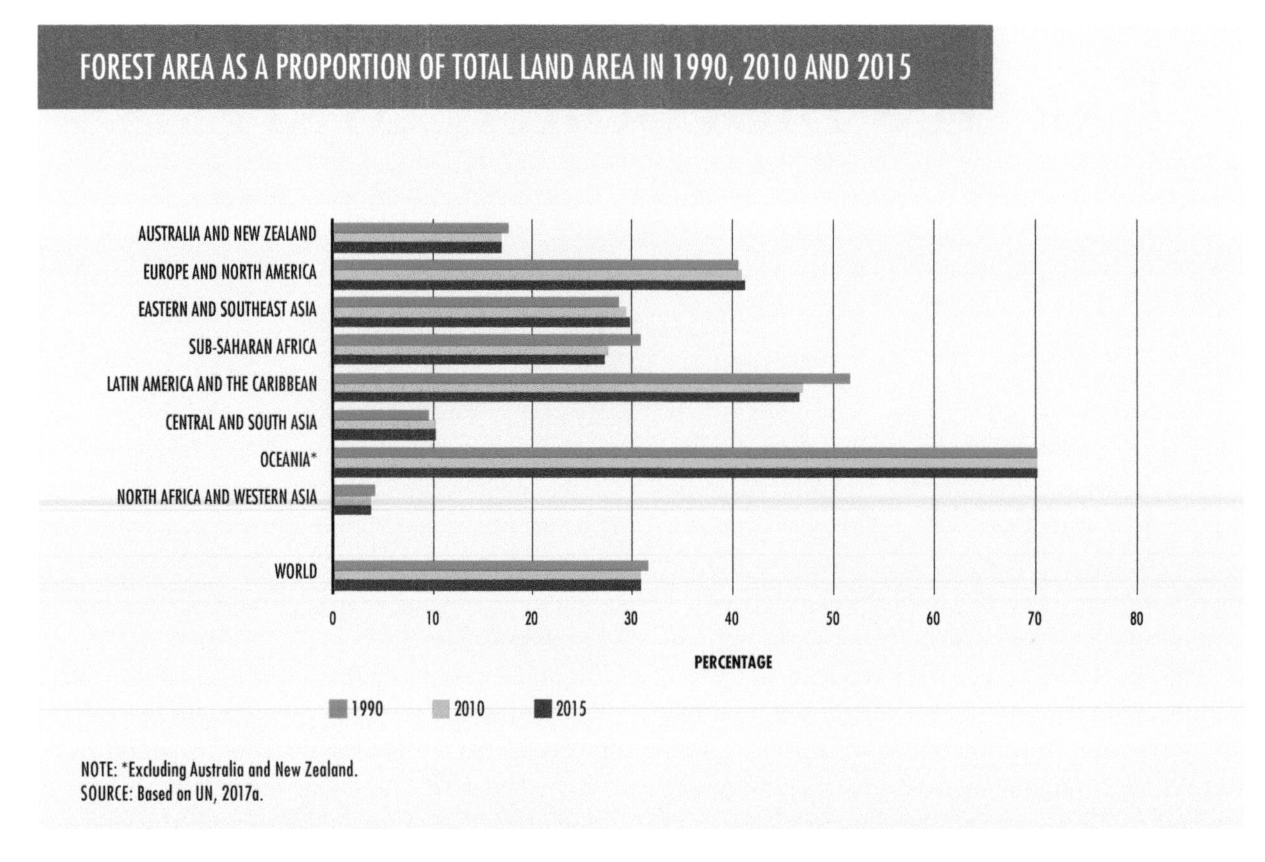

dominate this zone, which also contains larch, balsam, pine, and fir, and some deciduous hardwoods such as birch and aspen. The acidic needles from the evergreens make the leaf litter that is changed into soil humus. The acidic soil limits the plants that develop.

Animals in boreal forests include deer, caribou, bear, and wolves. Birds in this zone include goshawks, red-tailed hawks, sapsuckers, grouse, and nuthatches. Relatively few animals emigrate from this habitat during winter. Conifer seeds are the basic winter food. The disappearing aspen habitat of the beaver has decreased their numbers and has reduced the size of wetlands.

Tundra

About 5 percent of the earth's surface is covered with Arctic tundra, and 3 percent with alpine tundra. The Arctic tundra is the area of Europe, Asia, and North America north of the boreal coniferous forest zone, where the soils remain frozen most of the year. Arctic tundra has a permanent frozen subsoil, called permafrost. Deep snow and low temperatures slow the soil-forming process. The area is bounded by a 50 degrees Fahrenheit circumpolar isotherm, known as the summer isotherm. The cold temperature north of this line prevents normal tree growth.

The tundra landscape is covered by mosses, lichens, and low shrubs, which are eaten by caribou, reindeer, and musk oxen. Wolves eat these herbivores. Bear, fox, and lemming also live here. The larger mammals, including marine mammals and the overwintering birds, have large fat layers beneath the skin and long dense fur or dense feathers that provide protection. The small mammals burrow beneath the ground to avoid the harsh winter climate. The most common Arctic bird is the old squaw duck. Ptarmigans and eider ducks are also very common. Geese, falcons, and loons are some of the nesting birds of the area.

The alpine tundra, which exists at high altitude in all latitudes, is acted upon by winds, cold temperatures, and snow. The plant growth is mostly cushion and mat-forming plants.

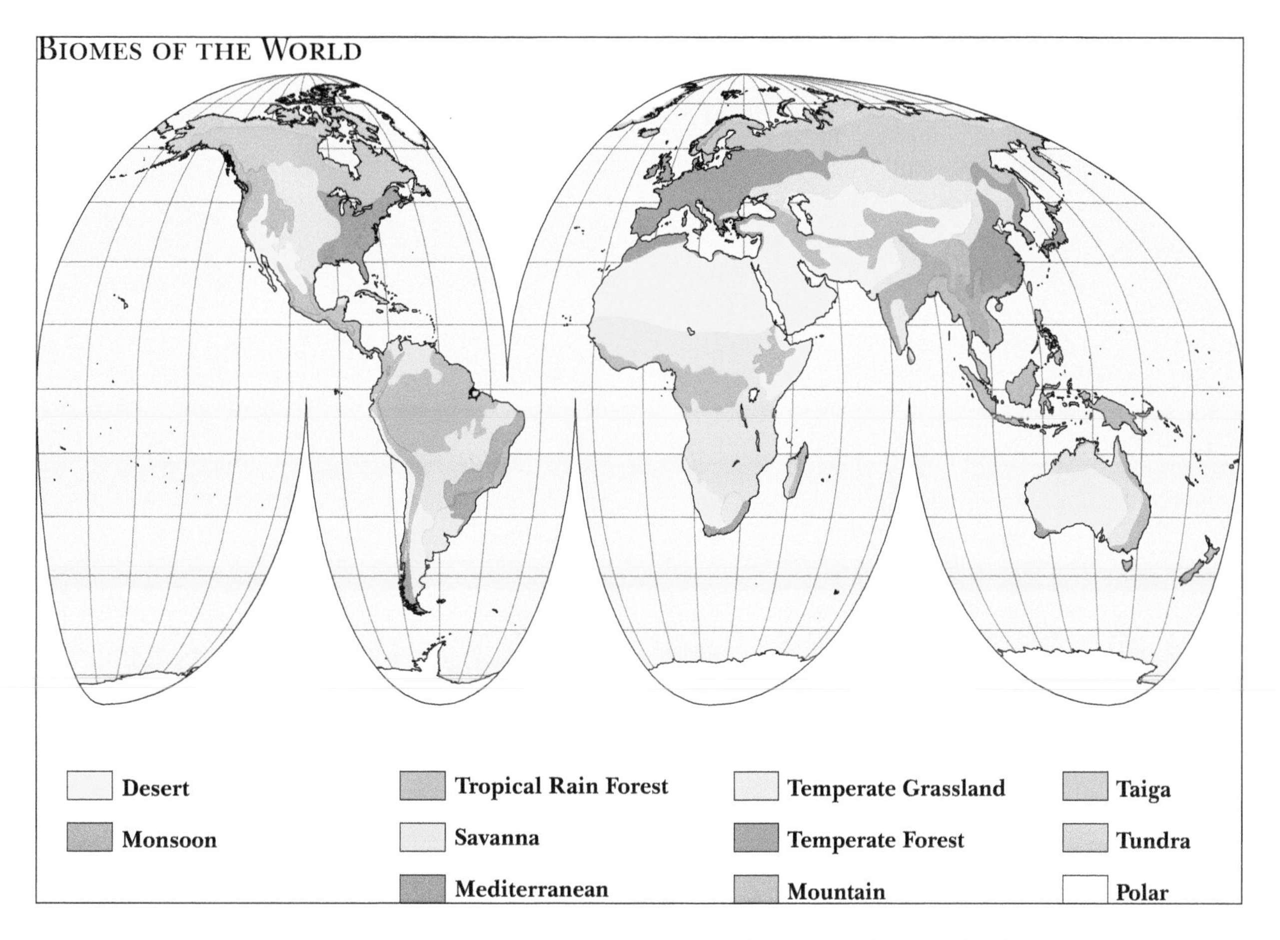

Desert

The desert biome covers about one-seventh of the earth's surface. Deserts typically receive no more than 10 inches (25 centimeters) of rainfall a year, but evaporation generally exceeds rainfall. Deserts are found around the Tropic of Cancer and the Tropic of Capricorn. As the warm air rises over the equator, it cools and loses its water content. This dry air descends in the two subtropical zones on each side of the equator; as it warms, it picks up moisture, resulting in drying the land.

Rainfall is a key agent in shaping the desert. The lack of sufficient plant cover removes the natural protection that prevents soil erosion during storms. High winds also cut away the ground.

Some desert plants obtain water from deep below the surface, for example, the mesquite tree, which has roots that are 40 feet (13 meters) deep. Other plants, such as the barrel cactus, store large amounts of water in their leaves, roots, or stems. Other plants slow the loss of water by having tiny leaves or shedding their leaves. Desert plants have very short growth periods, because they cannot grow during the long drought periods.

Desert animals protect themselves from the Sun's heat by eating at night, staying in the shade during the day, and digging burrows in the ground. Among the world's large desert animals are the camel, coyote, mule deer, Australian dingo, and Asian saiga. The digestive process of some desert animals produces water. A method used by some animals to conserve water is the reabsorption of water from their feces and urine.

Grassland

Grasslands cover about a quarter of the earth's surface, and can be found between forests and deserts. Treeless grasslands grow in parts of central North America, Central America, and eastern South America that have between 10 and 40 inches (250-1,000 millimeters) of erratic rainfall. The climate has a high rate of evaporation and periodic major droughts. The biome is also subject to fire.

Some grassland plants survive droughts by growing deep roots, while others survive by being dormant. Grass seeds feed the lizards and rodents that become the food for hawks and eagles. Large animals include bison, coyotes, mule deer, and wolves. The grasslands produce more food than any other biome. Poor grazing and agricultural practices and mining destroy the natural stability and fertility of these lands. The reduced carrying capacity of these lands causes an increase in water pollution and erosion of the soil. Diverse natural grasslands appear to be more capable of surviving drought than are simplified manipulated grass systems. This may be due to slower soil mineralization and nitrogen turnover of plant residues in the simplified system.

Savannas are open grasslands containing deciduous trees and shrubs. They are near the equator and are associated with deserts. Grasses grow in clumps and do not form a continuous layer. The northern savanna bushlands are inhabited by oryx and gazelles. The southern savanna supports springbuck and eland. Elephants, antelope, giraffe, zebras, and black rhinoceros are found on the savannas. Lions, leopards, cheetah, and hunting dogs are the primary predators here. Kangaroos are found in the savannas of Australia. Savannas cover South America north and south of the Amazon rainforest, where jaguar and deer can be found.

Chaparral

The chaparral or Mediterranean biome is found in the Mediterranean Basin, California, southern Australia, middle Chile, and Cape Province of South America. This region has a climate of wet winters and summer drought. The plants have tough leathery leaves and may contain thorns. Regional fires clear the area of dense and dead vegetation. Fire, heat, and drought shape the region. The vegetation dwarfing is due to the severe drought and extreme climate changes. The seeds from some plants, such as the California manzanita and South African fire lily, are protected by the soil during a fire and later germinate and rapidly grow to form new plants.

Ocean

The ocean biome covers more than 70 percent of the earth's surface and includes 90 percent of its volume. The ocean has four zones. The intertidal zone is shallow and lies at the land's edge. The continental shelf, which begins where the intertidal zone ends, is a plain that slopes gently seaward. The neritic zone (continental slope) begins at a depth of about 600 feet (180 meters), where the gradual slant of the continental shelf becomes a sharp tilt toward the ocean floor, plunging about 12,000 feet (3,660 meters) to the ocean bottom, which is known as the abyss. The abyssal zone is so deep that it does not have light.

Plankton are animals that float in the ocean. They include algae and copepods, which are microscopic crustaceans. Jellyfish and animal larva are also considered plankton. The nekton are animals that move freely through the water by means of their muscles. These include fish, whales, and squid. The benthos are animals that are attached to or crawl along the ocean's floor. Clams are examples of benthos. Bacteria decompose the dead organic materials on the ocean floor.

The circulation of materials from the ocean's floor to the surface is caused by winds and water temperature. Runoff from the land contains polluting chemicals such as pesticides, nitrogen fertilizers, and animal wastes. Rivers carry loose soil to the ocean, where it builds up the bottom areas. Overfishing has caused fisheries to collapse in every world sector. In some parts of the northwestern Atlantic Ocean, there has been a shift from bony fish to cartilaginous fish dominating the fisheries.

Human Impact on Biomes

Human interaction with biomes has increased biotic invasions, reduced the numbers of species, changed the quality of land and water resources, and caused the proliferation of toxic compounds. Managed care of biomes may not be capable of undoing these problems.

Ronald J. Raven

NATURAL RESOURCES

SOILS

Soils are the loose masses of broken and chemically weathered rock mixed with organic matter that cover much of the world's land surface, except in polar regions and most deserts. The two major solid components of soil—minerals and organic matter—occupy about half the volume of a soil. Pore spaces filled with air and water account for the other half. A soil's organic material comes from the remains of dead plants and animals, its minerals from weathered fragments of bedrock. Soil is also an active, dynamic, ever-changing environment. Tiny pores in soil fill with air, water, bacteria, algae, and fungi working to alter the soil's chemistry and speed up the decay of organic material, making the soil a better living environment for larger plants and animals.

Soil Formation

The natural process of forming new soil is slow. Exactly how long it takes depends on how fast the bedrock below is weathered. This weathering process is a direct result of a region's climate and topography, because these factors influence the rate at which exposed bedrock erodes and vegetation is distributed. Global variations in these factors account for the worldwide differences in soil types.

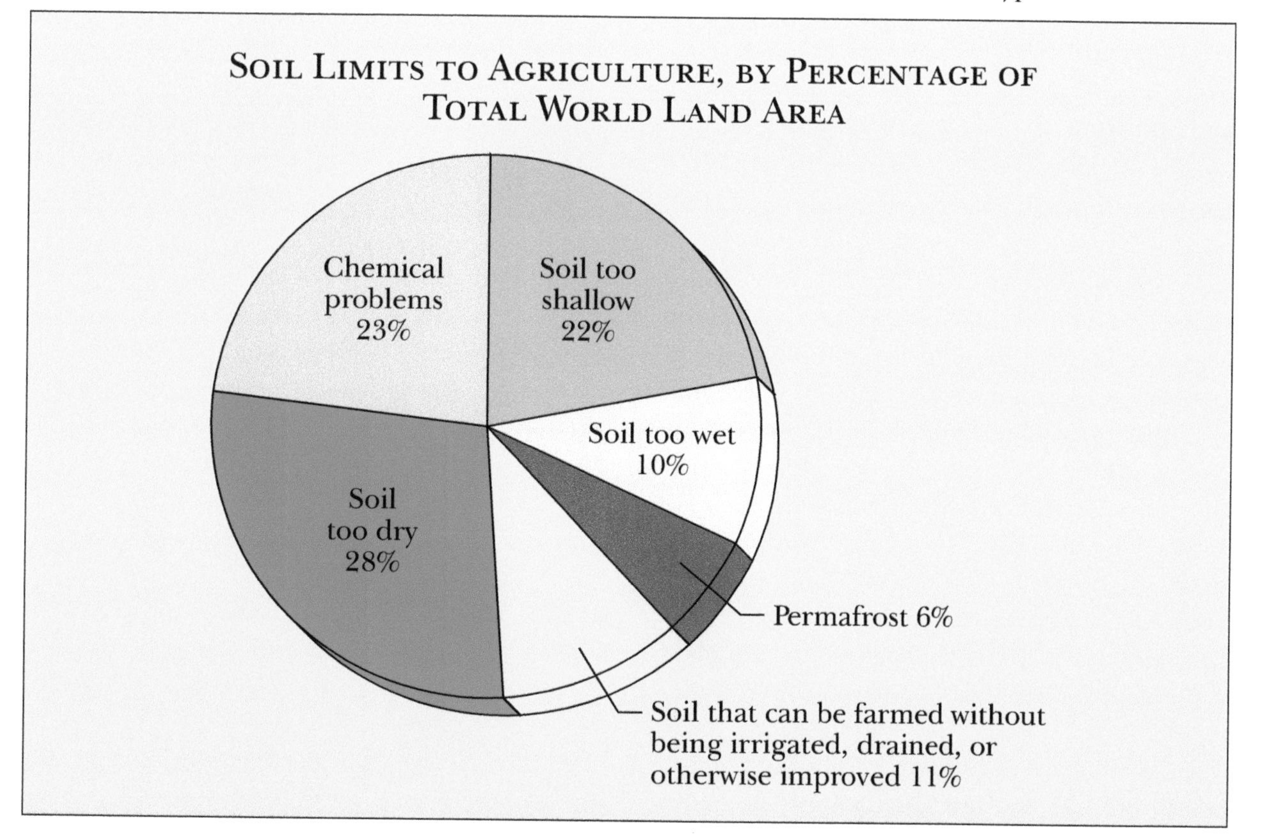

Climate is the principal factor in determining the type and rate of soil formation. Temperature and precipitation are the two main climatic factors that influence soil formation, and they vary with elevation and latitude. Water is the main agent of weathering, and the amount of water available depends on how much falls and how much runs off. The amount of precipitation and its distribution during the year influence the kind of soil formed and the rate at which it is formed. Increased precipitation usually results in increased rates of soil formation and deep soils. Temperature and precipitation also determine the kind and amount of vegetation in a region, which determines the amount of available organics.

Topography is a characteristic of the landscape involving slope angle and slope length. Topographic relief governs the amount of water that runs off or enters a soil. On flat or gently sloping land, soil tends to stay in place and may become thick, but as the slope increases so does the potential for erosion. On steep slopes, soil cover may be very thin, possibly only a few inches, because precipitation washes it downhill; on level plains, soil profiles may be several feet thick.

Types of Soil

Typically, bedrock first weathers to form regolith, a protosoil devoid of organic material. Rain, wind, snow, roots growing into cracks, freezing and thawing, uneven heating, abrasion, and shrinking and swelling break large rock particles into smaller ones. Weathered rock particles may range in size from clay to silt, sand, and gravel, with the texture and particle size depending largely on the type of bedrock. For example, shale yields finer-textured soils than sandstone. Soils formed from eroded limestone are rich in base minerals; others tend to be acidic. Generally, rates of soil formation are largely determined by the rates at which silicate minerals in the bedrock weather: the more silicates, the longer the formation time.

In regions where organic materials, such as plant and animal remains, may be deposited on top of regolith, rudimentary soils can begin to form. When waste material is excreted, or a plant or animal dies, the material usually ends up on the earth's surface. Organisms that cause decomposition, such as bacteria and fungi, begin breaking down the remains into a beneficial substance known as humus. Humus restores minerals and nutrients to the soil. It also improves the soil's structure, helping it to retain water. Over time, a skeletal soil of coarse, sandy material with trace amounts of organics gradually forms. Even in a region with good weathering rates and adequate organic material, it can take as long as fifty years to form 12 inches (30 centimeters) of soil. When new soil is formed from weathering bedrock, it can take from 100 to 1,000 years for less than an inch of soil to accumulate.

Water moves continually through most soils, transporting minerals and organics downward by a process called leaching. As these materials travel downward, they are filtered and deposited to form distinct soil horizons. Each soil horizon has its own color, texture, and mineral and humus content. The O-horizon is a thin layer of rotting organics covering the soil. The A-horizon, commonly called topsoil, is rich in humus and minerals. The B-horizon is a subsoil rich in minerals but poor in humus. The C-horizon consists of weathered bedrock; the D-horizon is the bedrock itself.

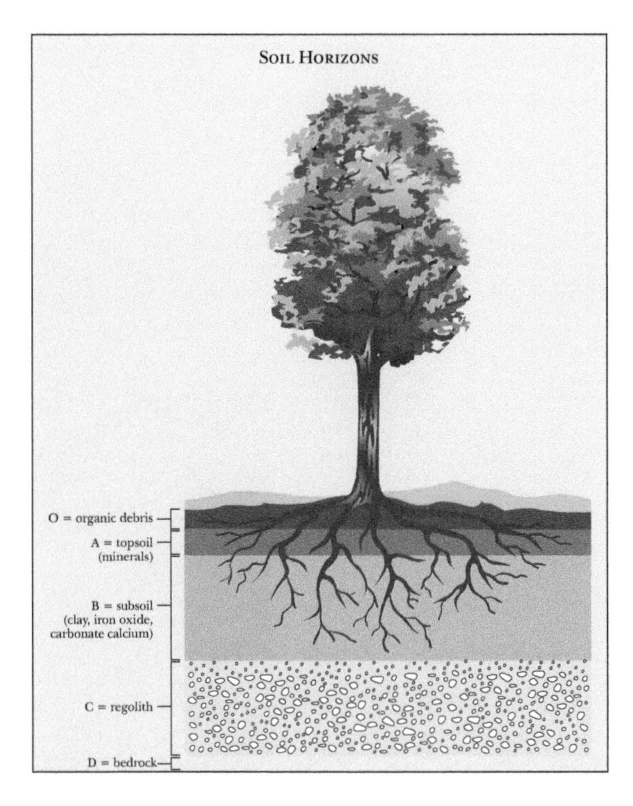

Because Earth's surface is made of many different rock types exposed at differing amounts and weathering at different rates at different locations, and because the availability of organic matter varies greatly around the planet due to climatic and seasonal conditions, soil is very diverse and fertile soil is unevenly distributed. Structure and composition are key factors in determining soil fertility. In a fertile soil, plant roots are able to penetrate easily to obtain water and dissolved nutrients. A loam is a naturally fertile soil, consisting of masses of particles from clays (less than 0.002 mm across), through silts (ten times larger) to sands (100 times larger), interspersed with pores, cracks, and crevices.

The Roles of Soil

In any ecosystem, soils play six key roles. First, soil serves as a medium for plant growth by mechanically supporting plant roots and supplying the eighteen nutrients essential for plants to survive. Different types of soil contain differing amounts of these eighteen nutrients; their combination often determines the types of vegetation present in a region, and as a result, influences the number and types of animals the vegetation can support, including humans. Humans rely on soil for crops necessary for food and fiber.

Second, the property of a particular soil is the controlling factor in how the hydrologic system in a region retains and transports water, how contaminants are stored or flushed, and at what rate water is naturally purified. Water enters the soil in the form of precipitation, irrigation, or snowmelt that falls or runs off soil. When it reaches the soil, it will either be surface water, which evaporates or runs into streams, or subsurface water, which soaks into the soil where it is either taken up by plant roots or percolates downward to enter the groundwater system. Passing through soil, organic and inorganic pollutants are filtered out, producing pure groundwater.

Soil also functions as an air-storage facility. Air is pushed into and drawn out of the soil by changes in barometric pressure, high winds, percolating water, and diffusion. Pore spaces within soil provide access to oxygen to organisms living underground as well as to plant roots. Soil pore spaces also contain carbon dioxide, which many bacteria use as a source of carbon.

Soil is nature's recycling system, through which organic waste products and decaying plants and animals are assimilated and their elements made available for reuse. The production and assimilation of humus within soil converts mineral nutrients into forms that can be used by plants and animals, who return carbon to the atmosphere as carbon dioxide. While dead organic matter amounts to only about 1 percent of the soil by weight, it is a vital component as a source of minerals.

Soil provides a habitat for many living things, from insects to burrowing animals, from single microscopic organisms to massive colonies of subterranean fungi. Soils contain much of the earth's genetic diversity, and a handful of soil may contain billions of organisms, belonging to thousands of species. Although living organisms only account for about 0.1 percent of soil by weight, 2.5 acres (one hectare) of good-quality soil can contain at least 300 million small invertebrates—mites, millipedes, insects, and worms. Just 1 ounce (30 grams) of fertile soil can contain 1 million bacteria of a single type, 100 million yeast cells, and 50,000 fungus mycelium. Without these, soil could not convert nitrogen, phosphorus, and sulphur to forms available to plants.

Finally, soil is an important factor in human culture and civilization. Soil is a building material used to make bricks, adobe, plaster, and pottery, and often provides the foundation for roads and buildings. Most important, soil resources are the basis for agriculture, providing people with their dietary needs.

Because the human use of soils has been haphazard and unchecked for millennia, soil resources in many parts of the world have been harmed severely. Human activities, such as overcultivation, inexpert irrigation, overgrazing of livestock, elimination of tree cover, and cultivating steep slopes, have caused natural erosion rates to increase many times over. As a result of mismanaged farm and forest lands, escalated erosional processes wash off or blow away an estimated 75 billion tons of soil annually, eroding away one of civilization's crucial resources.

Randall L. Milstein

WATER

Life on Earth requires water—without it, life on Earth would cease. As human populations grow, the freshwater resources of the world become scarcer and more polluted, while the need for clean water increases. Although nearly three-quarters of Earth's surface is covered with water, only about 0.3 percent of that water is freshwater suitable for consumption and irrigation. This is because more than 97 percent of Earth's water is ocean salt water, and most of the remaining freshwater is frozen in the Antarctic ice cap. Only the small amounts that remain in lakes, rivers, and groundwater is available for human use.

All of earth's water cycles between the ocean, land, atmosphere, plants, and animals over and over. On average, a molecule of surface water cycles from the ocean, to the atmosphere, to the land and back again in less than two weeks. Water consumed by plants or animals takes longer to return to the oceans, but eventually the cycle is completed.

Water's Uses

Water supports the lives of all living creatures. People use for drinking, cooking, cleaning, and bathing. Water also plays a key role in society since humans can travel on it, make electricity with it, fish in it, irrigate crops with it, and use it for recreation. Globally, more than 4 trillion cubic meters of freshwater is used each day. Agriculture accounts for about 70 percent, industry uses 20 percent, and domestic and municipal activities use 10 percent. To produce beef requires between 1,320 and 5,283 gallons (5,000 and 20,000 liters) of water for every 35 ounces (1 kg). A similar amount of wheat requires between 660 and 1056 gallons (2,500 and 4,000 liters) of water. Manufactured goods also require significant amounts of freshwater; a car consumes between 13,000 and 20,000 gallons (49,210 to 75,708 liters), a smartphone has a water footprint of nearly 3,100 gallons (11,734 liters) and a teeshirt uses around 660 gallons (2,498 liters).

The average American family uses more than 300 gallons of drinking quality water per day at home. Roughly 70 percent of this use occurs indoors for drinking, bathing and showering, flushing the toilet, and washing dishes. Outdoor water use for landscape watering, washing cars, and cleaning windows, etc., accounts for 30 percent of household use, although it can be much higher in drier parts of the country. For example, the arid West has some of the highest per capita residential water use because of landscape irrigation.

As the world's population grows, the demand for fresh water will also increase. A study by the World Bank concluded that approximately 80 percent of human illness results from insufficient water supplies and poor water quality caused by lack of sanitation, so careful management of water resources is essential for improving the health of people in the twenty-first century.

Groundwater Supply and Quality

The amount of groundwater in the Earth is seventy times greater than all of the freshwater lakes combined. Groundwater is held within the rocks below the ground surface and is the primary source of water in many parts of the world. In the United States, approximately 50 percent of the population uses some groundwater. However, problems with both groundwater supplies and its quality threaten its future use.

The U.S. Environmental Protection Agency (EPA) found that 45 percent of the large public water systems in the United States that use groundwater were contaminated with synthetic organic chemicals that posed potential health threats. Another major problem occurs when groundwater is used faster than it is replaced by precipitation infiltrating through the ground surface. Many of the arid regions of earth are already suffering from this problem. For example, one-third of the wells in Beijing, China, have gone dry due to overuse. In the United States, the Ogallala Aquifer of the Great Plains, the

The World and North America's Greatest Rivers and Lakes

Longest river	
Nile (North Africa)	4,130 miles (6,600 km.)
Missouri-Mississippi (United States)	3,740 miles ((6,000 km.)
Largest river by average discharge	
Amazon (South America)	6,181,000 cubic feet/second (175,000 cubic meters/second)
Missouri-Mississippi (United States)	600,440 cubic feet/second (17,000 cubic meters/second)
Largest freshwater lake by volume	
Lake Baikal (Russia)	5,280 cubic miles (22,000 cubic km.)
Lake Superior (United States)	3,000 cubic miles (12,500 cubic km.)

largest in North America, is being severely overused. This aquifer irrigates 30 percent of U.S. farmland, but some areas of the aquifer have declined by up to 60 percent. In one part of Texas, so much water has been pumped out that the aquifer has essentially dried up there. Once depleted, the aquifer will take over 6,000 years to replenish naturally through rainfall.

Surface Water Supply and Quality

Surface water is used for transportation, recreation, electrical generation, and consumption. Ships use rivers and lakes as transport routes, people fish and boat on rivers and lakes, and dams on rivers often are used to generate electricity. The largest river on earth is the Amazon in South America, which has an average flow of 212,500 cubic meters per second, more than twelve times greater than North America's Mississippi River. Earth's largest lake by volume—Lake Baikal in Russia—holds 5,521 cubic miles of water (23,013 cubic kilometers), or approximately 20 percent of Earth's fresh surface water. This is a volume of water approximately equivalent to all five of the North American Great Lakes combined.

Although surface water has more uses, it is more prone to pollution than groundwater. Almost every human activity affects surface water quality. For example, water is used to create paper for books, and some of the chemicals used in the paper process are discharged into surface water sources. Most foods are grown with agricultural chemicals, which can contaminate water sources. According to 2018 surveys on national water quality from the U.S. Environmental Protection Agency, nearly half of U.S. rivers and streams and more than one-third of lakes are polluted and unfit for swimming, fishing, and drinking.

Earth's Future Water Supply

Inadequate water supplies and water quality problems threaten the lives of more than 1.5 million people worldwide. The World Health Organization estimates that polluted water causes the death of 361,000 children under five years of age each year and affects the health of 20 percent of Earth's population. As the world's population grows, these problems are likely to worsen.

The United Nations estimates that if current consumption patterns continue, 52 percent of the world's people will live in water-stressed conditions by 2050. Since access to clean freshwater is essential to health and a decent standard of living, efforts must be made to clean up and conserve the planet's freshwater.

Mark M. Van Steeter

Exploration and Transportation

Exploration and Historical Trade Routes

The world's exploration was shaped and influenced substantially by economic needs. Lacking certain resources and outlets for trade, many societies built ships, organized caravans, and conducted military expeditions to protect their frontiers and obtain new markets.

Over the last 5,000 years, the world evolved from a cluster of isolated communities into a firmly integrated global community and capitalist world system. By the beginning of the twentieth century, explorers had successfully navigated the oceans, seas, and landmasses and gathered many regional economies into the beginnings of a global economy.

Early Trade Systems

Trade and exploration accompanied the rise of civilization in the Middle East. Egyptian pharaohs, looking for timber for shipbuilding, established trade relations with Mediterranean merchants. Phoenicians probed for new markets off the coast of North Africa and built a permanent settlement at Carthage. By 513 BCE, the Persian Empire stretched from the Indus River in India to the Libyan coast, and it controlled the pivotal trade routes in Iran and Anatolia. A regional economy was taking shape, linking Africa, Asia, and Europe into a blended economic system.

Alexander the Great's victory against the Persian Empire in 330 BCE thrust Greece into a dominant position in the Middle Eastern economy. Trade between the Mediterranean and the Middle East increased, new roads and harbors were constructed, and merchants expanded into sub-Saharan Africa, Arabia, and India. The Romans later benefited from the Greek foundation. Through military and political conquest, Rome consolidated its control over such diverse areas as Arabia and Britain and built a system of roads and highways that facilitated the growth of an expanding world economy. At the apex of Roman power in 200 CE, trade routes provided the empire with Greek marble, Egyptian cloth, seafood from Black Sea fisheries, African slaves, and Chinese silk.

The emergence of a profitable Eurasian trade route linked people, customs, and economies from the South China Sea to the Roman Empire. Although some limited activity occurred during the Hellenistic period, East-West trade flourished following the rise of the Han Dynasty in China. With the opening of the Great Silk Road from 139 BCE to 200 CE, goods and services were exchanged between people from three different continents.

The Great Silk Road was an intricate network of middlemen stretching from China to the Mediterranean Sea. Eastern merchants sold their products at markets in Afghanistan, Iran, and even Syria, and exchanged a variety of commodities through the use of camel caravans. Chinese spices, perfumes, metals, and especially silk were in high demand. The Parthians from central Asia added their own sprinkling of merchandise, introducing both the East and the West to various exotic fruits, rare birds, and ostrich eggs.

Romans peddled glassware, statuettes, and acrobatic performing slaves. Since communication lines were virtually nonexistent during this period, trade routes were the only means by which ideas regarding art, religion, and culture could mix. The contacts and exchanges enacted along the Great Silk Road initiated a process of cultural diffusion among a diversity of cultures and increased each culture's knowledge of the vast frontiers of world geography.

The Atlantic Slave Trade

Beginning in the fifteenth century, European navigators explored the West African coastline seeking gold. Supplies were difficult to procure, because most of the gold mines were located in the interior along the Senegal River and in the Ashanti forests. Because mining required costly investments in time, labor, and security, the Europeans quickly shifted their focus toward the slave trade. Although slavery had existed since antiquity, the Atlantic slave trade generated one of the most significant movements of people in world history. It led to the forced migration of more than 10 million Africans to South America, the Caribbean islands, and North America. It ensured the success of several imperial conquests, and it transformed the demographic, cultural, and political landscape on four continents.

Originally driven by their quest to circumnavigate Africa and open a lucrative trade route with India, the Portuguese initiated a systematic exploration of the West African coastline. The architect of this system, Henry the Navigator, pioneered the use of military force and naval superiority to annex African islands and open up new trade routes, and he increased Portugal's southern frontier with every acquisition. In 1415 his ships captured Ceuta, a prosperous trade center located on the Mediterranean coast overlooking North African trade routes. Over the next four decades, Henry laid claim to the Madeira Islands, the Canary Islands, the Azores, and Cape Verde. After his death, other Portuguese explorers continued his pursuit of circumnavigation of Africa.

Diego Cão reached the Congo River in 1483 and sent several excursions up the river before returning to Lisbon. Two explorers completed the Portuguese mission at the end of the fifteenth century. Vasco da Gama, sailing from 1497 to 1499, and Bartholomeu Dias, from 1498 to 1499, sailed past the southern tip of Africa and eventually reached India. Since Muslims had already created a number of trade links between East Africa, Arabia, and India, Portuguese exploration furthered the integration of various regions into an emerging capitalist world system.

When the Portuguese shifted their trading from gold to slaves, the other European powers followed suit. The Netherlands, Spain, France, and England used their expanding naval technology to explore the Atlantic Ocean and ship millions of slaves across the ocean. A highly efficient and organized trade route quickly materialized. Since the Europeans were unwilling to venture beyond the walls of their coastal fortresses, merchants relied on African sources for slaves, supplying local kings and chiefs with the means to conduct profitable slave-raiding parties in the interior. In both the Congo and the Gold Coast region, many Africans became quite wealthy trading slaves.

In 1750 merchants paid King Tegbessou of Dahomey 250,000 pounds for 9,000 slaves, and his income exceeded the earnings of many in England's merchant and landowning class. After purchasing slaves, dealers sold them in the Americas to work in the mines or on plantations. Commodities such as coffee and sugar were exported back to Europe for home consumption. Merchants then sold alcohol, tobacco, textiles, and firearms to Africans in exchange for more slaves. This practice was abolished by the end of the nineteenth century, but not before more than 10 million Africans had been violently removed from their homeland. The Atlantic slave trade, however, joined port cities from the Gold Coast and Guinea in Africa with Rio de Janeiro, Hispaniola, Havana, Virginia, Charleston, and Liverpool, and constituted a pivotal step toward the rise of a unified global economy.

Magellan and Zheng He

The Portuguese explorer Ferdinand Magellan generated considerable interest in the Asian markets when he led an expedition that sailed around the world from 1519 to 1522. Looking for a quick route to Asia and the Spice Islands, he secured financial backing from the king of Spain. Magellan sailed from Spain in 1519, canvassed the eastern coastline of South America, and visited Argentina. He ultimately traversed the narrow straits along the southern tip of the continent and ventured into the uncharted waters of the Pacific Ocean.

Magellan explored the islands of Guam and the Philippines but was killed in a skirmish on Mactan in 1521. Some of his crew managed to return to Spain in 1522, and one member subsequently published a journal of the expedition that dramatically enhanced the world's understanding of the major sea lanes that connected the continents.

China also opened up new avenues of trade and exploration in Southeast Asia during the fifteenth century. Under the direction of Chinese emperor Yongle, explorer Zheng He organized seven overseas trips from 1405 to 1433 and investigated economic opportunities in Korea, Vietnam, the Indian Ocean, and Egypt. His first voyage consisted of more than 28,000 men and 400 ships and represented the largest naval force assembled prior to World War I.

Zheng's armada carried porcelains, silks, lacquerware, and artifacts to Malacca, the vital port city in Indonesia. He purchased an Arab medical text on drug therapy and had it translated into Chinese. He introduced giraffes and mahogany wood into the mainland's economy, and his efforts helped spread Chinese ideas, customs, diet, calendars, scales and measures, and music throughout the global economy. Zheng He's discoveries, coupled with all the material gathered by the European explorers, provided cartographers and geographers with a credible store of knowledge concerning world geography.

Emerging Global Trade Networks

From 1400 to 1900, several regional economic systems facilitated the exchange of goods and services throughout a growing world system. Building on the triangular relationships produced by the slave trade, the Atlantic region helped spread new foodstuffs around the globe. Plants and plantation crops provided societies with a plentiful supply of sweet potatoes, squash, beans, and maize. This system, often referred to as the Columbian exchange, also assisted development in other regions by supplying the global economy with an ample money supply in gold and silver. Europeans sent textiles and other manufactures to the Americas. In return, they received minerals from Mexico; sugar and molasses from the Caribbean; money, rum, and tobacco from North America; and foodstuffs from South America. Trade routes also closed the distance between the Pacific coastline in the Americas and the Pacific Rim.

Additional thriving trade routes existed in the African-West Asian region. Linking Europe and Africa with Arabia and India, this area experienced a considerable amount of trade over land and through the sea lanes in the Persian Gulf and Red Sea. Europeans received grains, timber, furs, iron, and hemp from Russia in exchange for wool textiles and silver. Central Asians secured stores of cotton textiles, silk, wheat, rice, and tobacco from India and sold silver, horses, camel, and sheep to the Indians. Ivory, blankets, paper, saltpeter, fruits, dates, incense, coffee, and wine were regularly exchanged among merchants situated along the trade route connecting India, Persia, the Ottoman Empire, and Europe.

Finally, a Russian-Asian-Chinese market provided Russia's ruling czars with arms, sugar, tobacco, and grain, and a sufficient supply of drugs, medicines, livestock, paper money, and silver moved eastward. Overall, this system linked the economies of three continents and guaranteed that a nation could acquire essential foodstuffs, resources, and money from a variety of sources.

Several profitable trade routes existed in the Indian Ocean sector. After Malacca emerged as a key trading port in the sixteenth century, this territory served as an international clearinghouse for the global economy. Indians sent tin, elephants, and wood into Burma and Siam. Rice, silk, and sugar were sold to Bengal. Pepper and other spices were shipped westward across the Arabian Sea, while Ceylon furnished India with vast quantities of jewels, cinnamon, pearls, and elephants. The booming interregional trade routes positioned along the Indian coastline ensured that many of the vast commodities produced in the world system could be obtained in India.

The final region of crucial trade routes was between Southeast Asia and China. While the extent of Asian overseas trade prior to the twentieth century is usually downplayed, an abundance of products flowed across the Bay of Bengal and the South China Sea. Japan procured silver, copper, iron, swords, and sulphur from Cantonese merchants, and Japanese-finished textiles, dyes, tea, lead, and manufactures were in high demand on the mainland. The Chinese also purchased silk and ceramics from the Philippines in exchange for silver. Burma and Siam traded pepper, sappan wood, tin, lead, and saltpeter to China for satin, velvet, thread, and labor. As goods increasingly moved from the Malabar coast in India to the northern boundaries of Korea and Japan, the Pacific Rim played a prominent role in the global economy.

Robert D. Ubriaco, Jr.

ROAD TRANSPORTATION

Roads—the most common surfaces on which people and vehicles move—are a key part of human and economic geography. Transportation activities form part of a nation's economic product: They strengthen regional economy, influence land and natural resource use, facilitate communication and commerce, expand choices, support industry, aid agriculture, and increase human mobility. The need for roads closely correlates with the relative location of centers of population, commerce, industry, and other transportation.

History of Road Making

The great highway systems of modern civilization have their origin in the remote past. The earliest travel was by foot on paths and trails. Later, pack animals and crude sleds were used. The development of the wheel opened new options. As various ancient civilizations reached a higher level, many of them realized the importance of improved roads.

The most advanced highway system of the ancient world was that of the Romans. When Roman civilization was at its peak, a great system of military roads reached to the limits of the empire. The typical Roman road was bold in conception and construction, built in a straight line when possible, with a deep multilayer foundation, perfect for wheeled vehicles.

After the decline of the Roman Empire, rural road building in Europe practically ceased, and roads fell into centuries of disrepair. Commerce traveled by water or on pack trains that could negotiate the badly maintained roads. Eventually, a commercial revival set in, and roads and wheeled vehicles increased.

Interest in the art of road building was revived in Europe in the late eighteenth century. P. Trésaguet, a noted French engineer, developed a new method of lightweight road building. The regime of French dictator Napoleon Bonaparte (1800–1814) encouraged road construction, chiefly for military purposes. At about the same time, two Scottish engineers, Thomas Telford and John McAdam, also developed road-building techniques.

Roads in the United States

Toward the end of the eighteenth century, public demand in the United States led to the improvement of some roads by private enterprise. These improvements generally took the form of toll roads, called "turnpikes" because a pike was rotated in each road to allow entry after the fee was paid, and generally were located in areas adjacent to larger cities. In the early nineteenth century, the federal government paid for an 800-mile-long macadam road from Cumberland, Maryland, to Vandalia, Illinois.

With the development of railroads, interest in road building began to wane. By 1900, however, demand for better roads came from farmers, who wanted to move their agricultural products to market more easily. The bicycle craze of the 1890s and the advent of motorized vehicles also added to the demand for more and better roads. Asphalt and concrete technology was well developed by then; now, the problem was financing. Roads had been primarily a local issue, but the growing demand led to greater state and federal involvement in funding.

The Federal-Aid Highway Act of 1956 was a milestone in the development of highway transportation in the United States; it marked the beginning of the largest peacetime public works program in the history of the world, creating a 41,000-mile National System of Interstate and Defense Highways, built to high standards. Later legislation expanded funding, improved planning, addressed environmental concerns, and provided for more balanced transportation. Other developed countries also developed highway programs but were more restrained in construction.

Highway Classification

Modern roads can be classified by roadway design or traffic function. The basic type of roadway is the conventional, undivided two-way road. Divided highways have median strips or other physical barriers separating the lanes going in opposite directions.

Another quality of a roadway is its right-of-way control. The least expensive type of system controls most side access and some minor at-grade intersections; the more expensive type has side access fully controlled and no at-grade intersections. The amount of traffic determines the number of lanes. Two or three lanes in each direction is typical, but some roads in Los Angeles have five lanes, while some sections of the Trans-Canada Highway have only one lane. Some highways are paid for entirely from public funds; if users pay directly when they use the road, they are called tollways or turnpikes.

Roads are classified as expressway, arterial, collector, and local in urban areas, with a similar hierarchy in rural areas. The highest level—expressway—is intended for long-distance travel.

Roads and Development

Transportation presents a severe challenge for sustainable development. The number of motor vehicles at the end of the second decade of the twenty-first century—estimated at more than 1.2 billion worldwide—is growing almost everywhere at higher rates than either population or the gross domestic product. Overall road traffic grows even more quickly. The tiny nation of San Marino has nearly 1.2 cars per person. Americans own one car for every 1.88 residents. In Great Britain, there is one car for every 5.3 people.

Highways around the world have been built to help strengthen national unity. The Trans-Canada Highway, the world's longest national road, for example, extends east-west across the breadth of the country. Completed in the 1960s, it had the same goal as the Canadian Pacific Railroad a century before, to improve east-west commerce within Canada.

Sometimes, existing highways need to be upgraded; in less-developed countries, this can simply mean paving a road for all-weather operation. An example of a late-1990s project of this nature was the Brazil-Venezuela Highway project, which had this description: Improve the Brazil-Venezuela highway link by completion of paving along the BR-174, which runs northward from Manaus in the Amazon, through Boa Vista and up to the frontier, so opening a route to the Caribbean. Besides the investment opportunities in building the road itself, the highway would result in investment opportunities in mining, tourism, telecommunications, soy and rice production, trade with Venezuela, manufacturing in the Manaus Free Trade Zone, ecotourism in the Amazon, and energy integration.

Growing road traffic has required increasingly significant national contributions to road construction. Beginning in the 1960s, the World Bank began to finance road construction in several countries. It required that projects be organized to the highest technical and economic standards, with private contracting and international competitive bidding rather than government workers. Still, there were questions as to whether these economic assessments had a road-sector bias and properly incorporated environmental costs. Sustainability was also a question—could the facilities be maintained once they were built?

In the 1990s, the World Bank financed a program to build an asphalt road network in Mozambique. Asphalt makes very smooth roads but is very maintenance-intensive, requiring expensive imported equipment and raw materials. By the end of the decade, the roads required resurfacing but the debt was still outstanding. Alternative materials would have given a rougher road, but it could have been built with local materials and labor.

The European Investment Bank has become a major player in the construction of highways linking Eastern and Western Europe to further European integration. Some of the fastest growth in the world in ownership of autos has been in Eastern Europe. There is a two-way feedback effect between highway construction and auto ownership.

Environment Consequences

Highways and highway vehicles have social, economic, and environmental consequences. Compromise is often necessary to balance transportation needs against these constraints. For example, in Israel, there has been a debate over construction of the Trans-Israel highway, a US$1.3 billion, six-lane highway stretching 180 miles (300 km.) from Galilee to the Negev.

Demand on resources for worldwide road infrastructure far exceeds available funds; governments increasingly are looking to external sources such as tolls. Private toll roads, common in the nineteenth century, are making a comeback. This has spread from the United States to Europe, where private and government-owned highway operators have begun to sell shares on the stock market. Private companies are not only operating and financing roads in Europe, they are also designing and building them. In Eastern Europe, where road construction languished under communism, private financing and toll collecting are seen as the means of supporting badly needed construction.

Industrial development in poor countries is adversely affected by limited transportation. Costs are high-unreliable delivery schedules make it necessary to maintain excessive inventories of raw materials and finished goods. Poor transport limits the radius of trade and makes it difficult for manufacturers to realize the economies of large-scale operations to compete internationally.

In more difficult terrain, roads become more expensive because of a need for cuts and fills, bridges, and tunnels. To save money, such roads often have steeper grades, sharper curves, and reduced width than might be desired. Severe weather changes also damage roads, further increasing maintenance costs.

Stephen B. Dobrow

RAILWAYS

Railroads were the first successful attempts by early industrial societies to develop integrated communication systems. Today, global societies are linked by Internet systems dependent upon communication satellites orbiting around Earth. The speed by which information and ideas can reach remote places breaks down isolation and aids in the developing of a world community. In the nineteenth century, railroads had a similar impact. Railroads were critical for the creation of an urban-industrial society: They linked regions and remote places together, were important contributors in developing nation-states, and revolutionized the way business was conducted through the creation of corporations. Although alternative forms of transportation exist, railroads remain important in the twenty-first century.

The Industrial Revolution and the Railroad

Development of the steam engine gave birth to the railroad. Late in the eighteenth century, James Watt perfected his steam engine in England. Water was superheated by a boiler and vaporized into steam, which was confined to a cylinder behind a piston. Pressure from expanding steam pushes the cylinder forward, causing it to do work if it is attached to wheels. Watt's engine was used in the manufacturing of textiles, thus beginning the Industrial Revolution whereby machine technology mass produced goods for mass consumption. Robert Fulton was the first innovator to commercially apply the steam engine to water transportation. His steamboat *Clermont* made its maiden voyage up the Hudson River in 1807.

Not until the 1820s was a steam engine used for land transportation. Rivers and lakes were natural features where no road needed to be built. Applying steam to land movement required some type of roadbed. In England, George Stephenson ran a locomotive over iron strips attached to wooden rails. Within a short time, England's forges were able to roll rails made completely of iron shaped like an inverted "U."

The amount of profit a manufacturer could make was determined partially by the cost of transportation. The lower the cost of moving cargo and people, the higher the profitability. Several alternatives existed before the emergence of railroads, although there were drawbacks compared to rail transportation. Toll roads were too slow. A loaded wagon pulled by four horses could average 15 miles (25 km.) a day. Canals were more efficient than early railroads, because barges pulled by mules moved faster over waterways. However, canals could not be built everywhere, especially over mountains.

The application of railroad technology, using steam as a power source, made it possible to overcome obstacles in moving goods and people over considerable distances and at profitable costs. Railroads transformed the way goods were purchased by reducing the costs for consumers, thus raising the living standards in industrial societies. Railroads transformed the human landscape by strengthening the link between farm and city, changed commercial cities into industrial centers, and started early forms of suburban growth well before automobiles arrived.

Financing Railroads

Constructing railroads was costly. Tunnels had to be blasted through mountains, and rivers had to be crossed by bridges. Early in the building of U.S. railroads, the nation's iron foundries could not meet the demands for rolled rails. Rails had to be imported from England until local forges developed more efficient technologies. Once a railway was completed, there was a constant need to maintain the right-of-way so that traffic flow would not be disrupted. Accidents were frequent, and it was an early practice to burn damaged cars because salvaging them was too expensive.

In some countries, railroads were built and operated by national governments. In the United States, railroads were privately owned; however, it was impossible for any single individual to finance and operate a rail system with miles of track. Businessmen raised money by selling stocks and bonds. Just as investors buy stocks in modern high-technology companies, investors purchased stocks and bonds in railroads.

Investing in railroads was good as long as they earned profits and returned money to their investors, but not all railroads made sufficient profits to reward their investors. Competition among railroads was heavy in the United States, and some railroads charged artificially low fares to attract as much business as they could. When ambitious investment schemes collapsed, railroads went bankrupt and were taken over by financiers.

Selling shares of common stock and bonds was made possible by creating corporations. Railroads were granted permission from state governments to organize a corporation. Every investor owned a portion of the railroad. Stockholders' interests were served by boards of directors, and all business transactions were opened for public inspection. One important factor of the corporation was that it relieved individuals from the responsibilities associated with accidents. The railroad, as a corporation, was held accountable, and any compensation for claims made against the company came out of corporate funds, not from individual pockets. This had an impact on the law profession, as law schools began specializing in legal matters relevant to railroads and interstate commerce.

The Success of Railroads

Railroads usually began by radiating outward from port cities where merchants engaged in transoceanic trade. A classic example, in the United States, is the country's first regional railroad—the Baltimore and Ohio. Construction commenced from Baltimore in 1828; by 1850, the railroad had crossed the Appalachian Mountains and was on the Ohio River at Wheeling, Virginia.

Once trunk lines were established, rail networks became more intensive as branch lines were built to link smaller cities and towns. Countries with extremely large continental dimensions developed interior articulating cities where railroads from all directions converged. Chicago and Atlanta are two such cities in the United States. Chicago was surrounded by three circular railroads (belts) whose only function was to interchange cars. Railroads from the Pacific Coast converged with lines from the Atlantic Coast as well as routes moving north from the Gulf Coast.

Mechanized farms and heavy industries developed within the network. Railroads made possible the extraction of fossil fuels and metallic ores, the necessary ingredients for industrial growth. Extension of railroads deep into Eastern Europe helped to generate massive waves of immigration into both North and South America, creating multicultural societies.

Building railroads in Africa and South Asia made it possible for Europe to increase its political control over native populations. The ultimate aim of the colonial railroad was to develop a colony's economy according to the needs of the mother country. Railroads were usually single-line routes transhipping commodities from interior centers to coastal ports for exportation. Nairobi, Kenya, began as a rail hub linking British interests in Uganda with Kenya's port city of Mombasa. Similar examples existed in Malaysia and Indonesia.

Railroads generated conflicts among colonial powers as nations attempted to acquire strategic resources. In 1904–1905 Russia and Japan fought a war in the Chinese province of Manchuria over railroad rights; Imperial Germany attempted to get around British interests in the Middle East by building a railroad linking Berlin with Baghdad to give Germany access to lucrative oil fields. India was a region of loosely connected provinces until British railroads helped establish unification. The resulting sense of national unity led to the termination of British rule in 1947 and independence for India and Pakistan.

In the United States, private railroads discontinued passenger service among cities early in the 1970s and the responsibility was assumed by the federal government (Amtrak). Most Americans riding trains do so as commuters traveling from the suburbs to jobs in the city. The U.S. has no true high-speed trains, aside from sections of Amtrak's Acela line in the Northeast Corridor, where it can reach 150 mph for only 34 miles of its 457-mile span. Passenger service remains popular in Japan and Europe. France, Germany, and Japan operate high-speed luxury trains with speeds averaging above 100 miles (160 km.) per hour.

Railroads are no longer the exclusive means of land transportation as they were early in the twentieth century. Although competition from motor vehicles and air freight provide alternate choices, railroads have remained important. France and England have direct rail linkage beneath the English Channel. In the United States, great railroad mergers and the application of computer technology have reduced operating costs while increasing profits. Transoceanic container traffic has been aided by railroads hauling trailers on flatcars. Railroads began the process of bringing regions within a nation together in the nineteenth century just as the computer and the World Wide Web began uniting nations throughout the world at the end of the twentieth century.

Sherman E. Silverman

Air Transportation

The movement of goods and people among places is an important field of geographic study. Transportation routes form part of an intricate global network through which commodities flow. Speed and cost determine the nature and volume of the materials transported, so air transportation has both advantages and disadvantages when compared with road, rail, or water transport.

Early Flying Machines

The transport of people and freight by air is less than a century old. Although hot-air balloons were used in the late eighteenth century for military purposes, aerial mapping, and even early photography, they were never commercially important as a means of transportation. In the late nineteenth century, the German count Ferdinand von Zeppelin began experimenting with dirigibles, which added self-propulsion to lighter-than-air craft. These aircraft were used for military purposes, such as the bombing of Paris in World War I. However, by the 1920s zeppelins had become a successful means of passenger transportation. They carried thousands of passengers on trips in Europe or across the Atlantic Ocean and also were used for exploration. Nevertheless, they had major problems and were soon superseded by flying machines heavier than air. The early term for such a machine was an aeroplane, which is still the word used for airplane in Great Britain.

Following pioneering advances with the internal combustion engine and in aerodynamic theory using gliders, the development of powered flight in a heavier-than-air machine was achieved by Wilbur and Orville Wright in December, 1903. From that time, the United States moved to the forefront of aviation, with Great Britain and Germany also making significant contributions to air transport. World War I saw the further development of aviation for military purposes, evidenced by the infamous bombing of Guernica.

Early Commercial Service

Two decades after the Wright brothers' brief flight, the world's first commercial air service began, covering the short distance from Tampa to St. Petersburg in Florida. The introduction of airmail service by the U.S. Post Office provided a new, regular source of income for commercial airlines in the United States, and from these beginnings arose the modern Boeing Company, United Airlines, and American Airlines. Europe, however, was the home of the world's first commercial airlines. These include the Deutsche Luftreederie in Germany, which connected Berlin, Leipzig, and Weimar in 1919; Farman in France, which flew from Paris to London; and KLM in the Netherlands (Amsterdam to London), followed by Qantas—the Queensland and Northern Territory Aerial Services, Limited—in Australia. The last two are the world's oldest still operating airlines.

Aircraft played a vital role in World War II, as a means of attacking enemy territory, defending territory, and transporting people and equipment. A humanitarian use of air power was the Berlin Airlift of 1948, when Western nations used airplanes to deliver food and medical supplies to the people of West Berlin, which the Soviet Union briefly blockaded on the ground.

Cargo and Passenger Service

The jet engine was developed and used for fighter aircraft during World War II by the Germans, the British, and the United States. Further research led to civil jet transport, and by the 1970s, jet planes accounted for most of the world's air transportation. Air travel in the early days was extremely expensive, but technological advances enabled longer flights with heavier loads, so commercial air travel became both faster and more economical.

Most air travel is made for business purposes. Of business trips between 750 and 1,500 miles (1,207 and 2,414 km.), air captures almost 85 percent, and of trips more than 1,500 miles (2,414 km.), 90 percent are made by air. The United States had 5,087 public airports in 2018, a slight decrease from the 5,145 public airports operating in 2014. Conversely, the number of private airports increased over this period from 13,863 to 14,549.

The biggest air cargo carriers in 2019 were Federal Express, which carried more than 15.71 billion freight tonne kilometres (FTK), Emirates Skycargo (12.27 billion FTK), and United Parcel Service (11.26 billion FTK).

The first commercial supersonic airliner, the British-French Concorde, which could fly at more than twice the speed of sound, began regular service in early 1976. However, the fleet was grounded after a Concorde crash in France in mid-2000. The first space shuttle flew in 1981. There have been 135 shuttle missions since then, ending with the successful landing of Space Shuttle Orbiter Atlantis on July 21, 2011. The shuttles have transported 600 people and 3 million pounds (1.36 million kilograms) of cargo into space.

Health Problems Transported by Air

The high speed of intercontinental air travel and the increasing numbers of air travelers have increased the risk of exotic diseases being carried into destination countries, thereby globalizing diseases previously restricted to certain parts of the world. Passengers traveling by air might be unaware that they are carrying infections or viruses. The worldwide spread of HIV/AIDS after the 1980s was accelerated by international air travel.

Disease vectors such as flies or mosquitoes can also make air journeys unnoticed inside airplanes. At some airports, both airplane interiors and passengers are subjected to spraying with insecticide upon arrival and before deplaning. The West Nile virus (West Nile encephalitis) was previously found only in Africa, Eastern Europe, and West Asia, but in the 1990s it appeared in the northeastern United States, transported there by birds, mosquitos, or people.

It was feared in the mid-1990s that the highly infectious and deadly Ebola virus, which originated in tropical Africa, might spread to Europe and the United States, by air passengers or through the importing of monkeys. The devastation of native bird communities on the island of Guam has been traced to the emergence there of a large population of brown tree snakes, whose ancestors are thought to have arrived as accidental stowaways on a military airplane in the late 1940s.

Ray Sumner

Energy and Engineering

Energy Sources

Energy is essential for powering the processes of modern industrial society: refining ores, manufacturing products, moving vehicles, heating buildings, and powering appliances. In 1999 energy costs were half a trillion dollars in the United States alone. All technological progress has been based on harnessing more energy and using it more effectively. Energy use has been shaped by geography and also has shaped economic and political geography.

Ancient to Modern Energy

Energy use in traditional tribal societies illustrates all aspects of energy use that apply in modern human societies. Early Stone Age peoples had only their own muscle power, fueled by meat and raw vegetable matter. Warmth for living came from tropical or subtropical climates. Then a new energy source, fire, came into use. It made cold climates livable. It enabled the cooking of roots, grains, and heavy animal bones, vastly increasing the edible food supply. Its heat also hardened wood tools, cured pottery, and eventually allowed metalworking.

Nearly as important as fire was the domestication of animals, which multiplied available muscle energy. Domestic animals carried and pulled heavy loads. Domesticated horses could move as fast as the game to be hunted or large animals to be herded.

Increased energy efficiency was as important as new energy sources in making tribal societies more successful. Cured animal hides and woven cloth were additional factors enabling people to move to cooler climates. Cooking fires also allowed drying meat into jerky to preserve it against times of limited supply. Fire-cured pottery helped protect food against pests and kept water close by. However, energy benefits had costs. Fire drives for hunting may have caused major animal extinctions. Periodic burning of areas for primitive agriculture caused erosion. Trees became scarce near the best campsites because they had been used for camp fires—the first fuel shortage.

Energy Fundamentals

Human use of energy revolves about four interrelated factors: energy sources, methods of harnessing the sources, means of transporting or storing energy, and methods of using energy. The potential energies and energy flows that might be harnessed are many times greater than present use.

The Sun is the primary source of most energy on Earth. Sunlight warms the planet. Plants use photosynthesis to transform water and carbon dioxide into the sugars that power their growth and indirectly power plant-eating and meat-eating animals. Many other energies come indirectly from the Sun. Remains of plants and animals become fossil fuels. Solar heat evaporates water, which then falls as rain, causing water flow in rivers. Regional differences in the amount of sunlight received and reflected cause temperature differences that generate winds, ocean currents, and temperature differences between different ocean layers. Food for muscle power of humans and animals is the most basic energy system.

Energy Sources

Biomass—wood or other vegetable matter that can be burned—is still the most important energy source in much of the world. Its basic use is to provide heat for cooking and warmth. Biomass fuels are often agricultural or forestry wastes. The advantage of biomass is that it is grown, so it can be replaced. However, it has several limitations. Its low energy content per unit volume and unit mass makes it unprofitable to ship, so its use is limited to the amount nearby. Collecting and processing biomass fuels costs energy, so the net energy is less. Biomass energy production may compete with food production, since both come from the soil. Finally, other fuels can be cheaper.

Greater concentration of biomass energy or more efficient use would enable it to better compete against other energy sources. For example, fermenting sugars into fuel alcohol is one means of concentrating energy, but energy losses in processing make it expensive.

Fossil fuels have more concentrated chemical energy than biomass. Underground heat and pressure compacts trees and swampy brush into the progressively more energy-concentrated peat, lignite coal, bituminous coal, and anthracite or black coal, which is mostly carbon. Industrializing regions turned to coal when they had exhausted their firewood. Like wood, coal could be stored and shoveled into the fire box as needed. Large deposits of coal are still available, but growth in the use of coal slowed by the mid-twentieth century because of two competing fossil fuels, petroleum and natural gas.

Petroleum includes gasoline, diesel fuel, and fuel oil. It forms from remains of one-celled plants and animals in the ocean that decompose from sugars into simpler hydrogen and carbon compounds (hydrocarbons). Petroleum yields more energy per unit than coal, and it is pumped rather than shoveled. These advantages mean that an oil-fired vehicle can be cheaper and have greater range than a coal-fired vehicle.

There are also hydrocarbon gases associated with petroleum and coal. The most common is the natural gas methane. Methane does not have the energy density of hydrocarbon liquids, but it burns cleanly and is a fuel of choice for end uses such as homes and businesses.

Petroleum and natural gas deposits are widely scattered throughout the world, but the greatest known deposits are in an area extending from Saudi Arabia north through the Caucasus Mountains. Deposits extend out to sea in areas such as the Persian Gulf, the North Sea, and the Gulf of Mexico. Other sources, such as oil tar sands and shale oil, are currently seen as a potentially important source of energy, but controversies surrounding the extraction, refining, and delivery processes make these energy sources a matter of significant debate and concern.

Heat engines transform the potential of chemical energies. James Watt's steam engine (1782) takes heat from burning wood or coal (external combustion), boils water to steam, and expands it through pistons to make mechanical motion. In the twentieth century, propeller-like steam turbines were developed to increase efficiency and decrease complexity. Auto and diesel engines burn fuel inside the engine (internal combustion), and the hot gases expand through pistons to make mechanical motion. Expanding them through a gas turbine is a jet engine. Heat engines can create energy from other sources, such as concentrated sunlight, nuclear fission, or nuclear fusion. The electrical generator transforms mechanical motion into electricity that can move by wire to uses far away. Such transportation (or wheeling) of electricity means that one power plant can serve many customers in different locations.

Flowing water and wind are two of the oldest sources of industrial power. The Industrial Revolution began with water power and wind power, but they could only be used in certain locations, and they were not as dependable as steam engines. In the early twentieth century, electricity made river power practical again. Large dams along river valleys with adequate water and steep enough slopes enabled areas like the Tennessee Valley to be industrial centers. In the 1970s wind power began to be used again, this time for generating electricity.

Solar energy can be tapped directly for heat or to make electricity. Although sunlight is free, it is not concentrated energy, so getting usable energy requires more equipment cost. Consequently, fossil-fueled heat is cheaper than solar heat, and power from the conventional utility grid has been much less expensive than solar-generated electricity. However, prices of solar equipment continue to drop as technologies improve.

Future Energy Sources

Possible future energy sources are nuclear fission, nuclear fusion, geothermal heat, and tides. Fission reactors contain a critical mass of radioactive heavy elements that sustains a chain reaction of atoms splitting (fissioning) into lighter elements—releasing heat to run a steam turbine. Tremendous amounts of fission energy are available, but reactor costs and safety issues have kept nuclear prices higher than that of coal.

Nuclear fusion involves the same reaction that powers the Sun: four hydrogen atoms fusing into one helium atom. However, duplicating the Sun's heat in a small area without damaging the surrounding reactor may be too expensive to allow profitable fusion reactors.

Geothermal power plants, tapping heat energy from within the earth, have operated since 1904, but widespread use depends on cheaper drilling to make them practical in more than highly volcanic areas. Tidal power is limited to the few bays that concentrate tidal energy.

Energy and Warfare

Much of ancient energy use revolved about herding animals and conducting warfare. Horse riders moved faster and hit harder than warriors on foot. The bow and arrow did not change appreciably for thousands of years. Herders on the plains rode horses and used the bow and arrow as part of tending their flocks, and the small amounts of metal needed for weapons was easily acquired. Consequently, the herders could invade and plunder much more advanced peoples. From Scythians to Parthians to Mongols, these people consistently destroyed the more advanced civilizations.

The geographical effect was that ancient civilizations generally developed only if they had physical barriers separating them from the flat plains of herding peoples. Egypt had deserts and seas. The Greeks and Romans lived on mountainous peninsulas, safe from easy attack. The Chinese built the Great Wall along their northern frontier to block invasions.

Nomadic riders dominated until the advent of an energy system of gunpowder and steel barrels began delivering lead bullets. With them, the Russians broke the power of the Tartars in Eurasia in the late fifteenth century, and various peoples from Europe conquered most of the world. Energy and industrial might became progressively more important in war with automatic weapons, high explosives, aircraft, rockets, and nuclear weapons.

By World War II, oil had become a reason for war and a crucial input for war. The Germans attempted to seize petroleum fields around Baku on the Caspian. Later in the war, major Allied attacks targeted oil fields in Romania and plants in Germany synthesizing liquid fuels. During the Arab-Israeli War of 1973, Arabs countered Western support of Israel with an oil boycott that rocked Western economies. In 1990 Iraq attempted to solve a border dispute with its oil-rich neighbor, Kuwait, by seizing all of Kuwait. An alliance, led by the United States, ejected the Iraqis.

Other wars occur over petroleum deposits that extend out to sea. European nations bordering on the North Sea negotiated a complete demarcation of economic rights throughout that body. Tensions between China and other Asian countries continue to mount over rights to the resources available in the South China Sea. Current estimates suggest that there may be 90 trillion cubic feet of natural gas and 11 billion barrels of oil in proved and probable reserves, with much more potentially undiscovered. The area is claimed by China, Vietnam, Malaysia, and the Philippines. Turkey and Greece have not resolved ownership division of Aegean waters that might have oil deposits.

Energy, Development, and Energy Efficiency

Ancient civilizations tended to grow and use locally available food and firewood. Soils and wood supplies often were depleted at the same time, which often coincided with declines in those civilizations.

The Industrial Revolution caused development to concentrate in new wooded areas where rivers suitable for power, iron ore, and coal were close together, for example, England, Silesia, and the Pittsburgh area. The iron ore of Alsace in France combined with nearby coal from the Ruhr in Germany fueled tremendous growth, not always peacefully.

By the late nineteenth century, the development of Birmingham, Alabama, demonstrated that railroads enabled a wider spread between coal deposits, iron ore deposits, and existing population centers. By the 1920s, the Soviet Union developed entirely new cities to connect with resources. By the 1970s, unit trains and ore-carrying ships transported coal from the thick coal beds in Montana and Wyoming to the United States' East Coast and to countries in Asia.

The mechanized transport of electrical distribution and distribution of natural gas in pipelines also changed settlement patterns. Trains and subway trains allowed cities to spread along rail corridors in the late nineteenth century and early twentieth century. By the 1940s, cars and trucks enabled cities such as Los Angeles and Phoenix to spread into suburbs. The trend continues with independent solar power that allows houses to be sited anywhere.

Advances in technology have allowed people to get more while using less energy. For example, early peoples stampeded herds of animals over cliffs for food, which was mostly wasted. Horseback hunting was vastly more efficient. Likewise, fireplaces in colonial North America were inefficient, sending most of their heat up the chimney. In the late eighteenth century, inventor and statesman Benjamin Franklin developed a metallic cylinder radiating heat in all directions, which saved firewood.

The ancient Greeks and others pioneered the use of passive solar energy and efficiency after they exhausted available firewood. They sited buildings to absorb as much low winter sun as possible and constructed overhanging roofs to shade buildings from the high summer sun. That siting was augmented by heavy masonry building materials that buffered the buildings from extremes of heat and cold. Later, metal pipes and glass meant that solar energy could be used for water and space heating.

The first seven decades of the twentieth century saw major declines in energy prices, and cars and appliances became less efficient. That changed abruptly with the energy crises and high prices of the 1970s. Since then, countries such as Japan, with few local energy resources, have worked to increase efficiency so they will be less sensitive to energy shocks and be able to thrive with minimal energy inputs. This trend could lead eventually to economies functioning on only solar and biomass inputs.

Solid-state electronics, use of light emitting diode (LED) or compact fluorescent lamps (CFLs) rather than incandescent bulbs, and fuel cells, which convert fuel directly into electricity more efficiently than combustion engines, all could lead to less energy use. The speed of their adoption depends on the price of competing energies. According to the U.S. Energy Information Administration's (EIA) International Energy Outlook 2019 (IEO2019), the global supply of crude oil, other liquid hydrocarbons, and biofuels is expected to be adequate to meet world demand through 2050. However, many have noted that continuing to burn fossil fuels at our current rate is not sustainable, not because reserves will disappear, but because the damage to the climate would be unacceptable.

Energy and Environment

Energy affects the environment in three major ways. First, firewood gathering in underdeveloped countries contributes to deforestation and resulting erosion. Although more efficient stoves and small solar cookers have been designed, efficiency increases are competing against population increases.

Energy production also frequently causes toxic pollutant by-products. Sulfur dioxide (from sulfur impurities in coal and oil) and nitrogen oxides (from nitrogen being formed during combustion) damage lungs and corrode the surfaces of buildings. Lead additives in gasoline make internal combustion engines run more efficiently, but they cause low-grade lead poisoning. Spent radioactive fuel from nuclear fission reactors is so poisonous that it must be guarded for centuries.

Finally, carbon dioxide from the burning of fossil fuels may be accelerating the greenhouse effect, whereby atmospheric carbon dioxide slows the planetary loss of heat. If the effect is as strong as some research suggests, global temperatures may increase several degrees on average in the twenty-first century, with unknown effects on climate and sea level.

Roger V. Carlson

ALTERNATIVE ENERGIES

The energy that lights homes and powers industry is indispensable in modern societies. This energy usually comes from mechanical energy that is converted into electrical energy by means of generators—complex machines that harness basic energy captured when such sources as coal, oil, or wood are burned under controlled conditions. This energy, in turn, provides the thermal energy used for heating, cooling, and lighting and for powering automobiles, locomotives, steamships, and airplanes. Because such natural resources as coal, oil, and wood are being used up, it is vital that these nonrenewable sources of energy be replaced by sources that are renewable and abundant. It is also desirable that alternative sources of energy be developed in order to cut down on the pollution that results from the combustion of the hydrocarbons that make the nonrenewable fuels burn.

The Sun as an Energy Source

Energy is heat. The Sun provides the heat that makes Earth habitable. As today's commonly used fuel resources are used less, solar energy will be used increasingly to provide the power that societies need in order to function and flourish.

There are two forms of solar energy: passive and active. Humankind has long employed passive solar energy, which requires no special equipment. Ancient cave dwellers soon realized that if they inhabited caves that faced the Sun, those caves would be warmer than those that faced away from the Sun. They also observed that dark surfaces retained heat and that dark rocks heated by the Sun would radiate the heat they contained after the Sun had set. Modern builders often capitalize on this same knowledge by constructing structures that face south in the Northern Hemisphere and north in the Southern Hemisphere. The windows that face the Sun are often large and unobstructed by draperies and curtains. Sunlight beats through the glass and, in passive solar houses, usually heats a dark stone or brick floor that will emit heat during the hours when there is no sunlight. Just as an automobile parked in the sunlight will become hot and retain its heat, so do passive solar buildings become hot and retain their heat.

Active solar energy is derived by placing specially designed panels so that they face the Sun. These panels, called flat plate collectors, have a flat glass top beneath which is a panel, often made of copper with a black overlay of paint, that retains heat. These panels are constructed so that heat cannot escape from them easily. When water circulated through pipes in the panels becomes hot, it is either pumped into tanks where it can be stored or circulated through a central heating system.

Some active solar devices are quite complex and best suited to industrial use. Among these is the focusing collector, a saucer-shaped mirror that centers the Sun's rays on a small area that becomes extremely hot. A power plant at Odeillo in the French Pyrenees Mountains uses such a system to concentrate the Sun's rays on a concave mirror. The mirror directs its incredible heat to an enormous, confined body of water that the heat turns to steam, which is then used to generate electricity.

Another active solar device is the solar or photovoltaic cell, which gathers heat from the Sun and turns it into energy directly. Such cells help to power spacecraft that cannot carry enough conventional fuel to sustain them through long missions in outer space.

Geothermal Heating

The earth's core is incredibly hot. Its heat extends far into the lower surfaces of the planet, at times causing eruptions in the form of geysers or volcanoes. Many places on Earth have springs that are warmed by heat from the earth's core.

In some countries, such as Iceland, warm springs are so abundant that people throughout the country bathe in them through the coldest of winters. In Iceland, geothermal energy is used to heat and light homes, making the use of fossil fuels unnecessary.

Hot areas exist beneath every acre of land on Earth. When such areas are near the surface, it is easy to use them to produce the energy that humans require. As dependence on fossil fuels decreases, means will increasingly be found of drawing on Earth's subterranean heat as a major source of energy.

Wind Power

Anyone who has watched a sailboat move effortlessly through the water has observed how the wind can be used as a source of kinetic energy—the kind of energy that involves motion—whose movement is transferred to objects that it touches. Wind power has been used throughout human history. In its more refined aspects, it has been employed to power windmills that cause turbines to rotate, providing generators with the power they require to produce electricity.

Windmills typically have from two to twenty blades made of wood or of heavy cloth such as canvas. Windmills are most effective when they are located in places where the wind regularly blows with considerable velocity. As their blades turn, they cause the shafts of turbines to rotate, thus powering generators. The electricity created is usually transmitted over metal cables for immediate use or for storage.

Modern vertical-axis wind turbines have two or three strips of curved metal that are attached at both ends to a vertical pole. They can operate efficiently even if they are not turned toward the wind. These windmills are a great improvement over the old horizontal axis windmills that have been in use for many years. From 2000 to 2015, cumulative wind capacity around the world increased from 17,000 megawatts to more than 430,000 megawatts. In 2015, China also surpassed the EU in the number of installed wind turbines and continues to lead installation efforts. Production of wind electricity in 2016 accounted for 16 percent of the electricity generated by renewables.

Oceans as Energy Sources

Seventy percent of the earth's surface is covered by oceans. Their tides, which rise and fall with predictable regularity twice a day, would offer a ready source of energy once it becomes economically feasible to harness them and store the electrical energy they can provide. The most promising spots to build facilities to create electrical energy from the tides are places where the tides are regularly quite dramatic, such as Nova Scotia's Bay of Fundy, where the difference between high and low tides averages about 55 feet (17 meters).

Some tidal power stations that currently exist were created by building dams across estuaries. The sluices of these dams are opened when the tide comes in and closed after the resulting reservoir fills. The water captured in the reservoir is held for several hours until the tide is low enough to create a considerable difference between the level of the wa-

Ocean Energy

The oceans have tremendous untapped energy flows in currents and tremendous potential energy in the temperature differences between warmer tropical surface waters and colder deep waters, known as ocean thermal energy conversion. In both cases, the insurmountable cost has been in transporting energy to users on shore.

ter in the reservoir and that outside it. Then the sluice gates are opened and, as the water rushes out at a high rate of speed, it turns turbines that generate electricity.

The world's first large-scale tidal power plant was the Rance Tidal Power Station in France, which became operational in 1966. It was the largest tidal power station in terms of output until Sihwa Lake Tidal Power Station opened in South Korea in August 2011.

Future of Renewable Energy

As pollution becomes a huge problem throughout the world, the race to find nonpolluting sources of energy is accelerating rapidly. Scientists are working on unlocking the potential of the electricity generated by microbes as a fuel source, for example. New technologies are making renewable energy sources economically practical. As supplies of fossil fuels have diminished, pressure to become less dependent on them has grown worldwide. Alternative energy sources are the wave of the future.

R. Baird Shuman

Engineering Projects

Human beings attempt to overcome the physical landscape by building forms and structures on the earth. Most structures are small-scale, like houses, telephone poles, and schools. Other structures are great engineering works, such as hydroelectric projects, dams, canals, tunnels, bridges, and buildings.

Hydroelectric Projects

The potential for hydroelectricity generation is greatest in rapidly flowing rivers in mountainous or hilly terrain. The moving water turns turbines that, in turn, generate electricity. Hydroelectric power projects also can be built on escarpments and fall lines, where there is tremendous untapped energy in the falling water.

Most of the potential for hydroelectricity remains untapped. Only about one-sixth of the suitable rivers and falls are used for hydroelectric power. Certain areas of the world have used more of their potential than others. The percent of potential hydropower capacity that has not been developed is 71 percent in Europe, 75 percent in North America, 79 percent in South America, 95 percent in Africa, 95 percent in the Middle East, and 82 percent in Asia-Pacific. China, Brazil, Canada, and the United States currently produce the most hydroelectric power.

In Africa, only Zambia, Zimbabwe, and Ghana produce significant hydroelectricity. The region's total generating capacity needs to increase by 6 percent per year to 2040 from the current total of 125 GW to keep pace with rising electricity demand. In Southeast Asia, countries continue to grapple with the need to build up their hydroelectric plants without causing harm to the rivers that are used to supply food, water, and transportation.

Dams

Dams serve several purposes. One purpose is the generation of hydroelectric power, as discussed above. Dams also provide flood control and irrigation. Rivers in their natural state tend to rise and fall with the seasons. This can cause serious problems for people living in downstream valleys. Flood-control dams also can be used to regulate the flow of water used for irrigation and other projects. A final reason to build dams is to reduce swampland, in order to control insects and the diseases they carry.

Famous dams are found in all regions of the world. In North America, two of the most notable dams are Hoover Dam, completed in 1936, on the Colorado River between Arizona and Nevada; and the Grand Coulee Dam, completed in 1942, on the Columbia River in Washington State.

In South America, the most famous dam is the Itaipu Dam, completed in 1983, on the Paraná River between Brazil and Paraguay. In Africa, the Aswan High Dam was completed in 1970, on the Nile River in Egypt, and the Kariba Dam was completed in 1958, on the Zambezi River between Zambia and Zimbabwe. In Asia, the Three Gorges Dam spans the Yangtze River by the town of Sandouping in Hubei province, China. The Three Gorges Dam has been the world's largest power station in terms of installed capacity (22,500 MW) since 2012.

Bridges

Bridges are built to span low-lying land between two high places. Most commonly, there is a river or other body of water in the way, but other features that might be spanned include ravines, deep valleys and trenches, and swamps. A related engineering project is the causeway, in which land in a low-lying area is built up and a road is then constructed on it.

The longest bridge in the world is the Akashi Kaikyo in Japan near Osaka. It was built in 1998 and spans 6,529 feet (1,990 meters), connecting the island of Hōnshū to the small island of Awaji. The Storebælt Bridge in Denmark, also completed in 1998, spans 5,328 feet (1,624 meters), connecting the island of Sjaelland, on which Copenhagen is situated, with the rest of Denmark. Another bridge spanning more than 5,300 feet is the Osman Gazi Bridge in Turkey. The bridge was opened on 1 July 1, 2016, ad to become the longest bridge in Turkey and the fourth-longest suspension bridge in the world by the length of its central span. The length of the bridge is expected to be surpassed by the Çanakkale 1915 Bridge, which is currently under construction across the Dardanelles strait.

Other long bridges can be found across the Humber River in Hull, England; across the Chiang Jiang (Yangtze River) in China; in Hong Kong, Norway, Sweden, and Turkey and elsewhere in Japan.

The longest bridge in the United States is the Lake Pontchartrain Causeway, Louisiana, which spans 24 miles (38.5 km.), the Verrazano-Narrows Bridge in New York City between Staten Island and Brooklyn was once the longest suspension bridge in the world. Completed in 1964, its main span measures 4,260 feet (1,298 meters).

Canals

Moving goods and people by water is generally cheaper and easier, if a bit slower, than moving them by land. Before the twentieth century, that cost savings overwhelmed the advantages of land travel—speed and versatility. Therefore, human beings have wanted to move things by water whenever possible. To do so, they had two choices: locate factories and people near water, such as rivers, lakes, and oceans, or bring water to where the factories and people are, by digging canals.

One of the most famous canals in the world is the Erie Canal, which runs from Albany to Buffalo in New York State. Built in 1825 and running a length of 363 miles (584 km.), the Erie Canal opened up the Great Lakes region of North America to devel-

ENGINEERING WORKS AND ENVIRONMENTAL PROBLEMS

Although engineering allows humans to overcome natural obstacles, works of engineering often have unintended consequences. Many engineering projects have caused unanticipated environmental problems.

Dams, for instance, create large lakes behind them by trapping water that is released slowly. This water typically contains silt and other material that eventually would have formed soil downstream had the water been allowed to flow naturally. Instead, the silt builds up behind the dam, eventually diminishing the lake's usefulness. As an additional consequence, there is less silt available for soil-building downstream.

Canals also can cause environmental harm by diverting water from its natural course. The river from which water is diverted may dry up, negatively affecting fish, animals, and the people who live downstream.

The benefits of engineering works must be weighed against the damage they do to the environment. They may be worthwhile, but they are neither all good nor all bad: There are benefits and drawbacks in building any engineering project.

opment and led to the rise of New York City as one of the world's dominant cities.

Two other important canals in world history are the Panama Canal and the Suez Canal. The Panama Canal connects the Atlantic and Pacific Oceans over a length of 50.7 miles (81.6 km.) on the isthmus of Panama in Central America. Completed in 1914, the Panama Canal eliminated the long and dangerous sea journey around the tip of South America. The Suez Canal in Egypt, which runs for 100 miles (162 km.) and was completed in 1856, eliminates a similar journey around the Cape of Good Hope in South Africa.

The longest canal in the world is the Grand Canal in China, which was built in the seventh century and stretches a length of 1,085 miles (2,904 km.). It connects Tianjin, near Beijing in the north of China, with Nanjing on the Chang Jiang (Yangtze River) in Central China. The Karakum Canal runs across the Central Asian desert in Turkmenistan from the Amu Darya River westward to Ashkhabad. It was begun in the 1954, and completed in 1988 and is navigable over much of its 854-mile (1,375-km.) length. The Karakum Canal and carries 13 cubic kilometres (3.1 cu mi) of water annually from the Amu-Darya River across the Karakum Desert to irrigate the dry lands of Turkmenistan.

Many canals are found in Europe, particularly in England, France, Belgium, the Netherlands, and Germany, and in the United States and Canada, especially connecting the Great Lakes to each other and to the Ohio and Mississippi Rivers.

Tunnels

Tunnels connect two places separated by physical features that would make it extremely difficult, if not impossible, for them to be connected without cutting directly through them. Tunnels can be used in place of bridges over water bodies so that water traffic is not impeded by a bridge span. Tunnels of this type are often found in port cities, and cities with them include Montreal, Quebec; New York City; Hampton Roads, Virginia; Liverpool, England; or Rio de Janeiro, Brazil.

Tunnels are often used to go through mountains that might be too tall to climb over. Trains especially are sensitive to changes in slope, and train tunnels are found all over the world. Less common are automobile and truck tunnels, although these are also found in many places. Train and automotive tunnels through mountains are common in the Appalachian Mountains in Pennsylvania, the Rockies in the United States and Canada, Japan, and the Alps in Italy, France, Switzerland, and Austria.

The Chunnel

Arguably the most famous—and one of the most ambitious—tunnels in the world goes by the name Chunnel. Completed in 1994, it connects Dover, England, to Calais, France, and runs 31 miles (50 km.). "Chunnel" is short for the Channel Tunnel, named for the English Channel, the body of water that it goes under. It was built as a train tunnel, but cars and trucks can be carried through it on trains. In the year 2000 plans were underway to cut a second tunnel, to carry automobiles and trucks, that would run parallel to the first Chunnel.

The Seikan Tunnel in Japan, connects the large island of Hōnshū with the northern island of Hōkkaidō. The Seikan Tunnel is nearly 2.4 miles (4 km.) longer than Europe's Chunnel; however, the undersea portion of the tunnel is not as long as that of the Chunnel.

Buildings

Historically, North America has been home to the tallest buildings in the world. Chicago has been called the birthplace of the skyscraper and was at one time home to the world's tallest building. In 1998, however, the two Petronas Towers (each 1,483 feet/452 meters tall) were completed in Kuala Lumpur, Malaysia, surpassing the height of the world's tallest building, Chicago's Sears Tower (1,450 feet/442 meters), which had been completed in 1974. In 2019, the tallest completed building in the world is the 2,717-foot (828-metre) tall Burj Khalifa in Dubai, the tallest building since 2008.

Of the twenty tallest buildings standing in the year 2019, China is home to ten (Shanghai Tower, Ping An Finance Center, Goldin Finance 117, Guangzhou CTF Finance Center, Tianjin CFT Finance Center, China Zun, Shanghai World Financial Center, International Commerce Center, Wuhan Greenland Center, Changsha); Malaysia (the Petronas towers) and the United States (One World Trade Center and Central Park Towers) boast two each; Vietnam has one (Landmark 81 in Ho Chi Minh City), as does Russia (Lakhta Center), Taiwan (Taipei 101), South Korea (Lotte World Trade Center), Saudi Aragia (Abraj Al-Bait Clock Tower in Mecca).

Timothy C. Pitts

INDUSTRY AND TRADE

MANUFACTURING

Manufacturing is the process by which value is added to materials by changing their physical form—shape, function, or composition. For example, an automobile is manufactured by piecing together thousands of different component parts, such as seats, bumpers, and tires. The component parts in unassembled form have little or no utility, but pieced together to produce a fully functional automobile, the resulting product has significant utility. The more utility something has, the greater its value. In other words, the value of the component parts increases when they are combined with the other parts to produce a useful product.

Employment in Manufacturing

On a global scale, 28 percent of the world's working population had jobs in the manufacturing sector in the third decade of the century. The rest worked in agriculture (28 percent) and services (49 percent). The importance of each of these sectors varies from country to country and from time period to time period. High-income countries have a higher percentage of their labor force employed in manufacturing than low-income countries do. For example, in the United States 19 percent of the labor force worked in manufacturing by 2019, whereas the African country of Tanzania had only 7 percent of its labor force employed in the manufacturing sector at that time.

At the end of the twentieth century, the vast majority of the U.S. labor force (74 percent) worked in services, a sector that includes jobs such as computer programmers, lawyers, and teachers. By the end of the second decade of the twenty-first century, the percentage has risen to slightly more than 79 percent. Only 1 percent worked in agriculture and mining. This employment structure is typical for a high-income country. In low-income countries, in contrast, the majority of the labor force have agricultural jobs. In Tanzania, for example, 66 percent of the labor force worked in agriculture, while services accounted for 27 percent of the jobs.

The importance of manufacturing as an employer changes over time. In 1950 manufacturing accounted for 38 percent of all jobs in the United States. The percentage of jobs accounted for by the manufacturing sector in high-income countries has decreased in the post-World War II period. The decreasing share of manufacturing jobs in high-income countries is partly attributable to the fact that many manufacturing companies have replaced people with machines on assembly lines. Because one machine can do the work of many people, manufacturing has become less labor-intensive (uses fewer people to perform a particular task) and more capital-intensive (uses machines to perform tasks formerly done by people). In the future, manufacturing in high-income countries is expected to become increasingly capital-intensive. It is not inconceivable that manufacturing's share of the U.S. labor force could fall below 10 percent over the course of the twenty-first century.

Geography of Manufacturing

Every country produces manufactured goods, but the vast bulk of manufacturing activity is concentrated geographically. Four countries—China, the United States, Japan, and Germany—produce almost 60 percent of the world's manufactured goods. The concentration of manufacturing activity in a small number of regions means that there are other regions where very little manufacturing occurs. Africa is a prime example of a region with little manufacturing.

Different countries tend to specialize in the production of different products. For example, 50 percent of the automobiles that were produced in that late 1990s were produced in three countries—Germany, Japan, and the United States. In the production of television sets, the top three countries were China, Japan, and South Korea, which together produced 48 percent of the world's television sets. It is important to note that these patterns change over time. For example, in 1960 the top three automobile-producing countries were Germany, the United Kingdom, and the United States, which together produced 76 percent of the world's automobiles.

Multinational Corporations

A multinational corporation is a corporation that is headquartered in one country but owns business facilities, for example, manufacturing plants, in other countries. Some examples of multinational corporations from the manufacturing sector include the automobile maker Ford, whose headquarters are the in the United States, the pharmaceutical company Bayer, whose headquarters are in Germany, and the candy manufacturer Nestlé, whose headquarters are in Switzerland. Since the end of World War II, multinational corporations have become increasingly important in the world economy. Most multinational corporations are headquartered in high-income countries, such as Japan, the United Kingdom, and the United States.

Companies open manufacturing plants in other countries for a variety of reasons. One of the most common reasons is that it allows them to circumvent barriers to trade that are imposed by foreign governments, especially tariffs and quotas. A tariff is an import tax that is imposed upon foreign-manufactured goods as they enter a country. A quota is a limitation imposed on the volume of a particular good that a particular country can export to another country. The net effect of tariffs and quotas is to increase the cost of imported goods for consumers.

Governments impose tariffs and quotas partly to raise revenue and partly to encourage consumers to purchase goods manufactured in their own country. Foreign manufacturers faced with tariffs and quotas often begin manufacturing their product in the country imposing the tariffs and quotas. As tariffs and quotas apply to imported goods only, producing in the country imposing the quotas or tariffs effectively makes these trade barriers obsolete.

Companies also open manufacturing plants in other countries because of differences in labor costs among countries. While most manufacturing takes place in high-income countries, some low-income countries have become increasingly attractive as production locations because their workers can be hired much more cheaply than in high-income countries. For example, in late 2019, the average manufacturing job in the United States paid more than US$22.50 per hour. By comparison, manufacturing employees in the Philippines earned a few cents more than US$2.50 per hour.

This dramatic differences in labor costs have prompted some companies to close down their manufacturing plants in high-income countries and open up new plants in low-income countries. This has resulted in high-income countries purchasing more manufactured goods from low-income countries.

More than half the clothing imported into the United States came from Asian countries, for example, China, Taiwan, and South Korea, where labor costs were much lower than in the United States. Much of this clothing was made in factories where workers were paid by companies headquartered in the United States. For example, most of the Nike sports shoes that were sold in the United States were made in China, Indonesia, Vietnam, and Pakistan.

Transportation and Communications Technology

The ability of companies to have manufacturing plants in other countries stems from the fact that the world has a sophisticated and efficient transportation and communications system. An advanced transportation and communications system makes it relatively easy and relatively cheap to transfer information and goods between geographically distant locations. Thus, Nike can manufacture soccer balls in Pakistan and transport them quickly and cheaply to customers in the United States.

The extent to which transportation and communications systems have improved during the last two centuries can be illustrated by a few simple examples. In 1800, when the stagecoach was the primary method of overland transportation, it took twenty hours to travel the ninety miles from Lansing, Michigan, to Detroit, Michigan. Today, with the automobile, the same journey takes approximately ninety minutes. In 1800 sailing ships traveling at an average speed of ten miles per hour were used to transport people and goods between geographically distant countries. In the year 2019 jet-engine aircraft could traverse the globe at speeds in excess of 600 miles per hour. Communications technology has also improved over time.

In 1930 a three-minute telephone call between New York and London, England, cost more than US$385 in 2018 dollars. In the year 2019 the same telephone call could be made for less than a dime. In addition to modern telephones, there are fax machines, email, videoconferencing capabilities, and a host of other technologies that make communication with other parts of the world both inexpensive and swift.

Future Prospects

The global economy of the twenty-first century presents a wide variety of opportunities and challenges. Sophisticated communications and transportation networks provide increasing numbers of manufacturing companies with more choices as to where to locate their factories. However, high-income countries like the United States are increasingly in competition with other countries (both high- and low-income) to maintain existing and manufacturing investments and attract new ones. Persuading existing companies to keep their U.S. factories open and not move overseas has been a major challenge. Likewise, making the United States as an attractive place for foreign companies to locate their manufacturing plants is an equally challenging task.

Neil Reid

Globalization of Manufacturing and Trade

Why are most of the patents issued worldwide assigned to Asian corporations? How did a Taiwanese earthquake prevent millions of Americans from purchasing memory upgrades for their computers? Why have personal incomes in Beijing nearly doubled in less than a decade?

Answers to these questions can be found in the geography of globalization. Globalization is an economic, political, and social process characterized by the integration of the world's many systems of manufacturing and trade into a single and increasingly seamless marketplace. The result: a new world geography.

This new geography is associated with the expansion of manufacturing and trade as capitalist principles replace old ideologies and state-controlled economies. With expanded free markets, the process of manufacturing and trading is constantly changing. Globalization delivers economic growth through improved manufacturing processes, newly developed goods, foreign investment in overseas manufacturing, and expanded employment.

The economies of developing countries are slowly transitioning from agricultural to industrial activities. Nevertheless, more than 65 percent of workers in these countries continue to work in agriculture. Meanwhile, developed countries, such as Australia and Germany, are experiencing high-technology service sector growth and reduced manufacturing employment. In the United States, nearly 30 percent of all workers were employed in manufacturing during the 1950s, but by 2019, less than 8.5 percent were.

In between these extremes, former state-controlled economies, like Romania, are adopting more efficient economic development strategies. Other nations and economic models, such as Indonesia and China, are pulled into the global marketplace by the growth and expansion of market economies. Despite the different economic paths of developing, transitioning, and developed nations, manufacturing and trade link all nations together and represent an economic convergence with important implications for political, business, and labor leaders—as well as all the world's citizens.

The geographies of manufacturing and trade can be examined as the distribution and location of economic activities in response to technological change and political and economic change.

Distribution and Location

Questions about where people live, work, and spend their money can be answered by reading product labels in any shopping mall, supermarket, or automobile dealership. They reveal the fact that manufacturing is a multistage process of component fabrication and final product assembly that can occur continents apart. For example, a shirt may be designed in New Jersey, assembled in Costa Rica from North Carolina fabric, and sold in British Columbia. To understand how goods produced in faraway locations are sold at neighborhood stores, geographers investigate the spatial, or geographic, distribution of natural resources, manufacturing plants, trading patterns, and consumption.

Historically, the geography of manufacturing and trade has been closely linked to the distribution of raw materials, workers, and buyers. In earlier times, this meant that manufacturing and trade were highly localized functions. In the eighteenth century, every North American town had cobblers or blacksmiths who produced goods from local resources for sale in local markets. By the start of the Industrial Revolution, improved transportation and manufacturing techniques had significantly enlarged the geography of manufacturing and trade. As distances increased, new manufacturing and trading centers developed. The location of these centers was contingent upon site and situation. Site and situation refer to a physical location, or site, relative to needed materials, transportation networks, and markets. For example, Pittsburgh, Pennsylvania, became the site of a major steel industry because it was near coal and iron resources. Pittsburgh also benefited from its historical role as a port town on a major river system that provided access to both western and eastern markets.

While relative location and transportation costs continue to be important factors, the geographic distribution of production and movement of goods across space is more complex than the simple calculus of site and situation. New global and local geographies of manufacturing and trade have been fueled by two major factors: technology and political change.

Technological Change

The old saying that time is money partially explains where goods are manufactured and traded. By compressing time and space, technology has enabled people, goods, and information to go farther more quickly. In the process, technology has reduced interaction costs, such as telecommunications. Just as steel enabled railroads to push farther westward, new technologies reduce the distance between places and people.

By increasing physical and virtual access to people, places, and things, technology has eliminated many barriers to global trade. However, improved telecommunications and transportation are only part of technology's contribution to globalization. If time is money, new efficient manufacturing processes also have reduced costs and facilitated globalization.

Armed with more efficient production processes, reliable telecommunications infrastructures, and transportation improvements, businesses can increase profits and remain competitive by seeking out lower-cost labor markets thousands of miles from consumers. As trade and manufacturing are increasingly spatially separate activities, the geographic distribution of manufacturing promotes an uneven distribution of income. The global distribution of manufacturing plants is closely related to industry-specific skill and wage requirements. For example, low-wage and low-skill jobs tend to concentrate in the developing regions of Asia, South America, and Africa. Alternately, high-technology and high-wage manufacturing activities concentrate in more developed regions.

In some cases, high wages and global competition force corporations to move their manufacturing plants to save costs and remain competitive. During the early 1990s, this byproduct of globalization was a major issue during the U.S. and Canadian debates to ratify the North American Free Trade Agreement (NAFTA). Focusing on primarily U.S. and Canadian companies that moved jobs to Mexico, the debate contributed to growing anxiety over job security as plants relocate to low-cost labor markets in South America and around the world.

As global competition increases, the geography of manufacturing and trade is increasingly global and rapidly changing. One company that has adapted to the shifting nature of global trade and manufacturing is Nike. Based in Beaverton, Oregon, Nike designs and develops new products at its Oregon world headquarters. However, Nike has internationalized much of its manufacturing capacity to compete in an aggressive athletic apparel industry. Over the last twenty-five years, Nike's strategy has meant shifts in production from high-wage U.S. locations to numerous low-wage labor markets around Pacific Rim.

Political and Economic Change: A New World Order

In order for companies such as Nike to successfully adapt to changing global dynamics, a stable international, or multilateral, trading system must be in place. In 1948 the General Agreement on Tariffs and Trade (GATT) was the first major step toward developing this stable global trading infrastructure. During that same period, the World Bank and International Monetary Fund were created to stabilize and standardize financial markets and practices. However, Cold War politics postponed complete economic integration for nearly half a century. Since the collapse of communism, globalization has accelerated as economies coalesce around the principles of free markets and capitalism. These important changes have become institutionalized through multilateral trade agreements and international trading organizations.

International trading organizations try to minimize or eliminate barriers to free and fair trade between nations. Trade barriers include tariffs (taxes levied on imported goods), product quotas, government subsidies to domestic industry, domestic content rules, and other regulations. Barriers prevent competitive access to domestic markets by artificially raising the prices of imported goods too high or preventing foreign firms from achieving economies of scale. In some cases, tariffs can also be used to promote fair trade by effectively leveling the playing field.

Because tariffs can be used both to promote fair trade and to unfairly protect markets, trading organizations are responsible for distinguishing between the two. For example, the Asian Pacific Economic Cooperation (APEC) forum has established guidelines to promote fair trade and attract foreign investment. APEC initiatives include a public Web-based database of member state tariff schedules and related links. Through programs such as the APEC information-sharing project, trading organizations are streamlining the international business process and promoting the overall stability of international markets.

The Future

As the globalization of manufacturing and trade continues, a new world geography is emerging. Unlike the Cold War's east-west geography and politics of ideology, an economic politics divides the developed and developing world along a north-south axis. While the types of conflicts associated with these new politics and the rules of engagement are unclear, it is evident that a new hierarchy of nations is emerging.

Globalization will raise the economic standard of living in most nations, but it has also widened the gap between richer and poorer countries. A small group of nations generates and controls most of the world's wealth. Conversely, the poorest countries account for roughly two-thirds of the world's population and less than 10 percent of its wealth.

This fundamental question of economic justice was a motive behind globalization's first major political clash. During the 1999 World Trade Organization (WTO) meetings in Seattle, Washington, approximately 50,000 environmentalists, labor unionists, and human and animal rights activists protested against numerous issues, including cultural intolerance, economic injustice, environmental degradation, political repression, and unfair labor practices they attribute to free trade. While the protesters managed to cancel the opening ceremonies, the United Nations secretary-general, Kofi Annan, expressed the general sentiment of most WTO member states. Agreeing that the protesters' concerns were important, Annan also asserted that the globalization of manufacturing and trade should not be used as a scapegoat for domestic failures to protect individual rights. More important, the secretary-general feared that those issues could be little more than a pretext for a return to unilateral trade policies, or protectionism.

Like the Seattle protesters, supporters of multilateral trade advocate political and economic reforms. Proponents emphasize that open markets promote open societies. Free traders earnestly believe economic engagement encourages rogue nations to improve poor human rights, environmental, and labor records. It is argued that economic engagement raises the expectations of citizens, thereby promoting change.

Conclusion

Technological and political change have made global labor and consumer markets more accessible and established an economic world hierarchy. At the top, one-fifth of the world's population consumes the vast majority of produced goods and controls more than 80 percent of the wealth. At the bottom of this hierarchy, poor nations are industrializing but possess less than 10 percent of the world's wealth. In political, social, and cultural terms, this global economic reality defines the contours and cleavages of a changing world geography. Whether geographers calculate the economic and political costs of a widening gap between rich and poor or chart the flow of funds from Tokyo to Toronto, the globalization of manufacturing and trade will remain central to the study of geography well into the twenty-first century.

Jay D. Gatrell

Modern World Trade Patterns

Trade, its routes, and its patterns are an integral part of modern society. Trade is primarily based on need. People trade the goods that they have, including money, to obtain the goods that they don't have. Some nations are very rich in agriculture or natural resources, while others are centers of industrial or technical activity. Because nations' needs change only slowly, trade routes and trading patterns develop that last for long periods of time.

Types of Trade

The movement of goods can occur among neighboring countries, such as the United States and Mexico, or across the globe, as between Japan and

Italy. Some trade routes are well established with regularly scheduled service connecting points. Such service is called liner service. Liners may also serve intermediate points along a trade route to increase their revenue.

Some trade occurs only seasonally, such as the movement of fresh fruits from Chile to California. Some trade occurs only when certain goods are demanded, such as special orders of industrial goods. This type of service is provided by operators called tramps. They go where the business of trade takes them, rather than along fixed liner schedules and routes.

Many people think of international trade as being carried on great ships plying the oceans of the world. Such trade is important; however, a considerable amount of trade is carried by other modes of transportation. Ships and airplanes carry large volumes of freight over large distances, while trucks, trains, barges, and even animal transport are used to move goods over trade routes among neighboring or landlocked countries.

Trade Routes

Through much of human history, trade routes were limited. Shipping trade carried on sailing vessels, for example, was limited by the prevailing winds that powered the ships. Land routes were limited by the location of water, mountain ranges, and the slow development of roads through thick forests and difficult terrain. The mechanization of transportation eventually freed ships and other forms of transport to follow more direct trade routes. Also, the development of canals and transcontinental highway systems allowed trade routes to develop based solely upon economic requirements.

Other changes in trade routes have occurred with industrialization of transport systems. The world began to have a great need for coal. Trade routes ran to the countries in which coal was mined. Ships and trains delivered coal to the power industry worldwide. Later, trade shifted to locations where oil (petroleum) was drilled. Now, oil is delivered to those same powerplants and industrial sites around the world.

Noneconomic Factors

Some trade is not purely economic in nature. Political relationships among countries can play an important part in their trade relations. For example, many national governments try to protect their countries' automobile and electronics industries from outside competition by not allowing foreign goods to be imported easily. Governments control imports by assessing duties, or tariffs, on selected imports.

Some national governments use the concept of cabotage to protect their home transportation industries by requiring that certain percentages of imported and exported trade goods be carried by their own carriers. For example, the U.S. government might require that 50 percent of its trade use American ships, planes, or trucks. The government might also require that all American carriers employ only American citizens.

Nations also can exert pressure on their trading partners by limiting access to port or airport facilities. Stronger nations may force weaker nations into accepting unequal trade agreements. For example, the United States once had an agreement with Germany concerning air passenger service between the two countries. The agreement allowed United States carriers to carry 80 percent of the passengers, while German carriers were permitted to carry only 20 percent of the passengers.

Multilateral Trade

In situations in which pairs of trading nations do not have direct diplomatic contact with each other, they make their trade arrangements through other nations. Such trade is referred to as multilateral. Certain carriers cater to this type of trade. They operate their ships or planes in around-the-world service. They literally travel around the globe picking up and depositing cargo along the way for a variety of nations.

Trade Patterns

For many years, world populations were coast centered. This means that most of the people in the country lived close to the coast. This was due primarily to the availability of water transportation systems to move both goods and people. At this time, major railroad, highway and airline systems did not exist. As railway and highway systems pushed into the interiors of nations, the population followed, and goods were needed as well as produced in these areas. Thus, over the years many inland population centers have developed that require transportation systems to move goods into and away from this area.

In these cases, international trade to these inland centers required the use of a number of different modes of transportation. Each of the different modes required additional paperwork and time for repackaging and securing of the cargo. For example, cargo coming off ships from overseas was unloaded and placed in warehouse storage. At some later time, it was loaded onto trucks that carried it to railyards. There it would be unloaded, stored, and then loaded onto railcars. At the destination, the cargo would once again be shifted to trucks for the final delivery. During the course of the trip, the cargo would have been handled a number of times, with the possibility of damage or loss occurring each time.

Containerization

As more goods began to move in international trade, the systems for packaging and securing of cargo became more standardized. In the 1960s, shipments began to move in containers. These are highway truck trailers which have been removed from the chassis leaving only the box. Container packaging has become the standard for most cargos moving today in both domestic and international trade. With the advent of containerization of cargo in international trade, cargo movements could quickly move intermodally. Intermodal shipping involves the movement of cargo by using more than a single mode of transportation.

Land, water, and air carriers have attempted to make the intermodal movement of cargo in international trade as seamless as possible. They have not only standardized the box for carrying cargo, but they have also standardized the handling equipment, so that containers move quickly from one mode to another. Advances in communications and electronic banking allow the paperwork and payments also to be completed and transferred rapidly.

As the demands for products have grown and as the size of industrial plants has grown, the size of movements of raw materials and containerized cargo has also grown. Thus, the sizes of the ships and trains required to move these large volumes of cargo have also increased.

The development of VLCC's (very large crude carriers) has allowed shippers to move large volumes of oil products. The development of large bulk carriers has allowed for the carriage of large volumes of dry raw materials such as grains or iron ore. These large vessels take advantage of what is known as economies of scale. Goods can be moved more cheaply when large volumes of them are moved at the same time. This is because the doubling of the volume of cargo moved does not double the cost to build or operate the vessels in which it is carried. This savings reduces the cost to move large volumes of cargo.

THE WORLD TRADE ORGANIZATION AND GLOBAL TRADING

In 1998 domestic political pressures and an expected domestic surplus of rice prompted the Japanese government to unilaterally implement a 355-percent tariff on foreign rice, violating the United Nations' General Agreement on Tariffs and Trade (GATT). On April 1, 1999, Japan agreed to return to GATT import levels and imposed new over-quota tariffs. While domestic Japanese politics could have prompted a trade war with rice-exporting countries, the crisis demonstrates how multilateral trading initiatives promote stability. Without an agreement, rice exporters might not have gained access to Japanese markets. By returning to GATT minimum quotas and implementing over-quota taxes, the compromise addressed the interests of both domestic and foreign rice growers.

Intermodal Transportation

Intermodal transportation has allowed cargo to move seamlessly across both international boundaries and through different modes of transportation. This seamless movement has changed ocean trade routes over recent years.

The development of the Pacific Rim nations created a demand for trade between East Asia and both the United States and Europe. This trade has usually taken the all-water routes between Asia and Europe. Ships moving from East Asia across the Pacific Ocean pass through the Panama Canal and cross the Atlantic Ocean to reach Western Europe. This journey is in excess of 10,000 miles (16,000 km.) and usually takes about thirty days for most ships to complete. The all-water route from Asia to New York is similar. The distance is almost as great as that to Europe and requires about twenty-one to twenty-four days to complete.

Intermodal transportation has given shippers alternatives to all-water routes. A great volume of Asian goods is now shipped to such western U.S. ports as Seattle, Oakland, and Los Angeles, from which these goods are carried by trains across the United States to New York. The overall lengths of these routes to New York are only about 7,400 miles (12,000 km.) and take between only fifteen and nineteen days to complete. Cargos continuing to Europe are put back on ships in New York and complete their journeys in an additional seven to ten days. Such intermodal shipping can save as much as a week in delivery time.

Airfreight

Another changing trend in trade patterns is the development of airfreight as an international competitor. Modern aircraft have improved dramatically both in their ability to lift large weights of cargo as well as their ability to carry cargos over long distances. Because of the speed at which aircraft travel in comparison to other modes of transportation, goods can be moved quickly over large distances. Thus, high-value cargos or very fragile cargos can move very quickly by aircraft.

The drawback to airfreight movement of cargo is that it is more expensive than other modes of travel. However, for businesses that need to move perishable commodities, such as flowers of the Netherlands, or expensive commodities, such as Paris fashions or Singapore-made computer chips, airfreight has become both economic and essential.

Robert J. Stewart

Political Geography

Forms of Government

Philosophers and political scientists have studied forms of government for many centuries. Ancient Greek philosophers such as Plato and Aristotle wrote about what they believed to be good and bad forms of government. According to Plato's famous work, *The Republic*, the best form of government was one ruled by philosopher-kings. Aristotle wrote that good governments, whether headed by one person (a kingship), a few people (an aristocracy), or many people (a polity), were those that ruled for the benefit of all. Those that were based on narrow, selfish interests were considered bad forms of government, whether ruled by an individual (a tyranny), a few people (an oligarchy), or many people (a democracy). Thus, democracy was not always considered a good form of government.

Constitutions and Political Institutions

All governments have certain things in common: institutions that carry out legislative, executive, and judicial functions. How these institutions are supposed to function is usually spelled out in a country's constitution, which is a guide to organizing a country's political system. Most, but not all, countries have written constitutions. Great Britain, for example, has an unwritten constitution based on documents such as the Magna Carta, the English Bill of Rights, and the Treaty of Rome, and on unwritten codes of behavior expected of politicians and members of the royal family.

The world's oldest written constitution still in use is that of the United States. All countries have written or unwritten constitutions, and most follow them most of the time. Some countries do not follow their constitutions—for example, the Soviet Union did not; other countries, for example France, change their constitutions frequently.

Constitutions usually first specify if the country is to be a monarchy or a republic. Few countries still have monarchies, and those that do usually grant the monarch only ceremonial powers and duties. Countries with monarchies at the beginning of the twenty-first century included Spain, Great Britain, Lesotho, Swaziland, Sweden, Saudi Arabia, and Jordan. Most countries that do not have monarchies are republics.

Constitutions also specify if power is to be concentrated in the hands of a strong national government, which is a unitary system; if it is to be divided between a national and various subnational governments such as states, provinces, or territories, which is a federal system; or if it is to be spread among various subnational governments that might delegate some power to a weak national government, which is a confederate system.

Examples of countries with unitary systems include Great Britain, France, and China; federal systems include the United States, Germany, Russia, Canada, India, and Brazil. There were no confederate systems by the third decade of the twenty-first century, although there are examples from history as well as confederations of various groups and nations. The United States under its eighteenth-century Articles of Confederation and the nineteenth-century Confederate States of America, made up of the rebelling Southern states, were confederate systems. Switzerland was a confederation for much of the nineteenth century. The concept of dividing power between the national and subnational governments is called the vertical axis of power.

MONARCHIES OF THE WORLD

Realm/Kingdom	*Monarch*	*Type*
Principality of Andorra	Co-Prince Emmanuel Macron; Co-Prince Archbishop Joan Enric Vives Sicília	Constitutional
Antigua and Barbuda	Queen Elizabeth II	Constitutional
Commonwealth of Australia	Queen Elizabeth II	Constitutional
Commonwealth of the Bahamas	Queen Elizabeth II	Constitutional
Barbados	Queen Elizabeth II	Constitutional
Belize	Queen Elizabeth II	Constitutional
Canada	Queen Elizabeth II	Constitutional
Grenada	Queen Elizabeth II	Constitutional
Jamaica	Queen Elizabeth II	Constitutional
New Zealand	Queen Elizabeth II	Constitutional
Independent State of Papua New Guinea	Queen Elizabeth II	Constitutional
Federation of Saint Kitts and Nevis	Queen Elizabeth II	Constitutional
Saint Lucia	Queen Elizabeth II	Constitutional
Saint Vincent and the Grenadines	Queen Elizabeth II	Constitutional
Solomon Islands	Queen Elizabeth II	Constitutional
Tuvalu	Queen Elizabeth II	Constitutional
United Kingdom of Great Britain and Northern Ireland	Queen Elizabeth II	Constitutional
Kingdom of Bahrain	King Hamad bin Isa	Mixed
Kingdom of Belgium	King Philippe	Constitutional
Kingdom of Bhutan	King Jigme Khesar Namgyel	Constitutional
Brunei Darussalam	Sultan Hassanal Bolkiah	Absolute
Kingdom of Cambodia	King Norodom Sihamoni	Constitutional
Kingdom of Denmark	Queen Margrethe II	Constitutional
Kingdom of Eswatini	King Mswati III	Absolute
Japan	Emperor Naruhito	Constitutional
Hashemite Kingdom of Jordan	King Abdullah II	Constitutional
State of Kuwait	Emir Sabah al-Ahmad	Constitutional
Kingdom of Lesotho	King Letsie III	Constitutional
Principality of Liechtenstein	Prince Regnant Hans-Adam II (Regent: The Hereditary Prince Alois)	Constitutional
Grand Duchy of Luxembourg	Grand Duke Henri	Constitutional
Malaysia	Yang di-Pertuan Agong Abdullah	Constitutional
Principality of Monaco	Sovereign Prince Albert II	Constitutional

Monarchies of the World *(continued)*

Realm/Kingdom	*Monarch*	*Type*
Kingdom of Morocco	King Mohammed VI	Constitutional
Kingdom of the Netherlands	King Willem-Alexander	Constitutional
Kingdom of Norway	King Harald V	Constitutional
Sultanate of Oman	Sultan Haitham bin Tariq	Absolute
State of Qatar	Emir Tamim bin Hamad	Mixed
Kingdom of Saudi Arabia	King Salman bin Abdulaziz	Absolute theocracy
Kingdom of Spain	King Felipe VI	Constitutional
Kingdom of Sweden	King Carl XVI Gustaf	Constitutional
Kingdom of Thailand	King Vajiralongkorn	Constitutional
Kingdom of Tonga	King Tupou VI	Constitutional
United Arab Emirates	President Khalifa bin Zayed	Mixed
Vatican City State	Pope Francis	Absolute theocracy

Whether governments share power with subnational governments or not, there must be institutions to make laws, enforce laws, and interpret laws: the legislative, executive, and judicial branches of government. How these branches interact is what determines whether governments are parliamentary, presidential, or mixed parliamentary-presidential. In a presidential system, such as in the United States, the three branches—legislative, executive, and judicial—are separate, independent, and designed to check and balance each other according to a constitution. In a parliamentary system, the three branches are not entirely separate, and the legislative branch is much more powerful than the executive and judicial branches.

Great Britain is a good example of a parliamentary system. Some countries, such as France and Russia, have created a mixed parliamentary-presidential system, wherein the three branches are separate but are not designed to check and balance each other. In a mixed parliamentary-presidential system, the executive (led by a president) is the most powerful branch of government.

Looking at political systems in this way—how the legislative, executive, and judicial branches of government interact—is to examine the horizontal axis of power. All governments are unitary, federal, or confederate, and all are parliamentary, presidential, or mixed parliamentary-presidential. One can find examples of different combinations. Great Britain is unitary and parliamentary. Germany is federal and parliamentary. The United States is federal and presidential. France is unitary and mixed parliamentary-presidential. Russia is federal and mixed parliamentary-presidential. Furthermore, virtually all countries are either republics or monarchies.

Types of Government

Constitutions describe how the country's political institutions are supposed to interact and provide a guide to the relationship between the government and its citizens. Thus, while governments may have similar political institutions—for example, Germany and India are both federal, parliamentary republics—how the leaders treat their citizens can vary widely. However, governments may have political systems that function similarly although they have different forms of constitutions and institutions. For example, Great Britain, a unitary, parliamentary monarchy with an unwritten constitution, treats its citizens very similarly to the United States, which is a federal, presidential republic with a written constitution.

The three most common terms used to describe the relationships between those who govern and those who are governed are democratic, authoritarian, and totalitarian. Characteristics of democracies are free, fair, and meaningfully contested elections; majority rule and respect for minority rights and opinions; a willingness to hand power to the opposition after an election; the rule of law; and civil rights and liberties, including freedom of speech and press, freedom of association, and freedom to travel. The United States, Canada, Japan, and most European countries are democratic.

An authoritarian system is one that curtails some or all of the characteristics of a democratic regime. For example, authoritarian regimes might permit token electoral opposition by allowing other political parties to run in elections, but they do not allow the opposition to win those elections. If the opposition did win, the authoritarian regime would not hand over power. Authoritarian regimes do not respect the rule of law, the rights of minorities to dissent, or freedom of the press, speech, or association. Authoritarian governments use the police, courts, prisons, and the military to intimidate and threaten their citizens, thus preventing people from uniting to challenge the existing political rulers. Afghanistan, Cuba, Iran, Uzbekistan, Saudi Arabia, Chad, Syria, Libya, Sudan, Belarus, and China are examples of countries with authoritarian regimes.

Totalitarian regimes are similar to authoritarian regimes but are even more extreme. Under a totalitarian regime, there is no legal opposition, no freedom of speech, and no rule of law whatsoever. Totalitarian regimes attempt to control totally all members of the society to the point where everyone always must actively demonstrate their loyalty to and support for the regime. Nazi Germany under Adolf Hitler's rule (1933-1945) and the Soviet Union under Joseph Stalin's rule (1928-1953) are examples of totalitarian regimes. As of 2019, only Eritrea and North Korea are the still have governments classified as totalitarian dictatorships.

Forms of Government: Putting it All Together

In *The Republic*, Plato asserts that people have varied dispositions, and, therefore, there are various types of governments. In recent years, regimes have been created that some call mafiacracies (rule by criminal mafias), narcocracies (rule by narcotics gangs), gerontocracies (rule by very old people), theocracies (rule by religious leaders), and so forth. Such variations show the ingenuity of the human mind in devising forms of government.

Whatever labels that are given to a political system, there remain basic questions to be asked about that regime: Is it a monarchy or a republic? Is all power concentrated in the hands of a national government, or is power shared between a national government and the states or provinces? Are its institutions those of a parliamentary, presidential, or mixed parliamentary-presidential system? Is it democratic, authoritarian, or totalitarian? Finally, does it live up to its constitution, both in terms of how power is supposed to be distributed among institutions and in its relationship between the government and the people? To paraphrase Aristotle, how many rulers are there, and in whose interests do they rule?

Nathaniel Richmond

Political Geography

Students of politics have been aware that there is a significant relationship between physical and political geography since the time of ancient Greece. The ancient Greek philosopher Plato argued that a *polis* (politically organized society) must be of limited geographical size and limited population or it would lack cohesion. The ideal *polis* would be only as geographically large as required to feed about 5,000 people, its maximum population.

Plato's illustrious pupil, Aristotle, agreed that stable states must be small. "One can build a wall around the Hellespont," the main territory of ancient Greece, he wrote in his treatise *Politics*, "but that will not make it a polis." Today human ideas differ about the maximum area of a successful state or nation-state, but the close influence of physical geography on political geography and their profound mutual effects on politics itself are not in question.

Geographical Influences on Politics

The physical shape and contours of states may be called their physical geography; the political shape and contours of states, starting with their basic structure as unified state, federation, or confederation, are primary features of their political geography. The idea of "political geography" also can refer to variations in a population's political attitudes and behavior that are influenced by geographical features. Thus, the combination of plentiful land and sparse population tend toward an independent spirit, especially where the economy is agriculturally based. This has historically been the case in the western United States; in the Pampas region of Argentina, where cattle are raised by independent-minded gauchos (cowboys); and on the Brazilian frontier, where government regulation is routinely resisted.

Likewise, where physical geography presents significant difficulties for inhabitants in earning a living or associating, as where there is rough terrain and poor soil or inhospitable climate, the populace is likely to exhibit a hardy, self-reliant character that strongly influences political preferences. Thus, physical geography helps to shape national character, including aspects of a nation's politics.

Furthermore, it is well known that where physical geography isolates one part of a country's population from the rest, political radicalism may take root. This tendency is found in coastal cities and remote regions, where labor union radicalism has often been pronounced. Populations in coastal locations with access to foreign trade often show a more liberal, tolerant, and outgoing spirit, as reflected in their political opinions. In ancient Greece, the coastal access enjoyed by Athens through a nearby port in the fifth century BCE had a strong influence on its liberal and democratic political order. In modern times, China's coastal cities, such as Tientsin, and North American cities such as San Francisco, show similar influences.

The Geographical Imperative

In many instances, political geography is shaped by what may be called the "geographical imperative." Physical geography in these instances demands, or at least strongly suggests, that political geography follow its course. The numerous valleys of mountainous Greece strongly influenced the emergence of the small, often fiercely independent, polis of ancient times. The formation and borders of Asian states such as Bhutan, Nepal, and Tibet have been strongly influenced by the Himalaya Mountains, and the Alps, which shape Switzerland.

As another example, physical geography demands that the land between the Pacific Ocean and the Andes Mountains along the western edge of South America be organized as a separate country—Chile. Island geography often plays a decisive role in its political geography. The qualified political unity of Great Britain can be directly traced to its insular status. Small islands often find themselves combined into larger units, such as the Hawaiian Islands.

The absence of the geographical imperative, however, leaves political geography an open question. For example, Indonesia comprises some 1,300 islands stretching 3,000 miles in bodies of water such as the Indian Ocean and the Celebes Sea. With so many islands, Indonesia lacks a geographical imperative to be a unified state. It also lacks the imperative of ethnic and cultural homogeneity and cohesion, a circumstance mirrored in its political life, since it has remained unified only through military force. As control by the military waned after the fall of the authoritarian General Suharto in 1998, conflicts among the nation's diverse peoples have threatened its breakup. No such threat, however, confronts Australia, an immense island continent where a European majority dominates a fragmented and primitive aboriginal minority. In Australia, the geographical imperative suggests a unity supported by the cultural unity of the majority.

As many examples show, the geographical imperative is not absolute. For example, mountainous Greece is politically united in the twenty-first century. Although long shielded geographically, Tibet lost its political independence after it was successfully invaded by China. The formerly independent Himalayan state Sikkim was taken over by India. Thus, political will trumps physical geography.

The frequency of exceptions to the geographical imperative illustrates that human freedom, while not unlimited, often plays a key role in shaping political geography. As one example, the Baltic Republics—Lithuania, Latvia, and Estonia—historically have been dominated, or largely swallowed up, by neighboring Russia. By the start of the twenty-first century, however, they had regained their independence through the political will to self-rule and the drive for cultural survival.

Strategically Significant Locations

Locations of great economic or military significance become focal points of political attention and, potentially, of military conflict. There are innumerable such places in the world, but several stand out as models of how important physical geography can be for political geography in the context of international politics.

One significant example is the Panama Canal, without which ships must sail around South America. The Suez Canal, which connects European and Asian shipping, is a similar waterway, saving passage around Africa. The canal's significance was reduced after 1956, however, when its blockage after the Arab-Israeli war of that year led to the building of supertankers too large to traverse it. Another example is Gibraltar, whose fortifications command the entrance to the Mediterranean Sea from the Atlantic Ocean. A final example is the Bosporus, the tiny entrance from the Black Sea to waters leading to the Mediterranean Sea. It is the only warm-water route to and from Eastern Russia and therefore is of great military and economic importance for regional and world power politics.

Charles F. Bahmueller

Geopolitics

Geopolitics is a concept pertaining to the role of purely geographical features in the relations among states in international politics. Geopolitics is especially concerned with the geographical locations of the states in relationship to one another. Geopolitical relationships incorporate social, economic, political, and historical features of the states that interact with purely geographical elements to influence the strategic thinking and behavior of nations in the international sphere.

Coined in 1899 by the Swedish theorist Rudolf Kjellen, the term "geopolitics" combines the logic of the search for security and competition for dominance among states with geographical methodology. *Geopolitics* must not, however, be confused with *political geography*, which focuses on individual states' territorial sizes, boundaries, resources, internal political relations, and relations with other states.

Geopolitical is a term frequently used by military and political strategists, politicians and diplomats, political scientists, journalists, statesmen, and a variety of other government officials, such as policy planners and intelligence analysts.

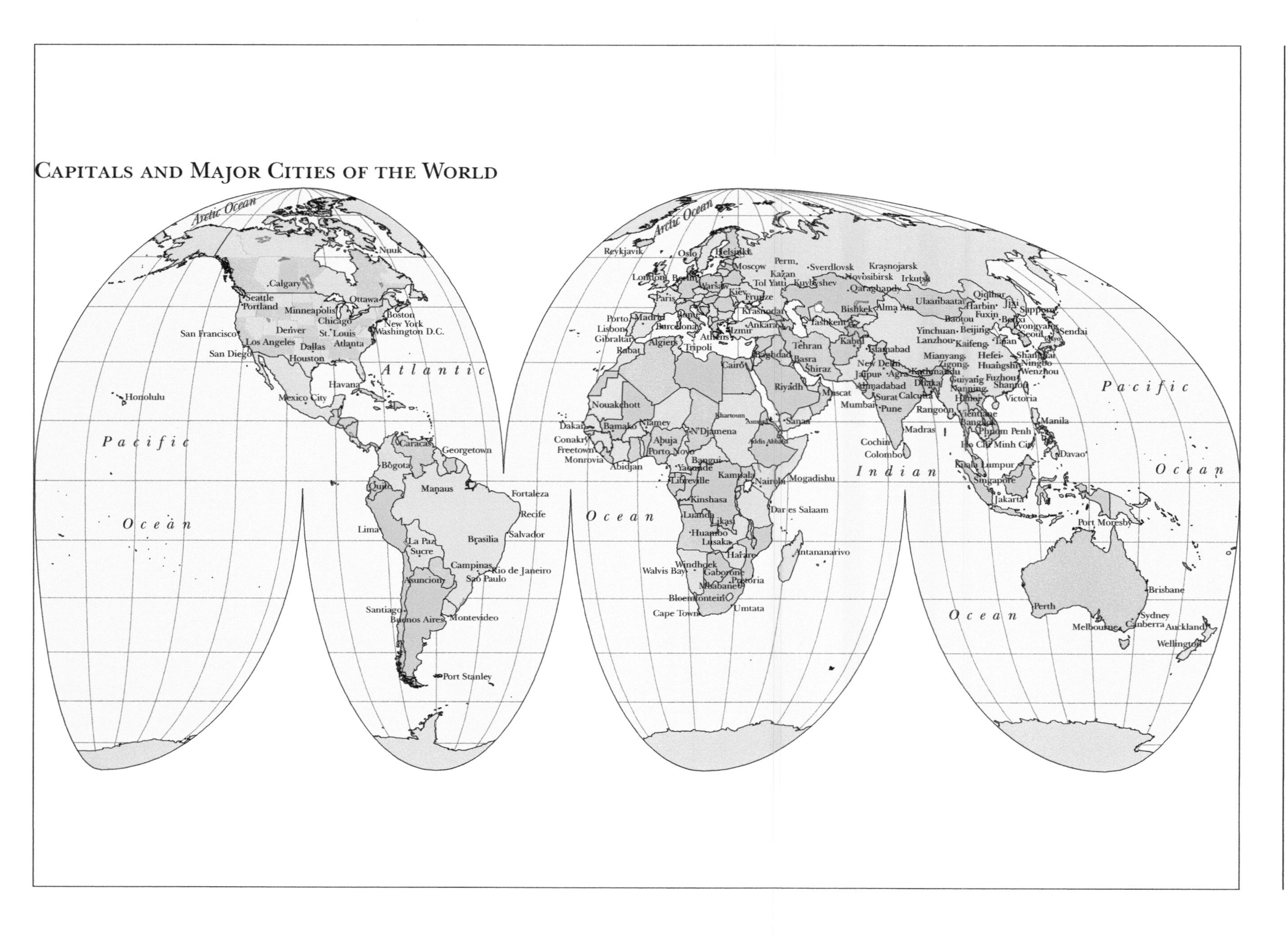
CAPITALS AND MAJOR CITIES OF THE WORLD
Arctic Ocean
Arctic Ocean
Pacific Ocean
Atlantic
Ocean
Indian
Pacific
Ocean
Ocean
Honolulu
Nuuk
Calgary
Seattle
Portland
Minneapolis
Ottawa
Chicago
Boston
New York
Washington D.C.
San Francisco
Denver
St. Louis
Los Angeles
Dallas
Atlanta
San Diego
Houston
Havana
Mexico City
Caracas
Georgetown
Bogota
Quito
Manaus
Fortaleza
Recife
Lima
Brasilia
Salvador
La Paz
Sucre
Campinas
Rio de Janeiro
São Paulo
Asuncion
Santiago
Buenos Aires
Montevideo
Port Stanley
Reykjavik
Oslo
Helsinki
Moscow
Perm
Kazan
Sverdlovsk
Krasnojarsk
Novosibirsk
Irkutsk
London
Berlin
Tol Yatti
Kuybyshev
Qaraghandy
Warsaw
Kiev
Paris
Frunze
Bishkek
Alma Ata
Ulaanbaatar
Qiqihar
Harbin
Jixi
Sapporo
Porto
Madrid
Rome
Krasnodar
Tashkent
Baotou
Fuxin
Lisbon
Barcelona
Ankara
Yinchuan
Beijing
Seoul
Sendai
Gibraltar
Athens
Izmir
Tehran
Kabul
Lanzhou
Kaifeng
Tianjin
Rabat
Algiers
Tripoli
Baghdad
Islamabad
Mianyang
Hefei
Shanghai
Basra
Shiraz
New Delhi
Zigong
Huangshi
Ningbo
Cairo
Jaipur
Agra
Kathmandu
Wenzhou
Guiyang
Fuzhou
Riyadh
Muscat
Ahmadabad
Dhaka
Nanning
Shantou
Pacific
Surat
Calcutta
Hanoi
Victoria
Mumbai
Pune
Rangoon
Nouakchott
Khartoum
Sanaa
Vientiane
Bangkok
Manila
Dakar
Bamako
Niamey
N'Djamena
Madras
Phnom Penh
Conakry
Abuja
Cochin
Ho Chi Minh City
Freetown
Porto Novo
Addis Ababa
Colombo
Davao
Monrovia
Abidjan
Bangui
Yaounde
Kuala Lumpur
Kampala
Libreville
Nairobi
Mogadishu
Singapore
Kinshasa
Jakarta
Dar es Salaam
Luanda
Likasi
Port Moresby
Huambo
Lusaka
Harare
Antananarivo
Windhoek
Walvis Bay
Gaborone
Pretoria
Mbabane
Bloemfontein
Brisbane
Cape Town
Umtata
Perth
Sydney
Melbourne
Canberra
Auckland
Wellington

Power Struggles Among States

The idea of geopolitics arises in the course of what might be considered the universal struggle for power among the world's most powerful nations, which compete for political and military leadership. How one state can threaten another, for example, is often influenced by geographical factors in combination with technological, social, economic and other factors. The extent to which individual states can threaten each other depends in no small measure on purely geographical considerations.

By the close of twentieth century the Cold War that had dominated world security concerns was over. Nevertheless, the United States still worried about the danger of being attacked by nuclear missiles fired, not by the former Soviet Union, but by irresponsible, fanatical, or suicidal states. American political leaders and military planners were concerned with the geographical position of so-called "rogue states." or "states of concern." In 1994, North Korea, Cuba, Iran, Libya under Muammar Gaddafi, and Ba'athist Iraq were listed as states of concern. By 2019, a list of state sponsors of terrorism included Iran, North Korea, Sudan, and Syria.

Geographical factors play prominent roles in assessments of the different threats that those states presented to American interests. How far those states are located from American territory determines whether their missiles might pose a serious threat. A missile may be able to reach only the periphery of U.S. soil, or it might be able to carry only a small payload. Similar considerations determine the threat such states pose for U.S. forces stationed abroad, as well as for such important U.S. allies as Japan, Western Europe, or Israel. Such questions are thus said to constitute geopolitical, or geostrategic, considerations.

There are many examples of the influence of geopolitical factors on international relations among nations in the past. For example, the Bosporus, the narrow sea lane linking the Black Sea and the Mediterranean where Istanbul is situated, has long been considered of great strategic importance. In the nineteenth century, the Bosporus was the only direct route through which the Russian navy could reach southern Europe and the Mediterranean Sea.

Because of Russia's nineteenth century history of expansionism and its integration into the pre-World War I European state system, with its networks of competing military alliances, the Bosporus took on added geopolitical meaning. It was the congested (and therefore vulnerable) space through which Russian naval power had to pass to reach the Mediterranean.

A Peacefully Resolved Border Dispute

The peaceful resolution of the border dispute between the Southern African states of Botswana and Namibia was hailed by observers of African politics. Instead of resorting to the armed warfare that so often has marked similar disputes on the continent, the two states chose a different course in 1996, when they found negotiations stalemated. They submitted their claims to the International Court of Justice in The Hague and agreed to accept the court's ruling. Late in 1999, by an eleven-to-four vote, the court ruled for Botswana, and Namibia kept its word to embrace the decision. At issue was a tiny island in the Chobe River on Botswana's northern border. An 1890 treaty between colonial rulers Great Britain and Germany had described the border at the disputed point vaguely, as the river's "main channel." The court took the course of the deepest channel to mark the agreed boundary, giving Botswana title to the 1.4-square-mile (3.5-sq. km.) territory.

Historical Origins of Geopolitics

Although political geography was a well-established field by the late nineteenth century, geopolitics was just beginning to emerge as a field of study and political analysis at the end of the century. In 1896 the German theorist Friedrich Ratzel published his *Political Geography*, which put forward the idea of the state as territory occupied by a people bound together by an idea of the state. Ratzel's theory embraced Social Darwinist notions that justified the current boundaries of nations. Ratzel viewed the state as a biological organism in competition for land with other states. The ethical implication of his theory seemed to be that "might makes right."

That theme set the stage for later German geopolitical thought, especially the notion of the need for *Lebensraum* (living room)—space into which the people of a nation could expand. German dictator Adolf Hitler justified his attack on Russia during World War II partly upon his claim that the German people needed more *Lebensraum* to the east. To some modern geographers, the use of geopolitical theories to serve German fascism and to justify other instances of military aggression tarnished geopolitics itself as a field of study.

Historical Development of Geopolitics

Modern geopolitics has further origins in the work of the Scottish geographer Sir Halford John Mackinder. In 1904 he published a seminal article, "The Geographical Pivot of History," in which he argued that the world is made up of a Eurasian "heartland" and a secondary hinterland (the remainder of the world), which he called the "marginal crescent." According to his theory, international politics is the struggle to gain control of the heartland. Any state that managed that feat would dominate the world.

A major proposition of Mackinder's theory was that geographical factors are not merely causative factors, but coercive. He tried to describe the physical features of the world that he believed directed human actions. In his view, "Man and not nature initiates, but nature in large measure controls." Geopolitical factors were therefore to a great extent determinants of the behavior of states. If this were true, geopolitics as a science could have deep relevance and corresponding influence among governments.

After Mackinder's time, the concept of geopolitics had a double significance. On the one hand it was a purely descriptive theory of geographic causation in history. On the other hand, its purveyors also believed, as Mackinder argued in 1904, that geopolitics has "a practical value as setting into perspective some of the competing forces in current international politics." Mackinder sought to promote this field of study as a companion to British statecraft, a tool to further Britain's national interest. By extension, geopolitical theory could assist any government in forming its political/military strategy.

As applied to the early twentieth-century world of international politics, however, Mackinder's theory had major weaknesses. Among his most glaring oversights were his failure to appreciate the rise of the United States, which attained considerable naval power after the turn of the century. Also, he failed to foresee the crucial strategic role that air power would play in warfare—and with it the immense change that air power could make in geopolitical considerations. Air power moves continents closer together, revolutionizing their geopolitical relationships.

One of Mackinder's chief critics was Nicolas John Spykman. Spykman argued that Mackinder had overvalued the potential economic, and therefore political, power of the Eurasian heartland, which could never reach its full potential because it could not overcome the obstacles to internal transportation. Moreover, the weaknesses of the remainder of the world—in effect, northern, western and southern Europe—could be overcome through forging alliances.

The dark side of geopolitical thought as handmaiden to political and military strategy became apparent in the Germany of the 1920s. At that time German theorists sought the resurrection of a German state broken by failure in World War I, the harsh terms of the Versailles Treaty that ended the war, and the hyperinflation that followed, wiping out the German middle class. In his 1925 article "Why Geopolitik?" Karl Haushofer urged the practical applications of *Geopolitik.* He urged that this form of analysis had not only "come to stay" but could also form important services for German political leaders, who should use all available tools "to carry on the fight for Germany's existence."

Haushofer ominously suggested that the "struggle" for German existence was becoming increasingly difficult because of the growth of the country's population. A people, he wrote, should study the living spaces of other nations so it could be prepared to "seize any possibility to recover lost ground." This discussion clearly implied that, from geopolitical necessity, Germany should seek additional territory to feed itself—a view carried into effect by Hitler in his quest for *Lebensraum* in attacking the Soviet Union, including its wheat-producing breadbasket, the Ukraine.

After World War II, a chastened Haushofer sought to soft-pedal both the direction and influence of his prewar writings. However, Hitler's morally heinous use of *Geopolitik* left geopolitical theorizing permanently tainted, in some eyes. Nevertheless, there is no necessary connection between geopolitics as a purely analytic description and geopolitics as the basis for a selfish search for power and advantage.

Geopolitics in the Twenty-first Century

Geopolitical considerations were unquestionably of profound relevance to the principal states of the post-World War II Cold War period. After the fall of the Berlin Wall in 1989, however, some theorists thought that the age of geopolitics had passed. In 1990 American strategic theorist Edward N. Luttwak, for example, argued that the importance of military power in international affairs had declined precipitously with the winding down of the Cold War. Military power had been overtaken in significance by economic prowess. Consequently, geopolitics had been eclipsed by what Luttwak called "geoeconomics," the waging of geopolitical struggle by economic means.

The view of Luttwak and various geographers of the declining significance of military power and geopolitical analysis, however, was soon proved to be overdrawn by events. As early as the first months of 1991, before the Soviet Union was officially dismantled, military power asserted itself as a key determinant on the international scene. Led by the United States, a far-flung alliance of nations participated in a war to remove Iraqi dictator Saddam Hussein's forces from neighboring Kuwait, which Iraq had illegally occupied. The decisive and successful use of military power in that war dramatically disproved assertions of its growing irrelevance.

Similarly, in the first three decades of the twenty-first century, military power retained its preeminence in the dynamics of international politics, even as economic forces were seen to gather momentum. To states throughout Asia and the West (especially Western Europe and the United States), the relative military capability of potential adversaries, and therefore geopolitics, remained a vital feature of the international order. Central to this view of the world scene is the growing military rivalry of the United States and China in East Asia. As China modernizes and expands its nuclear and conventional forces, it may feel itself capable of challenging America's predominant military power and prestige in East Asia. This possibility heightens the use of geopolitical thinking, giving it currency in analyzing this emerging situation.

Geopolitics as Civilizational Clash

A sometimes controversial expression of geopolitical analysis has been offered by Samuel Huntington of Harvard University. In his *The Clash of Civilizations and the Remaking of World Order* (1996) Huntington constructs a theory to explain certain tendencies of international behavior. He divides the world into a number of cultural groupings, or "civilizations," and argues that the character of various international conflicts can best be explained as conflicts or clashes of civilizations. In his view, Western civilization differs from the civilization of Orthodox Christianity, with a variety of conflicts erupting between the two. An example is the attack by the North Atlantic Treaty Organization (NATO), the bastion of the West, on Serbia, which is part of the Orthodox East.

Huntington's other civilizations include Islamic, Jewish, Eastern Caribbean, Hindu, Sinic (Chinese), and Japanese. The clash between Israel and its neighbors, the struggle between Pakistan and India over Kashmir, the rivalries between the United States and China and between China and India, for example, can be viewed as civilizational conflicts. Huntington has stated, however, that his theory is not intended to explain all of the historical past, and he does not expect it to remain valid long into the future.

Charles F. Bahmueller

NATIONAL PARK SYSTEMS

The world's first national parks were established as a response to the exploitation of natural resources, disappearance of wildlife, and destruction of natural landscapes that took place during the late nineteenth century. Government efforts to preserve natural areas as parks began with the establishment of Yellowstone National Park in the United States in 1872 and were soon adopted in other countries, including Australia, Canada, and New Zealand.

While the preservation of nature continues to be an important benefit provided by national parks, worldwide increases in population and the pressures of urban living have raised public interest in setting aside places that provide opportunities for solitude and interaction with nature.

Because national parks have been established by nations with diverse cultural values, land resources, and management philosophies, there is no single definition of what constitutes a national park. In some countries, areas used principally for recreational purposes are designated as national parks; other countries emphasize preservation of outstanding scenic, geologic, or biological resources. The terminology used for national parks also varies among countries. For example, protected areas that are similar to national parks may be called reserves, preserves, or sanctuaries.

Diverse landscapes are protected within national parks, including swamps, river deltas, dune areas, mountains, prairies, tropical rainforests, temperate forests, arid lands, and marine environments. Individual parks within nations form networks that vary with respect to size, accessibility, function, and the type of natural landscapes preserved. Some national park areas are isolated and sparsely populated, such as Greenland National Park; others, such as Peak District National Park in Great Britain, contain numerous small towns and are easily accessible to urban populations.

The functions of national parks include the preservation of scenic landscapes, geological features, wilderness, and plants and animals within their natural habitats. National parks also serve as outdoor laboratories for education and scientific research and as reservoirs for genetic information. Many are components of the United Nations International Biosphere Reserve Program.

National parks also play important roles in preserving cultures, by protecting archaeological, cultural, and historical sites. The United Nations recognizes several national parks that possess important cultural attributes as World Heritage Sites. Tourism to national parks has become important to the economies of many developing nations, especially in Eastern and Southern Africa, India, Nepal, Ecuador, and Indonesia. Parks are sources of local employment and can stimulate improvements to transportation and other types of infrastructure while encouraging productive use of lands that are of marginal agricultural use.

The International Union for Conservation of Nature has developed a system for classifying the world's protected areas, with Category II areas designated as national parks. Using this definition, there are 3,044 national parks in the world, with a mean average size of 457 square miles (1,183 sq. km.) each. Together, they cover an area of about 1.5 million square miles (4 million sq. km.), accounting for about 2.7 percent of the total land area on Earth.

STEPHEN T. MATHER AND THE U.S. NATIONAL PARK SERVICE

In 1914 businessman and conservationist Stephen T. Mather wrote to Secretary of the Interior Franklin K. Lane about the poor condition of California's Yosemite and Sequoia National Parks. Lane wrote back, "if you don't like the way the national parks are being run, come on down to Washington and run them yourself." Mather accepted the challenge and became an assistant to Lane and later the first director of the U.S. National Park Service, from 1917 to 1929.

North America

In 1916 management of U.S. national parks and monuments was shifted from the U.S. Army to the newly established National Park Service (NPS). The system has since grown in size to protect sixty-one national parks, as well as other natural areas including national monuments, seashores, and preserves.

North America's second-largest system of national parks is Parks Canada, created in 1930. Among the best-known Canadian parks is Banff, established in southern Alberta in 1885. Preserved within this area are glacially carved valleys, evergreen forests, and turquoise lakes. Parks Canada has the goal of protecting representative examples of each of Canada's vegetation and physiographic regions.

Mexico began providing protection for natural areas in the late nineteenth century. Among its system of sixty-seven national parks is Dzibilchaltún, an important Mayan archaeological site on the Yucatán Peninsula. With fewer resources available for park management, the emphasis in Mexico remains the preservation of scenic beauty for public use.

South America

Two of South America's best-known national parks are located within Argentina's park system. Nahuel Huapi National Park preserves two rare deer species of the Andes, while Iguazú National Park, located on the border with Brazil, is home to tapir, ocelot, and jaguar.

Located on a plateau of the western slope of the Andes Mountains in Chile, Lauca National Park is one of the world's highest parks, with an average elevation of more than 14,000 feet (4,267 meters)-an altitude nearly as high as the tallest mountains in the continental United States. Huascarán, another mountain park located in western Peru, boasts twenty peaks that exceed 19,000 feet (5,791 meters) in elevation. The volcanic islands of Galapagos Islands National Park, managed by Ecuador, have been of interest to biologists since British naturalist Charles Darwin studied variation and adaptation in animal species there in 1835.

Australia and New Zealand

Established in 1886, Royal was Australia's first national park. Perhaps better known to tourists, Uluru National Park in Australia's Northern Territory protects two rock domes, Ayer's Rock and Mount Olga, that rise above the plains 15 miles (40 km.) apart.

Along with Australia and other former colonies of Great Britain, New Zealand was a leader in establishing early national parks. The first of these was Tongagiro, created in 1887 to protect sacred lands of the Maori people on the North Island. New Zealand's South Island features several national parks including Fiordland, created in 1904 to preserve high mountains, forests, rivers, waterfalls, and other spectacular features of glacial origin.

Africa

Game poaching continues to be a severe problem in Africa, where animals are slaughtered for ivory, meat, and hides. Many African national parks were established to protect large game. South Africa's national park system began in 1926, when the Sabie Game Preserve of the eastern Transvaal region became Kruger National Park. Among South Africa's greatest attractions to foreign visitors, Kruger is famous for its population of lions and elephants.

East Africa is also known for outstanding game sanctuaries, such as Serengeti National Park, created prior to Tanzania's independence from Great Britain. Another national park in Tanzania, Kilimanjaro, protects Africa's highest and best-known mountain. Other African countries with well-developed park systems include Kenya, the Democratic Republic of the Congo (formerly Zaire), and Zambia. Although there is now a network of national parks in Africa that protects a wide range of habitats in various regions, there remains a need to protect additional areas in the arid northern part of the continent that includes the Sahara Desert.

Europe

In comparison with the United States, the national park concept spread more slowly within Europe. In 1910 Germany set aside Luneburger Heide National Park near the Elbe River, and in 1913, Sweden established Sarek, Stora Sjöfallet, Peljekasje, and Abisko National Parks. Swiss National Park was founded in Switzerland in 1914, in the Lower Engadine region. Great Britain has several national parks, including Lake District, a favorite recreation destination for English poet William Wordsworth. Spain's Doñana National Park, located on its southwestern coast, preserves the largest dune area on the European continent.

Asia

The system of land tenure and rural economy in many Asian countries has made it difficult for national governments to set aside large areas free from human exploitation. Many national parks established by colonial powers prior to World War II were maintained or expanded by countries following independence. For example, Kaziranga National Park is a refuge for the largest heard of rhinoceros in India. Established in 1962, Thailand's Khao Yai National Park protects a sample of the country's wildlife, while Indonesia's Komodo Island National Park preserves the habitat for the large lizards known as Komodo dragons.

In Japan, high population density has made it difficult to limit human activities within large areas. Some Japanese national parks are principally recreation areas rather than wildlife sanctuaries and may contain cultural features such as Shinto shrines. One of the best known national parks in Japan is Fuji-Hakone-Izu, which contains world-famous Mount Fuji, a volcano with a nearly symmetrical shape.

The Future

National parks serve as relatively undisturbed enclaves that protect examples of the world's most outstanding natural and cultural resources. The movement to establish these areas is a relatively recent attempt to achieve an improved balance between human activities and the earth. In recent years, rising incomes and lower costs for international travel have improved the accessibility of national parks to a larger number of persons, meaning that park visitation is likely to continue to rise.

Thomas A. Wikle

Boundaries and Time Zones

International Boundaries

International boundaries are the marked or imaginary lines traversing natural terrain of land or water that mark off the territory of one politically organized society—a state or nation-state—from other states. In addition, states claim "air boundaries." While satellites circumnavigate the earth without nations' permission, airplanes and other air vessels that fly much lower must gain the permission of states over whose territory they travel.

The existence of international boundaries is a consequence of the "territoriality" that is a feature of modern human societies. All politically organized societies, except for nomadic tribes, claim to rule some exactly defined geographical territory. International boundaries provide the limits that define this territory.

International boundaries have ancient origins. For example, the oldest sections of the Great Wall of China date back to the Ch'in Dynasty of the second century BCE. The Roman Empire also maintained boundaries to its territories, such as Hadrian's Wall in the north of England, built by the Romans in 122 CE as a defensive barrier against marauders. In these and other ancient instances, however, there was little thought that borders must be exact.

The existence of precisely drawn boundaries among states is relatively recent. The modern state has existed for no more than a few hundred years. In addition, means to determine many boundaries have come into existence only in the nineteenth and twentieth centuries, with the invention of scientific methods and instruments, along with accompanying vocabulary, for determining exact boundaries. The most basic terms of this vocabulary begin with "latitude" and "longitude" and their subdivisions into the "minutes" and "seconds" used in determining boundaries. In modern times, a new attitude toward states' territory was born, especially with the nineteenth century forms of nationalism, which tend to regard every acre of territory as sacred.

Types of Boundaries

There are several types of international boundaries. Some are geographical features, including rivers, lakes, oceans, and seas. Thus boundaries of the United States include the Great Lakes, which border Canada to the north; the Rio Grande, a river that forms part of the U.S. boundary with Mexico to the south; the Atlantic and Pacific Oceans, to the east and west, respectively; and the Gulf of Mexico, to the south. In Africa, Lake Victoria bounds parts of Tanzania, Uganda, and Kenya; and rivers, such as sections of the Congo and the Zambezi, form natural boundaries among many of the continent's states.

Other geographical features, such as mountains, often form international boundaries. The Pyrenees, for example, separate France and Spain and cradle the tiny state of Andorra. In South America, the Andes frequently serve as a boundary, such as between Argentina and Chile. The Himalayas in South Central Asia create a number of borders, such as between India, China, and Tibet and between Nepal, Butan, and their neighbors. When there are no clear geographical barriers between states, boundaries must be decided by mutual consent or the threat of force. In the 2016 presidential campaign, Donald Trump repeatedly called for a wall to be built between the United States and Mexico, claim-

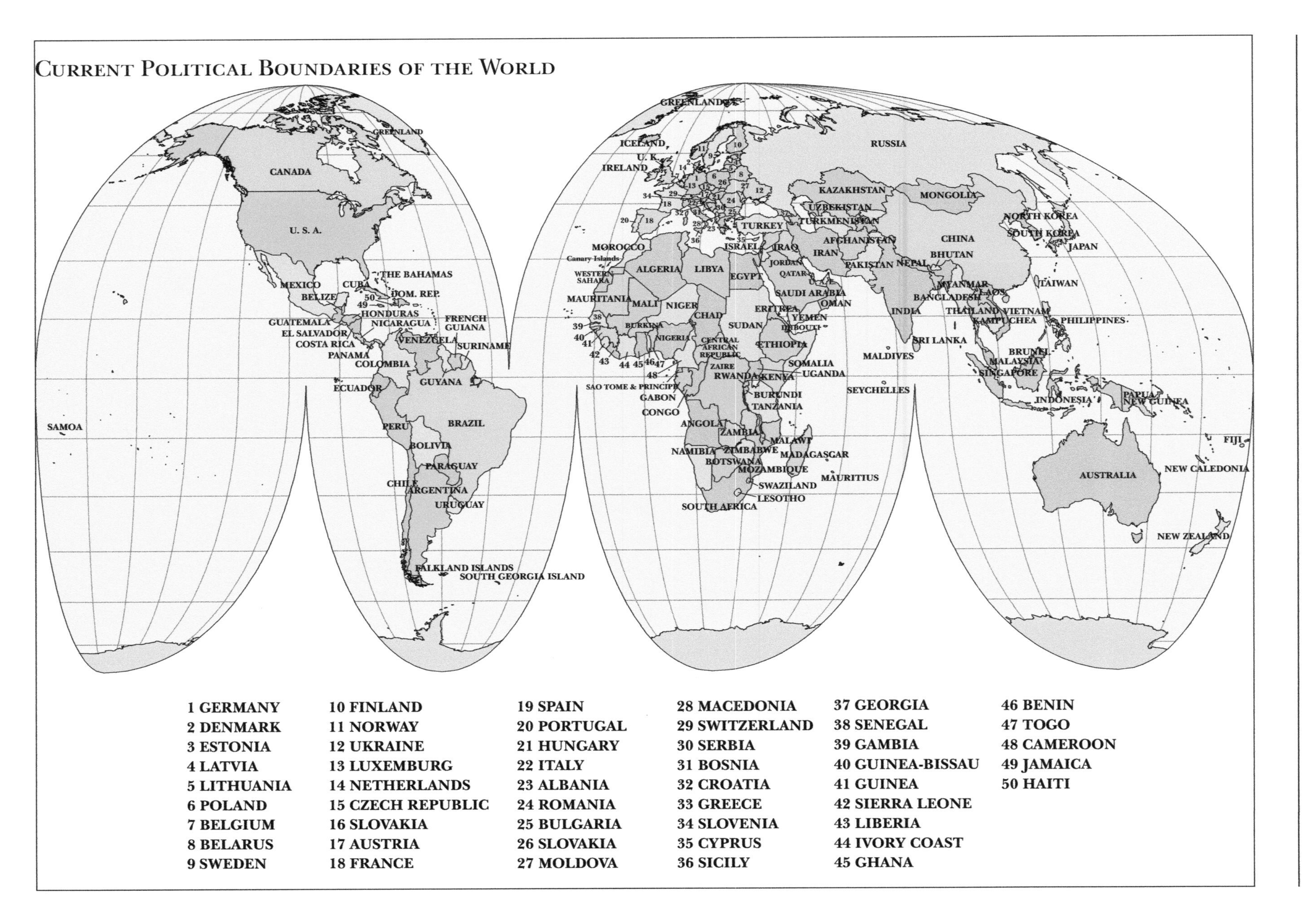
CURRENT POLITICAL BOUNDARIES OF THE WORLD
SAMOA
CANADA
U. S. A.
GREENLAND
MEXICO
BELIZE
GUATEMALA
EL SALVADOR
COSTA RICA
PANAMA
NICARAGUA
HONDURAS
CUBA
THE BAHAMAS
DOM. REP.
COLOMBIA
VENEZUELA
GUYANA
SURINAME
FRENCH GUIANA
ECUADOR
PERU
BRAZIL
BOLIVIA
PARAGUAY
CHILE
ARGENTINA
URUGUAY
FALKLAND ISLANDS
SOUTH GEORGIA ISLAND
ICELAND
IRELAND
U. K.
MOROCCO
Canary Islands
WESTERN SAHARA
MAURITANIA
MALI
ALGERIA
LIBYA
EGYPT
NIGER
CHAD
SUDAN
NIGERIA
BURKINA
CENTRAL AFRICAN REPUBLIC
SAO TOME & PRINCIPE
GABON
CONGO
ZAIRE
RWANDA
BURUNDI
KENYA
UGANDA
TANZANIA
ANGOLA
ZAMBIA
MALAWI
NAMIBIA
BOTSWANA
ZIMBABWE
MOZAMBIQUE
MADAGASCAR
SWAZILAND
LESOTHO
SOUTH AFRICA
ERITREA
ETHIOPIA
DJIBOUTI
SOMALIA
MAURITIUS
SEYCHELLES
TURKEY
ISRAEL
JORDAN
IRAQ
SAUDI ARABIA
QATAR
YEMEN
OMAN
IRAN
AFGHANISTAN
PAKISTAN
TURKMENISTAN
UZBEKISTAN
KAZAKHSTAN
RUSSIA
MONGOLIA
CHINA
NEPAL
BHUTAN
INDIA
BANGLADESH
MYANMAR
SRI LANKA
MALDIVES
THAILAND
LAOS
VIETNAM
KAMPUCHEA
MALAYSIA
SINGAPORE
BRUNEI
INDONESIA
PHILIPPINES
TAIWAN
NORTH KOREA
SOUTH KOREA
JAPAN
PAPUA NEW GUINEA
AUSTRALIA
NEW CALEDONIA
FIJI
NEW ZEALAND
1 GERMANY
2 DENMARK
3 ESTONIA
4 LATVIA
5 LITHUANIA
6 POLAND
7 BELGIUM
8 BELARUS
9 SWEDEN
10 FINLAND
11 NORWAY
12 UKRAINE
13 LUXEMBURG
14 NETHERLANDS
15 CZECH REPUBLIC
16 SLOVAKIA
17 AUSTRIA
18 FRANCE
19 SPAIN
20 PORTUGAL
21 HUNGARY
22 ITALY
23 ALBANIA
24 ROMANIA
25 BULGARIA
26 SLOVAKIA
27 MOLDOVA
28 MACEDONIA
29 SWITZERLAND
30 SERBIA
31 BOSNIA
32 CROATIA
33 GREECE
34 SLOVENIA
35 CYPRUS
36 SICILY
37 GEORGIA
38 SENEGAL
39 GAMBIA
40 GUINEA-BISSAU
41 GUINEA
42 SIERRA LEONE
43 LIBERIA
44 IVORY COAST
45 GHANA
46 BENIN
47 TOGO
48 CAMEROON
49 JAMAICA
50 HAITI

ing that Mexico would pay for it. As of 2019, the wall has not been completed, however.

Creation and Change of International Boundaries

War and conquest often have been used to determine borders. Such wars, however, historically have created hostility among losers. Political pressures to recover lost lands build up among aggrieved losers, and such irredentist claims provide fuel for future wars. A classic example is the fate of the regions of Alsace and Lorraine between France and Germany. Although natural resources in the form of coal played a substantial role in the dispute over this area, national pride was also a potent element.

Whether boundaries are fixed through compelling geographical imperatives or in their absence, states typically sign treaties agreeing to their location. These may be treaties that conclude wars, or boundary commissions set up by those involved may draw up borders to which states give formal agreement. In 1846, for example, negotiators for Great Britain and the United States settled on the forty-ninth parallel as the boundary between the western United States and Canada, although in the United States, "Fifty-four [degrees latitude] Forty [minutes] or Fight" had been a popular motto in the presidential election campaign of 1844.

Sometimes no accepted borders exist because of chronic hostility between states. Thus, maps of the Kashmir region between India and Pakistan, claimed by both countries, show only a "line of control" or cease-fire line to divide the two warring states. Similarly, only a cease-fire line, drawn at the armistice of the Korean War of 1950-1953, divides North and South Korea; a mutually agreed-upon border remains unfixed.

In rare instances, no true boundary exists to mark where a state's territory begins and ends. Classic cases are found on the Arabian Peninsula, where the land borders of principalities, known as the Gulf Sheikdoms, are vague lines in the sand. Such circumstances usually create no difficulties where nothing is at stake, but when oil is discovered, states must come to agreement or risk coming to blows.

In other instances, negotiations and international arbitration have been effective for determining borders. Perhaps the most important principle for determining the borders of newly created states is found in the Latin phrase, *Uti possidetis iurus*. This principle is used when states become independent after having been colonies or constituent parts of a larger state that has broken up. The principle holds that states shall respect the borders in place when they were colonies. *Uti possidetis* was first extensively used in South America in the nineteenth century, when European colonial powers withdrew, leaving several newly born states to determine their own boundaries. The principle may be used as a basis for border agreements among the fifteen states of the former Soviet Union.

Besides war and negotiation, purchase has sometimes been a means of creating international boundaries. For example, in 1853 the United States purchased territory from Mexico in the southwest; in 1867, it purchased Alaska from Russia.

In rare cases, natural boundaries may change naturally or be changed deliberately by one side, incurring resentment among victims. An example occurred in 1997, when Vietnam complained that China had built an embankment on a border river embankment that caused the river to change its course; China countered that Vietnam had built a dam altering the river's course.

Other border difficulties among states include conflicts over water that flows from one country to another. In the 1990s, for example, Mexico complained of excessive U.S. use of Colorado River waters and demanded adjustment.

Border Disputes

Border disputes among states in the past two centuries have been numerous and lethal. In the twentieth century, numerous such controversies degenerated into violence. In Asia, India and Pakistan fought over Kashmir, beginning in 1947-1949 and recurring in 1965 and 1999. China has been involved in violent border disputes with India, especially in 1962; Vietnam in 1979; and Russia in 1969. In South America, border wars between Ecuador and Peru broke out in 1941, 1981, and 1995. This dispute was settled by negotiation in 1998. In Africa, among numerous recent armed conflicts, the bloody border conflict between Eritrea and Ethiopia in the 1990s was notable.

Other recent disputes have ended peacefully. Eritrea avoided violence with Yemen over several Red Sea islands by accepting arbitration by an international tribunal. In 1995 Saudi Arabia and the United Arab Emirates negotiated a peaceful agreement to their border dispute involving oil rights.

As of 2019, there are four ongoing border conflicts: Israeli-Syrian ceasefire line incidents during the Syrian Civil War, the War in Donbass, India-Pakistan military confrontation, and the 2019 Turkish offensive into north-eastern Syria, code-named by Turkey as Operation Peace Spring. Many unresolved boundary disputes might yet lead to conflicts. Among the most complex is the multinational dispute over the 600 tiny Spratly Islands in the South China Sea. Uninhabited but potentially valuable because of oil, the Spratlys are claimed by China, Brunei, Malaysia, Indonesia, the Philippines, Taiwan, and Vietnam.

Border Policies

Problems with international borders are not limited to territorial disputes. Policies regarding how borders should be operated—including the key questions of who and what should be allowed entrance and exit under what conditions—can be expected to continue as long as independent states exist. While the members of the European Union have agreed to allow free passage of people and goods among themselves, this policy does not extent to nonmembers.

The most important purpose of states is to protect the lives and property of their citizens. One of the principal purposes of international boundaries is to further this purpose. Most states insist on controlling their borders, although borders seem increasingly porous. Given the imperatives of control and the increasing difficulties of maintaining it, issues surrounding international borders are expected to continue indefinitely in the twenty-first century.

Charles F. Bahmueller

Global Time and Time Zones

Before the nineteenth century, people kept time by local reckoning of the position of the Sun; consequently, thousands of local times existed. In medieval Europe, "hours" varied in length, depending upon the seasons: Each hour was determined by the Roman Catholic Church. In the sixteenth century, Holy Roman emperor Charles V was the first secular ruler to decree hours to be of equal length. As the industrial and scientific revolutions swept Europe, North America, and other areas, some form of time standardization became necessary as communities and regions increasingly interacted. In 1780 Geneva, Switzerland, was the first locality known to employ a standard time, set by the town-hall clockkeeper, throughout the town and its immediate vicinity.

The growth and expansion of railroads, providing the first relatively fast movement of people and goods from city to city, underscored the need for a standard system in Great Britain. As early as 1828, Sir John Herschel, Astronomer Royal, called for a national standard time system based on instruments at the Royal Observatory at Greenwich. That practice began in 1852, when the British telegraph system had developed sufficiently for the Greenwich time signals to be sent instantly to any point in the country.

As railroads expanded through North America, they exposed a problem of local time variation similar to that in Great Britain but on a far larger scale, since the distances between the East and West Coasts were much greater than in Great Britain. In order for long-distance train schedules to work, different parts of the country had to coordinate their clocks. The first to suggest a standard time framework for the United States was Charles F. Dowd, president of Temple Grove Seminary for Women in Saratoga Springs, New York. Initially, Dowd proposed putting all U.S. railroads on a single standard time, based on the time in Washington, D.C. When he realized that the time in California would be behind such a standard by almost four hours, he produced a revised system, establishing four time zones in the United States. Dowd's plan, published in 1870, included the first known map of a time zone system for the country.

Not everyone was happy with the designation of Washington, D.C., as the administrative center of time in the United States. Northeastern railroad executives urged that New York, the commercial capital of the nation, be used instead: Many cities and towns in the region already had standardized to New York time out of practical necessity. Dowd proposed a compromise: to set the entire national time zone system in the United States using the Greenwich prime meridian, already in use in many parts of the world for maritime and scientific purposes. In 1873 the American Association of Railways (AAR) flatly rejected the proposal.

In the end, Dowd proved to be a visionary. In 1878 Sandford Fleming, chief engineer of the government of Canada, proposed a worldwide system of twenty-four time zones, each fifteen degrees of longitude in width, and each bisected by a meridian, beginning with the prime meridian of Greenwich. William F. Allen, general secretary of the AAR and armed with a deep knowledge of railroad practices and politics, took up the crusade and persuaded the railroads to agree to a system. At noon on Sunday, November 18, 1883, most of the more than six hundred U.S. railroad lines dropped the fifty-three arbitrary times they had been using and adopted Greenwich-indexed meridians that defined the times in each of four times zones: eastern, central, mountain, and Pacific. Most major cities in the United States and Canada followed suit.

Time System for the World

Almost at the same time that American railroads adopted a standard time zone system, the State Department, authorized by the United States Congress, invited governments from around the world to assemble delegates in Washington, D.C., to adopt a global system. The International Meridian Conference assembled in the autumn of 1884, attended by representatives of twenty-five countries. Led by Great Britain and the United States, most favored adoption of Greenwich as the official prime meridian and Greenwich mean time as universal time.

There were other contenders: The French wanted the prime meridian to be set in Paris, and the Germans wanted it in Berlin; others proposed a mountaintop in the Azores or the tip of the Great Pyramid in Egypt. Greenwich won handily. The conference also agreed officially to start the universal day at midnight, rather than at noon or at sunrise, as practiced in many parts of the world. Each time zone in the world eventually came to have a local name, although technically, each goes by a letter in the alphabet in order eastward from Greenwich.

Once a global system was in place, there was a new issue: Many jurisdictions wanted to adjust their clocks for part of the year to account for differences in the number of hours of daylight between summer and winter months. In 1918 Congress decreed a system of daylight saving time for the United States but almost immediately abolished it, leaving state governments and communities to their local options. Daylight saving time, or a form of it, returned in the United States and many Allied nations during World War II. In the Uniform Time Act of 1966, Congress finally established a national system of daylight saving time, although with an option for states to abstain.

To the extent that it indicates how human communities want to manipulate time for social, political, or economic reasons, the issue of daylight saving time, rather than the establishment of a system of world time zones, is a better clue to the geographical issues involved in time administration. Both the history and the present format of the world time zone system show that the mathematically precise arrangement envisioned by many of the pioneers of time zones is not as important as things on the ground.

In the United States, the railroad time system adopted in 1883 drew the boundary between eastern time and central time more or less between the thirteen original states and the trans-Appalachian West: The entire Midwest, including Ohio, Indiana, and Michigan, fell in the central time zone. As the center of population migrated westward, train speeds increased, highways developed, and New York emerged as the center of mass media in the United States, the boundary between the eastern and central time zones marched steadily westward. In 1918 it ran down the middle of Ohio; by the 1960s, it was at the outskirts of Chicago.

One of the principal reasons for the popularity of Greenwich as the site of the prime meridian (zero degrees longitude), is that it places the international date line (180 degrees longitude)—where, in effect, time has to move forward to the next day rather than the next hour—far out in the Pacific Ocean where few people are affected by what otherwise would be an awkward arrangement. However, even this line is somewhat irregular, to avoid placing a small section of eastern Russia and some of the Aleutian Islands of the United States in different days.

Coordinated Universal Time Coordinated Universal Time (or UTC) is the primary time standard by which the world regulates clocks and time. It is within about 1 second of mean solar time at 0° longitude, and is not adjusted for daylight saving time. In some countries, the term Greenwich Mean Time is used. The co-ordination of time and frequency transmissions around the world began on January 1, 1960. UTC was first officially adopted as CCIR Recommendation 374, Standard-Frequency and Time-Signal Emissions, in 1963, but the official abbreviation of UTC and the official English name of Coordinated Universal Time (along with the French equivalent) were not adopted until 1967. UTC uses a *slightly* different second called the *SI second*. That is based on *atomic clocks*. Atomic clocks are more regular than the slightly variable Earth's rotation period. Hence, the essential difference between GMT and UTC is that they use different definitions of exactly how long one second of time is.

By 1950 most nations had adopted the universal time zone system, although a few followed later: Saudi Arabia in 1962, Liberia in 1972. Despite adhering to the system in principle, many nations take considerable liberties with the zones, especially if their territory spans several. All of Western Europe, despite covering an area equivalent to two zones, remains on a single standard. The People's Republic of China, which stretches across five different time zones, arbitrarily sets the entire country officially on Beijing time, eight hours behind Greenwich. Iran, Afghanistan, India, and Myanmar, each of which straddle time zone boundaries, operate on half-hour compromise systems as their time standards (as does Newfoundland). As late as 1978, Guyana's standard time was three hours, forty-five minutes in advance of Greenwich.

It can be argued that adoption of a worldwide system of time zones in the late nineteenth century was one of the earliest manifestations of the emergence of a global economy and society, and has been a crucial factor in the unfolding of this process throughout the twentieth century and beyond.

Ronald W. Davis

Global Education

Themes and Standards in Geography Education

Many people believe that the study of geography consists of little more than knowing the locations of places. Indeed, in the past, whole generations of students grew up memorizing states, capitals, rivers, seas, mountains, and countries. Most students found that approach boring and irrelevant. During the 1990s, however, geography education in the United States underwent a remarkable transformation.

While it remains important to know the locations of places, geography educators know that place name recognition is just the beginning of geographic understanding. Geography classes now place greater emphasis on understanding the characteristics of and the connections between places. Three things have led to the renewal of geography education: the five themes of geography, the national geography standards, and the establishment of a network of geographic alliances.

The Five Themes of Geography

One of the first efforts to move geography education beyond simple memorization was the National Geographic Society's publication of five themes of geography in 1984: location, place, human-environment interactions, movement, and regions. Not intended to be a checklist or recipe for understanding the world, these themes merely provided a framework for teachers—many of whom did not have a background in the subject—to incorporate geography throughout a social studies curriculum. The five themes were promoted widely by the National Geographic Society and are still used by some teachers to organize their classes.

Location is about knowing where things are. Both the absolute location (where a place is on earth's surface) and relative location (the connections between places) are important. The concept of place involves the physical and human characteristics that distinguish one place from another. The theme of human/environment interaction recognizes that people have relationships within defined places and are influenced by their surroundings. For example, many different types of housing have been created as adaptations to the world's diverse climates. The theme of movement involves the flow of people, goods, and ideas around the world. Finally, regions are human creations to help organize and understand Earth, and geography studies how they form and change.

The National Geography Standards

Geography was one of six subjects identified by President George H. W. Bush and the governors of the U.S. states when they formulated the National Education Goals in 1989. While the goals themselves foundered amid the political debate that followed their adoption, one tangible result of the initiative was the creation of Geography for Life: The National Geography Standards. More than 1,000 teachers, professors, business people, and government officials were involved in the writing of Geography for Life. The project was supported by four geography organizations: the American Geographical Society, the Association of American Geographers, the National Council for Geographic Education, and the National Geographic Society. The resulting book defines what every U.S. student

GEOGRAPHY STANDARDS

The geographically informed person knows and understands the following:

- how to use maps and other geographic representations, tools, and technologies to acquire, process, and report information from a spatial perspective;
- how to use mental maps to organize information about people, places, and environments in a spatial context;
- how to analyze the spatial organization of people, places, and environments on Earth's surface;
- the physical and human characteristics of places;
- that people create regions to interpret Earth's complexity;
- how culture and experience influence people's perceptions of places and regions;
- the physical processes that shape the patterns of Earth's surface;
- the characteristics and spatial distribution of ecosystems on Earth's surface;
- the characteristics, distribution, and migration of human populations on Earth's surface;
- the characteristics, distribution, and complexity of Earth's cultural mosaics;
- the patterns and networks of economic interdependence on Earth's surface;
- the processes, patterns, and functions of human settlement;
- how the forces of cooperation and conflict among people influence the division and control of Earth's surface;
- how human actions modify the physical environment;
- how physical systems affect human systems;
- the changes that occur in the meaning, use, distribution, and importance of resources;
- how to apply geography to interpret the past;
- how to apply geography to interpret the present and plan for the future.

Source: National Geography Standards Project. Geography for Life: National Geography Standards, Second Edition. Washington, D.C.: National Geographics Research and Exploration, 2012.

should know and be able to accomplish in geography.

Each of the eighteen standards is designed to develop students' geographic skills, including asking geographic questions; acquiring, organizing, and analyzing geographic information; and answering the questions. Each standard features explanations, examples, and specific requirements for students in grades four, eight, and twelve.

Geography Alliances and the Future of Geography Education

To publicize efforts in geography education, a network of geography alliances was established between 1986 and 1993. Today, each U.S. state has a geography alliance that links teachers and organizations such as the National Geographic Society and the National Council for Geographic Education to sponsor workshops, teacher training sessions, field experiences, and other ways of sharing the best in geographic teaching and learning.

A 2013 executive summary prepared by the National Geographic Society for the *Road Map for 21st Century Geography Education Project* continues to champion the goal of better geography education in K–12 schools. The Road Map Project represents the collaborative effort of four national organizations: the American Geographical Society (AGS), the Association of American Geographers (AAG), the National Council for Geographic Education (NCGE), and the National Geographic Society (NGS). The project partners share belief that geography education is essential for student success in all aspects of their adult lives—careers, civic lives, and personal decision making. It also is essential for the education of specialists who can help society addressing critical issues in the areas of social welfare, economic stability, environmental health, and international relations.

Eric J. Fournier

Global Data

World Gazetteer of Oceans and Continents

Places whose names are printed in SMALL CAPS *are subjects of their own entries in this gazetteer.*

Aden, Gulf of. Deep-water area between the Red and Arabian Seas, bounded by Somalia, Africa, on the south and Yemen on the north. Water is warmer and saltier in the Gulf of Aden than in the Red and Arabian Seas, because little water enters from rain or land runoff.

Africa. Second-largest continent, connected to Asia by the narrow isthmus of Suez. Bounded on the east by the Indian Ocean and on the west by the Atlantic Ocean. Countries of Africa are Algeria, Angola, Benin, Botswana, Burkina Faso, Burundi, Cameroon, Central African Republic, Chad, Congo, Côte d'Ivoire (Ivory Coast), the Democratic Republic of Congo, Egypt, Ethiopia, Gabon, Gambia, Ghana, Guinea, Kenya, Liberia, Libya, Madagascar, Malawi, Mali, Mauritania, Morocco, Mozambique, Namibia, Niger, Nigeria, Rio Muni (Mbini), Rwanda, Senegal, Sierra Leone, Somalia, South Africa, Sudan, Tanzania, Togo, Tunisia, Uganda, Western Sahara, Zambia, and Zimbabwe. Climate ranges from hot and rainy near the equator, to hot and dry in the huge Sahara Desert in the north and the Kalahari Desert in the south, to warm and fairly mild at the northern and southern extremes. Paleontological evidence indicates that humans originally evolved in Africa.

Agulhas Current. Warm, swift ocean current moving south along East Africa's coast. Part moves between Africa and Madagascar to form the Mozambique Current. The warm water of the

Oceans and Continents

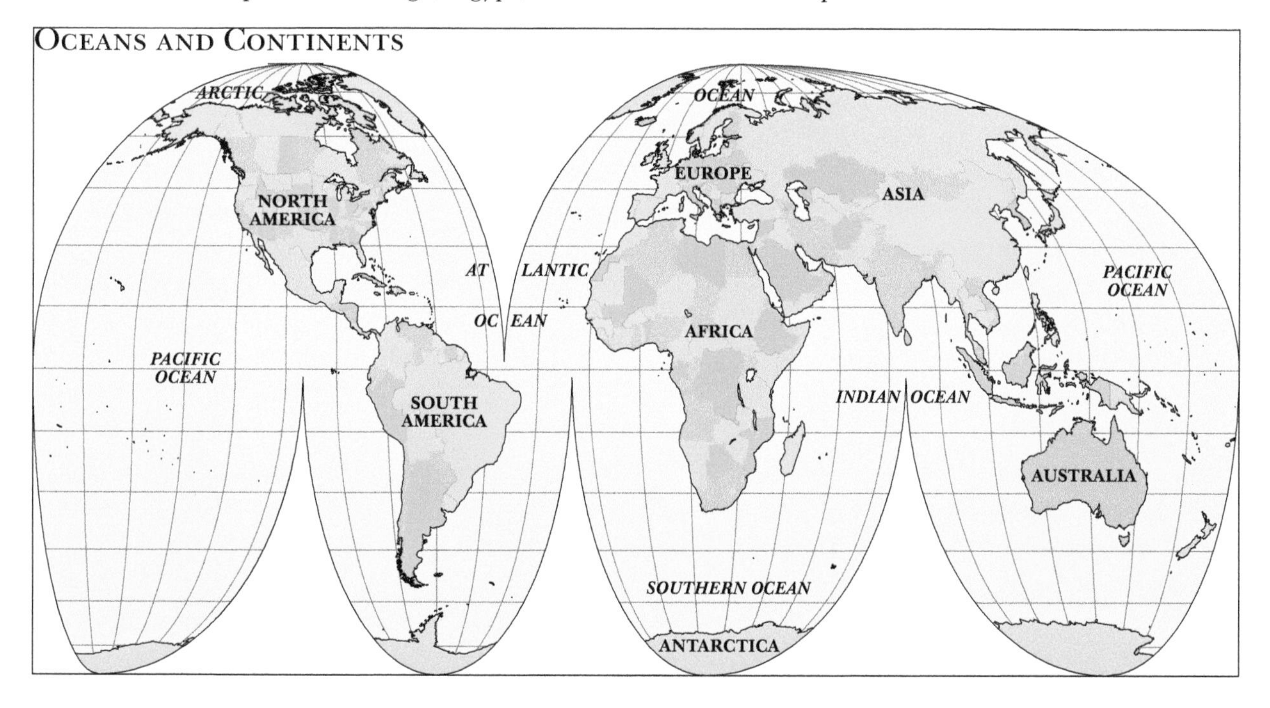

Agulhas Current increases the average temperatures in the eastern part of South Africa.

Agulhas Plateau. Relatively small ocean-bottom plateau that lies south of South Africa, at the area where the Indian and Atlantic Oceans meet.

Aleutian Islands. Chain of volcanic islands that extends 1,100 miles (1,770 km.) from the tip of the Alaska Peninsula to the Kamchatka Peninsula in Russia and forms the boundary between the North Pacific Ocean and the Bering Sea. The area is hazardous to navigation and has been called the "Home of Storms."

Aleutian Trench. Located on the northern margin of the Pacific Ocean, stretching 3,666 miles (5,900 km.) from the western edge of the Aleutian Island chain to Prince William Sound, Alaska. Depth is 25,263 feet (7,700 meters).

American Highlands. Elevated region on the Antarctic coast between Enderby Land and Wilkes Land, located far south of India. The Lambert and Fisher glaciers originate in the American Highlands and move down to feed the Amery Ice Shelf.

Amery Ice Shelf. Year-round shelf of relatively flat ice in a bay of Antarctica, located at approximately longitude 70 degrees east, between Mac. Robertson Land and the American Highlands. The ice shelf is fed by the Lambert and Fisher glaciers.

Amundsen Sea. Portion of the southernmost Pacific Ocean off the Wahlgreen Coast of Antarctica, approximately longitude 100 to 120 degrees west. Named for the Norwegian explorer Roald Amundsen, who became the first person to reach the South Pole in 1911.

Antarctic Circle. Latitude of 66.3 degrees south. South of this line, the Sun does not set on the day of the summer solstice, about December 22 in the Southern Hemisphere, and does not rise on the day of the winter solstice, about June 21.

Antarctic Circumpolar Current. Eastward-flowing current that circles Antarctica and extends from the surface to the deep ocean floor. The largest-volume current in the oceans. Extends northward to approximately 40 degrees south latitude and is driven by westerly winds.

Antarctic Convergence. Meeting place where cold Antarctic water sinks below the warmer sub-Antarctic water.

Antarctic Ocean. See Southern Ocean.

Antarctica. Fifth-largest continent, located at the southernmost part of the world. There are two major regions; western Antarctica, which includes the mountainous Antarctic peninsula, and eastern Antarctica, which is mostly a low continental shield area. An ice cap up to 13,000 feet (4,000 meters) thick covers 95 percent of the continent's surface. Temperatures in the austral summer (December, January, and February) rarely rise above 0 degrees Fahrenheit (-18 degrees Celsius) except on the peninsula. By international treaty, the continent is not owned by any single country, and human access is largely regulated. There has never been a self-supporting human habitation on Antarctica.

Arabian Sea. Portion of the Indian Ocean bounded by India on the east, Pakistan on the north, and Oman and Yemen of the Arabian Peninsula on the west.

Arctic Circle. Latitude of 66.3 degrees north. North of this line, the Sun does not set on the day of the summer solstice, about June 21 in the Northern Hemisphere, and does not rise on the day of the winter solstice, about December 22.

Arctic Ocean. World's smallest ocean. It centers on the geographic North Pole and connects to the Pacific Ocean through the Bering Sea, and to the Atlantic Ocean through the Greenland Sea. The Arctic Ocean is covered with ice up to 13 feet (4 meters) thick all year, except at its edges. Norwegian explorers on the ship *Fram* stayed locked in the icepack from 1893 to 1896, in order to study the movement of polar ice. They drifted in the ice a total of 1,028 miles (1,658 km.), from the Bering Sea to the Greenland Sea, proving that there was no land mass under the Arctic ice at the top of the world. Also known as Arctic Sea or Arctic Mediterranean Sea.

Argentine Basin. Basin on the floor of the western Atlantic Ocean, off the coast of Argentina in South America. Among ocean basins, this one is unusually circular.

Ascension Island. Isolated volcanic island in the South Atlantic Ocean, about midway between South America and Africa. One of the islands visited by British biologist Charles Darwin during his five-year voyage on the *Beagle*.

Asia. Largest continent; joins with Europe to form the great Eurasian landmass. Asia is bounded by the Arctic Ocean on the north, the western Pacific Ocean on the east, and the Indian Ocean on the south. Its countries include Afghanistan, Bahrain, Bangladesh, Bhutan, Cambodia, China, India, Iran, Iraq, Irian Jaya, Israel, Japan, Jordan, Kalimantan, Kazakhstan, North and South Korea, Kyrgyzstan, Laos, Lebanon, Malaysia, Myanmar, Mongolia, Nepal, Oman, Pakistan, the Philippines, Russia, Sarawak, Saudi Arabia, Sri Lanka, Sumatra, Syria, Tajikistan, Thailand, Asian Turkey, Turkmenistan, United Arab Emirates, Uzbekistan, Vietnam, and Yemen. Climates include virtually all types on earth, from arctic to tropical, desert to rainforest. Asia has the highest (Mount Everest) and lowest (Dead Sea) surface points in the world. Nearly 60 percent of the world's people live in Asia.

Atlantic Ocean. Second-largest body of water in the world, covering more than 25 percent of Earth's surface. Bordered by North and South America on the west, and Europe and East Africa on the east. The widest part (5,500 miles/8,800 km.) lies between West Africa and Mexico, along 20 degrees latitude. Scientists disagree on the north-south boundaries of the Atlantic; if one includes the Arctic Ocean and the Southern Ocean, the Atlantic Ocean extends about 13,300 miles (21,400 km.). The deepest spot (28,374 feet/8,648 meters) is found in the Puerto Rico Trench. The Atlantic Ocean has been a major route for trade and communications, especially between North America and Europe, for hundreds of years. This is because of its relatively narrow size and favorable currents, such as the Gulf Stream.

Australasia. Loosely defined term for the region, which, at the least, includes Australia and New Zealand; at the most, it also includes other South Pacific Islands in the region.

Australia. Smallest continent, sometimes called the "island continent." Located between the Indian and Pacific Oceans. It is the only continent occupied by a single nation, the Commonwealth of Australia. Australia is the flattest and driest continent; two-thirds is either desert or semiarid. Geologically, it is the oldest and most isolated continent. Unlike any other place on Earth, large mammals never evolved in Australia. Marsupials (pouched, warm-blooded animals) and unusual birds developed in their place.

Azores. Archipelago (group of islands) in the eastern Atlantic Ocean lying about 994 miles (1,600 km.) west of Portugal. The islands are of volcanic origin and have been known, fought over, and used by the Europeans since before the fourteenth century. Spanish explorer Christopher Columbus stopped in the Azores to wait for favorable winds before his first trip across the Atlantic Ocean.

Barents Sea. Partially enclosed section of the Arctic Ocean, bounded by Russia and Norway on the south and the Russian island of Navaya Zemlaya on the east. The Barents Sea was important in World War II because Allied convoys had to cross it, through storms and submarine patrols, to deliver war supplies to Murmansk, the only ice-free port in western Russia. It was named for the Dutch explorer Willem Barents.

Bays. See under individual names.

Beaufort Sea. Area of the Arctic Ocean located off the northern coast of Alaska and western Canada. It is usually frozen over and has no islands. Named for British admiral Sir Francis Beaufort, who devised the Beaufort Wind Scale as a means of classifying wind force at sea.

Bengal, Bay of. Northeast arm of the Indian Ocean, bounded by India on the west and Myanmar on the east. The Ganges River empties into the Bay of Bengal. The great ports of Calcutta and Madras in India, and Rangoon in Myanmar lie in the bay, making it a busy and important area for shipping for centuries.

Benguela Current. Northward-flowing current along the western coast of Southern Africa. Normally, the Benguela Current carries cold, rich water that wells up from the ocean depths and supports a large fishing industry. A change in winds can reduce the oxygen supply and kill huge numbers of fish, similar to what may happen off the coast of Peru during El Niño weather conditions.

Bering Sea. Portion of the northernmost Pacific Ocean that is bounded by the state of Alaska on the east, Russia and the Kamchatka Peninsula on the west, and the Bering Strait on the north. It is a valuable fishing ground, rich in shrimp, crabs, and fish. Whales, fur seals, sea otters, and walrus are also found there.

Bering Strait. Narrowest point of connection between the Bering Sea and the Arctic Ocean, located between the easternmost point of Siberia on the west and Alaska on the east. The Bering Strait is 52 miles (84 km.) wide. During the Ice Age, when the sea level was lower, humans and animals were able to walk from the Asian continent across a land bridge—now known as Beringia—to the North American continent across the frozen strait, providing the first human access to the Americas.

Bikini Atoll. Small atoll in the Marshall Islands group in the western Pacific Ocean. In the 1940s, the United States began testing nuclear bombs on Bikini and neighboring atolls. The U.S. Army removed the inhabitants of Bikini, and testing occurred from 1946 to 1958. The Bikini inhabitants were allowed to return in 1969, then removed again in 1978 when high levels of radioactivity were found to remain.

Black Sea. Large inland sea situated where southeastern Europe meets Asia; connected to the Mediterranean Sea through Turkey's Bosporus strait. The sea covers an area of about 178,000 square miles (461,000 sq. km.), with a maximum depth of more than 7,250 feet (2,210 meters).

Brazil Current. Extension of part of the warm, westward-flowing South Equatorial Current, which turns south to the coast of Brazil. The Brazil Current has very salty water because of its long flow across the equator. It joins the West Wind Drift and moves eastward across the South Atlantic Ocean as part of the South Atlantic Gyre.

California, Gulf of. Branch of the eastern Pacific Ocean that separates Baja California from mainland Mexico. Warm, nutrient-rich water supports a variety of fish, oysters, and sponges. California gray whales migrate to the gulf to give birth and breed, January through March. Fisheries and tourism are important industries in the Gulf of California. Also known as the Sea of Cortés.

California Current. Cool water that flows southeast along the western coast of North America from Washington State to Baja California. The eastern portion of the North Pacific Gyre.

Canada Basin. Part of the ocean floor that lies north of northeastern Canada and Alaska. The Beaufort Sea lies above the Canada Basin.

Cape Horn. Southernmost tip of South America. It is the site of notoriously severe storms and is hazardous to shipping.

Cape Verde Plateau. Atlantic Ocean plateau lying off the western bulge of the African continent. The volcanic Cape Verde Islands lie on the plateau.

Caribbean Sea. Portion of the western Atlantic Ocean bounded by Central and South America to the west and south, and the islands of the Antilles chain on the north and east. Mostly tropical in climate, the Caribbean Sea supports a large variety of plant and animal life. Its islands, including Puerto Rico, the Cayman Islands, and the Virgin Islands, are popular tourist sites.

Caspian Sea. World's largest inland sea. Located east of the Caucasus Mountains at Europe's south easternmost extremity, it dominates the expanses of western Central Asia. Its basin is 750 miles (1,200 kilometers) long, and its average width is 200 miles (320 kilometers). It covers 149,200 square miles (386,400 sq. km.).

Central America. Region generally understood to constitute the irregularly shaped neck of land linking North and South America, containing Belize, Guatemala, Honduras, El Salvador, Nicaragua, Costa Rica, and Panama.

Chukchi Sea. Portion of the Arctic Ocean, bounded by the Bering Strait on the south, Siberia on the southwest, and Alaska on the southeast. The Chukchi Sea is the area of exchange between waters and sea life of the Pacific and Arctic Oceans, and so is an area of interest to oceanographers and fishermen.

Clarion Fracture Zone. East-west-running fracture zone that begins off the west coast of Mexico and extends approximately 2.500 miles (4,023 km.) to the southwest.

Cocos Basin. Relatively small ocean basin located off the west coast of Sumatra in the northeast Indian Ocean.

Coral Sea. Area of the Pacific Ocean off the northeast coast of Australia, between Australia on the southwest, Papua New Guinea and the Solomon Islands on the northeast, and New Caledonia on the east. Site of a naval battle in 1942 that prevented the Japanese invasion of Australia.

Cortés, Sea of. See California, Gulf of.

Denmark Strait. Channel that separates Greenland and Iceland and connects the North Atlantic Ocean with the Arctic Ocean.

Dover, Strait of. Body of water between England and the European continent, separating the North Sea from the English Channel. It is 33 miles (53 km.) wide at its narrowest point. The tunnel between England and France (known as the "Chunnel") was cut into the rock under the Strait of Dover.

Drake Passage. Narrow part of the Southern Ocean that connects the Atlantic and Pacific Oceans between the southern tip of South America and the Antarctic peninsula. Named for sixteenth-century English navigator and explorer Sir Francis Drake, who discovered the passage when his ship was blown into it during a violent storm. Also called Drake Strait.

East China Sea. Area of the western Pacific Ocean bounded by China on the west, the Yellow Sea on the north, and Japan on the northeast. Large oil deposits were found under the East China Sea floor in the 1980s.

East Pacific Rise. Broad, nearly continuous undersea mountain range that extends from the southern end of Baja California southward, then curves east near Antarctica. It is formed along the southeast side of the Pacific Plate and is part of the Ring of Fire, a nearly continuous ring of volcanic and tectonic activity around the rim of the Pacific Ocean. Also called East Pacific Ridge.

East Siberian Sea. Portion of the Arctic Ocean bounded by the Chukchi Sea on the east, Siberia on the south, and the Laptev Sea on the west. Much of the East Siberian Sea is covered with ice year-round.

Eastern Hemisphere. The half of the earth containing Europe, Asia, and Africa; generally understood to fall between longitudes 20 degrees west and 160 degrees east.

El Niño. Conditions—also known as El Niño-Southern Oscillation (ENSO) events—that occur every two to ten years and cause weather and ocean temperature changes off the coast of Ecuador and Peru. Most of the time, the Peru Current causes cold, nutrient-rich water to well up off the coast of Ecuador and Peru. During ENSO years, the cold upwelling is replaced by warmer surface water that does not support plankton and fish. Fisheries decline and seabirds starve. Climatic changes of El Niño can bring floods to normally dry areas and drought to wet areas. Effects can extend across North and South America, and to the western Pacific Ocean. During the 1990s, the ENSO event fluctuated but did not go completely away, which caused tremendous damage to fisheries and agriculture, storms and droughts in North America, and numerous hurricanes.

Emperor Seamount Chain. Largest known example of submerged underwater volcanic ridges, located in the northern Pacific Ocean and extending southward from the Kamchatka Peninsula in Russia for about 2,500 miles (4,023 km.).

Enderby Land. Section of Antarctica that lies between the Indian Ocean and the South Polar Plateau, east of Queen Maud Land. Enderby Land lies between approximately longitude 45 and 60 degrees east.

English Channel. Strait water separating continental France from Great Britain. Runs for roughly 350 miles (560 km.), from the Atlantic Ocean in the west to the Strait of Dover in the east.

Equatorial Current. Currents just north and south of the equator that flow from east to west. Equatorial currents are found in the Pacific and Atlantic Oceans. The equatorial currents and the trade winds, which move in the same direction, greatly aid oceangoing traffic.

Eurasia. Term for the combined landmass of Europe and Asia.

Europe. Sixth-largest continent, actually a large peninsula of the Eurasian landmass. Europe is densely populated and includes the countries of Albania, Andorra, Austria, Belarus, Belgium, Bulgaria, Bosnia-Herzegovina, Croatia, the Czech Republic, Denmark, Estonia, Finland, France, Germany, Greece, Hungary, Iceland, Ireland, Italy, Latvia, Lithuania, Macedonia, Malta, Moldova, Monaco, the Netherlands, Norway, Poland, Portugal, Romania, Slovakia, Spain, Switzerland, Turkey, and the United Kingdom (England, Northern Ireland, Scotland, and Wales). Climate ranges from near arctic in the north, to temperate and Mediterranean in the south.

Florida Current. Water moving northward along the east coast of Florida to Cape Hatteras, North Carolina, where it joins the Gulf Stream.

Fundy, Bay of. Large inlet on the North American Atlantic coast, northwest of Maine, separating New Brunswick and Nova Scotia in Canada. Renowned for having the largest tidal change in the world, more than 56 feet (17 meters).

Galápagos Islands. Located directly on the equator, 600 miles (965 km.) west of Ecuador. The islands are volcanic in origin and sit directly in the cold Peru Current, which cools the islands and creates unusual microclimates and fogs. The extreme isolation of the islands allowed unique species to develop. Biologist Charles Darwin visited the Galápagos in the 1830s, and the unusual organisms he observed helped him to conceive the theory of evolution.

Galápagos Rift. Divergent plate boundary extending between the Galápagos Islands and South America. The first hydrothermal vent community was discovered in 1977 in the Galápagos Rift. This unusual type of biological habitat is based on energy from bacteria that use heat and chemicals to make food, instead of sunlight.

Grand Banks. Portion of the northwest Atlantic Ocean southeast of Nova Scotia and Newfoundland. The Grand Banks are extremely rich fishing grounds, although in the 1980s and 1990s catches of cod, flounder, and many other fish dropped dramatically due to overfishing and pollution.

Great Barrier Reef. Largest coral reef in the world, lying in the Coral Sea off the east coast of Australia. The reef system and its small islands stretch for more than 1,100 miles (1,750 km.) and is difficult to navigate through. The reefs are home to an incredible variety of tropical marine life, including large numbers of sharks.

Greenland. Largest island in the world that is not rated as a continent; lies between the northernmost part of the Atlantic Ocean and the Arctic Ocean, northeast of the North American continent. About 90 percent of Greenland is permanently covered with an ice sheet and glaciers. Residents engage in limited agriculture, growing potatoes, turnips, and cabbages. Most people live along the southwest coast, where the climate is warmed by the North Atlantic current.

Greenland Sea. Body of water bounded by Greenland on the west, Iceland on the north, and Spitsbergen on the east. It is often ice-covered.

Guinea, Gulf of. Arm of the North Atlantic Ocean below the great bulge of West Africa.

Gulf Stream. Current of westward-moving warm water originating along the equator in the Atlantic Ocean. The mass of water moves along the east coast of Florida as the Florida Current, then turns in a northeasterly direction off North Carolina to become the Gulf Stream. The Gulf Stream flows northeast past Newfoundland and the western edge of the British Isles. The warmer water of the Gulf Stream moderates the climate of northwestern Europe, causing temperatures in winter to be several degrees warmer than in areas of North America at the same latitudes. The Gulf Stream decreases the time required for ships to travel from North America to Europe. This was an important factor in trade and communication in American Colonial times and has continued to be significant.

Gulfs. See under individual names.

Hatteras Abyssal Plain. Part of the floor of the northwest Atlantic Ocean Basin, east of North Carolina. It rises to form shallow sandbars around Cape Hatteras, which are a notorious navigational hazard. In the seventeenth and eighteenth centuries, so many ships were lost in the area that Cape Hatteras became known as "The Graveyard of the Atlantic."

horse latitudes. Latitude belts between 30 and 35 degrees north and south latitude, where winds are usually light and variable and the climate mostly hot and dry.

Humboldt Current. See Peru Current

Iceland. Island country bounded by the Greenland Sea on the north, the Norwegian Sea on the east, and the Atlantic Ocean on the south and west. Total area of 39,768 square miles (103,000 sq. km.). The nearest land mass is Greenland, 200 miles (320 km.) to the northwest. Situated on top of the northern part of the Atlantic Mid-Oceanic Ridge, it is characterized by major volcanic activities, geothermal springs, and glaciers.

Idzu-Bonin Trench. Ocean trench in the western Pacific Ocean, about 6,082 miles (9,810 km.) long and 2,624 feet (800 meters) deep.

Indian Ocean. Third-largest of the world's oceans, bounded by the continents of Africa to the west, Asia to the north, Australia to the east, and Antarctica to the south. Most of the Indian Ocean lies below the equator. It has an approximate area of 33 million square miles (76 million sq. km.) and an average depth of about 13,120 feet (4,000 meters). Its deepest point is 24,442 feet (7,450 meters), in the Java Trench. The Indian Ocean was the first major ocean to be used as a trade route, particularly by the Egyptians. About 600 BCE, the Egyptian ruler Necho sent an expedition into the Indian Ocean, and the ship circumnavigated Africa, probably the first time this feat was accomplished. Warm winds blowing over the northern part of the Indian Ocean from May to September pick up huge amounts of moisture, which falls on India and Sri Lanka as monsoons. Fishing is important and mostly is done by small, family boats. About 40 percent of the world's offshore oil production comes from the Indian Ocean.

Indonesian Trench. See Java Trench.

Intracoastal Waterway. Series of bays, sounds, and channels, part natural and part human-made, that extends along the eastern coast of the United States from the Delaware River in New Jersey, south to the tip of Florida, then around the west coast of Florida. It extends around the Gulf Coast to the Rio Grande in Texas. It runs 2,455 miles (3,951 km.) and is an important, protected route for commercial and pleasure boat traffic.

Japan, Sea of. Marginal sea of the western Pacific Ocean that is bounded by Japan on the east and the Russian mainland on the west. Its surface area is approximately 377,600 square miles (978,000 sq. km.). It has an average depth of 5,750 feet (1,750 meters) and a maximum depth of 12,276 feet (3,742 meters).

Japan Trench. Ocean trench approximately 497 miles (800 km.) long, beginning at the eastern edge of the Japanese islands and stretching southward toward the Mariana Trench. Depth is 27,560 feet (8,400 meters).

Java Sea. Portion of the western Pacific Ocean between the islands of Java and Borneo. The sea has a total surface area of 167,000 square miles (433,000 sq. km.) and a comparatively shallow average depth of 151 feet (46 meters).

Java Trench. Ocean trench in the Indian Ocean, 2,790 miles (4,500 km.) long and 24,443 feet (7,450 meters) deep. Also called the Indonesian Trench.

Kermadec Trench. Ocean trench approximately 930 miles (1,500 km.) long, located in the southwest Pacific Ocean, beginning northeast of New Zealand. It has a depth of 32,800 feet (10,000 meters). Its northern end connects with the Tonga Trench.

Kurile Trench. Ocean trench approximately 1,367 miles (2,200 km.) long along the northeast rim of the Pacific Ocean, beginning at the north end of the Japanese island chain and extending northeastward. Depth is 34,451 feet (10,500 meters).

Labrador Current. Cold current that begins in Baffin Bay between Greenland and northeastern Canada and flows southward. The Labrador Current sometimes carries icebergs into North Atlantic shipping channels; such an iceberg caused the famous sinking of the great passenger ship *Titanic* in 1912.

Laptev Sea. Marginal sea of the Arctic Ocean off the coast of northern Siberia. The Taymyr Peninsula bounds it on the west and the New Siberian Islands on the east. Its area is about 276,000 square miles (714,000 sq. km.). Its average depth is 1,896 feet (578 meters), and the greatest depth is 9,774 feet (2,980 meters).

Lord Howe Rise. Elevation of the floor of the western Pacific Ocean that lies between Australia and New Guinea and under the Tasman Sea.

Mac. Robertson Land. Land near the coast of Antarctica, located between the Indian Ocean and the south Polar Plateau, east of Enderby Land. Mac. Robertson Land lies between approximately longitude 60 and 65 degrees east.

Macronesia. Loose grouping of islands in the Atlantic Ocean that includes the Azores, Madeira, the Canary Islands and Cape Verde. The term is derived from Greek words meaning "large" and "island" and should not be confused with Micronesia, small islands in the central and North Pacific Ocean.

Madagascar. Large island nation, officially called the Malagasy Republic, located in the Indian Ocean about 200 miles from the southeast coast of Africa. Although geographically tied to the African continent, it has a culture more closely tied to those of France and Southeast Asia. Area is 226,657 square miles (587,042 sq. km.).

Magellan, Strait of. Waterway connecting the south Atlantic Ocean with the South Pacific. Ships passing through the strait, north of Tierra del Fuego Island, avoid some of the world's roughest seas around Cape Horn.

magnetic poles. The two points on the earth, one in the Northern Hemisphere and one in the Southern Hemisphere, which are defined by the internal magnetism of the earth. Each point attracts one end of a compass needle and repels the opposite end.

Malacca, Strait of. Relatively narrow passage (200 miles/322 kilometers wide) bordered by Malaysia and Sumatra and linking the South China Sea and the Java Sea. It is one of the most heavily traveled waterways in the world, with more than one thousand ships every week.

Mariana Trench. Lowest point on Earth's surface, with a maximum depth of 36,150 feet (11,022 meters) in the Challenger Deep. The Mariana Trench is located on the western margin of the Pacific Ocean southeast of Japan, and is approximately 1,584 miles (2,550 km.) long.

Marie Byrd Land. Section of Antarctica located at the base of the Antarctic peninsula and shaped like a large peninsula itself. It is bounded at its base by the Ross Ice Shelf and the Ronne Ice Shelf.

Mediterranean Sea. Large sea that separates the continents of Europe, Africa, and Asia. It takes its name from Latin words meaning "in the middle of land"—a reference to its nearly land-locked nature. Covers about 969,100 square miles (2.5 million sq. km.) and extends 2,200 miles (3,540 km.) from west to east and about 1,000 miles (1,600 km.) from north to south at its widest. Its greatest depth is 16,897 feet (5,150 meters).

Melanesia. One of three divisions of the Pacific Islands, along with Micronesia and Polynesia; located in the western Pacific. The name Melanesia, for "dark islands," was given to the area because of its inhabitants' dark skins

Mexico, Gulf of. Nearly enclosed arm of the western Atlantic Ocean, bounded by the states of Florida, Alabama, Mississippi, Louisiana, and Texas, and Mexico and the Yucatan Peninsula. Cuba is located in the gap between the Yucatan Peninsula and Florida. Most ocean water enters through the Yucatan passage and exits the Gulf of Mexico around the tip of Florida, becoming the Florida Current. Fisheries, tourism, and oil production are important activities.

Micronesia. One of three divisions of the Pacific Islands, along with Melanesia and Polynesia. Micronesia means "small islands." Micronesia's islands are mostly atolls and coral islands, but some are of volcanic origin. The more than 2,000 islands of Micronesia are located in the Pacific Ocean east of the Philippines, mostly north of the equator.

Mid-Atlantic Ridge. Steep-sided, underwater mountain range running down the middle of the Atlantic Ocean. Formed by the divergent boundaries, or region where tectonic plates are separating.

Mozambique Current. See Agulhas Current.

New Britain Trench. Ocean trench in the southwest Pacific Ocean, about 5,158 miles (8,320 km.) long and 2,460 feet (750 meters) deep.

New Hebrides Basin. Part of the Coral Sea, located east of Australia and west of the New Hebrides island chain. The basin contains volcanic islands, both old and recent.

New Hebrides Trench. Ocean trench in the southwest Pacific Ocean, about 5,682 miles (9,165 km.) long and 3,936 feet (1,200 meters) deep.

North America. Third-largest continent, usually considered to contain all land and nearby islands in the Western Hemisphere north of the Isthmus of Panama, which connects it to South America. The major mainland countries are Canada, the United States, Mexico, Guatemala, El Salvador, Honduras, Nicaragua, Costa Rica, and Panama. Island countries include the islands of the Caribbean Sea and Greenland. Climate ranges from arctic to tropical.

North Atlantic Current. Continuation of the Gulf Stream, originating near the Grand Banks off Newfoundland. It curves eastward and divides into a northern branch, which flows into the Norwegian Sea, a southern branch, which flows eastward, and a branch that forms the Canary Current and flows south along the coast of Europe.

North Atlantic Gyre. Large mass of water, located in the Atlantic Ocean in the Northern Hemisphere, that rotates clockwise. Warm water moves toward the pole and cold water moves toward the equator.

North Pacific Current. Eastward flow of water in the Pacific Ocean in the Northern Hemisphere. It originates as the Kuroshio Current and moves from Japan toward North America.

North Pacific Gyre. Large mass of water, located in the Pacific Ocean in the Northern Hemisphere, that rotates clockwise. Warm water moves toward the pole and cold water moves toward the equator.

North Pole. Northern end of the earth's geographic axis, located at 90 degrees north latitude and longitude zero degrees. The North Pole itself is located on the Polar Abyssal Plain, about 14,000 feet (4,000 meters) deep in the Arctic Ocean. U.S. explorer Robert Edwin is credited with being the first person to reach the North Pole, in 1909, although there is historical dispute over the claim. The North Pole is different from the North magnetic pole.

North Sea. Arm of the northeastern Atlantic Ocean, bounded by Great Britain on the west and Norway, Denmark, and Germany on the east and south. The North Sea is one of the great fishing areas of the world and an important source of oil.

Northern Hemisphere. The half of the earth above the equator.

Norwegian Sea. Section of the North Atlantic Ocean. Norway borders it on the east and Iceland on the west. A submarine ridge linking

Greenland, Iceland, the Faroe Islands, and northern Scotland separates the Norwegian Sea from the open Atlantic Ocean. Cut by the Arctic Circle, the sea is often associated with the Arctic Ocean to the north. Reaches a maximum depth of about 13,020 feet (3,970 meters).

Oceania. Loosely applied term for the large island groups of the central and South Pacific; sometimes used to include Australia and New Zealand.

Okhotsk, Sea of. Nearly enclosed area of the northwestern Pacific Ocean bounded by Russia's Kamchatka Peninsula on the east and Siberia on the west. It is open to the Pacific Ocean on the south side only through Japan and the Kuril Islands, a string of islands belonging to Russia.

Pacific Ocean. Largest body of water in the world, covering more than one-third of Earth's surface—an area of about 70 million square miles (181 million sq. km.), more than the entire land area of the world. At its widest point, between Panama in Central America and the Philippines, it stretches 10,700 miles (17,200 km.). It runs 9,600 miles (15,450 km.) from the Bering Strait in the north to Antarctica in the south. Bordered by North and South America in the east, and Asia and Australia in the west. The average depth is about 12,900 feet (3,900 meters). It contains the deepest point on Earth (36,150 feet/11,022 meters), in the Challenger Deep of the Mariana Trench, southwest of Japan. The Pacific Ocean bottom is more geologically varied than the Indian or Atlantic Oceans; it has more volcanoes, ridges, trenches, seamounts, and islands. The vast size of the Pacific Ocean was a formidable barrier to travel, communications, and trade well into the nineteenth century. However, evidence shows that people crossed the Pacific Ocean in rafts or canoes as early as 3,000 BCE.

Pacific Rim. Modern term for the nations of Asia and North and South America that border, or are in, the Pacific Ocean. Used mostly in discussions of economic growth.

Palau Trench. Ocean trench in the western Pacific Ocean, about 250 miles (400 km.) long and 26,425 feet (8,054 meters) deep.

Palmer Land. Section of Antarctica that occupies the base of the Antarctic peninsula.

Panama, Isthmus of. Narrow neck of land that joins Central and South America. In 1914 the Panama Canal was opened through the isthmus, creating a direct sea link between the Pacific Ocean and the Caribbean Sea. The canal stretches about 50 miles (80 km.) from Panama City on the Pacific to Colón on the Caribbean. More than 12,000 ships pass through the canal annually.

Persian Gulf. Large extension of the Arabian Sea that separates Iran from the Arabian Peninsula in the Middle East. It covers about 88,000 square miles (226,000 sq. km.) and is about 620 miles (1,000 km.) long and 125–185 miles (200–300 km.) wide.

Peru-Chile Trench. Ocean trench that runs along the eastern boundary of the Pacific Ocean, off the western edge of South America. It is 3,666 miles (5,900 km.) long and 26,576 feet (8,100 meters) deep.

Peru Current. Cold, broad current that originates in the southernmost part of the South Pacific Gyre and flows up the west coast of South America. Off the coast of Peru, prevailing winds usually push the warmer surface water to the west. This causes the nutrient-rich, colder water of the Peru Current to well up to the surface, which provides excellent feeding for fish. At times, the upwelling ceases and biological, economic, and climatic catastrophe can result in El Niño weather conditions. Also known as the Humboldt Current.

Philippine Trench. Ocean trench located on the western rim of the Pacific Ocean, at the eastern margin of the Philippine islands. It is about 870 miles (1,400 km.) long and 34,451 feet (10,500 meters) deep.

Polynesia. One of three main divisions of the Pacific Islands, along with Melanesia and Micronesia. The islands are spread through the central and South Pacific. Polynesia means "many islands." Mostly small, the islands are predominantly coral atolls, but some are of volcanic origin.

Puerto Rico Trench. Ocean trench in the western Atlantic Ocean, about 27,500 feet (8,385 meters) deep and 963 miles (1,550 km.) long.

Queen Maud Land. Section of Antarctica that lies between the Atlantic Ocean and the south Polar Plateau, between approximately longitude 15 and 45 degrees east.

Red Sea. Narrow arm of water separating Africa from the Arabian Peninsula. One of the saltiest bodies of ocean water on Earth, as a result of high evaporation and little freshwater input. It was used as a trade route for Mediterranean, Indian, and Chinese peoples for centuries before the Europeans discovered it in the fifteenth century. The Suez Canal was opened in 1869 between the Mediterranean Sea and the Red Sea, cutting the distance from the northern Indian Ocean to northern Europe by about 5,590 miles (9,000 km.). This greatly increased the economic and military importance of the Red Sea.

Ring of Fire. Nearly continuous ring of volcanic and tectonic activity around the margins of the Pacific Ocean.

Ross Ice Shelf. Thick layer of ice in the Ross Sea off the coast of Antarctica. The relatively flat ice is attached to and nourished by a continental glacier.

Ross Sea. Bay in the Southern Ocean off the coast of Antarctica, located south of New Zealand. Named for English explorer James Clark Ross, the first person to break through the Antarctic ice pack in a ship, in 1841.

St. Peter and St. Paul. Cluster of rocks showing above the surface of the Atlantic Ocean between Brazil and West Africa. Important landmarks in the days of slave ships.

Sargasso Sea. Warm, salty area of water located in the Atlantic Ocean south and east of Bermuda, formed from water that circulates around the center of the North Atlantic Gyre. Named for the seaweed, *Sargassum*, that floats on the surface in large amounts.

Scotia Sea. Area of the southernmost Atlantic Ocean between the southern tip of South America and the Antarctic peninsula. The area is known for severe storms.

Seas. See under individual names.

Siam, Gulf of. See Thailand, Gulf of.

South America. Fourth-largest continent, usually considered to contain all land and nearby islands in the Western Hemisphere south of the Isthmus of Panama, which connects it to North America. Countries are Argentina, Bolivia, Brazil, Chile, Colombia, Ecuador, French Guiana, Guyana, Paraguay, Peru, Suriname, Uruguay, and Venezuela. Climate ranges from tropical to cold, nearly sub-Antarctic.

South Atlantic Gyre. Large mass of water, located in the Atlantic Ocean in the Southern Hemisphere, that rotates counterclockwise. Warm water moves toward the pole and cold water moves toward the equator.

South China Sea. Portion of the western Pacific Ocean that lies along the east coast of China, Vietnam, and the southeastern part of the Gulf of Thailand. The eastern and southern edges are defined by the Philippine and Indonesian Islands.

South Equatorial Current. Part of the South Atlantic Gyre that is split in two by the eastern prominence of Brazil. One part moves along the northeastern coast of South America toward the Caribbean Sea and the North Atlantic Ocean; the other turns southward and forms the Brazil Current.

South Pacific Gyre. Large mass of water, located in the Pacific Ocean in the Southern Hemisphere, that rotates counterclockwise. Warm water moves toward the pole and cold water moves toward the equator.

South Pole. Southern end of the earth's geographic axis, located at 90 degrees south latitude and longitude zero degrees. The first person to reach the South Pole was Norwegian explorer Roald Amundsen, in 1911. The South Pole is different from the South magnetic pole.

Southeastern Pacific Plateau. Portion of the Pacific Ocean floor closest to South America.

Southern Hemisphere. The half of the earth below the equator.

Southern Ocean. Not officially recognized as one of the major oceans, but a commonly used term for water surrounding Antarctica and extending northward to 50 degrees south latitude. Also known as the Antarctic Ocean.

Straits. See under individual names.

Sunda Shelf. One of the largest continental shelves in the world, nearly 772,000 square miles (2 million sq. km.). Located in the Java Sea, South China Sea, and Gulf of Thailand. The area was above water in the Quaternary period, enabling large animals such as elephants and rhinoceros to migrate to Sumatra, Java, and Borneo.

Surtsey Island. Island formed by a volcanic explosion off the coast of Iceland in 1963. It is valuable to scientists studying how island flora and fauna develop and is a popular tourist site.

Tashima Current. See Tsushima Current.

Tasman Sea. Area of the Pacific Ocean off the southeast coast of Australia, between Australia and Tasmania on the west and New Zealand on the east. First crossed by the Morioris people sometime before 1300 CE Also called the Tasmanian Sea.

Tasmanian Sea. See Tasman Sea.

Thailand, Gulf of. Also known as the Gulf of Siam, inlet of the South China Sea, located between the Malay Archipelago and the Southeast Asian mainland. Bounded by Thailand, Cambodia, and Vietnam.

Tonga Trench. Ocean trench in the Pacific Ocean, northeast of New Zealand. It stretches for 870 miles (1,400 km.), beginning at the northern end of the Kermadec Trench. Depth is 32,810 feet (10,000 meters).

Tsushima Current. Warm current in the western Pacific Ocean that flows out of the Yellow Sea into the Sea of Japan in the spring and summer. Also called Tashima Current.

Walvis Ridge (Walfisch Ridge). Long, narrow undersea elevation near the southwestern coast of Africa, which extends about 1,900 miles (3,000 km.) in a southwesterly direction under the Atlantic Ocean.

Weddell Sea. Bay in the Southern Ocean bounded by the Antarctic peninsula on the west and a northward bulge of Antarctica on the east, stretching from approximately longitude 60 to 10 degrees west. One of the harshest environments on Earth; surface water temperatures stay near 32 degrees Fahrenheit (0 degrees Celsius) all year. The Weddell Sea was the site of much whaling and seal hunting in the nineteenth and twentieth centuries.

West Caroline Trench. See Yap Trench.

West Wind Drift. Surface portion of the Antarctic Circumpolar Current, driven by westerly winds. Often extremely rough; seas as high as 98 feet (30 meters) have been reported.

Western Hemisphere. The half of the earth containing North and South America; generally understood to fall between longitudes 160 degrees east and 20 degrees west.

Wilkes Land. Broad section near the coast of Antarctica, which lies south of Australia and east of the American Highlands. Wilkes Land is the nearest landmass to the South magnetic pole.

Yap Trench. Ocean trench in the western Pacific Ocean, about 435 miles (700 km.) long and 27,900 feet (8,527 meters) deep. Also called the West Caroline Trench.

Yellow Sea. Area of the Pacific Ocean bounded by China on the north and west and Korea on the east. Named for the large amounts of yellow dust carried into it from central China by winds and by the Yangtze, Yalu, and Yellow Rivers. Parts of the sea often show a yellow color from the dust.

Kelly Howard

World's Oceans and Seas

Name	Approximate Area		Average Depth	
	Sq. Miles	Sq. Km.	Feet	Meters
Pacific Ocean	64,000,000	165,760,000	13,215	4,028
Atlantic Ocean	31,815,000	82,400,000	12,880	3,926
Indian Ocean	25,300,000	65,526,700	13,002	3,963
Arctic Ocean	5,440,200	14,090,000	3,953	1,205
Mediterranean & Black Seas	1,145,100	2,965,800	4,688	1,429
Caribbean Sea	1,049,500	2,718,200	8,685	2,647
South China Sea	895,400	2,319,000	5,419	1,652
Bering Sea	884,900	2,291,900	5,075	1,547
Gulf of Mexico	615,000	1,592,800	4,874	1,486
Okhotsk Sea	613,800	1,589,700	2,749	838
East China Sea	482,300	1,249,200	617	188
Hudson Bay	475,800	1,232,300	420	128
Japan Sea	389,100	1,007,800	4,429	1,350
Andaman Sea	308,100	797,700	2,854	870
North Sea	222,100	575,200	308	94
Red Sea	169,100	438,000	1,611	491
Baltic Sea	163,000	422,200	180	55

MAJOR LAND AREAS OF THE WORLD

	Approximate Land Area		
Area	*Sq. Mi.*	*Sq. Km.*	*Percent of World Total*
World	57,308,738	148,429,000	100.0
Asia (including Middle East)	17,212,041	44,579,000	30.0
Africa	11,608,156	30,065,000	20.3
North America	9,365,290	24,256,000	16.3
Central America, South America, & Caribbean	6,879,952	17,819,000	8.9
Antarctica	5,100,021	13,209,000	8.9
Europe	3,837,082	9,938,000	6.7
Oceania, including Australia	2,967,966	7,687,000	5.2

Major Islands of the World

Island	Location	Area	
		Sq. Mi.	Sq. Km.
Greenland	North Atlantic Ocean	839,999	2,175,597
New Guinea	Western Pacific Ocean	316,615	820,033
Borneo	Western Pacific Ocean	286,914	743,107
Madagascar	Western Indian Ocean	226,657	587,042
Baffin	Canada, North Atlantic Ocean	183,810	476,068
Sumatra	Indonesia, northeast Indian Ocean	182,859	473,605
Hōnshū	Japan, western Pacific Ocean	88,925	230,316
Great Britain	North Atlantic Ocean	88,758	229,883
Ellesmere	Canada, Arctic Ocean	82,119	212,688
Victoria	Canada, Arctic Ocean	81,930	212,199
Sulawesi (Celebes)	Indonesia, western Pacific Ocean	72,986	189,034
South Island	New Zealand, South Pacific Ocean	58,093	150,461
Java	Indonesia, Indian Ocean	48,990	126,884
North Island	New Zealand, South Pacific Ocean	44,281	114,688
Cuba	Caribbean Sea	44,218	114,525
Newfoundland	Canada, North Atlantic Ocean	42,734	110,681
Luzon	Philippines, western Pacific Ocean	40,420	104,688
Iceland	North Atlantic Ocean	39,768	102,999
Mindanao	Philippines, western Pacific Ocean	36,537	94,631
Ireland	North Atlantic Ocean	32,597	84,426
Hōkkaidō	Japan, western Pacific Ocean	30,372	78,663
Hispaniola	Caribbean Sea	29,355	76,029
Tasmania	Australia, South Pacific Ocean	26,215	67,897
Sri Lanka	Indian Ocean	25,332	65,610
Sakhalin (Karafuto)	Russia, western Pacific Ocean	24,560	63,610
Banks	Canada, Arctic Ocean	23,230	60,166
Devon	Canada, Arctic Ocean	20,861	54,030
Tierra del Fuego	Southern tip of South America	18,605	48,187
Kyūshū	Japan, western Pacific Ocean	16,223	42,018
Melville	Canada, Arctic Ocean	16,141	41,805
Axel Heiberg	Canada, Arctic Ocean	15,779	40,868
Southampton	Hudson Bay, Canada	15,700	40,663

Countries of the World

			Area		Population Density	
Country	*Region*	*Population*	*Square Miles*	*Square Kilometers*	*Persons/ Sq. Mi.*	*Persons/ Sq. Km.*
Afghanistan	Asia	31,575,018	249,347	645,807	127	49
Albania	Europe	2,862,427	11,082	28,703	259	100
Algeria	Africa	42,545,964	919,595	2,381,741	47	18
Andorra	Europe	76,177	179	464	425	164
Angola	Africa	29,250,009	481,354	1,246,700	60	23
Antigua and Barbuda	Caribbean	104,084	171	442	609	235
Argentina	South America	44,938,712	1,073,518	2,780,400	41	16
Armenia	Europe	2,962,100	11,484	29,743	259	100
Australia	Australia	25,576,880	2,969,907	7,692,024	9	3
Austria	Europe	8,877,036	32,386	83,879	275	106
Azerbaijan	Asia	10,027,874	33,436	86,600	300	116
Bahamas	Caribbean	386,870	5,382	13,940	73	28
Bahrain	Asia	1,543,300	300	778	5,136	1,983
Bangladesh	Asia	167,888,084	55,598	143,998	3,020	1,166
Barbados	Caribbean	287,025	166	430	1,730	668
Belarus	Europe	9,465,300	80,155	207,600	119	46
Belgium	Europe	11,515,793	11,787	30,528	976	377
Belize	Central America	398,050	8,867	22,965	44	17
Benin	Africa	11,733,059	43,484	112,622	269	104
Bhutan	Asia	821,592	14,824	38,394	55	21
Bolivia	South America	11,307,314	424,164	1,098,581	26	10
Bosnia and Herzegovina	Europe	3,511,372	19,772	51,209	179	69
Botswana	Africa	2,302,878	224,607	581,730	10.4	4
Brazil	South America	210,951,255	3,287,956	8,515,767	64	25
Brunei	Asia	421,300	2,226	5,765	189	73
Bulgaria	Europe	7,000,039	42,858	111,002	163	63
Burkina Faso	Africa	20,244,080	104,543	270,764	194	75
Burundi	Africa	10,953,317	10,740	27,816	1,020	394
Cambodia	Asia	16,289,270	69,898	181,035	233	90
Cameroon	Africa	24,348,251	179,943	466,050	135	52
Canada	North America	37,878,499	3,855,103	9,984,670	10	4
Cape Verde	Africa	550,483	1,557	4,033	352	136
Central African Republic	Africa	4,737,423	240,324	622,436	21	8
Chad	Africa	15,353,184	495,755	1,284,000	31	12
Chile	South America	17,373,831	291,930	756,096	60	23
China, People's Republic of	Asia	1,400,781,440	3,722,342	9,640,821	376	145
Colombia	South America	46,103,400	440,831	1,141,748	105	40
Comoros	Africa	873,724	719	1,861	1,215	469

Country	*Region*	*Population*	*Area*		*Population Density*	
			Square Miles	*Square Kilometers*	*Persons/ Sq. Mi.*	*Persons/ Sq. Km.*
Costa Rica	Central America	5,058,007	19,730	51,100	256	99
Côte d'Ivoire (Ivory Coast)	Africa	25,823,071	124,680	322,921	207	80
Croatia	Europe	4,087,843	21,831	56,542	186	72
Cuba	Caribbean	11,209,628	42,426	109,884	264	102
Cyprus	Europe	864,200	2,276	5,896	381	147
Czech Republic	Europe	10,681,161	30,451	78,867	350	135
Dem. Republic of the Congo	Africa	86,790,567	905,446	2,345,095	96	37
Denmark	Europe	5,814,461	16,640	43,098	350	135
Djibouti	Africa	1,078,373	8,880	23,000	122	47
Dominica	Caribbean	71,808	285	739	251	97
Dominican Republic	Caribbean	10,358,320	18,485	47,875	559	216
Ecuador	South America	17,398,588	106,889	276,841	163	63
Egypt	Africa	99,873,587	387,048	1,002,450	258	100
El Salvador	Central America	6,704,864	8,124	21,040	826	319
Equatorial Guinea	Africa	1,358,276	10,831	28,051	124	48
Eritrea	Africa	3,497,117	46,757	121,100	75	29
Estonia	Europe	1,324,820	17,505	45,339	75	29
Eswatini (Swaziland)	Africa	1,159,250	6,704	17,364	174	67
Ethiopia	Africa	107,534,882	410,678	1,063,652	262	101
Fed. States of Micronesia	Pacific Islands	105,300	271	701	388	150
Fiji	Pacific Islands	884,887	7,078	18,333	124	48
Finland	Europe	5,527,405	130,666	338,424	41	16
France	Europe	67,022,000	210,026	543,965	319	123
Gabon	Africa	2,067,561	103,347	267,667	21	8
Gambia	Africa	2,228,075	4,127	10,690	539	208
Georgia	Europe	3,729,600	26,911	69,700	140	54
Germany	Europe	83,073,100	137,903	357,168	603	233
Ghana	Africa	30,280,811	92,098	238,533	329	127
Greece	Europe	10,724,599	50,949	131,957	210	81
Grenada	Caribbean	108,825	133	344	818	316
Guatemala	Central America	17,679,735	42,042	108,889	420	162
Guinea	Africa	12,218,357	94,926	245,857	129	50
Guinea-Bissau	Africa	1,604,528	13,948	36,125	114	44
Guyana	South America	782,225	83,012	214,999	9.3	3.6
Haiti	Caribbean	11,263,077	10,450	27,065	1,077	416
Honduras	Central America	9,158,345	43,433	112,492	210	81
Hungary	Europe	9,764,000	35,919	93,029	272	105
Iceland	Europe	360,390	39,682	102,775	9.1	3.5
India	Asia	1,357,041,500	1,269,211	3,287,240	1,069	413
Indonesia	Asia	268,074,600	735,358	1,904,569	365	141
Iran	Asia	83,096,438	636,372	1,648,195	131	50

Country	Region	Population	Area Square Miles	Area Square Kilometers	Population Density Persons/ Sq. Mi.	Population Density Persons/ Sq. Km.
Iraq	Asia	39,309,783	169,235	438,317	233	90
Ireland	Europe	4,921,500	27,133	70,273	181	70
Israel	Asia	9,141,680	8,522	22,072	1,073	414
Italy	Europe	60,252,824	116,336	301,308	518	200
Jamaica	Caribbean	2,726,667	4,244	10,991	642	248
Japan	Asia	126,140,000	145,937	377,975	865	334
Jordan	Asia	10,587,132	34,495	89,342	307	119
Kazakhstan	Asia	18,592,700	1,052,090	2,724,900	18	7
Kenya	Africa	47,564,296	224,647	581,834	212	82
Kiribati	Pacific Islands	120,100	313	811	383	148
Kuwait	Asia	4,420,110	6,880	17,818	642	248
Kyrgyzstan	Asia	6,309,300	77,199	199,945	83	32
Laos	Asia	6,492,400	91,429	236,800	70	27
Latvia	Europe	1,910,400	24,928	64,562	78	30
Lebanon	Asia	6,855,713	4,036	10,452	1,740	672
Lesotho	Africa	2,263,010	11,720	30,355	194	75
Liberia	Africa	4,475,353	37,466	97,036	119	46
Libya	Africa	6,470,956	683,424	1,770,060	9.6	3.7
Liechtenstein	Europe	38,380	62	160	622	240
Lithuania	Europe	2,793,466	25,212	65,300	111	43
Luxembourg	Europe	613,894	998	2,586	614	237
Madagascar	Africa	25,680,342	226,658	587,041	114	44
Malawi	Africa	17,563,749	45,747	118,484	383	148
Malaysia	Asia	32,715,210	127,724	330,803	256	99
Maldives	Asia	378,114	115	298	3,287	1,269
Mali	Africa	19,107,706	482,077	1,248,574	39	15
Malta	Europe	493,559	122	315	3,911	1,510
Marshall Islands	Pacific Islands	55,500	70	181	795	307
Mauritania	Africa	3,984,233	397,955	1,030,700	10.4	4
Mauritius	Africa	1,265,577	788	2,040	1,606	620
Mexico	North America	126,577,691	759,516	1,967,138	166	64
Moldova	Europe	2,681,735	13,067	33,843	205	79
Monaco	Europe	38,300	0.78	2.02	49,106	18,960
Mongolia	Asia	3,000,000	603,902	1,564,100	4.9	1.9
Montenegro	Europe	622,182	5,333	13,812	117	45
Morocco	Africa	35,773,773	172,414	446,550	207	80
Mozambique	Africa	28,571,310	308,642	799,380	93	36
Myanmar (Burma)	Asia	54,339,766	261,228	676,577	207	80
Namibia	Africa	2,413,643	318,580	825,118	7.5	2.9
Nauru	Pacific Islands	11,000	8	21	1,357	524
Nepal	Asia	29,609,623	56,827	147,181	521	201

Country	*Region*	*Population*	*Area*		*Population Density*	
			Square Miles	*Square Kilometers*	*Persons/ Sq. Mi.*	*Persons/ Sq. Km.*
Netherlands	Europe	17,370,348	16,033	41,526	1,083	418
New Zealand	Pacific Islands	4,952,186	104,428	270,467	47	18
Nicaragua	Central America	6,393,824	46,884	121,428	137	53
Niger	Africa	21,466,863	458,075	1,186,408	47	18
Nigeria	Africa	200,962,000	356,669	923,768	565	218
North Korea	Asia	25,450,000	46,541	120,540	546	211
North Macedonia	Europe	2,077,132	9,928	25,713	210	81
Norway	Europe	5,328,212	125,013	323,782	41	16
Oman	Asia	4,183,841	119,499	309,500	36	14
Pakistan	Asia	218,198,000	310,403	803,940	703	271
Palau	Pacific Islands	17,900	171	444	104	40
Panama	Central America	4,158,783	28,640	74,177	145	56
Papua New Guinea	Pacific Islands	8,558,800	178,704	462,840	47	18
Paraguay	South America	7,052,983	157,048	406,752	44	17
Peru	South America	32,162,184	496,225	1,285,216	65	25
Philippines	Asia	108,785,760	115,831	300,000	939	363
Poland	Europe	38,386,000	120,728	312,685	319	123
Portugal	Europe	10,276,617	35,556	92,090	290	112
Qatar	Asia	2,740,479	4,468	11,571	614	237
Republic of the Congo	Africa	5,399,895	132,047	342,000	41	16
Romania	Europe	19,405,156	92,043	238,391	210	81
Russia[1]	Europe/Asia	146,877,088	6,612,093	17,125,242	23	9
Rwanda	Africa	12,374,397	10,169	26,338	1,217	470
St Kitts and Nevis	Caribbean	56,345	104	270	541	209
St Lucia	Caribbean	180,454	238	617	756	292
St Vincent and Grenadines	Caribbean	110,520	150	389	736	284
Samoa	Pacific Islands	199,052	1,093	2,831	181	70
San Marino	Europe	34,641	24	61	1,471	568
Sahrawi Arab Dem. Rep.[2]	Africa	567,421	97,344	252,120	6	2.3
São Tomé and Príncipe	Africa	201,784	386	1,001	523	202
Saudi Arabia	Asia	34,218,169	830,000	2,149,690	41	16
Senegal	Africa	16,209,125	75,955	196,722	212	82
Serbia	Europe	6,901,188	29,913	77,474	231	89
Seychelles	Africa	96,762	176	455	552	213
Sierra Leone	Africa	7,901,454	27,699	71,740	285	110
Singapore	Asia	5,638,700	279	722.5	20,212	7,804
Slovakia	Europe	5,450,421	18,933	49,036	287	111
Slovenia	Europe	2,084,301	7,827	20,273	267	103
Solomon Islands	Pacific Islands	682,500	10,954	28,370	62	24
Somalia	Africa	15,181,925	246,201	637,657	62	24
South Africa	Africa	58,775,022	471,359	1,220,813	124	48

Country	Region	Population	Area		Population Density	
			Square Miles	Square Kilometers	Persons/ Sq. Mi.	Persons/ Sq. Km.
South Korea	Asia	51,811,167	38,691	100,210	1,339	517
South Sudan	Africa	12,575,714	248,777	644,329	52	20
Spain	Europe	46,934,632	195,364	505,990	241	93
Sri Lanka	Asia	21,803,000	25,332	65,610	860	332
Sudan	Africa	40,782,742	710,251	1,839,542	57	22
Suriname	South America	568,301	63,251	163,820	9.1	3.5
Sweden	Europe	10,344,405	173,860	450,295	59	23
Switzerland	Europe	8,586,550	15,940	41,285	539	208
Syria	Asia	17,070,135	71,498	185,180	238	92
Taiwan	Asia	23,596,266	13,976	36,197	1,689	652
Tajikistan	Asia	9,127,000	55,251	143,100	166	64
Tanzania	Africa	55,890,747	364,900	945,087	153	59
Thailand	Asia	66,455,280	198,117	513,120	335	130
Timor-Leste	Asia	1,167,242	5,760	14,919	202	78
Togo	Africa	7,538,000	21,853	56,600	344	133
Tonga	Pacific Islands	100,651	278	720	362	140
Trinidad and Tobago	Caribbean	1,359,193	1,990	5,155	683	264
Tunisia	Africa	11,551,448	63,170	163,610	183	71
Turkey	Europe/Asia	82,003,882	302,535	783,562	271	105
Turkmenistan	Asia	5,851,466	189,657	491,210	31	12
Tuvalu	Pacific Islands	10,200	10	26	1,020	392
Uganda	Africa	40,006,700	93,263	241,551	429	166
Ukraine[3]	Europe	41,990,278	232,820	603,000	180	70
United Arab Emirates	Asia	9,770,529	32,278	83,600	303	117
United Kingdom	Europe	66,435,600	93,788	242,910	708	273
United States	North America	330,546,475	3,796,742	9,833,517	87	34
Uruguay	South America	3,518,553	68,037	176,215	52	20
Uzbekistan	Asia	32,653,900	172,742	447,400	189	73
Vanuatu	Pacific Islands	304,500	4,742	12,281	65	25
Vatican City	Europe	1,000	0.17	0.44	5,887	2,273
Venezuela	South America	32,219,521	353,841	916,445	91	35
Vietnam	Asia	96,208,984	127,882	331,212	751	290
Yemen	Asia	28,915,284	175,676	455,000	166	64
Zambia	Africa	16,405,229	290,585	752,612	57	22
Zimbabwe	Africa	15,159,624	150,872	390,757	101	39

Notes: (1) Including the population and area of Autonomous Republic of Crimea and City of Sevastopol, Ukraine's administrative areas on the Crimean Peninsula, which are claimed by Russia; (2) Administration is split between Morocco and the Sahrawi Arab Democratic Republic (Western Sahara), both of which claim the entire territory; (3) Excludes Crimea.
Source: U.S. Census Bureau, International Data Base

Past and Projected World Population Growth, 1950-2050

Year	*Approximate World Population*	*Ten-Year Growth Rate (%)*
1950	2,556,000,053	18.9
1960	3,039,451,023	22.0
1970	3,706,618,163	20.2
1980	4,453,831,714	18.5
1990	5,278,639,789	15.2
2000	6,082,966,429	12.6
2010	6,848,932,929	10.7
2020	7,584,821,144	8.7
2030	8,246,619,341	7.3
2040	8,850,045,889	5.6
2050	9,346,399,468	—

Note: The listed years are the baselines for the estimated ten-year growth rate figures; for example, the rate for 1950-1960 was 18.9%.
Source: U.S. Census Bureau, International Data Base

WORLD'S LARGEST COUNTRIES BY AREA

			Area	
Rank	*Country*	*Region*	*Sq. Miles*	*Sq. Km.*
1	Russia	Europe/Asia	6,612,093	17,125,242
2	Canada	North America	3,855,103	9,984,670
3	United States	North America	3,796,742	9,833,517
4	China	Asia	3,722,342	9,640,821
5	Brazil	South America	3,287,956	8,515,767
6	Australia	Australia	2,969,907	7,692,024
7	India	Asia	1,269,211	3,287,240
8	Argentina	South America	1,073,518	2,780,400
9	Kazakhstan	Asia	1,052,090	2,724,900
10	Algeria	Africa	919,595	2,381,741
11	Democratic Rep. of the Congo	Africa	905,446	2,345,095
12	Saudi Arabia	Asia	830,000	2,149,690
13	Mexico	North America	759,516	1,967,138
14	Indonesia	Asia	735,358	1,904,569
15	Sudan	Africa	710,251	1,839,542
16	Libya	Africa	683,424	1,770,060
17	Iran	Asia	636,372	1,648,195
18	Mongolia	Asia	603,902	1,564,100
19	Peru	South America	496,225	1,285,216
20	Chad	Africa	495,755	1,284,000
21	Mali	Africa	482,077	1,248,574
22	Angola	Africa	481,354	1,246,700
23	South Africa	Africa	471,359	1,220,813
24	Niger	Africa	458,075	1,186,408
25	Colombia	South America	440,831	1,141,748
26	Bolivia	South America	424,164	1,098,581
27	Ethiopia	Africa	410,678	1,063,652
28	Mauritania	Africa	397,955	1,030,700
29	Egypt	Africa	387,048	1,002,450
30	Tanzania	Africa	364,900	945,087

Source: U.S. Census Bureau, International Data Base

World's Smallest Countries by Area

Rank	*Country*	*Region*	*Area*	
			Sq. Miles	*Sq. Km.*
1	Vatican City*	Europe	0.17	0.44
2	Monaco*	Europe	0.78	2.02
3	Nauru	Pacific Islands	8	21
4	Tuvalu	Pacific Islands	10	26
5	San Marino*	Europe	24	61
6	Liechtenstein*	Europe	62	160
7	Marshall Islands	Pacific Islands	70	181
8	Saint Kitts and Nevis	Central America	104	270
9	Maldives	Asia	115	298
10	Malta	Europe	122	315
11	Grenada	Caribbean	133	344
12	Saint Vincent and the Grenadines	Central America	150	389
13	Barbados	Caribbean	166	430
14	Antigua and Barbuda	Caribbean	171	442
15	Palau	Pacific Islands	171	444
16	Seychelles	Africa	176	455
17	Andorra*	Europe	179	464
18	Saint Lucia	Central America	238	617
19	Federated States of Micronesia	Pacific Islands	271	701
20	Tonga	Pacific Islands	278	720

Note: Asterisks (*) denote countries on continents; all other countries are islands or island groups.
Source: U.S. Census Bureau, International Data Base

World's Largest Countries by Population

Rank	*Country*	*Region*	*Population*
1	China	Asia	1,401,028,280
2	India	Asia	1,357,746,150
3	United States	North America	329,229,067
4	Indonesia	Asia	268,074,600
5	Pakistan	Asia	218,385,000
6	Brazil	South America	211,032,216
7	Nigeria	Africa	200,962,000
8	Bangladesh	Asia	167,983,726
9	Russia	Europe/Asia	146,877,088
10	Mexico	North America	126,577,691
11	Japan	Asia	126,020,000
12	Philippines	Asia	108,210,625
13	Ethiopia	Africa	107,534,882
14	Egypt	Africa	99,930,038
15	Vietnam	Asia	96,208,984
16	Democratic Rep. of the Congo	Africa	86,790,567
17	Germany	Europe	83,149,300
18	Iran	Asia	83,142,818
19	Turkey	Europe/Asia	82,003,882
20	France	Europe	67,060,000
21	Thailand	Asia	66,461,867
22	United Kingdom	Europe	66,435,600
23	Italy	Europe	60,252,824
24	South Africa	Africa	58,775,022
25	Tanzania	Africa	55,890,747
26	Myanmar	Asia	54,339,766
27	South Korea	Asia	51,811,167
28	Kenya	Africa	47,564,296
29	Spain	Europe	46,934,632
30	Colombia	South America	46,127,200

Source: U.S. Census Bureau, International Data Base

World's Smallest Countries by Population

Rank	*Country*	*Region*	*Population*
1	Vatican City	Europe	1,000
2	Tuvalu	Pacific Islands	10,200
3	Nauru	Pacific Islands	11,000
4	Palau	Pacific Islands	17,900
5	San Marino	Europe	34,641
6	Monaco	Europe	38,300
7	Liechtenstein	Europe	38,380
8	Marshall Islands	Pacific Islands	55,500
9	Saint Kitts and Nevis	Central America	56,345
10	Dominica	Caribbean	71,808
11	Andorra	Europe	76,177
12	Seychelles	Africa	96,762
13	Tonga	Pacific Islands	100,651
14	Antigua and Barbuda	Caribbean	104,084
15	Federated States of Micronesia	Pacific Islands	105,300
16	Grenada	Caribbean	108,825
17	Saint Vincent and the Grenadines	Central America	110,520
18	Kiribati	Pacific Islands	120,100
19	Saint Lucia	Central America	180,454
20	Samoa	Pacific Islands	199,052
21	São Tomé and Príncipe	Africa	201,784
22	Barbados	Caribbean	287,025
23	Vanuatu	Pacific Islands	304,500
24	Iceland	Europe	360,390
25	Maldives	Asia	378,114
26	Bahamas	Caribbean	386,870
27	Belize	Central America	398,050
28	Brunei	Asia	421,300
29	Malta	Europe	493,559
30	Cape Verde	Africa	550,483

Source: U.S. Census Bureau, International Data Base.

WORLD'S MOST DENSELY POPULATED COUNTRIES

				Area		*Persons Per Square*	
Rank	*Country*	*Region*	*Population*	*Sq. Miles*	*Sq. Km.*	*Mile*	*Km.*
1	Monaco	Europe	38,300	0.78	2.02	49,106	18,960
2	Singapore	Asia	5,638,700	279	722.5	20,212	7,804
3	Vatican City	Europe	1,000	0.17	0.44	5,887	2,273
4	Bahrain	Asia	1,543,300	300	778	5,136	1,983
5	Malta	Europe	493,559	122	315	3,911	1,510
6	Maldives	Asia	378,114	115	298	3,287	1,269
7	Bangladesh	Asia	167,983,726	55,598	143,998	3,021	1,167
8	Lebanon	Asia	6,855,713	4,036	10,452	1,740	672
9	Barbados	Caribbean	287,025	166	430	1,730	668
10	Taiwan	Asia	23,596,266	13,976	36,197	1,689	652
11	Mauritius	Africa	1,265,577	788	2,040	1,606	620
12	San Marino	Europe	34,641	24	61	1,471	568
13	Nauru	Pacific Islands	11,000	8	21	1,344	519
14	South Korea	Asia	51,811,167	38,691	100,210	1,339	517
15	Rwanda	Africa	12,374,397	10,169	26,338	1,217	470
16	Comoros	Africa	873,724	719	1,861	1,215	469
17	Netherlands	Europe	17,426,881	16,033	41,526	1,087	420
18	Haiti	Caribbean	11,263,077	10,450	27,065	1,077	416
19	Israel	Asia	9,149,500	8,522	22,072	1,074	415
20	India	Asia	1,357,746,150	1,269,211	3,287,240	1,070	413
21	Burundi	Africa	10,953,317	10,740	27,816	1,020	394
22	Tuvalu	Pacific Islands	10,200	10	26	1,015	392
23	Belgium	Europe	11,515,793	11,849	30,689	974	376
24	Philippines	Asia	108,210,625	115,831	300,000	934	361
25	Japan	Asia	126,020,000	145,937	377,975	862	333
26	Sri Lanka	Asia	21,803,000	25,332	65,610	860	332
27	El Salvador	Central America	6,704,864	8,124	21,040	826	319
28	Grenada	Caribbean	108,825	133	344	818	316
29	Marshall Islands	Pacific Islands	55,500	70	181	795	307
30	Saint Lucia	Central America	180,454	238	617	756	292

Source: U.S. Census Bureau, International Data Base

WORLD'S LEAST DENSELY POPULATED COUNTRIES

				Area		Persons Per Square	
Rank	*Country*	*Region*	*Population*	*Sq. Miles*	*Sq. Km.*	*Mile*	*Km.*
1	Mongolia	Asia	3,000,000	603,902	1,564,100	4.9	1.9
2	Western Sahara	Africa	567,421	97,344	252,120	6	2.3
3	Namibia	Africa	2,413,643	318,580	825,118	7.5	2.9
4	Australia	Australia	25,594,366	2,969,907	7,692,024	9	3
5	Suriname	South America	568,301	63,251	163,820	9.1	3.5
6	Iceland	Europe	360,390	39,682	102,775	9.1	3.5
7	Guyana	South America	782,225	83,012	214,999	9.3	3.6
8	Libya	Africa	6,470,956	683,424	1,770,060	9.6	3.7
9	Canada	North America	37,898,384	3,855,103	9,984,670	10	4
10	Mauritania	Africa	3,984,233	397,955	1,030,700	10.4	4
11	Botswana	Africa	2,302,878	224,607	581,730	10.4	4
12	Kazakhstan	Asia	18,592,700	1,052,090	2,724,900	18	7
13	Central African Republic	Africa	4,737,423	240,324	622,436	21	8
14	Gabon	Africa	2,067,561	103,347	267,667	21	8
15	Russia	Europe/Asia	146,877,088	6,612,093	17,125,242	23	9
16	Bolivia	South America	11,307,314	424,164	1,098,581	26	10
17	Chad	Africa	15,353,184	495,755	1,284,000	31	12
18	Turkmenistan	Asia	5,851,466	189,657	491,210	31	12
19	Oman	Asia	4,183,841	119,499	309,500	36	14
20	Mali	Africa	19,107,706	482,077	1,248,574	39	15
21	Argentina	South America	44,938,712	1,073,518	2,780,400	41	16
22	Saudi Arabia	Asia	34,218,169	830,000	2,149,690	41	16
23	Republic of the Congo	Africa	5,399,895	132,047	342,000	41	16
24	Finland	Europe	5,527,405	130,666	338,424	41	16
25	Norway	Europe	5,328,212	125,013	323,782	41	16
26	Paraguay	South America	7,052,983	157,048	406,752	44	17
27	Belize	Central America	398,050	8,867	22,965	44	17
28	Algeria	Africa	42,545,964	919,595	2,381,741	47	18
29	Niger	Africa	21,466,863	458,075	1,186,408	47	18
30	Papua New Guinea	Pacific Islands	8,558,800	178,704	462,840	47	18

Source: U.S. Census Bureau, International Data Base

WORLD'S MOST POPULOUS CITIES

Rank	*City*	*Country*	*Region*	*Population*
1	Chongqing	China	Asia	30,484,300
2	Shanghai	China	Asia	24,256,800
3	Beijing	China	Asia	21,516,000
4	Chengdu	China	Asia	16,044,700
5	Karachi	Pakistan	Asia	14,910,352
6	Guangzhou	China	Asia	14,043,500
7	Istanbul	Turkey	Europe	14,025,000
8	Tokyo	Japan	Asia	13,839,910
9	Tianjin	China	Asia	12,784,000
10	Mumbai	India	Asia	12,478,447
11	São Paulo	Brazil	South America	12,252,023
12	Moscow	Russia	Europe/Asia	12,197,596
13	Kinshasa	Dem. Rep. of Congo	Africa	11,855,000
14	Baoding	China	Asia	11,194,372
15	Lahore	Pakistan	Asia	11,126,285
16	Wuhan	China	Asia	11,081,000
17	Delhi	India	Asia	11,034,555
18	Harbin	China	Asia	10,635,971
19	Suzhou	China	Asia	10,459,890
20	Cairo	Egypt	Africa	10,230,350
21	Seoul	South Korea	Asia	10,197,604
22	Jakarta	Indonesia	Asia	10,075,310
23	Lima	Peru	South America	9,174,855
24	Mexico City	Mexico	North America	9,041,395
25	Ho Chi Minh City	Vietnam	Asia	8,993,082
26	Dhaka	Bangladesh	Africa	8,906,039
27	London	United Kingdom	Europe	8,825,001
28	Bangkok	Thailand	Asia	8,750,600
29	Xi'an	China	Asia	8,705,600
30	New York	United States	North America	8,622,698
31	Bangalore	India	Asia	8,425,970
32	Shenzhen	China	Asia	8,378,900
33	Nanjing	China	Asia	8,230,000
34	Tehran	Iran	Asia	8,154,051
35	Rio de Janeiro	Brazil	South America	6,718,903
36	Shantou	China	Asia	5,391,028
37	Kolkata	India	Asia	4,486,679
38	Shijiazhuang	China	Asia	4,303,700
39	Los Angeles	United States	North America	3,884,307
40	Buenos Aires	Argentina	South America	3,054,300

MAJOR LAKES OF THE WORLD

		Surface Area		*Maximum Depth*	
Lake	*Location*	*Sq. Mi.*	*Sq. Km.*	*Feet*	*Meters*
Caspian Sea	Central Asia	152,239	394,299	3,104	946
Superior	North America	31,820	82,414	1,333	406
Victoria	East Africa	26,828	69,485	270	82
Huron	North America	23,010	59,596	750	229
Michigan	North America	22,400	58,016	923	281
Aral	Central Asia	13,000	33,800	223	68
Tanganyika	East Africa	12,700	32,893	4,708	1,435
Baikal	Russia	12,162	31,500	5,712	1,741
Great Bear	North America	12,000	31,080	270	82
Nyasa	East Africa	11,600	30,044	2,316	706
Great Slave	North America	11,170	28,930	2,015	614
Chad	West Africa	9,946	25,760	23	7
Erie	North America	9,930	25,719	210	64
Winnipeg	North America	9,094	23,553	204	62
Ontario	North America	7,520	19,477	778	237
Balkhash	Central Asia	7,115	18,428	87	27
Ladoga	Russia	7,000	18,130	738	225
Onega	Russia	3,819	9,891	361	110
Titicaca	South America	3,141	8,135	1,214	370
Nicaragua	Central America	3,089	8,001	230	70
Athabasca	North America	3,058	7,920	407	124
Rudolf	Kenya, East Africa	2,473	6,405	—	—
Reindeer	North America	2,444	6,330	—	—
Eyre	South Australia	2,400	6,216	varies	varies
Issyk-Kul	Central Asia	2,394	6,200	2,297	700
Urmia	Southwest Asia	2,317	6,001	49	15
Torrens	Australia	2,200	5,698	—	—
Vänern	Sweden	2,141	5,545	322	98
Winnipegosis	North America	2,086	5,403	59	18
Mobutu Sese Seko	East Africa	2,046	5,299	180	55
Nettilling	North America	1,950	5,051	—	—

Note: The sizes of some lakes vary with the seasons.

MAJOR RIVERS OF THE WORLD

River	Region	Source	Outflow	Approximate Length Miles	Approximate Length Km.
Nile	N. Africa	Tributaries of Lake Victoria	Mediterranean Sea	4,180	6,690
Mississippi-Missouri-Red Rock	N. America	Montana	Gulf of Mexico	3,710	5,970
Yangtze Kiang	East Asia	Tibetan Plateau	China Sea	3,602	5,797
Ob	Russia	Altai Mountains	Gulf of Ob	3,459	5,567
Yellow (Huang He)	East Asia	Kunlun Mountains, west China	Gulf of Chihli	2,900	4,667
Yenisei	Russia	Tannu-Ola Mountains, western Tuva, Russia	Arctic Ocean	2,800	4,506
Paraná	S. America	Confluence of Paranaiba and Grande Rivers	Río de la Plata	2,795	4,498
Irtysh	Russia	Altai Mountains, Russia	Ob River	2,758	4,438
Congo	Africa	Confluence of Lualaba and Luapula Rivers, Congo	Atlantic Ocean	2,716	4,371
Heilong (Amur)	East Asia	Confluence of Shilka and Argun Rivers	Tatar Strait	2,704	4,352
Lena	Russia	Baikal Mountains, Russia	Arctic Ocean	2,652	4,268
Mackenzie	N. America	Head of Finlay River, British Columbia, Canada	Beaufort Sea	2,635	4,241
Niger	West Africa	Guinea	Gulf of Guinea	2,600	4,184
Mekong	Asia	Tibetan Plateau	South China Sea	2,500	4,023
Mississippi	N. America	Lake Itasca, Minnesota	Gulf of Mexico	2,348	3,779
Missouri	N. America	Confluence of Jefferson, Gallatin, and Madison Rivers, Montana	Mississippi River	2,315	3,726
Volga	Russia	Valdai Plateau, Russia	Caspian Sea	2,291	3,687
Madeira	S. America	Confluence of Beni and Maumoré Rivers, Bolivia-Brazil boundary	Amazon River	2,012	3,238
Purus	S. America	Peruvian Andes	Amazon River	1,993	3,207
São Francisco	S. America	S.W. Minas Gerais, Brazil	Atlantic Ocean	1,987	3,198

River	*Region*	*Source*	*Outflow*	*Approximate Length*	
				Miles	*Km.*
Yukon	N. America	Junction of Lewes and Pelly Rivers, Yukon Terr., Canada	Bering Sea	1,979	3,185
St. Lawrence	N. America	Lake Ontario	Gulf of St. Lawrence	1,900	3,058
Rio Grande	N. America	San Juan Mountains, Colorado	Gulf of Mexico	1,885	3,034
Brahmaputra	Asia	Himalayas	Ganges River	1,800	2,897
Indus	Asia	Himalayas	Arabian Sea	1,800	2,897
Danube	Europe	Black Forest, Germany	Black Sea	1,766	2,842
Euphrates	Asia	Confluence of Murat Nehri and Kara Su Rivers, Turkey	Shatt-al-Arab	1,739	2,799
Darling	Australia	Eastern Highlands, Australia	Murray River	1,702	2,739
Zambezi	Africa	Western Zambia	Mozambique Channel	1,700	2,736
Tocantins	S. America	Goiás, Brazil	Pará River	1,677	2,699
Murray	Australia	Australian Alps, New S. Wales	Indian Ocean	1,609	2,589
Nelson	N. America	Head of Bow River, western Alberta, Canada	Hudson Bay	1,600	2,575
Paraguay	S. America	Mato Grosso, Brazil	Paraná River	1,584	2,549
Ural	Russia	Southern Ural Mountains, Russia	Caspian Sea	1,574	2,533
Ganges	Asia	Himalayas	Bay of Bengal	1,557	2,506
Amu Darya (Oxus)	Asia	Nicholas Range, Pamir Mountains, Turkmenistan	Aral Sea	1,500	2,414
Japurá	S. America	Andes, Colombia	Amazon River	1,500	2,414
Salween	Asia	Tibet, south of Kunlun Mountains	Gulf of Martaban	1,500	2,414
Arkansas	N. America	Central Colorado	Mississippi River	1,459	2,348
Colorado	N. America	Grand County, Colorado	Gulf of California	1,450	2,333
Dnieper	Russia	Valdai Hills, Russia	Black Sea	1,419	2,284
Ohio-Allegheny	N. America	Potter County, Pennsylvania	Mississippi River	1,306	2,102
Irrawaddy	Asia	Confluence of Nmai and Mali rivers, northeast Burma	Bay of Bengal	1,300	2,092
Orange	Africa	Lesotho	Atlantic Ocean	1,300	2,092

River	Region	Source	Outflow	Approximate Length Miles	Km.
Orinoco	S. America	Serra Parima Mountains, Venezuela	Atlantic Ocean	1,281	2,062
Pilcomayo	S. America	Andes Mountains, Bolivia	Paraguay River	1,242	1,999
Xi Jiang	East Asia	Eastern Yunnan Province, China	China Sea	1,236	1,989
Columbia	N. America	Columbia Lake, British Columbia, Canada	Pacific Ocean	1,232	1,983
Don	Russia	Tula, Russia	Sea of Azov	1,223	1,968
Sungari	East Asia	China-North Korea boundary	Amur River	1,215	1,955
Saskatchewan	N. America	Canadian Rocky Mountains	Lake Winnipeg	1,205	1,939
Peace	N. America	Stikine Mountains, British Columbia, Canada	Great Slave River	1,195	1,923
Tigris	Asia	Taurus Mountains, Turkey	Shatt-al-Arab	1,180	1,899

HIGHEST PEAKS IN EACH CONTINENT

			Height	
Continent	*Mountain*	*Location*	*Feet*	*Meters*
Asia	Everest	Tibet & Nepal	29,028	8,848
South America	Aconcagua	Argentina	22,834	6,960
North America	McKinley	Alaska	20,320	6,194
Africa	Kilimanjaro	Tanzania	19,340	5,895
Europe	Elbrus	Russia & Georgia	18,510	5,642
Antarctica	Vinson Massif	Ellsworth Mountains	16,066	4,897
Australia	Kosciusko	New South Wales	7,316	2,228

Note: The world's highest sixty-six mountains are all in Asia.

MAJOR DESERTS OF THE WORLD

Desert	Location	Approximate Area		Type
		Sq. Miles	Sq. Km.	
Antarctic	Antarctica	5,400,000	14,002,200	polar
Sahara	North Africa	3,500,000	9,075,500	subtropical
Arabian	Southwest Asia	1,000,000	2,593,000	subtropical
Great Western (Gibson, Great Sandy, and Great Victoria)	Australia	520,000	1,348,360	subtropical
Gobi	East Asia	500,000	1,296,500	cold winter
Patagonian	Argentina, South America	260,000	674,180	cold winter
Kalahari	Southern Africa	220,000	570,460	subtropical
Great Basin	Western United States	190,000	492,670	cold winter
Thar	South Asia	175,000	453,775	subtropical
Chihuahuan	Mexico	175,000	453,775	subtropical
Karakum	Central Asia	135,000	350,055	cold winter
Colorado Plateau	Southwestern United States	130,000	337,090	cold winter
Sonoran	United States and Mexico	120,000	311,160	subtropical
Kyzylkum	Central Asia	115,000	298,195	cold winter
Taklimakan	China	105,000	272,265	cold winter
Iranian	Iran	100,000	259,300	cold winter
Simpson	Eastern Australia	56,000	145,208	subtropical
Mojave	Western United States	54,000	140,022	subtropical
Atacama	Chile, South America	54,000	140,022	cold coastal
Namib	Southern Africa	13,000	33,709	cold coastal
Arctic	Arctic Circle			polar

Highest Waterfalls of the World

Waterfall	*Location*	*Source*	*Height*	
			Feet	*Meters*
Angel	Canaima National Park, Venezuela	Rio Caroni	3,212	979
Tugela	Natal National Park, South Africa	Tugela River	3,110	948
Utigord	Norway	glacier	2,625	800
Monge	Marstein, Norway	Mongebeck	2,540	774
Mutarazi	Nyanga National Park, Zimbabwe	Mutarazi River	2,499	762
Yosemite	Yosemite National Park, California, U.S.	Yosemite Creek	2,425	739
Espelands	Hardanger Fjord, Norway	Opo River	2,307	703
Lower Mar Valley	Eikesdal, Norway	Mardals Stream	2,151	655
Tyssestrengene	Odda, Norway	Tyssa River	2,123	647
Cuquenan	Kukenan Tepuy, Venezuela	Cuquenan River	2,000	610
Sutherland	Milford Sound, New Zealand	Arthur River	1,904	580
Kjell	Gudvanger, Norway	Gudvangen Glacier	1,841	561
Takkakaw	Yoho Natl Park, British Columbia, Canada	Takkakaw Creek	1,650	503
Ribbon	Yosemite National Park, California, U.S.	Ribbon Stream	1,612	491
Upper Mar Valley	near Eikesdal, Norway	Mardals Stream	1,536	468
Gavarnie	near Lourdes, France	Gave de Pau	1,388	423
Vettis	Jotunheimen, Norway	Utla River	1,215	370
Hunlen	British Columbia, Canada	Hunlen River	1,198	365
Tin Mine	Kosciusko National Park, Australia	Tin Mine Creek	1,182	360
Silver Strand	Yosemite National Park, California, U.S.	Silver Strand Creek	1,170	357
Basaseachic	Baranca del Cobre, Mexico	Piedra Volada Creek	1,120	311
Spray Stream	Lauterburnnental, Switzerland	Staubbach Brook	985	300
Fachoda	Tahiti, French Polynesia	Fautaua River	985	300
King Edward VIII	Guyana	Courantyne River	850	259
Wallaman	near Ingham, Australia	Wallaman Creek	844	257
Gersoppa	Western Ghats, India	Sharavati River	828	253
Kaieteur	Guyana	Rio Potaro	822	251
Montezuma	near Rosebery, Tasmania	Montezuma River	800	240
Wollomombi	near Armidale, Australia	Wollomombi River	722	2203

Source: Fifth Continent Australia Pty Limited

GLOSSARY

Places whose names are printed in SMALL CAPS *are subjects of their own entries in this glossary.*

Ablation. Loss of ice volume or mass by a GLACIER. Ablation includes melting of ice, SUBLIMATION, DEFLATION (removal by WIND), EVAPORATION, and CALVING. Ablation occurs in the lower portions of glaciers.

Abrasion. Wearing away of ROCKS in STREAMS by grinding, especially when rocks and SEDIMENT are carried along by stream water. The STREAMBED and VALLEY are carved out and eroded, and the rocks become rounded and smoothed by abrasion.

Absolute location. Position of any PLACE on the earth's surface. The absolute location can be given precisely in terms of DEGREES, MINUTES, and SECONDS of LATITUDE (0 to 90 degrees north or south) and of LONGITUDE (0 to 180 degrees east or west). The EQUATOR is 0 degrees latitude; the PRIME MERIDIAN, which runs through Greenwich in England, is 0 degrees longitude.

Abyss. Deepest part of the OCEAN. Modern TECHNOLOGY—especially sonar—has enabled accurate mapping of the ocean floors, showing that there are MOUNTAIN CHAINS, or RIDGES, in all the oceans, as well as deep CANYONS or TRENCHES closer to the edges of the oceans.

Acid rain. PRECIPITATION containing high levels of nitric or sulfuric acid; a major environmental problem in parts of North America, Europe, and Asia. Natural precipitation is slightly acidic (about 5.6 on the pH SCALE), because CARBON DIOXIDE—which occurs naturally in the ATMOSPHERE—is dissolved to form a weak carbonic acid.

Adiabatic. Change of TEMPERATURE within the ATMOSPHERE that is caused by compression or expansion without addition or loss of heat.

Advection. Horizontal movement of AIR from one PLACE to another in the ATMOSPHERE, associated with WINDS.

Advection fog. FOG that forms when a moist AIR mass moves over a colder surface. Commonly, warm moist air moves over a cool OCEAN CURRENT, so the air cools to SATURATION POINT and fog forms. This phenomenon, known as sea fog, occurs along subtropical west COASTS.

Aerosol. Substances held in SUSPENSION in the ATMOSPHERE, as solid particles or liquid droplets.

Aftershock. EARTHQUAKE that follows a larger earthquake and originates at or near the focus of the latter; many aftershocks may follow a major earthquake, decreasing in frequency and magnitude with time.

Agglomerate. Type of ROCK composed of volcanic fragments, usually of different sizes and rough or angular.

Aggradation. Accumulation of SEDIMENT in a STREAMBED. Aggradation often results from reduced flow in the channel during dry periods. It also occurs when the STREAM's load (BEDLOAD and SUSPENDED LOAD) is greater than the stream capacity. A BRAIDED STREAM pattern often results.

Air current. Air currents are caused by differential heating of the earth's surface, which causes heated air to rise. This causes WINDS at the surface as well as higher in the earth's ATMOSPHERE.

Air mass. Large body of air with distinctive homogeneous characteristics of TEMPERATURE, HUMIDITY, and stability. It forms when air remains stationary over a source REGION for a period of time, taking on the conditions of that region. An air mass can extend over a million square miles with a depth of more than a mile. Air masses are classified according to moisture content (*m* for maritime or *c* for continental) and temperature

(*A* for ARCTIC, *P* for polar, *T* for tropical, or *E* for equatorial). The air masses affecting North America are mP, cP, and mT. The interaction of AIR masses produces WEATHER. The line along which air masses meet is a FRONT.

Albedo. Measure of the reflective properties of a surface; the ratio of reflected ENERGY (INSOLATION) to the total incoming energy, expressed as a percentage. The albedo of Earth is 33 percent.

Alienation (land). Land alienation is the appropriation of land from its original owners by a more powerful force. In preindustrial societies, the ownership of agricultural land is of prime importance to subsistence farmers.

Alkali flat. Dry LAKEBED in an arid REGION, covered with a layer of SALTS. A well-known example is the Alkali Flat area of White Sands National Monument in New Mexico; it is the bed of a large lake that formed when the GLACIERS were melting. It is covered with a form of gypsum crystals called selenite. This material is blown off the surface into large SAND DUNES. Also called a salina. See also BITTER LAKE.

Allogenic sediment. SEDIMENT that originates outside the PLACE where it is finally deposited; SAND, SILT, and CLAY carried by a STREAM into a LAKE are examples.

Alluvial fan. Common LANDFORM at the mouth of a CANYON in arid REGIONS. Water flowing in a narrow canyon immediately slows as it leaves the canyon for the wider VALLEY floor, depositing the SEDIMENTS it was transporting. These spread out into a fan shape, usually with a BRAIDED STREAM pattern on its surface. When several alluvial fans grow side by side, they can merge into one continuous sloping surface between the HILLS and the valley. This is known by the Spanish word *bajada*, which means "slope."

Alluvial system. Any of various depositional systems, excluding DELTAS, that form from the activity of RIVERS and STREAMS. Much alluvial SEDIMENT is deposited when rivers top their BANKS and FLOOD the surrounding countryside. Buried alluvial sediments may be important water-bearing RESERVOIRS or may contain PETROLEUM.

Alluvium. Material deposited by running water. This includes not only fertile SOILS, but also CLAY, SILT, or SAND deposits resulting from FLUVIAL processes. FLOODPLAINS are covered in a thick layer of alluvium.

Altimeter. Instrument for measuring ALTITUDE, or height above the earth's surface, commonly used in airplanes. An altimeter is a type of ANEROID BAROMETER.

Altitudinal zonation. Existence of different ECOSYSTEMS at various ELEVATIONS above SEA LEVEL, due to TEMPERATURE and moisture differences. This is especially pronounced in Central America and South America. The hot and humid COASTAL PLAINS, where bananas and sugarcane thrive, is the *tierra caliente.* From about 2,500 to 6,000 feet (750–1,800 meters) is the *tierra templada*; crops grown here include coffee, wheat, and corn, and major cities are situated in this zone. From about 6,000 to 12,000 feet (1,800–3,600 meters) is the *tierra fria*; here only hardy crops such as potatoes and barley are grown, and large numbers of animals are kept. From about 12,000 to 15,000 feet (3,600 to 4,500 meters) lies the *tierra helada*, where hardy animals such as sheep and alpaca graze. Above 15,000 feet (4,500 meters) is the frozen *tierra nevada*; no permanent life is possible in the permanent SNOW and ICE FIELDS there.

Angle of repose. Maximum angle of steepness that a pile of loose materials such as SAND or ROCK can assume and remain stable; the angle varies with the size, shape, moisture, and angularity of the material.

Antecedent river. STREAM that was flowing before the land was uplifted and was able to erode at the pace of UPLIFT, thus creating a deep CANYON. Most deep canyons are attributed to antecedent rivers. In the Davisian CYCLE OF EROSION, this process was called REJUVENATION.

Anthropogeography. Branch of GEOGRAPHY founded in the late nineteenth century by German geographer Friedrich Ratzel. The field is closely related to human ECOLOGY—the study of humans, their DISTRIBUTION over the earth, and their interaction with their physical environment.

Anticline. Area where land has been upfolded symmetrically. Its center contains stratigraphically older rocks. See also Syncline.

Anticyclone. High-pressure system of rotating winds, descending and diverging, shown on a weather chart by a series of closed isobars, with a high in the center. In the Northern Hemisphere, the rotation is clockwise; in the Southern Hemisphere, the rotation is counterclockwise. An anticyclone brings warm weather.

Antidune. Undulatory upstream-moving bed form produced in free-surface flow of water over a sand bed in a certain range of high flow speeds and shallow flow depths.

Antipodes. Temperate zone of the Southern Hemisphere. The term is now usually applied to the countries of Australia and New Zealand. The ancient Greeks had believed that if humans existed there, they must walk upside down. This idea was supported by the Christian Church in the Middle Ages.

Antitrade winds. Winds in the upper atmosphere, or geostrophic winds, that blow in the opposite direction to the trade winds. Antitrade winds blow toward the northeast in the Northern Hemisphere and toward the southeast in the Southern Hemisphere.

Aperiodic. Irregularly occurring interval, such as found in most weather cycles, rendering them virtually unpredictable.

Aphelion. Point in the earth's 365-day revolution when it is at its greatest distance from the Sun. This is caused by Earth's elliptical orbit around the Sun. The distance at aphelion is 94,555,000 miles (152,171,500 km.) and usually falls on July 4. The opposite of perihelion.

Aposelene. Earth's farthest point from the Moon.

Aquifer. Underground body of porous rock that contains water and allows water percolation through it. The largest aquifer in the United States is the Ogallala Aquifer, which extends south from South Dakota to Texas.

Arête. Serrated or saw-toothed ridge, produced in glaciated mountain areas by cirques eroding on either side of a ridge or mountain range. From the French word for knife-edge.

Arroyo. Spanish word for a dry streambed in an arid area. Called a wadi in Arabic and a wash in English.

Artesian well. Well from which groundwater flows without mechanical pumping, because the water comes from a confined aquifer, and is therefore under pressure. The Great Artesian Basin of Australia has hundreds of artesian wells, called bores, that provide drinking water for sheep and cattle. The name comes from the Artois region of France, where the phenomenon is common. A subartesian well is sunk into an unconfined aquifer and requires a pump to raise water to the surface.

Asteroid belt. Region between the orbits of Mars and Jupiter containing the majority of asteroids.

Asthenosphere. Part of the earth's upper mantle, beneath the lithosphere, in which plate movement takes place. Also known as the low-velocity zone.

Astrobleme. Remnant of a large impact crater on Earth.

Astronomical unit (AU). Unit of measure used by astronomers that is equivalent to the average distance from the Sun to Earth (93 million miles/150 million km.).

Atmospheric pressure. Weight of the earth's atmosphere, equally distributed over earth's surface and pressing down as a result of gravity. On average, the atmosphere has a force of 14.7 pounds per square inch (1 kilogram per centimeter) squared at sea level, also expressed as 1013.2 millibars. Variations in atmospheric pressure, high or low, cause winds and weather changes that affect climate. Pressure decreases rapidly with altitude or distance from the surface: Half of the total atmosphere is found below 18,000 feet (5,500 meters); more than 99 percent of the atmosphere is below 30 miles (50 km.) of the surface. Atmospheric pressure is measured with a barometer.

Atoll. Ring-shaped growth of CORAL REEF, with a LAGOON in the middle. Charles Darwin, who observed many Pacific atolls during his voyage on the *Beagle* in the nineteenth century, suggested that they were created from FRINGING REEFS around volcanic ISLANDS. As such islands sank beneath the water (or as SEA LEVELS rose), the coral continued growing upward. SAND resting atop an atoll enables plants to grow, and small human societies have arisen on some atolls. The world's largest atoll, Kwajalein in the Marshall Islands, measures about 40 by 18 miles (65 by 30 km.), but perhaps the most famous atoll is Bikini Atoll—the SITE of nuclear-bomb testing during the 1950s.

Aurora. Glowing and shimmering displays of colored lights in the upper ATMOSPHERE, caused by interaction of the SOLAR WIND and the charged particles of the IONOSPHERE. Auroras occur at high LATITUDES. Near the North Pole they are called aurora borealis or northern lights; near the South Pole, aurora australis or southern lights.

Austral. Referring to an object or occurrence that is located in the SOUTHERN HEMISPHERE or related to Australia.

Australopithecines. Erect-walking early human ancestors with a cranial capacity and body size within the RANGE of modern apes rather than of humans.

Avalanche. Mass of SNOW and ice falling suddenly down a MOUNTAIN slope, often taking with it earth, ROCKS, and trees.

Bank. Elevated area of land beneath the surface of the OCEAN. The term is also used for elevated ground lining a body of water.

Bar (climate). Measure of ATMOSPHERIC PRESSURE per unit surface area of 1 million dynes per square centimeter. Millibars (thousandths of a bar) are the MEASUREMENT used in the United States. Other countries use kilopascals (kPa); one kilopascal is ten millibars.

Bar (land). RIDGE or long deposit of SAND or gravel formed by DEPOSITION in a RIVER or at the COAST. Offshore bars and baymouth bars are common coastal features.

Barometer. Instrument used for measuring ATMOSPHERIC PRESSURE. In the seventeenth century, Evangelista Torricelli devised the first barometer—a glass tube sealed at one end, filled with mercury, and upended into a bowl of mercury. He noticed how the height of the mercury column changed and realized this was a result of the pressure of air on the mercury in the bowl. Early MEASUREMENTS of atmospheric pressure were, therefore, expressed as centimeters of mercury, with average pressure at SEA LEVEL being 29.92 inches (760 millimeters). This cumbersome barometer was replaced with the ANEROID BAROMETER—a sealed and partially evacuated box connected to a needle and dial, which shows changes in atmospheric pressure. See also ALTIMETER.

Barrier island. Long chain of SAND islands that forms offshore, close to the COAST. LAGOONS or shallower MARSHES separate the barrier islands from the mainland. Such LOCATIONS are hazardous for SETTLEMENTS because they are easily swept away in STORMS and HURRICANES.

Basalt. IGNEOUS EXTRUSIVE ROCK formed when LAVA cools; often black in color. Sometimes basalt occurs in tall hexagonal columns, such as the Giant's Causeway in Ireland, or the Devils Postpile at Mammoth, California.

Basement. Crystalline, usually PRECAMBRIAN, IGNEOUS and METAMORPHIC ROCKS that occur beneath the SEDIMENTARY ROCK on the CONTINENTS.

Basin order. Approximate measure of the size of a STREAM BASIN, based on a numbering scheme applied to RIVER channels as they join together in their progress downstream.

Batholith. Large LANDFORM produced by IGNEOUS INTRUSION, composed of CRYSTALLINE ROCK, such as GRANITE; a large PLUTON with a surface area greater than 40 square miles (100 sq. km.). Most mountain RANGES have a batholith underneath.

Bathymetric contour. Line on a MAP of the OCEAN floor that connects points of equal depth.

Beaufort scale. Scale that measures wind force, expressed in numbers from 0 to 12. The original Beaufort scale was based on descriptions of the state of the sea. It was adapted to land conditions, using descriptions of chimney smoke, leaves of trees, and similar factors. The scale was devised in the early nineteenth century by Sir Francis Beaufort, a British naval officer.

Belt. Geographical region that is distinctive in some way.

Bergeron process. Precipitation formation in cold clouds whereby ice crystals grow at the expense of supercooled water droplets.

Bight. Wide or open bay formed by a curve in the coastline, such as the Great Australian Bight.

Biogenic sediment. Sediment particles formed from skeletons or shells of microscopic plants and animals living in seawater.

Biostratigraphy. Identification and organization of strata based on their fossil content and the use of fossils in stratigraphic correlation.

Bitter lake. Saline or brackish lake in an arid area, which may dry up in the summer or in periods of drought. The water is not suitable for drinking. Another name for this feature is "salina." See also Alkali flat.

Block lava. Lava flows whose surfaces are composed of large, angular blocks; these blocks are generally larger than those of aa flows and have smooth, not jagged, faces.

Block mountain. Mountain or mountain range with one side having a gentle slope to the crest, while the other slope, which is the exposed fault scarp, is quite steep. It is formed when a large block of the earth's crust is thrust upward on one side only, while the opposite side remains in place. The Sierra Nevada in California are a good example of block mountains. Also known as fault-block mountain.

Blowhole. Sea cave or tunnel formed on some rocky, rugged coastlines. The pressure of the seawater rushing into the opening can force a jet of seawater to rise or spout through an opening in the roof of the cave. Blowholes are found in Scotland, Tasmania, and Mexico, and on the Hawaiian islands of Kauai and Maui.

Bluff. Steep slope that marks the farthest edge of a floodplain.

Bog. Damp, spongy ground surface covered with decayed or decaying vegetation. Bogs usually are formed in cool climates through the in-filling, or silting up, of a lake. Moss and other plants grow outward toward the edge of the lake, which gradually becomes shallower, until the surface is completely covered. Bogs also can form on cold, damp mountain surfaces. Many bogs are filled with peat.

Bore. Standing wave, or wall, of water created in a narrow estuary when the strong incoming, or flood, tide meets the river water flowing outward; it moves upstream with the advancing tide, and downstream with the ebb tide. South America's Amazon River and Asia's Mekong River have large bores. In North America, the bore in the Bay of Fundy is visited by many tourists each year. Its St. Andrew's wharf is designed to handle changes in water level of as much as 53 feet (15 meters) in one day.

Boreal. Alluding to an item or event that is in the Northern Hemisphere.

Bottom current. Deep-sea current that flows parallel to bathymetric contours.

Brackish water. Water with salt content between that of salt water and fresh water; it is common in arid areas on the surface, in coastal marshes, and in salt-contaminated groundwater.

Braided stream. Stream having a channel consisting of a maze of interconnected small channels within a broader streambed. Braiding occurs when the stream's load exceeds its capacity, usually because of reduced flow.

Breaker. Wave that becomes oversteepened as it approaches the shore, reaching a point at which it cannot maintain its vertical shape. It then breaks, and the water washes toward the shore.

Breakwater. Large structure, usually of rock, built offshore and parallel to the coast, to absorb wave energy and thus protect the shore. Between the breakwater and the shore is an area of calm water, often used as a boat anchorage or

HARBOR. A similar but smaller structure is a seawall.

Breeze. Gentle WIND with a speed of 4 to 31 miles (6 to 50 km.) per hour. On the BEAUFORT SCALE, the numbers 2 through 6 represent breezes of increasing strength.

Butte. Flat-topped HILL, smaller than a MESA, found in arid REGIONS.

Caldera. Large circular depression with steep sides, formed when a VOLCANO explodes, blowing away its top. The ERUPTION of Mount St. Helens produced a caldera. Crater Lake in Oregon is a caldera that has filled with water. From the Spanish word for kettle.

Calms of Cancer. Subtropical BELT of high pressure and light WINDS, located over the OCEAN near 25 DEGREES north LATITUDE. Also known as the HORSE LATITUDES.

Calms of Capricorn. Subtropical BELT of high pressure and light WINDS, located over the OCEAN near 25 DEGREES south LATITUDE.

Calving. Loss of glacial mass when GLACIERS reach the SEA and large blocks of ice break off, forming ICEBERGS.

Cancer, tropic of. PARALLEL of LATITUDE at 23.5 DEGREES north; this line is the latitude farthest north on the earth where the noon SUN is ever directly overhead. The REGION between it and the tropic of CAPRICORN is known as the TROPICS.

Capricorn, tropic of. Line of LATITUDE at 23.5 DEGREES south; this line is the latitude farthest south on the earth where the noon SUN is ever directly overhead. The REGION between it and the tropic of CANCER is known as the TROPICS.

Carbon dating. Method employed by physicists to determine the age of organic matter—such as a piece of wood or animal tissue—to determine the age of an archaeological or paleontological SITE. The method works on the principle that the amount of radioactive carbon in living matter diminishes at a steady, and measurable, rate after the matter dies. Technique is also known as carbon-14 dating, after the radioactive carbon-14 isotope it uses. Also known as radiocarbon dating.

Carrying capacity. Number of animals that a given area of land can support, without additional feed being necessary. Lush GRASSLAND may have a carrying capacity of twenty sheep per acre, while more arid, SEMIDESERT land may support only two sheep per acre. The term sometimes is used to refer to the number of humans who can be supported in a given area.

Catastrophism. Theory, popular in the eighteenth and nineteenth centuries, that explained the shape of LANDFORMS and CONTINENTS and the EXTINCTION of species as the results of intense or catastrophic events. The biblical FLOOD of Noah was one such event, which supposedly explained many extinctions. Catastrophism is linked closely to the belief that the earth is only about 6,000 years old, and therefore tremendous forces must have acted swiftly to create present LANDSCAPES. An alternative or contrasting theory is UNIFORMITARIANISM.

Catchment basin. Area of land receiving the PRECIPITATION that flows into a STREAM. Also called catchment or catchment area.

Central place theory. Theory that explains why some SETTLEMENTS remain small while others grow to be middle-sized TOWNS, and a few become large cities or METROPOLISES. The explanation is based on the provision of goods and services and how far people will travel to acquire these. The German geographer Walter Christaller developed this theory in the 1930s.

Centrality. Measure of the number of functions, or services, offered by any CITY in a hierarchy of cities within a COUNTRY or a REGION. See also CENTRAL PLACE THEORY.

Chain, mountain. Another term for mountain RANGE.

Chemical farming. Application of artificial FERTILIZERS to the SOIL and the use of chemical products such as insecticides, fungicides, and herbicides to ensure crop success. Chemical farming is practiced mainly in high-income countries, because the cost of the chemical products is high. Farmers in low-income economies rely more on natural organic fertilizers such as animal waste.

Chemical weathering. Chemical decomposition of solid ROCK by processes involving water that change its original materials into new chemical combinations.

Chlorofluorocarbons (CFCs). Manufactured compounds, not occurring in nature, consisting of chlorine, fluorine, and carbon. CFCs are stable and have heat-absorbing properties, so they have been used extensively for cooling in refrigeration and air-conditioning units. Previously, they were used as propellants for aerosol products. CFCs rise into the STRATOSPHERE where ULTRAVIOLET RADIATION causes them to react with OZONE, changing it to oxygen and exposing the earth to higher levels of ultraviolet (UV) radiation. Therefore, the manufacture and use of CFCs was banned in many countries. The commercial name for CFCs is Freon.

Chorology. Description or mapping of a REGION. Also known as chorography.

Chronometer. Highly accurate CLOCK or timekeeping device. The first accurate and effective chronometers were constructed in the mid-eighteenth century by John Harrison, who realized that accurate timekeeping was the secret to NAVIGATION at SEA.

Cinder cone. Small conical HILL produced by PYROCLASTIC materials from a VOLCANO. The material of the cone is loose SCORIA.

Circle of illumination. Line separating the sunlit part of the earth from the part in darkness. The circle of illumination moves around the earth once in every approximately 24 hours. At the VERNAL and autumnal EQUINOXES, the circle of illumination passes through the POLES.

Cirque. Circular BASIN at the head of an ALPINE GLACIER, shaped like an armchair. Many cirques can be seen in MOUNTAIN areas where glaciers have completely melted since the last ICE AGE.

City Beautiful movement. Planning and architectural movement that was at its height from around 1890 to the 1920s in the United States. It was believed that classical architecture, wide and carefully laid-out streets, parks, and urban monuments would reflect the higher values of the society and be a civilizing, even uplifting, experience for the citizens of such cities. Civic pride was fostered through remodeling or modernizing older URBAN AREAS. Chicago, Illinois, and Pasadena, California, are cities where the planners of the City Beautiful movement left their imprint.

Clastic. ROCK or sedimentary matter formed from fragments of older rocks.

Climatology. Study of Earth CLIMATES by analysis of long-term WEATHER patterns over a minimum of thirty years of statistical records. Climatologists—scientists who study climate—seek similarities to enable grouping into climatic REGIONS. Climate patterns are closely related to natural VEGETATION. Computer TECHNOLOGY has enabled investigation of phenomena such as the EL NIÑO effect and global climate change. The KÖPPEN CLIMATE CLASSIFICATION system is the most commonly used scheme for climate classification.

Climograph. Graph that plots TEMPERATURE and PRECIPITATION for a selected LOCATION. The most commonly used climographs plot monthly temperatures and monthly precipitation, as used in the KÖPPEN CLIMATE CLASSIFICATION. Also spelled "climagraph." The term climagram is rarely used.

Clinometer. Instrument used by surveyors to measure the ELEVATION of land or the inclination (slope) of the land surface.

Cloud seeding. Injection of CLOUD-nucleating particles into likely clouds to enhance PRECIPITATION.

Cloudburst. Heavy rain that falls suddenly.

Coal. One of the FOSSIL FUELS. Coal was formed from fossilized plant material, which was originally FOREST. It was then buried and compacted, which led to chemical changes. Most coal was formed during the CARBONIFEROUS PERIOD (286 million to 360 million years ago) when the earth's CLIMATE was wetter and warmer than at present.

Coastal plain. Large area of flat land near the OCEAN. Coastal plains can form in various ways, but fluvial deposition is an important process. In the United States, the coastal plain extends from Texas to North Carolina.

Coastal wetlands. Shallow, wet, or flooded shelves that extend back from the freshwater-saltwater interface and may consist of marshes, bays, lagoons, tidal flats, or mangrove swamps.

Cognitive map. Mental image that each person has of the world, which includes locations and connections. These maps expand as children mature, from plans of their rooms, to their houses, to their neighborhoods. Adults know certain parts of the city and the streets connecting them.

Coke. Type of fuel produced by heating coal.

Col. Lower section of a ridge, usually formed by the headward erosion of two cirque glaciers at an arête. Sometimes called a saddle.

Colonialism. Control of one country over another state and its people. Many European countries have created colonial empires, including Great Britain, France, Spain, Portugal, the Netherlands, and Russia.

Columbian exchange. Interaction that occurred between the Americas and Europe after the voyages of Christopher Columbus. Food crops from the New World transformed the diet of many European countries.

Comet. Small body in the solar system, consisting of a solid head with a long gaseous tail. The elliptical orbit of a comet causes it to range from very close to the Sun to very far away. In ancient times, the appearance of a comet in the sky was thought to be an omen of great events or changes, such as war or the death of a king.

Comfort index. Number that expresses the combined effects of temperature and humidity on human bodily comfort. The index number is obtained by measuring ambient conditions and comparing these to a chart.

Commodity chain. Network linking labor, production, delivery, and sale for any product. The chain begins with the production of the raw material, such as the extraction of minerals by miners, and extends to the acquisition of the finished product by a consumer.

Complex crater. Impact crater of large diameter and low depth-to-diameter ratio caused by the presence of a central uplift or ring structure.

Composite cone. Cone or volcano formed by volcanic explosions in which the lava is of different composition, sometimes fluid, sometimes pyroclasts such as cinders. The alternation of layers allows a concave shape for the cone. These are generally regarded as the world's most beautiful volcanoes. Composite volcanoes are sometimes called stratovolcanoes.

Condensation nuclei. Microscopic particles that may have originated as dust, soot, ash from fires or volcanoes, or even sea salt; an essential part of cloud formation. When air rises and cools to the dew point (saturation), the moisture droplets condense around the nuclei, leading to the creation of raindrops or snowflakes. A typical air mass might contain 10 billion condensation nuclei in a single cubic yard (1 cubic meter) of air.

Cone of depression. Cone-shaped depression produced in the water table by pumping from a well.

Confined aquifer. Aquifer that is completely filled with water and whose upper boundary is a confining bed; it is also called an artesian aquifer.

Confining bed. Impermeable layer in the earth that inhibits vertical water movement.

Confluence. Place where two streams or rivers flow together and join. The smaller of the two streams is called a tributary.

Conglomerate. Type of sedimentary rock consisting of smaller rounded fragments naturally cemented together by another mineral. If the cemented fragments are jagged or angular, the rock is called breccia.

Conical projection. Map projection that can be imagined as a cone of paper resting like a witch's hat on a globe with a light source at its center; the images of the continents would be projected onto the paper. In reality, maps are constructed mathematically. A conic projection can show only part of one hemisphere. This projection is suitable for constructing a map of the United States, as a good equal-area representation can be achieved. Also called conic projection.

Consequent river. River that flows across a landscape because of gravity. Its direction is determined by the original slope of the land. Tributary streams, which develop later as erosion proceeds, are called subsequent streams.

Continental climate. Climate experienced over the central regions of large landmasses; drier and subject to greater seasonal extremes of temperature than at the continental margins.

Continental rift zones. Continental rift zones are places where the continental crust is stretched and thinned. Distinctive features include active volcanoes and long, straight valley systems formed by normal faults. Continental rifting in some cases has evolved into the breaking apart of a continent by seafloor spreading to form a new ocean.

Continental shelf. Shallow, gently sloping part of the seafloor adjacent to the mainland. The continental shelf is geologically part of the continent and is made of continental crust, whereas the ocean floor is oceanic crust. Although continental shelves vary greatly in width, on average they are about 45 miles (75 km.) wide and have slopes of 7 minutes (about one-tenth of a degree). The average depth of a continental shelf is about 200 feet (60 meters). The outer edge of the continental shelf is marked by a sharp change in angle where the continental slope begins. Most continental shelves were exposed above current sea level during the Pleistocene epoch and have been submerged by rising sea levels over the past 18,000 years.

Continental shield. Area of a continent that contains the oldest rocks on Earth, called cratons. These are areas of granitic rocks, part of the continental crust, where there are ancient mountains. The Canadian Shield in North America is an example.

Convectional rain. Type of precipitation caused when air over a warm surface is warmed and rises, leading to adiabatic cooling, condensation, and, if the air is moist enough, rain.

Convective overturn. Renewal of the bottom waters caused by the sinking of surface waters that have become denser, usually because of decreased temperature.

Convergence (climate). Air flowing in toward a central point.

Convergence (physiography). Process that occurs during the second half of a supercontinent cycle, whereby crustal plates collide and intervening oceans disappear as a result of plate subduction.

Convergent plate boundary. Compressional plate boundary at which an oceanic plate is subducted or two continental plates collide.

Convergent plate margin. Area where the earth's lithosphere is returned to the mantle at a subduction zone, forming volcanic "island arcs" and associated hydrothermal activity.

Conveyor belt current. Large cycle of water movement that carries warm water from the north Pacific westward across the Indian Ocean, around Southern Africa, and into the Atlantic, where it warms the atmosphere, then returns at a deeper ocean level to rise and begin the process again.

Coordinated universal time (UTC). International basis of time, introduced to the world in 1964. The basis for UTC is a small number of atomic clocks. Leap seconds are occasionally added to UTC to keep it synchronized with universal time.

Core-mantle boundary. Seismic discontinuity 1,790 miles (2,890 km.) below the earth's surface that separates the mantle from the outer core.

Core region. Area, generally around a country's capital city, that has a large, dense population and is the center of trade, financial services, and production. The rest of the country is referred to as the periphery. On a larger scale, the continent of Europe has a core region, which includes London, Paris, and Berlin; Iceland, Portugal, and Greece are peripheral locations.

Coriolis effect. Apparent deflection of moving objects above the earth because of the earth's rota-

TION. The deflection is to the right in the NORTHERN HEMISPHERE and to the left in the SOUTHERN HEMISPHERE. The deflection is inversely proportional to the speed of the earth's rotation, being negligible at the EQUATOR but at its maximum near the POLES. The Coriolis effect is a major influence on the direction of surface WINDS. Sometimes called Coriolis force.

Corrasion. EROSION and lowering of a STREAMBED by FLUVIAL action, especially by ABRASION of the bedload (material transported by the STREAM) but also including SOLUTION by the water.

Cosmogony. Study of the origin and nature of the SOLAR SYSTEM.

Cotton Belt. Part of the United States extending from South Carolina through Georgia, Alabama, Mississippi, Tennessee, Louisiana, Arkansas, Texas, and Oklahoma, where cotton was grown on PLANTATIONS using slave labor before the Civil War. After that war, the South stagnated for almost a century. Racial SEGREGATION contributed to cultural isolation from the rest of the United States. Cotton is still produced in this REGION, but California has overtaken the Southern STATES as a cotton producer, and other agricultural products, such as soybeans and poultry, have become dominant crops in the old Cotton Belt. In-migration, due to the SUN BELT attraction, has led to rapid urban growth.

Counterurbanization. Out-migration of people from URBAN AREAS to smaller TOWNS or RURAL areas. As large modern cities are perceived to be overcrowded, stressful, polluted, and dangerous, many of their residents move to areas they regard as more favorable. Such moves are often related to individuals' retirements; however, younger workers and families are also part of counterurbanization.

Crater morphology. Structure or form of CRATERS and the related processes that developed them.

Craton. Large, geologically old, relatively stable CORE of a continental LITHOSPHERIC PLATE, sometimes termed a CONTINENTAL SHIELD.

Creep. Slow, gradual downslope movement of SOIL materials under gravitational stress. Creep tests are experiments conducted to assess the effects of time on ROCK properties, in which environmental conditions (surrounding pressure, TEMPERATURE) and the deforming stress are held constant.

Crestal plane. Plane or surface that goes through the highest points of all beds in a fold; it is coincident with the axial plane when the axial plane is vertical.

Cross-bedding. Layers of ROCK or SAND that lie at an angle to horizontal bedding or to the ground.

Crown land. Land belonging to a NATION'S MONARCHY.

Crude oil. Unrefined OIL, as it occurs naturally. Also called PETROLEUM.

Crustal movements. PLATE TECTONICS theorizes that Earth's CRUST is not a single rigid shell, but comprises a number of large pieces that are in motion, separating or colliding. There are two types of crust—the older continental and the much younger OCEANIC CRUST. When PLATES diverge, at SEAFLOOR SPREADING zones, new (oceanic) crust is created from the MAGMA that flows out at the MID-OCEAN RIDGES. When plates converge and collide, denser oceanic crust is SUBDUCTED under the lighter CONTINENTAL CRUST. The boundaries at the areas where plates slide laterally, neither diverging nor converging, are called TRANSFORM FAULTS. The San Andreas Fault represents the world's best-known transform BOUNDARY. As a result of crustal movements, the earth can be deformed in several ways. Where PLATE BOUNDARIES converge, compression can occur, leading to FOLDING and the creation of SYNCLINES and ANTICLINES. Other stresses of the crust can lead to fracture, or faulting, and accompanying EARTHQUAKES. LANDFORMS created in this way include HORSTS, GRABEN, and BLOCK MOUNTAINS.

Culture hearth. LOCATION in which a CULTURE has developed; a CORE REGION from which the culture later spread or diffused outward through a larger REGION. Mesopotamia, the Nile Valley, and the Peruvian ALTIPLANO are examples of culture hearths.

Curie point. TEMPERATURE at which a magnetic MINERAL locks in its magnetization. Also known as Curie temperature.

Cycle of erosion. Influential MODEL of LANDSCAPE change proposed by William Morris Davis near the end of the nineteenth century. The UPLIFT of a relatively flat surface, or PLAIN, in an area of moderate RAINFALL and TEMPERATURE, led to gradual EROSION of the initial surface in a sequence Davis categorized as Youth, Maturity, and Old Age. The final landscape was called PENEPLAIN. Davis also recognized the stage of REJUVENATION, when a new uplift could give new ENERGY to the cycle, leading to further downcutting and erosion. The model also was used to explain the sequence of LANDFORMS developed in REGIONS of ALPINE GLACIERS. The model has been criticized as misleading, since CRUSTAL MOVEMENT is continuous and more frequent than Davis perhaps envisaged, but remained useful as a description of TOPOGRAPHY. Also known as the Davisian cycle or geomorphic cycle.

Cyclonic rain. In the NORTHERN HEMISPHERE winter, two low-pressure systems or CYCLONES—the Aleutian Low and the Icelandic Low—develop over the OCEAN near 60 DEGREES north LATITUDE. The polar FRONT forms where the cold and relatively dry ARCTIC AIR meets the warmer, moist air carried by westerly WINDS. The warm air is forced upward, cools, and condenses. These cyclonic STORMS often move south, bringing winter PRECIPITATION to North America, especially to the STATES of Washington and Oregon.

Cylindrical projection. MAP PROJECTION that represents the earth's surface as a rectangle. It can be imagined as a cylinder of paper wrapped around a globe with a light source at its center; the images of the CONTINENTS would be projected onto the paper. In reality, MAPS are constructed mathematically. It is impossible to show the North Pole or South Pole on a cylindrical projection. Although the map is conformal, distortion of area is extreme beyond 50 DEGREES north and south LATITUDES. The Mercator projection, developed in the sixteenth century by the Flemish cartographer Gerardus Mercator, is the best-known cylindrical projection. It has been popular with seamen because the shortest route between two PORTS (the GREAT CIRCLE route) can be plotted as straight lines that show the COMPASS direction that should be followed. Use of this projection for other purposes, however, can lead to misunderstandings about size; for example, compare Greenland on a globe and on a Mercator map.

Datum level. Baseline or level from which other heights are measured, above or below. MEAN SEA LEVEL is the datum commonly used in surveying and in the construction of TOPOGRAPHIC MAPS.

Daylight saving time. System of seasonal adjustments in CLOCK settings designed to increase hours of evening sunlight during summer months. In the spring, clocks are set ahead one hour; in the fall, they are put back to standard time. In North America, these changes are made on the first Sunday in April and the last Sunday in October. The U.S. Congress standardized daylight saving time in 1966; however, parts of Arizona, Indiana, and Hawaii do not follow the system.

Débâcle. In a scientific context, this French word means the sudden breaking up of ice in a RIVER in the spring, which can lead to serious, sudden flooding.

Debris avalanche. Large mass of SOIL and ROCK that falls and then slides on a cushion of AIR downhill rapidly as a unit.

Debris flow. Flowing mass consisting of water and a high concentration of SEDIMENT with a wide RANGE of size, from fine muds to coarse gravels.

Declination, magnetic. Measure of the difference, in DEGREES, between the earth's NORTH MAGNETIC POle and the North Pole on a MAP; this difference changes slightly each year. The needle of a magnetic COMPASS points to the earth's geomagnetic pole, which is not exactly the same as the North Pole of the geographic GRID or the set of lines of LATITUDE and LONGITUDE. The geomagnetic poles, north and south, mark the ends of the AXIS of the earth's MAGNETIC FIELD, but this field is not stationary. In fact, the geomagnetic poles have completely reversed hundreds of times throughout earth history. Lines of equal magnetic declination are called ISOGONIC LINES.

Declination of the Sun. LATITUDE of the SUBSOLAR POINT, the PLACE on the earth's surface where the SUN is directly overhead. In the course of a year, the declination of the Sun migrates from 23.5 DEGREES north LATITUDE, at the (northern) summer SOLSTICE, to 23.5 degrees south latitude, at the (northern) WINTER SOLSTICE. Hawaii is the only part of the United States that experiences the Sun directly overhead twice a year.

Deep-focus earthquakes. EARTHQUAKES occurring at depths ranging from 40 to 400 miles (70–700 km.) below the earth's surface. This RANGE of depths represents the zone from the base of the earth's CRUST to approximately one-quarter of the distance into Earth's MANTLE. Deep-focus earthquakes provide scientists information about the PLANET's interior structure, its composition, and SEISMICITY. Observation of deep-focus earthquakes has played a fundamental role in the discovery and understanding of PLATE TECTONICS.

Deep-ocean currents. Deep-ocean currents involve significant vertical and horizontal movements of seawater. They distribute oxygen- and nutrient-rich waters throughout the world's OCEANS, thereby enhancing biological productivity.

Defile. Narrow MOUNTAIN PASS or GORGE through which troops could march only in single file.

Deflation. EROSION by WIND, resulting in the removal of fine particles. The LANDFORM that typically results is a deflation hollow.

Deforestation. Removal or destruction of FORESTS. In the late twentieth century, there was widespread concern about tropical deforestation—destruction of the tropical RAINFOREST—especially that of Brazil. Forest clearing in the TROPICS is uneconomic because of low SOIL fertility. Deforestation causes severe EROSION and environmental damage; it also destroys habitat, which leads to the EXTINCTION of both plant and animal species.

Degradation. Process of CRATER EROSION from all processes, including WIND and other meteorological mechanisms.

Degree (geography). Unit of LATITUDE or LONGITUDE in the geographic GRID, used to determine ABSOLUTE LOCATION. One degree of latitude is about 69 miles (111 km.) on the earth's surface. It is not exactly the same everywhere, because the earth is not a perfect sphere. One degree of longitude varies greatly in length, because the MERIDIANS converge at the POLES. At the EQUATOR, it is 69 miles (111 km.), but at the North or South Pole it is zero.

Degree (temperature). Unit of MEASUREMENT of TEMPERATURE, based on the CELSIUS SCALE, except in the United States, which uses the FAHRENHEIT SCALE. On the Celsius scale, one degree is one-hundredth of the difference between the freezing point of water and the boiling point of water.

Demographic measure. Statistical data relating to POPULATION.

Demographic transition. MODEL of POPULATION change that fits the experience of many European countries, showing changes in birth and death rates. In the first stage, in preindustrial countries, population size was stable because both BIRTH RATES and DEATH RATES were high. Agricultural reforms, together with the INDUSTRIAL REVOLUTION and subsequent medical advances, led to a rapid fall in the death rate, so that the second and third stages of the model were periods of rapid population growth, often called the POPULATION EXPLOSION. In the fourth stage of the model, birth rates fall markedly, leading again to stable population size.

Dendritic drainage. Most common pattern of STREAMS and their TRIBUTARIES, occurring in areas of uniform ROCK type and regular slope. A MAP, or aerial photograph, shows a pattern like the veins on a leaf—smaller streams join the main stream at an acute angle.

Denudation. General word for all LANDFORM processes that lead to a lowering of the LANDSCAPE, including WEATHERING, mass movement, EROSION, and transport.

Deposition. Laying down of SEDIMENTS that have been transported by water, WIND, or ice.

Deranged drainage. LANDSCAPE whose integrated drainage network has been destroyed by irregular glacial DEPOSITION, yielding numerous shallow LAKE BASINS.

Derivative maps. MAPS that are prepared or derived by combining information from several other maps.

Desalinization. Process of removing SALT and MINERALS from seawater or from saline water occurring in AQUIFERS beneath the land surface to render it fit for AGRICULTURE or other human use.

Desert climate. Low PRECIPITATION, low HUMIDITY, high daytime TEMPERATURES, and abundant sunlight are characteristics of desert climates. The hot DESERTS of the world generally are located on the western sides of CONTINENTS, at LATITUDES from fifteen to thirty DEGREES north or south of the EQUATOR. One definition, based on precipitation, defines deserts as areas that receive between 0 and 9 inches (0 to 250 millimeters) of precipitation per year. REGIONS receiving more precipitation are considered to have a SEMIDESERT climate, in which some AGRICULTURE is possible.

Desert pavement. Surface covered with smoothed PEBBLES and gravels, found in arid areas where DEFLATION (WIND EROSION) has removed the smaller particles. Called a "gibber plain" in Australia.

Desertification. Increase in DESERT areas worldwide, largely as a result of overgrazing or poor agricultural practices in semiarid and marginal CLIMATES. DEFORESTATION, DROUGHT, and POPULATION increase also contribute to desertification. The REGION of Africa just south of the Sahara Desert, known as the SAHEL, is the largest and most dramatic demonstration of desertification.

Detrital rock. SEDIMENTARY ROCK composed mainly of grains of silicate MINERALS as opposed to grains of calcite or CLAYS.

Devolution. Breaking up of a large COUNTRY into smaller independent political units is the final and most extreme form of devolution. The Soviet Union devolved from one single country into fifteen separate countries in 1991. At an intermediate level, devolution refers to the granting of political autonomy or self-government to a REGION, without a complete split. The reopening of the Scottish Parliament in 1999 and the Northern Ireland parliament in 2000 are examples of devolution; the Parliament of the United Kingdom had previously met only in London and made laws there for all parts of the country. Canada experienced devolution with the creation of the new territory of Nunavut, whose residents elect the members of their own legislative assembly.

Dew point. TEMPERATURE at which an AIR mass becomes saturated and can hold no more moisture. Further cooling leads to CONDENSATION. At ground level, this produces DEW.

Diagenesis. Conversion of unconsolidated SEDIMENT into consolidated ROCK after burial by the processes of compaction, cementation, recrystallization, and replacement.

Diaspora. Dispersion of a group of people from one CULTURE to a variety of other REGIONS or to other lands. A Greek word, used originally to refer to the Jewish diaspora. Jewish people now live in many countries, although they have Israel as a HOMELAND. Similar to this are the diasporas of the Irish and the overseas Chinese.

Diastrophism. Deformation of the earth's CRUST by faulting or FOLDING.

Diatom ooze. Deposit of soft mud on the OCEAN floor consisting of the shells of diatoms, which are microscopic single-celled creatures with SILICA-rich shells. Diatom ooze deposits are located in the southern Pacific around Antarctica and in the northern Pacific. Other PELAGIC, or deep-ocean, SEDIMENTS include CLAYS and calcareous ooze.

Dike (geology). LANDFORM created by IGNEOUS intrusion when MAGMA or molten material within the earth forces its way in a narrow band through overlying ROCK. The dike can be exposed at the surface through EROSION.

Dike (water). Earth wall or DAM built to prevent flooding; an EMBANKMENT or artificial LEVEE. Sometimes specifically associated with structures built in the Netherlands to prevent the entry of seawater. The land behind the dikes was reclaimed for AGRICULTURE; these new fields are called POLDERS.

Distance-decay function. Rate at which an activity diminishes with increasing distance. The effect that distance has as a deterrent on human activity is sometimes described as the FRICTION OF DISTANCE. It occurs because of the time and cost of overcoming distances between people and their desired activity. An example of the distance-decay function is the rate of visitors to a football stadium. The farther people have to travel, the less likely they are to make this journey.

Distributary. STREAM that takes waters away from the main CHANNEL of a RIVER. A DELTA usually comprises many distributaries. Also called distributary channel.

Diurnal range. Difference between the highest and lowest TEMPERATURES registered in one twenty-four-hour period.

Diurnal tide. Having only one high tide and one low tide each lunar DAY; TIDES on some parts of the Gulf of Mexico are diurnal.

Divergent boundary. BOUNDARY that results where two PLATES are moving apart from each other, as is the case along MID-OCEANIC RIDGES.

Divergent margin. Area where the earth's CRUST and LITHOSPHERE form by SEAFLOOR SPREADING.

Doline. Large SINKHOLE or circular depression formed in LIMESTONE areas through the CHEMICAL WEATHERING process of carbonation.

Dolomite. MINERAL consisting of calcium and magnesium carbonate compounds that often forms from PRECIPITATION from seawater; it is abundant in ancient ROCKS.

Downwelling. Sinking of OCEAN water.

Drainage basin. Area of the earth's surface that is drained by a STREAM. Drainage basins vary greatly in size, but each is separated from the next by RIDGES, or drainage DIVIDES. The CATCHMENT of the drainage basin is the WATERSHED.

Drift ice. ARCTIC or ANTARCTIC ice floating in the open SEA.

Drumlin. Low HILL, shaped like half an egg, formed by DEPOSITION by CONTINENTAL GLACIERS. A drumlin is composed of TILL, or mixed-size materials. The wider end faces upstream of the glacier's movement; the tapered end points in the direction of the ice movement. Drumlins usually occur in groups or swarms.

Dust devil. Whirling cloud of DUST and small debris, formed when a small patch of the earth's surface becomes heated, causing hot AIR to rise; cooler air then flows in and begins to spin. The resulting dust devil can grow to heights of 150 feet (50 meters) and reach speeds of 35 miles (60 km.) per hour.

Dust dome. Dome of AIR POLLUTION, composed of industrial gases and particles, covering every large CITY in the world. The pollution sometimes is carried downwind to outlying areas.

Earth pillar. Formation produced when a boulder or caprock prevents EROSION of the material directly beneath it, usually CLAY. The clay is easily eroded away by water during RAINFALL, except where the overlying ROCK protects it. The result is a tall, slender column, as high as 20 feet (6.5 meters) in exceptional cases.

Earth radiation. Portion of the electromagnetic spectrum, from about 4 to 80 microns, in which the earth emits about 99 percent of its RADIATION.

Earth tide. Slight deformation of Earth resulting from the same forces that cause OCEAN TIDES, those that are exerted by the MOON and the SUN.

Earthflow. Term applied to both the process and the LANDFORM characterized by fluid downslope movement of SOIL and ROCK over a discrete plane of failure; the landform has a HUMMOCKY surface and usually terminates in discrete lobes.

Earth's heat budget. Balance between the incoming SOLAR RADIATION and the outgoing terrestrial reradiation.

Eclipse, lunar. Obscuring of all or part of the light of the Moon by the shadow of the earth. A lunar eclipse occurs at the full moon up to three times a year. The surface of the Moon changes from gray to a reddish color, then back to gray. The sequence may last several hours.

Eclipse, solar. At least twice a year, the Sun, Moon, and Earth are aligned in one straight line. At that time, the Moon obscures all the light of the Sun along a narrow band of the earth's surface, causing a total eclipse; in regions of Earth adjoining that area, there is a partial eclipse. A corona (halo of light) can be seen around the Sun at the total eclipse. Viewing a solar eclipse with naked eyes is extremely dangerous and can cause blindness.

Ecliptic, plane of. Imaginary plane that would touch all points in the earth's orbit as it moves around the Sun. The angle between the plane of the ecliptic and the earth's axis is 66.5 degrees.

Edge cities. Forms of suburban downtown in which there are nodal concentrations of office space and shopping facilities. Edge cities are located close to major freeways or highway intersections, on the outer edges of metropolitan areas.

Effective temperature. Temperature of a planet based solely on the amount of solar radiation that the planet's surface receives; the effective temperature of a planet does not include the greenhouse temperature enhancement effect.

Ejecta. Material ejected from the crater made by a meteoric impact.

Ekman layer. Region of the sea, from the surface to about 100 meters down, in which the wind directly affects water movement.

Eluviation. Removal of materials from the upper layers of a soil by water. Fine material may be removed by suspension in the water; other material is removed by solution. The removal by solution is called leaching. Eluviation from an upper layer leads to illuviation in a lower layer.

Enclave. Piece of territory completely surrounded by another country. Two examples are Lesotho, which is surrounded by the Republic of South Africa, and the Nagorno-Karabakh region, populated by Armenians but surrounded by Azerbaijan. The term is also used for smaller regions, such as ethnic neighborhoods within larger cities. See also Exclave.

Endemic species. Species confined to a restricted area in a restricted environment.

Endogenic sediment. Sediment produced within the water column of the body in which it is deposited; for example, calcite precipitated in a lake in summer.

Environmental degradation. Situation that occurs in slum areas and squatter settlements because of poverty and inadequate infrastructure. Too-rapid human population growth can lead to the accumulation of human waste and garbage, the pollution of groundwater, and denudation of nearby forests. As a result, life expectancy in such degraded areas is lower than in the rural communities from which many of the settlers came. Infant mortality is particularly high. When people leave an area because of such environmental degradation, that is referred to as ecomigration.

Environmental determinism. Theory that the major influence on human behavior is the physical environment. Some evidence suggests that temperature, precipitation, sunlight, and topography influence human activities. Originally espoused by early German geographers, this theory has led to some extreme stances, however, by authors who have sought to explain the dominance of Europeans as a result of a cool temperate climate.

Eolian (aeolian). Relating to, or caused by, wind. In Greek mythology, Aeolus was the ruler of the winds. Erosion, transport, and deposition are common eolian processes that produce landforms in desert regions.

Eolian deposits. Material transported by the wind.

Eolian erosion. Mechanism of erosion or crater degradation caused by wind.

Eon. Largest subdivision of geologic time; the two main eons are the Precambrian (c. 4.6 billion years ago to 544 million years ago) and the Phanerozoic (c. 544 million years ago to the present).

Ephemeral stream. Watercourse that has water for only a DAY or so.

Epicontinental sea. Shallow SEAS that are located on the CONTINENTAL SHELF, such as the North Sea or Hudson Bay. Also called an EPEIRIC SEA.

Epifauna. Organisms that live on the seafloor.

Epilimnion. Warmer surface layer of water that occurs in a LAKE during summer stratification; during spring, warmer water rises from great depths, and it heats up through the summer SEASON.

Equal-area projection. MAP PROJECTION that maintains the correct area of surfaces on7 a MAP, although shape distortion occurs. The property of such a map is called equivalence.

Erg. Sandy DESERT, sometimes called a SEA of SAND. Erg deserts account for less than 30 percent of the world's deserts. "Erg" is an Arabic word.

Eruption, volcanic. Emergence of MAGMA (molten material) at the earth's surface as LAVA. There are various types of volcanic eruptions, depending on the chemistry of the magma and its viscosity. Scientists refer to effusive and explosive eruptions. Low-viscosity magma generally produces effusive eruptions, where the lava emerges gently, as in Hawaii and Iceland, although explosive events can occur at those SITES as well. Gently sloping SHIELD VOLCANOES are formed by effusive eruptions; FLOODS, such as the Columbia Plateau, can also result. Explosive eruptions are generally associated with SUBDUCTION. Much gas, including steam, is associated with magma formed from OCEANIC CRUST, and the compressed gas helps propel the explosion. COMPOSITE CONES, such as Mount Saint Helens, are created by explosive eruptions.

Escarpment. Steep slope, often almost vertical, formed by faulting. Sometimes called a FAULT SCARP.

Esker. Deposit of coarse gravels that has a sinuous, winding shape. An esker is formed by a STREAM of MELTWATER that flowed through a tunnel it formed under a CONTINENTAL GLACIER. Now that the continental glaciers have melted, eskers can be found exposed at the surface in many PLACES in North America.

Estuarine zone. Area near the COASTLINE that consists of estuaries and coastal saltwater WETLANDS.

Etesian winds. WINDS that blow from the north over the Mediterranean during July and August.

Ethnocentrism. Belief that one's own ETHNIC GROUP and its CULTURE are superior to any other group.

Ethnography. Study of different CULTURES and human societies.

Eustacy. Any change in global SEA LEVEL resulting from a change in the absolute volume of available sea water. Also known as eustatic sea-level change.

Eustatic movement. Changes in SEA LEVEL.

Exclave. Territory that is part of one COUNTRY but separated from the main part of that country by another country. Alaska is an exclave of the United States; Kaliningrad is an exclave of Russia. See also ENCLAVE.

Exfoliation. When GRANITE rocks cooled and solidified, removal of the overlyingrock that was present reduced the pressure on the granite mass, allowing it to expand and causing sheets or layers of rock to break off. An exfoliation DOME, such as Half Dome in Yosemite National Park, is the resultant LANDFORM.

Exotic stream. RIVER that has its source in an area of high RAINFALL and then flows through an arid REGION or DESERT. The Nile River is the most famous exotic STREAM. In the United States, the Colorado River is a good example of an exotic stream.

Expansion-contraction cycles. Processes of wetting-drying, heating-cooling, or freezing-thawing, which affect SOIL particles differently according to their size.

Extrusive rock. Fine-grained, or glassy, ROCK which was formed from a MAGMA that cooled on the surface of the earth.

Fall line. Edge of an area of uplifted land, marked by WATERFALLS where STREAMS flow over the edge.

Fata morgana. Large mirage. Originally, the name given to a multiple mirage phenomenon often observed over the Straits of Messina and supposed to be the work of the fairy ("fata") Morgana. Another famous fata morgana is located in Antarctica.

Fathometer. Instrument that uses sound waves or sonar to determine the depth of water or the depth of an object below the water.

Fault drag. Bending of ROCKS adjacent to a FAULT.

Fault line. Line of breakage on the earth's surface. FAULTS may be quite short, but many are extremely long, even hundreds of miles. The origin of the faulting may lie at a considerable depth below the surface. Movement along the fault line generates EARTHQUAKES.

Fault plane. Angle of a FAULT. When fault blocks move on either side of a fault or fracture, the movement can be vertical, steeply inclined, or sometimes horizontal. In a NORMAL FAULT, the fault plane is steep to almost vertical. In a REVERSE FAULT, one block rides over the other, forming an overhanging FAULT SCARP. The angle of inclination of the fault plane from the horizontal is called the dip. The inclination of a fault plane is generally constant throughout the length of the fault, but there can be local variations in slope. In a STRIKE-SLIP FAULT the movement is horizontal, so no fault scarp is produced, although the FAULT LINE may be seen on the surface.

Fault scarp. FAULTS are produced through breaking or fracture of the surface ROCKS of the earth's CRUST as a result of stresses arising from tectonic movement. A NORMAL FAULT, one in which the earth movement is predominantly vertical, produces a steep fault scarp. A STRIKE-SLIP FAULT does not produce a fault scarp.

Feldspar. Family name for a group of common MINERALS found in such ROCKS as GRANITE and composed of silicates of aluminum together with potassium, sodium, and calcium. Feldspars are the most abundant group of minerals within the earth's CRUST. There are many varieties of feldspar, distinguished by variations in chemistry and crystal structure. Although feldspars have some economic uses, their principal importance lies in their role as rock-forming minerals.

Felsic rocks. IGNEOUS ROCKS rich in potassium, sodium, aluminum, and SILICA, including GRANITES and related rocks.

Fertility rate. DEMOGRAPHIC MEASURE of the average number of children per adult female in any given POPULATION. Religious beliefs, education, and other cultural considerations influence fertility rates.

Fetch. Distance along a large water surface over which a WIND of almost uniform direction and speed blows.

Feudalism. Social and economic system that prevailed in Europe before the INDUSTRIAL REVOLUTION. The land was owned and controlled by a minority comprising noblemen or lords; all other people were peasants or serfs, who worked as agricultural laborers on the lords' land. The peasants were not free to leave, or to do anything without their lord's permission. Other REGIONS such as China and Japan also had a feudal system in the past.

Firn. Intermediate stage between SNOW and glacial ice. Firn has a granular TEXTURE, due to compaction. Also called NÉVÉ.

Fission, nuclear. Splitting of an atomic nucleus into two lighter nuclei, resulting in the release of neutrons and some of the binding ENERGY that held the nucleus together.

Fissure. Fracture or crack in ROCK along which there is a distinct separation.

Flash flood. Sudden rush of water down a STREAM CHANNEL, usually in the DESERT after a short but intense STORM. Other causes, such as a DAM failure, could lead to a flash flood.

Flood control. Attempts by humans to prevent flooding of STREAMS. Humans have consistently settled on FLOODPLAINS and DELTAS because of the fertile SOIL for AGRICULTURE, and attempts at flood control date back thousands of years. In strictly agricultural societies such as ancient Egypt, people built VILLAGES above the FLOOD levels, but transport and industry made riverside LOCATIONS desirabl and engineers devised technological means to try to prevent flood damage. Artificial LEVEES, RESERVOIRS, and DAMS of ever-increasing size were built on rivers, as well as bypass channels leading to artificial floodplains. In many modern dam construction projects, the production of hydroelectric power was more important than flood control. Despite modern technology, floods cause the largest loss of human life of all natural disasters, especially in low-income countries such as Bangladesh.

Flood tide. Rising or incoming tide. Most parts of the world experience two flood tides in each 24-hour period.

Floodplain. Flat, low-lying land on either side of a stream, created by the deposition of alluvium from floods. Also called alluvial plain.

Fluvial. Pertaining to running water; for example, fluvial processes are those in which running water is the dominant agent.

Fog deserts. Coastal deserts where fog is an important source of moisture for plants, animals, and humans. The fog forms because of a cold ocean current close to the shore. The Namib Desert of southwestern Africa, the west coast of California, and the Atacama Desert of Peru are coastal deserts.

Föhn wind. Wind warmed and dried by descent, usually on the lee side of a mountain. In North America, these winds are called the Chinook.

Fold mountains. Rocks in the earth's crust can be bent by compression, producing folds. The Swiss Alps are an example of complex folding, accompanied by faulting. Simple upward folds are anticlines, downward folds are synclines; but subsequent erosion can produce landscapes with synclinal mountains.

Folding. Bending of rocks in the earth's crust, caused by compression. The rocks are deformed, sometimes pushed up to form mountain ranges.

Foliation. Texture or structure in which mineral grains are arranged in parallel planes.

Food web. Complex network of food chains. Food chains are interconnected, because many organisms feed on a variety of others, and in turn may be eaten by any of a number of predators.

Forced migration. Migration that occurs when people are moved against their will. The Atlantic slave trade is an example of forced migration. People were shipped from Africa to countries in Europe, Asia, and the New World as forced immigrants. Within the United States, some Native Americans were forced by the federal government to migrate to new reservations.

Ford. Short shallow section of a river, where a person can cross easily, usually by walking or riding a horse. To cross a stream in such a manner.

Formal region. Cultural region in which one trait, or group of traits, is uniform. Language might be the basis of delineation of a formal cultural region. For example, the Francophone region of Canada constitutes a formal region based on one single trait. One might also identify a formal Mormon region centered on the state of Utah, combining religion and landscape as defining traits. Cultural geographers generally identify formal regions using a combination of traits.

Fossil fuel. Deposit rich in hydrocarbons, formed from organic materials compressed in rock layers—coal, oil, and natural gas.

Fossil record. Fossil record provides evidence that addresses fundamental questions about the origin and history of life on the earth: When life evolved; how new groups of organisms originated; how major groups of organisms are related. This record is neither complete nor without biases, but as scientists' understanding of the limits and potential of the fossil record grows, the interpretations drawn from it are strengthened.

Fossilization. Processes by which the remains of an organism become preserved in the rock record.

Fracture zones. Large, linear zones of the seafloor characterized by steep cliffs, irregular topography, and faults; such zones commonly cross and displace oceanic ridges by faulting.

Free association. Relationship between sovereign nations in which one nation—invariably the larger—has responsibility for the other nation's defense. The Cook Islands in the South Pacific have such a relationship with New Zealand.

Friction of distance. Distance is of prime importance in social, political, economic, and other relationships. Large distance has a negative effect on human activity. The time and cost of overcoming distance can be a deterrent to various activities. This has been called the friction of distance.

Frigid zone. Coldest of the three CLIMATE zones proposed by the ancient Greeks on the basis of their theories about the earth. There were two frigid zones, one around each POLE. The Greeks believed that human life was possible only in the TEMPERATE ZONE.

Fringing reef. Type of CORAL REEF formed at the SHORELINE, extending out from the land in shallow water. The top of the coral may be exposed at low TIDE.

Frontier Thesis. Thesis first advanced by the U.S. historian Frederick Jackson Turner, who declared that U.S. history and the U.S. character were shaped by the existence of empty, FRONTIER lands that led to exploration and westward expansion and DEVELOPMENT. The closing of the frontier occurred when transcontinental railroads linked the East and West Coasts and SETTLEMENTS spread across the United States. This thesis was used by later historians to explain the history of South Africa, Canada, and Australia. Critics of the Frontier Thesis point out that minorities and women were excluded from this view of history.

Frost wedging. Powerful form of PHYSICAL WEATHERING of ROCK, in which the expansion of water as it freezes in JOINTS or cracks shatters the rock into smaller pieces. Also known as frost shattering.

Fumarole. Crack in the earth's surface from which steam and other gases emerge. Fumaroles are found in volcanic areas and areas of GEOTHERMAL activity, such as Yellowstone National Park.

Fusion energy. Heat derived from the natural or human-induced union of atomic nuclei; in effect, the opposite of FISSION energy.

Gall's projection. MAP PROJECTION constructed by projecting the earth onto a cylinder that intersects the sphere at 45 DEGREES north and 45 degrees south LATITUDE. The resulting map has less distortion of area than the more familiar CYLINDRICAL PROJECTION of Mercator.

Gangue. Apparently worthless ROCK or earth in which valuable gems or MINERALS are found.

Garigue. VEGETATION cover of small shrubs found in Mediterranean areas. Similar to the larger *maquis*.

Genus (plural, genera). Group of closely related species; for example, *Homo* is the genus of humans, and it includes the species *Homo sapiens* (modern humans) and *Homo erectus* (Peking Man, Java Man).

Geochronology. Study of the time SCALE of the earth; it attempts to develop methods that allow the scientist to reconstruct the past by dating events such as the formation of ROCKS.

Geodesy. Branch of applied mathematics that determines the exact positions of points on the earth's surface, the size and shape of the earth, and the variations of terrestrial GRAVITY and MAGNETISM.

Geoid. Figure of the earth considered as a MEAN SEA LEVEL surface extended continuously through the CONTINENTS.

Geologic terrane. Crustal block with a distinct group of ROCKS and structures resulting from a particular geologic history; assemblages of TERRANES form the CONTINENTS.

Geological column. Order of ROCK layers formed during the course of the earth's history.

Geomagnetic elements. MEASUREMENTS that describe the direction and intensity of the earth's MAGNETIC FIELD.

Geomagnetism. External MAGNETIC FIELD generated by forces within the earth; this force attracts materials having similar properties, inducing them to line up (point) along field lines of force.

Geostationary orbit. ORBIT in which a SATELLITE appears to hover over one spot on the PLANET'S EQUATOR; this procedure requires that the orbit be high enough that its period matches the planet's rotational period, and have no inclination relative to the equator; for Earth, the ALTITUDE is 22,260 miles (35,903 km.).

Geostrophic. Force that causes directional change because of the earth's ROTATION.

Geotherm. Curve on a TEMPERATURE-depth graph that describes how temperature changes in the subsurface.

Geothermal power. Power having its source in the earth's internal heat.

Glacial erratic. ROCK that has been moved from its original position and transported by becoming incorporated in the ice of a GLACIER. Deposited in a new LOCATION, the rock is noteworthy because its geology is completely different from that of the surrounding rocks. Glacial erratics provide information about the direction of glacial movement and strength of the flow. They can be as small as PEBBLES, but the most interesting erratics are large boulders. Erratics become smoothed and rounded by the transport and EROSION.

Glaciation. This term is used in two senses: first, in reference to the cyclic widespread growth and advance of ICE SHEETS over the polar and high- to mid-LATITUDE REGIONS of the CONTINENTS; second, in reference to the effect of a GLACIER on the TERRAIN it transverses as it advances and recedes.

Global Positioning System (GPS). Group of SATELLITES that ORBIT Earth every twenty-four hours, sending out signals that can be used to locate PLACES on Earth and in near-Earth orbits.

Global warming. Trend of Earth CLIMATES to grow increasingly warm as a result of the GREENHOUSE EFFECT. One of the most dramatic effects of global warming is the melting of the POLAR ICE CAPS and a consequent rise the level of the world's OCEANS.

Gondwanaland. Hypothesized ancient CONTINENT in the SOUTHERN HEMISPHERE that geologists theorize broke into at least two large segments; one segment became India and pushed northward to collide with the Eurasian LANDMASS, while the other, Africa, moved westward.

Graben. Roughly symmetrical crustal depression formed by the lowering of a crustal block between two NORMAL FAULTS that slope toward each other.

Granules. Small grains or pellets.

Gravimeter. Device that measures the attraction of GRAVITY.

Gravitational differentiation. Separation of MINERALS, elements, or both as a result of the influence of a gravitational field wherein heavy phases sink or light phases rise through a melt.

Great circle. Largest circle that goes around a sphere. On the earth, all lines of LONGITUDE are parts of great circles; however, the EQUATOR is the only line of LATITUDE that is a great circle.

Green mud. SOILS that develop under conditions of excess water, or waterlogged soils, can display colors of gray to blue to green, largely because of chemical reactions involving iron. Fine CLAY soils and muds in areas such as BOGS or ESTUARIES can be called green mud. This soil-forming process is called gleization.

Greenhouse effect. Trapping of the SUN's rays within the earth's ATMOSPHERE, with a consequence rise in TEMPERATURES that leads to GLOBAL WARMING.

Greenhouse gas. Atmospheric gas capable of absorbing electromagnetic radiation in the infrared part of the spectrum.

Greenwich mean time. Also known as universal time, the solar mean time on the MERIDIAN running through Greenwich, England—which is used as the basis for calculating time throughout most of the world.

Grid. Pattern of horizontal and vertical lines forming squares of uniform size.

Groundwater movement. Flow of water through the subsurface, known as groundwater movement, obeys set principles that allow hydrologists to predict flow directions and rates.

Groundwater recharge. Water that infiltrates from the surface of the earth downward through SOIL and ROCK pores to the WATER TABLE, causing its level to rise.

Growth pole. LOCATION where high-growth economic activity is deliberately encouraged and promoted. Governments often establish growth poles by creating industrial parks, open cities, special economic zones, new TOWNS, and other incentives. The plan is that the new industries will further stimulate economic growth in a cumulative trend. Automobile plants are a traditional form of growth industry but have been overtaken by high-tech industries and BIOTECHNOLOGY. In France, the term "technopole" is used for a high-tech growth pole. A related concept is SPREAD EFFECTS.

Guyot. Drowned volcanic ISLAND with a flat top caused by WAVE EROSION or coral growth. A type of SEAMOUNT.

Gyre. Large semiclosed circulation patterns of OCEAN CURRENTS in each of the major OCEAN BASINS that move in opposite directions in the Northern and Southern hemispheres.

Haff. Term used for various WETLANDS or LAGOONS located around the southern end of the Baltic Sea, from Latvia to Germany. Offshore BARS of SAND and shingle separate the haffs from the open SEA. One of the largest is the Stettiner Haff, which covers the BORDER REGION between Germany and Poland and is separated from the Baltic by the low-lying ISLAND of Usedom. The Kurisches Haff (in English, the Courtland Lagoon) is located on the Lithuanian border.

Harmonic tremor. Type of EARTHQUAKE activity in which the ground undergoes continuous shaking in response to subsurface movement of MAGMA.

Headland. Elevated land projecting into a body of water.

Headwaters. Source of a RIVER. Also called headstream.

Heat sink. Term applied to Antarctica, whose cold CLIMATE causes warm AIR masses flowing over it to chill quickly and lose ALTITUDE, affecting the entire world's WEATHER.

Heterosphere. Major realm of the ATMOSPHERE in which the gases hydrogen and helium become predominant.

High-frequency seismic waves. EARTHQUAKE WAVES that shake the ROCK through which they travel most rapidly.

Histogram. Bar graph in which vertical bars represent frequency and the horizontal axis represents categories. A POPULATION PYRAMID, or age-sex pyramid, is a histogram, as is a CLIMOGRAPH.

Historical inertia. Term used by economic geographers when heavy industries, such as steelmaking and large manufacture, that require huge capital investments in land and plant continue in operation for long periods, even after they become out of date, uncompetitive, or obsolete.

Hoar frost. Similar to DEW, except that moisture is deposited as ice crystals, not liquid dew, on surfaces such as grass or plant leaves. When moist AIR cools to saturation level at TEMPERATURES below the freezing point, CONDENSATION occurs directly as ice. Technically, hoar frost is not the same as frozen dew, but it is difficult to distinguish between the two.

Hogback. Steeply sloping homoclinal RIDGE, with a slope of 45 DEGREES or more. The angle of the slope is the same as the dip of the ROCK STRATA. These LANDFORMS develop in REGIONS where the underlying rocks, usually SEDIMENTARY, have been folded into anticlinal ridges and synclinal VALLEYS. Differential EROSION causes softer rock layers to wear away more rapidly than the harder layers of rock that form the hogback ridge. A similar feature with a gentler slope is called a CUESTA.

Homosphere. Lower part of the earth's ATMOSPHERE. In this area, 60 miles (100 km.) thick, the component gases are uniformly mixed together, largely through WINDS and turbulent AIR CURRENTS. Above the homosphere is the REGION of the atmosphere called the HETEROSPHERE. There, the individual gases separate out into layers on the basis of their molecular weight. The lighter gases, hydrogen and helium, are at the top of the heterosphere.

Hook. A long, narrow deposit of SAND and SILT that grows outward into the OCEAN from the land is called a SPIT or sandspit. A hook forms when currents or WAVES cause the deposited material to curve back toward the land. Cape Cod is the most famous spit and hook in the United States.

Horse latitudes. Parts of the OCEANS from about 30 to 35 DEGREES north or south of the EQUATOR. In these latitudes, AIR movement is usually light WINDS, or even complete calm, because there are semipermanent high-pressure cells called ANTICYCLONES, which are marked by dry subsiding air and fine clear WEATHER. The atmospheric circulation of an anticyclone is divergent and CLOCKWISE in the NORTHERN HEMISPHERE, so to the north of the horse latitudes are the westerly winds and to the south are the northeast TRADE WINDS. In the SOUTHERN HEMISPHERE, the circulation is reversed, producing the easterly winds and the southeast trade winds. It is believed that the name originated because when ships bringing immigrants to the Americas were becalmed for any length of time, horses were thrown overboard because they required too much FRESH WATER. Also called the CALMS OF CANCER.

Horst. FAULT block or piece of land that stands above the surrounding land. A horst usually has been uplifted by tectonic forces, but also could have originated by downward movement or lowering of the adjacent lands. Movement occurs along the parallel faults on either side of a horst. If the land is downthrown instead of uplifted, a VALLEY known as a GRABEN is formed. "Horst" comes from the German word for horse, because the flat-topped feature resembles a vaulting horse used in gymnastics.

Hot spot. PLACE on the earth's surface where heat and MAGMA rise from deep in the interior, perhaps from the lower MANTLE. Erupting VOLCANOES may be present, as in the formation of the Hawaiian Islands. More commonly, the heat from the rising magma causes GROUNDWATER to form HOT SPRINGS, GEYSERS, and other thermal and HYDROTHERMAL features. Yellowstone National Park is located on a hot spot. Also known as a MANTLE PLUME.

Hot spring. SPRING where hot water emerges at the earth's surface. The usual cause is that the GROUNDWATER is heated by MAGMA. A GEYSER is a special type of hot spring at which the water heats under pressure and that periodically spouts hot water and steam. Old Faithful is the best known of many geysers in Yellowstone National Park. In some countries, GEOTHERMAL ENERGY from hot springs is used to generate electricity. Also called thermal spring.

Humus. Uppermost layer of a SOIL, containing decaying and decomposing organic matter such as leaves. This produces nutrients, leading to a fertile soil. Tropical soils are low in humus, because the rate of decay is so rapid. Soils of GRASSLANDS and DECIDUOUS FOREST develop thick layers of humus. In a SOIL PROFILE, the layer containing humus is the O Horizon.

Hydroelectric power. Electricity generated when falling water turns the blades of a turbine that converts the water's potential ENERGY to mechanical energy. Natural WATERFALLS can be used, but most hydroelectric power is generated by water from DAMS, because the flow of water from a dam can be controlled. Hydroelectric generation is a RENEWABLE, clean, cheap way to produce power, but dam construction inundates land, often displacing people, who lose their homes, VILLAGES, and farmland. Aquatic life is altered and disrupted also; for example, Pacific salmon cannot return upstream on the Columbia River to their spawning REGION. In a few coastal PLACES, TIDAL ENERGY is used to generate hydroelectricity; La Rance in France is the oldest successful tidal power plant.

Hydrography. Surveying of underwater features or those parts of the earth that are covered by water, especially OCEAN depths and OCEAN CURRENTS. Hydrographers make MAPS and CHARTS of the ocean floor and COASTLINES, which are used by mariners for NAVIGATION. For centuries, mariners used a leadline, a long rope with a lead weight at the bottom. The line was thrown overboard and the depth of water measured. The unit of MEASUREMENT was FATHOMS (6 feet/1.8 meters), which is one-thousandth of a NAUTICAL MILE. The invention of sonar (underwater echo sounding) has enabled mapping of large areas, and hydrographers currently use both television cameras and SATELLITE data.

Hydrologic cycle. Continuous circulation of the earth's HYDROSPHERE, or waters, through EVAPORATION, CONDENSATION, and PRECIPITATION.

Other parts of the hydrologic cycle include RUNOFF, INFILTRATION, and TRANSPIRATION.

Hydrostatic pressure. Pressure imposed by the weight of an overlying column of water.

Hydrothermal vents. Areas on the OCEAN floor, typically along FAULT LINES or in the vicinity of undersea VOLCANOES, where water that has percolated into the ROCK reemerges much hotter than the surrounding water; such heated water carries various dissolved MINERALS, including metals and sulfides.

Hyetograph. Chart showing the DISTRIBUTION of RAINFALL over time. Typically, a hyetograph is constructed for a single STORM, showing the amount of total PRECIPITATION accumulating throughout the period. A hyetograph shows how rainfall intensity varies throughout the duration of a storm.

Hygrometer. Instrument for measuring the RELATIVE HUMIDITY of AIR, or the amount of water vapor in the ATMOSPHERE at any time.

Hypsometer. Instrument used for measuring ALTITUDE (height above SEA LEVEL), using boiling water that circulates around a THERMOMETER. Since ATMOSPHERIC PRESSURE falls with increased altitude, the boiling point of water is lower. The hypsometer relies on this difference in boiling point to calculate ELEVATION. A more common instrument for measuring altitude is the ALTIMETER.

Ice blink. Bright, usually yellowish-white glare or reflection on the underside of a CLOUD layer, produced by light reflected from an ice-covered surface such as pack ice. A similar phenomenon of reflection from a snow-covered surface is called snow blink.

Ice-cap climate. Earth's most severe CLIMATE, where the mean monthly TEMPERATURE is never above 32 DEGREES Fahrenheit (0 degrees Celsius). This climate is found in Greenland and Antarctica, which are high PLATEAUS, where KATABATIC WINDS blow strongly and frequently. At these high LATITUDES, INSOLATION (SOLAR ENERGY) is received for a short period in the summer months, but the high reflectivity of the ice and SNOW means that much is reflected back instead of being absorbed by the surface. No VEGETATION can grow, because the LANDSCAPE is permanently covered in ice and snow. Because AIR temperatures are so cold, PRECIPITATION is usually less than 5 inches (13 centimeters) annually. The POLES are REGIONS of stable, high-pressure air, where dry conditions prevail, but strong winds that blow the snow around are common. In the KÖPPEN CLIMATE CLASSIFICATION, the ice-cap climate is signified by the letters *EF*.

Ice sheet. Huge CONTINENTAL GLACIER. The only ice sheets remaining cover most of Antarctica and Greenland. At the peak of the last ICE AGE, around 18,000 years ago, ice covered as much as one-third of the earth's land surfaces. In the NORTHERN HEMISPHERE, there were two great ice sheets—the Laurentide ice sheet, covering North America, and the Scandinavian ice sheet, covering northwestern Europe and Scandinavia.

Ice shelf. Portion of an ICE SHEET extending into the OCEAN.

Ice storm. STORM characterized by a fall of freezing rain, with the formation of glaze on Earth objects.

Icefoot. Long, tapering extension of a GLACIER floating above the seawater where it enters the OCEAN. Eventually, it breaks away and forms an ICEBERG.

Igneous rock. ROCKS formed when molten material or MAGMA cools and crystallizes into solid rock. The type of rock varies with the composition of the magma and, more important, with the rate of cooling. Rocks that cool slowly, far beneath the earth's surface, are igneous INTRUSIVE ROCKS. These have large crystals and coarse grains. GRANITE is the most typical igneous intrusive rock. When cooling is more rapid, usually closer to or at the surface, finer-grained igneous EXTRUSIVE ROCKS such as rhyolite are formed. If the magma flows out to the surface as LAVA, it may cool quickly, forming a glassy rock called obsidian. If there is gas in the lava, rocks full of holes from bubbles of escaping gases form; PUMICE and BASALT are common igneous extrusive rocks.

Impact crater. Generally circular depression formed on the surface of a PLANET by the impact of a high-velocity projectile such as a METEORITE, ASTEROID, or COMET.

Impact volcanism. Process in which major impact events produce huge CRATERS along with MAGMA RESERVOIRS that subsequently produce volcanic activity. Such cratering is clearly visible on the MOON, Mars, Mercury, and probably Venus. It is assumed that Earth had similar craters, but EROSION has erased most of the evidence.

Import substitution. Economic process in which domestic producers manufacture or supply goods or services that were previously imported or purchased from overseas and foreign producers.

Index fossil. Remains of an ancient organism that are useful in establishing the age of ROCKS; index fossils are abundant and have a wide geographic DISTRIBUTION, a narrow stratigraphic RANGE, and a distinctive form.

Indian summer. Short period, usually not more than a week, of unusually warm WEATHER in late October or early November in the NORTHERN HEMISPHERE. Before the Indian summer, TEMPERATURES are cooler and there can be occurrences of FROST. Indian summer DAYS are marked by clear to hazy skies and calm to light WINDS, but nights are cool. The weather pattern is a high-pressure cell or ridge located for a few days over the East Coast of North America. The name originated in New England, referring to the practice of NATIVE AMERICANS gathering foods for winter storage over this brief spell. Similar weather in England is called an Old Wives' summer.

Infant mortality. DEMOGRAPHIC MEASURE calculated as the number of deaths in a year of infants, or children under one year of age, compared with the total number of live births in a COUNTRY for the same year. Low-income countries have high infant mortality rates, more than 100 infant deaths per thousand.

Infauna. Organisms that live in the seafloor.

Infiltration. Movement of water into and through the SOIL.

Initial advantage. In terms of economic DEVELOPMENT, not all LOCATIONS are suited for profitable investment. Some locations offer initial advantages, including an existing skilled labor pool, existing consumer markets, existing plants, and situational advantages. These advantages can also lead to clustering of a number of industries at a particular location and to further economic growth, which will provide the preconditions of initial advantage for further economic development.

Inlier. REGION of old ROCKS that is completely surrounded by younger rocks. These are often PLACES where ORES or MINERALS are found in commercial quantities.

Inner core. The innermost layer of the earth; the inner core is a solid ball with a radius of about 900 miles.

Inselberg. Exposed rocky HILL in a DESERT area, made of resistant ROCKS, rising steeply from the flat surrounding countryside. There are many inselbergs in Africa, but Uluru (Ayers Rock) in Australia is possibly the most famous inselberg. The word is German for "island mountain." A special type of inselberg is a bornhardt.

Insolation. ENERGY received by the earth from the SUN, which heats the earth's surface. The average insolation received at the top of the earth's ATMOSPHERE at an average distance from the Sun is called the SOLAR CONSTANT. Insolation is predominantly shortwave radiation, with wavelengths in the RANGE of 0.39 to 0.76 micrometers, which corresponds to the visible spectrum. Less than half of the incoming SOLAR ENERGY reaches the earth's surface-insolation is reflected back into space by CLOUDS; smaller amounts are reflected back by surfaces, absorbed, or scattered by the atmosphere. Insolation is not distributed evenly over the earth, because of Earth's curved surface. Where the rays are perpendicular, at the SUBSOLAR POINT, insolation is at the maximum. The word is a shortened form of incoming (or intercepted) SOLAR RADIATION.

Insular climate. Island climates are influenced by the fact that no PLACE is far from the SEA. Therefore, both the DIURNAL (daily) TEMPERATURE RANGE and the annual temperature range are small.

Insurgent state. STATE that arises when an uprising or guerrilla movement gains control of part of the territory of a COUNTRY, then establishes its own form of control or government. In effect, the insurgents create a state within a state. In Colombia, for example, the government and armed forces have been unable to control several REGIONS where insurgents have created their own domains. This is generally related to coca growing and the production of cocaine. Civilian farmers are unable to resist the drug-financed "armies."

Interfluve. Higher area between two STREAMS; the surface over which water flows into the stream. These surfaces are subject to RUNOFF and EROSION by RILL action and GULLYING. Over time, interfluves are lowered.

Interlocking spur. STREAM in a hilly or mountainous REGION that winds its way in a sinuous VALLEY between the different RIDGES, slowly eroding the ends of the spurs and straightening its course. The view of interlocking spurs looking upstream is a favorite of artists, as colors change with the receding distance of each interlocking spur.

Intermediate rock. IGNEOUS ROCK that is transitional between a basic and a silicic ROCK, having a SILICA content between 54 and 64 percent.

Internal migration. Movement of people within a COUNTRY, from one REGION to another. Internal MIGRATION in high-income economies is often urban-to-RURAL, such as the migration to the SUN BELT in the United States. In low-income economies, rural-to-URBAN migration is more common.

Intertillage. Mixed planting of different seeds and seedling crops within the same SWIDDEN or cleared patch of agricultural land. Potatoes, yams, corn, rice, and bananas might all be planted. The planting times are staggered throughout the year to increase the variety of crops or nutritional balance available to the subsistence farmer and his or her family.

Intrusive rock. IGNEOUS ROCK which was formed from a MAGMA that cooled below the surface of the earth; it is commonly coarse-grained.

Irredentism. Expansion of one COUNTRY into the territory of a nearby country, based on the residence of nationals in the neighboring country. Hitler used irredentist claims to invade Czechoslovakia, because small groups of German-speakers lived there in the Sudetenland. The term comes from Italian, referring to Italy's claims before World War I that all Italian-speaking territory should become part of Italy.

Isallobar. Imaginary line on a MAP or meteorological chart joining PLACES with an equal change in ATMOSPHERIC PRESSURE over a certain time, often three hours. Isallobars indicate a pressure tendency and are used in WEATHER FORECASTING.

Island arc. Chain of VOLCANOES next to an oceanic TRENCH in the OCEAN BASINS; an oceanic PLATE descends, or subducts, below another oceanic plate at ISLAND arcs.

Isobar. Imaginary line joining PLACES of equal ATMOSPHERIC PRESSURE. WEATHER MAPS show isobars encircling areas of high or low pressure. The spacing between isobars is related to the pressure gradient.

Isobath. Line on a MAP or CHART joining all PLACES where the water depth is the same; a kind of underwater CONTOUR LINE. This kind of map is a BATHYMETRIC CONTOUR.

Isoclinal folding. When the earth's CRUST is folded, the size and shape of the folds vary according to the force of compression and nature of the ROCKS. When the surface is compressed evenly so that the two sides of the fold are parallel, isoclinal folding results. When the sides or slopes of the fold are unequal or dissimilar in shape and angle, this can be an asymmetrical or overturned fold. See also ANTICLINE; SYNCLINE.

Isotherm. Line joining PLACES of equal TEMPERATURE. A world MAP with isotherms of average monthly temperature shows that over the OCEANS, temperature decreases uniformly from the EQUATOR to the POLES, and higher temperatures occur over the CONTINENTS in summer and lower temperatures in winter because of the unequal heating properties of land and water.

Isotropic surface. Hypothetical flat surface or PLAIN, with no variation in any physical attribute. An isotropic surface has uniform ELEVATION, SOIL type, CLIMATE, and VEGETATION. Economic geographic models study behavior on an isotropic surface before applying the results to the real world. For example, in an isotropic model, land value is highest at the CITY center and falls regularly with increasing distance from there. In the real world, land values are affected by elevation, water features, URBAN regulations, and other factors. The von Thuenen model of the Isolated State is based on a uniform plain or isotropic surface.

Isthmian links. Chains of ISLANDS between substantial LANDMASSES.

Isthmus. Narrow strip of land connecting two larger bodies of land. The Isthmus of Panama connects North and South America; the Isthmus of Suez connects Africa and Asia. Both of these have been cut by CANALS to shorten shipping routes.

Jet stream. WINDS that move from west to east in the upper ATMOSPHERE, 23,000 to 33,000 feet (7,000–10,000 meters) above the earth, at about 200 miles (300 km.) per hour. They are narrow bands, elliptical in cross section, traveling in irregular paths. Four jet streams of interest to earth scientists and meteorologists are the polar jet stream and the subtropical jet stream in the Northern and SOUTHERN HEMISPHERES. The polar jet stream is located at the TROPOPAUSE, the BOUNDARY between the TROPOSPHERE and the STRATOSPHERE, along the polar FRONT. There is a complex interaction between surface winds and jet streams. In winter the NORTHERN HEMISPHERE polar front can move as far south as Texas, bringing BLIZZARDS and extreme WEATHER conditions. In summer, the polar jet stream is located over Canada. The subtropical jet stream is located at the tropopause around 30 DEGREES north or south LATITUDE, but it also migrates north or south, depending on the SEASON. At times, the polar and subtropical jet streams merge for a few DAYS. Aircraft take advantage of the jet stream, or avoid it, depending on the direction of their flight. Upper atmosphere winds are also known as GEOSTROPHIC winds.

Joint. Naturally occurring fine crack in a ROCK, formed by cooling or by other stresses. SEDIMENTARY ROCKS can split along bedding planes; other joints form at right angles to the STRATA, running vertically through the rocks. In IGNEOUS ROCKS such as GRANITE, the stresses of cooling and contraction cause three sets of joints, two vertical and one parallel to the surface, which leads to the formation of distinctive LANDFORMS such as TORS. BASALT often demonstrates columnar jointing, producing tall columns that are mostly hexagonal in section. The presence of joints in BEDROCK hastens WEATHERING, because water can penetrate into the joints. This is particularly obvious in LIMESTONE, where joints are rapidly enlarged by SOLUTION. FROST WEDGING is a type of PHYSICAL WEATHERING that can split large boulders through the expansion when water in a joint freezes to form ice. Compare with FAULTS, which occur through tectonic activity.

Jurassic. Second of the three PERIODS that make up

Kame. Small HILL of gravel or mixed-size deposits, SAND, and gravel. Kames are found in areas previously covered by CONTINENTAL GLACIERS or ICE SHEETS, near what was the outer edge of the ice. They may have formed by materials dropping out of the melting ice, or in a deltalike deposit by a STREAM of MELTWATER. These deposits of which kames are made are called drift. Small LAKES called KETTLES are often found nearby. A closely spaced group of kames is called a kame field.

Karst. LANDSCAPE of SINKHOLES, underground STREAMS and caverns, and associated features created by CHEMICAL WEATHERING, especially SOLUTION, in REGIONS where the BEDROCK is LIMESTONE. The name comes from a region in the southwest of what is now Slovenia, the Krs (Kras) Plateau, but the karst region extends south through the Dinaric Alps bordering the

Adriatic Sea, into Bosnia-Herzegovina and Montenegro. Where limestone is well jointed, RAINFALL penetrates the JOINTS and enters the GROUNDWATER, carrying the MINERALS, especially calcium, away in solution. Most of the famous CAVES and caverns of the world are found in karst areas. The Carlsbad Caverns in New Mexico are a good example. Kentucky, Tennessee, and Florida also have well-known areas of karst. In some tropical countries, a form called tower karst is found. Tall conical or steep-sided HILLS of limestone rise above the flat surrounding landscape. Around 15 percent of the earth's land surface is karst TOPOGRAPHY.

Katabatic wind. GRAVITY DRAINAGE WINDS similar to MOUNTAIN BREEZES but stronger in force and over a larger area than a single VALLEY. Cold AIR collects over an elevated REGION, and the dense cold air flows strongly downslope. The ICE-SHEETS of Antarctica and Greenland produce fierce katabatic winds, but they can occur in smaller regions. The BORA is a strong, cold, squally downslope wind on the Dalmatian COAST of Yugoslavia in winter.

Kettle. Small depression, often a small LAKE, produced as a result of continental GLACIATION. It is formed by an isolated block of ice remaining in the ground MORAINE after a GLACIER has retreated. Deposited material accumulates around the ice, and when it finally melts, a steep hole remains, which often fills with water. Walden Pond, made famous by writer Henry David Thoreau, is a glacial kettle.

Khamsin. Hot, dry, DUST-laden WIND that blows in the eastern Sahara, in Egypt, and in Saudi Arabia, bringing high TEMPERATURES for three or four DAYS. Winds can reach GALE force in intensity. The word Khamsin is Arabic for "fifty" and refers to the period between March and June when the khamsin can occur.

Knickpoint. Abrupt change in gradient of the bed of a RIVER or STREAM. It is marked by a WATERFALL, which over time is eroded by FLUVIAL action, restoring the smooth profile of the riverbed. The knickpoint acts as a TEMPORARY BASE LEVEL for the upper part of the stream. Knickpoints can occur where a hard layer of ROCK is slower to erode than the rocks downstream, for example at Niagara Falls. Other knickpoints and waterfalls can develop as a result of tectonic forces. UPLIFT leads to new EROSION by a stream, creating a knickpoint that gradually moves upstream. The bed of a tributary GLACIER is often considerably higher than the VALLEY of the main glacier, so that after the glaciers have melted, a waterfall emerges over this knickpoint from the smaller hanging valley to join the main stream. Yosemite National Park has several such waterfalls.

Köppen climate classification. Commonly used scheme of CLIMATE classification that uses statistics of average monthly TEMPERATURE, average monthly PRECIPITATION, and total annual precipitation. The system was devised by Wladimir Köppen early in the twentieth century.

La Niña. WEATHER phenomenon that is the opposite part of EL NIÑO. When the SURFACE WATER in the eastern Pacific Ocean is cooler than average, the southeast TRADE WINDS blow strongly, bringing heavy rains to countries of the western Pacific. Scientists refer to the whole RANGE of TEMPERATURE, pressure, WIND, and SEA LEVEL changes as the SOUTHERN OSCILLATION (ENSO). The term "El Niño" gained wide currency in the U.S. media after a strong ENSO warm event in 1997–1998. A weak ENSO cold event, or La Niña, followed it in 1998. Means "the little girl" in Spanish. Alternative terms are "El Viejo" and "anti-El Niño."

Laccolith. LANDFORM of INTRUSIVE volcanism formed when viscous MAGMA is forced between overlying sedimentary STRATA, causing the surface to bulge upward in a domelike shape.

Lahar. Type of mass movement in which a MUDFLOW occurs because of a volcanic explosion or ERUPTION. The usual cause is that the heat from the LAVA or other pyroclastic material melts ice and SNOW at the VOLCANO's SUMMIT, causing a hot mudflow that can move downslope with great speed. The eruption of Mount Saint Helens in 1985 was accompanied by a lahar.

Lake basin. Enclosed depression on the surface of the land in which SURFACE WATERS collect; BASINS are created primarily by glacial activity and tectonic movement.

Lakebed. Floor of a LAKE.

Land bridge. Piece of land connecting two CONTINENTS, which permits the MIGRATION of humans, animals, or plants from one area to another. Many former land bridges are now under water, because of the rise in SEA LEVEL after the last ICE AGE. The Bering Strait connecting Asia and North America was an important land bridge for the latter continent.

Land hemisphere. Because the DISTRIBUTION of land and water surfaces on Earth is quite asymmetrical on either side of the EQUATOR, the NORTHERN HEMISPHERE might well be called the land hemisphere. For many centuries, Europeans refused to believe that there was not an equal area of land in the SOUTHERN HEMISPHERE. Explorers such as James Cook were dispatched to seek such a "Great South Land."

Landmass. Large area of land—an ISLAND or a CONTINENT.

Landsat. Space-exploration project begun in 1972 to MAP the earth continuously with SATELLITE imaging. The satellites have collected data about the earth: its AGRICULTURE, FORESTS, flat lands, MINERALS, waters, and ENVIRONMENT. These were the first satellites to aid in Earth sciences, helping to produce the best maps available and assisting farmers around the world to improve their crop yields.

Language family. Group of related LANGUAGES believed to have originated from a common prehistoric language. English belongs in the Indo-European language family, which includes the languages spoken by half of the world's peoples.

Lapilli. Small ROCK fragments that are ejected during volcanic ERUPTIONS. A lapillus ranges from about the size of a pea to not larger than a walnut. Some lapilli form by accretion of VOLCANIC ASH around moisture droplets, in a manner similar to hailstone formation. Lapilli sometimes form into a textured rock called lapillistone.

Laterite. Bright red CLAY SOIL, rich in iron oxide, that forms in tropical CLIMATES, where both TEMPERATURE and PRECIPITATION are high year-round, as ROCKS weather. It can be used in brick making and is a source of iron. When the soil is rich in aluminum, it is called BAUXITE. When laterite or bauxite forms a hard layer at the surface, it is called duricrust. Australia and sub-Saharan Africa have large areas of duricrust, some of which is thought to have formed under previous conditions during the TRIASSIC period.

Laurasia. Hypothetical SUPERCONTINENT made up of approximately the present CONTINENTS of the NORTHERN HEMISPHERE.

Lava tube. Cavern structure formed by the draining out of liquid LAVA in a pahoehoe flow.

Layered plains. Smooth, flat REGIONS believed to be composed of materials other than sulfur compounds.

Leaching. Removal of nutrients from the upper horizon or layer of a SOIL, especially in the humid TROPICS, because of heavy RAINFALL. The remaining soil is often bright red in color because iron is left behind. Despite their bright color, tropical soils are infertile.

Leeward. Rear or protected side of a MOUNTAIN or RANGE is the leeward side. Compare to WINDWARD.

Legend. Explanation of the different colors and symbols used on a MAP. For example, a map of the world might use different colors for high-income, middle-income, and low-income economies. A historical map might use different colors for countries that were once colonies of Britain, France, or Spain.

Light year. Distance traveled by light in one year; widely used for measuring stellar distances, it is equal to roughly 6 trillion miles (9.5 million km.).

Lignite. Low-grade COAL, often called brown coal. It is mined and used extensively in eastern Germany, Slovakia, and the Moscow Basin.

Liquefaction. Loss in cohesiveness of water-saturated SOIL as a result of ground shaking caused by an EARTHQUAKE.

Lithification. Process whereby loose material is transformed into solid ROCK by compaction or cementation.

Lithology. Description of ROCKS, such as rock type, MINERAL makeup, and fluid in rock pores.

Lithosphere. Solid outermost layer of the earth. It varies in thickness from a few miles to more than 120 miles (200 km.). It is broken into pieces known as TECTONIC PLATES, some of which are extremely large, while others are quite small. The upper layer of the lithosphere is the CRUST, which may be CONTINENTAL CRUST or OCEANIC CRUST. Below the crust is a layer called the ASTHENOSPHERE, which is weaker and plastic, enabling the motion of tectonic plates.

Littoral. Adjacent to or related to a SEA.

***Llanos*.** Grassy REGION in the Orinoco Basin of Venezuela and part of Colombia. SAVANNA VEGETATION gradually gives way to scrub at the outer edges of the *llanos*. The area is relatively undeveloped.

Loam. SOIL TEXTURE classification, indicating a soil that is approximately equal parts of SAND, SILT, and CLAY. Farmers generally consider a sandy loam to be the best soil texture because of its water-retaining qualities and the ease with which it can be cultivated.

Local sea-level change. Change in SEA LEVEL only in one area of the world, usually by land rising or sinking in that specific area.

Lode deposit. Primary deposit, generally a VEIN, formed by the filling of a FISSURE with MINERALS precipitated from a HYDROTHERMAL solution.

Loess. EOLIAN, or wind-blown, deposit of fine, silt-sized, light-colored material. Loess covers about 10 percent of the earth's land surface. The loess PLATEAU of China is good agricultural land, although susceptible to EROSION. Loess has the property of being able to form vertical CLIFFS or BLUFFS, and many people have built dwellings in the steep cliffs above the Huang He (Yellow) River. In the United States, loess deposits are found in the VALLEYS of the Platte, Missouri, Mississippi, and Ohio Rivers, and on the Columbia Plateau. A German word, meaning loose or unconsolidated, which comes from loess deposits along the Rhine River.

Longitudinal bar. Midchannel accumulation of SAND and gravel with its long end oriented roughly parallel to the RIVER flow.

Longshore current. Current in the OCEAN close to the SHORE, in the surf zone, produced by WAVES approaching the COAST at an angle. Also called a LITTORAL current. The longshore current combined with wave action can move large amounts of SAND and other BEACH materials down the coast, a process called LONGSHORE DRIFT.

Longshore drift. The movement of SEDIMENT parallel to the BEACH by a LONGSHORE CURRENT.

Maar. Explosion vent at the earth's surface where a volcanic cone has not formed. A small ring of pyroclastic materials surrounds the maar. Often a LAKE occupies the small CRATER of a maar. A larger form is called a TUFF RING.

Macroburst. Updrafts and downdrafts within a CUMULONIMBUS CLOUD or THUNDERSTORM can cause severe TURBULENCE. A DOWNBURST within a thunderstorm when windspeeds are greater than 130 miles (210 km.) per hour and over areas of 2.5 square miles (5 sq. km.) or more is called a macroburst. See also MICROBURST.

Magnetic poles. Locations on the earth's surface where the earth's MAGNETIC FIELD is perpendicular to the surface. The magnetic poles do not correspond exactly to the geographic North Pole and South Pole, or earth's AXIS; the difference is called magnetic variation or DECLINATION.

Magnetic reversal. Change in the earth's MAGNETIC FIELD from the North Pole to the South MAGNETIC POLE.

Magnetic storm. Rapid changes in the earth's MAGNETIC FIELD as a result of the bombardment of the earth by electrically charged particles from the SUN.

Magnetosphere. REGION surrounding a PLANET where the planet's own MAGNETIC FIELD predominates over magnetic influences from the SUN or other planets.

Mantle convection. Thermally driven flow in the earth's MANTLE thought to be the driving force of PLATE TECTONICS.

Mantle plume. Rising jet of hot MANTLE material that produces tremendous volumes of basaltic LAVA. See also HOT SPOT.

Map projection. Mathematical formula used to transform the curved surface of the earth onto a flat plane or sheet of paper. Projections are divided into three classes: CYLINDRICAL, CONICAL, and AZIMUTHAL.

Marchland. FRONTIER area where boundaries are poorly defined or absent. The marches themselves were a type of BOUNDARY REGION. Marchlands have changed hands frequently throughout history. The name is related to the fact that armies marched across them.

Mass balance. Summation of the net gain and loss of ice and SNOW mass on a GLACIER in a year.

Mass extinction. Die-off of a large percentage of species in a short time.

Mass wasting. Downslope movement of Earth materials under the direct influence of GRAVITY.

Massif. French term used in geology to describe very large, usually IGNEOUS INTRUSIVE bodies.

Meandering river. RIVER confined essentially to a single CHANNEL that transports much of its SEDIMENT load as fine-grained material in SUSPENSION.

Mechanical weathering. Another name for PHYSICAL WEATHERING, or the breaking down of ROCK into smaller pieces.

Mechanization. Replacement of human labor with machines. Mechanization occurred in AGRICULTURE as tractors, reapers, picking machinery, and similar technological inventions took the place of human farm labor. Mechanization in industry was part of the INDUSTRIAL REVOLUTION, as spinning and weaving machines were introduced into the textile industry.

Medical geography. Branch of geography specializing in the study of health and disease, with a particular emphasis on the areal spread or DIFFUSION of disease. The spatial perspective of geography can lead to new medical insights. Geographers working with medical researchers in Africa have made great contributions to understanding the role of disease on that CONTINENT. John Snow's studies of the origin and spread of cholera in London in 1854 mark the beginnings of medical geography.

Megalopolis. Conurbation formed when large cities coalesce physically into one huge built-up area. Originally coined by the French geographer Jean Gottman in the early 1960s for the northeastern part of the United States, from Boston to Washington, D.C.

Mesa. Flat-topped HILL with steep sides. EROSION removes the surrounding materials, while the mesa is protected by a cap of harder, more resistant ROCK. Usually found in arid REGIONS. A larger LANDFORM of this type is a PLATEAU; a smaller feature is a BUTTE. The Colorado Plateau and Grand Canyon in particular are rich in these landforms. From the Spanish word for table.

Mesosphere. Atmospheric layer above the STRATOSPHERE where TEMPERATURE drops rapidly.

Mestizo. Person of mixed European and Amerindian ancestry, especially in countries of LATIN AMERICA.

Metamorphic rock. Any ROCK whose mineralogy, MINERAL chemistry, or TEXTURE has been altered by heat, pressure, or changes in composition; metamorphic rocks may have IGNEOUS, SEDIMENTARY, or other, older metamorphic rocks as their precursors.

Metamorphic zone. Areas of ROCK affected by the same limited RANGE of TEMPERATURE and pressure conditions, commonly identified by the presence of a key individual MINERAL or group of minerals.

Meteor. METEOROID that enters the ATMOSPHERE of a PLANET and is destroyed through frictional heating as it comes in contact with the various gases present in the atmosphere.

Meteorite. Fragment of an ASTEROID that survives passage through the ATMOSPHERE and strikes the surface of the earth.

Meteoroid. Small planetary body that enters Earth's ATMOSPHERE because its path intersects the earth's ORBIT. Friction caused by the earth's atmosphere on the meteoroid creates a glowing METEOR, or "shooting star." This is a common phenomenon, and most meteors burn away completely. Those that are large enough to reach the ground are called METEORITES.

Microburst. Brief but intense downward WIND, lasting not more than fifteen minutes over an area of 0.6 to 0.9 square mile (1.5–8 sq. km.). Usually associated with THUNDERSTORMS, but are quite unpredictable. The sudden change in wind direction associated with a microburst can create wind shear that causes airplanes to crash, especially if it occurs during takeoff or landing. See also MACROBURST.

Microclimate. CLIMATE of a small area, at or within a few yards of the earth's surface. In this REGION, variations of TEMPERATURE, PRECIPITATION, and moisture can have a pronounced effect on the bioclimate, influencing the growth or well-being of plants and animals, including humans. DEW or FROST, RAIN SHADOW effects, wind-tunneling between tall buildings, and similar phenomena are studied by microclimatologists. Horticulturists know the variations in aspect that affect INSOLATION and temperature, so that certain plants grow best on south-facing walls, for example. The growing of grapes for wine production is a major industry where microclimatology is essential. The study of microclimatology was pioneered by the German meteorologist Rudolf Geiger.

Microcontinent. Independent LITHOSPHERIC PLATE that is smaller than a CONTINENT but possesses continental-type CRUST. Examples include Cuba and Japan.

Microstates. Tiny countries. In 2000, seventeen independent countries each had an area of less than 200 square miles (520 sq. km.). The smallest microstate is Vatican City, with an area of 0.2 square miles (0.5 sq. km.). Most of the world's microstates are island NATIONS, including Nauru, Tuvalu, Marshall Islands, Saint Kitts and Nevis, Seychelles, Maldives, Malta, Grenada, Saint Vincent and the Grenadines, Barbados, Antigua and Barbuda, and Palau.

Mineral species. Mineralogic division in which all the varieties in any one species have the same basic physical and chemical properties.

Monadnock. Isolated HILL far from a STREAM, composed of resistant BEDROCK. Monadnocks are found in humid temperate REGIONS. A similar LANDFORM in an arid region is an INSELBERG.

Monogenetic. Pertaining to a volcanic ERUPTION in which a single vent is used only once.

Moraine. Materials transported by a GLACIER, and often later deposited as a RIDGE of unsorted ROCKS and smaller material. Lateral moraine is found at the side of the glacier; medial moraine occurs when two glaciers join. Other types of moraine include ABLATION moraine, ground moraine, and push, RECESSIONAL, and TERMINAL MORAINE.

Mountain belts. Products of PLATE TECTONICS, produced by the CONVERGENCE of crustal PLATES. Topographic MOUNTAINS are only the surficial expression of processes that profoundly deform and modify the CRUST. Long after the mountains themselves have been worn away, their former existence is recognizable from the structures that mountain building forms within the ROCKS of the crust.

Nappe. Huge sheet of ROCK that was the upper part of an overthrust fold, and which has broken and traveled far from its original position due to the tremendous forces. The Swiss Alps have nappes in many LOCATIONS.

Narrows. STRAIT joining two bodies of water.

Nation-state. Political entity comprising a COUNTRY whose people are a national group occupying the area. The concept originated in eighteenth century France; in practice, such cultural homogeneity is rare today, even in France.

Natural increase, rate of. DEMOGRAPHIC MEASURE of POPULATION growth: the difference between births and deaths per year, expressed as a percentage of the POPULATION. The rate of natural increase for the United States in 2000 was 0.6 percent. In countries where the population is decreasing, the DEATH RATE is greater than the BIRTH RATE.

Natural selection. Main process of biological evolution; the production of the largest number of offspring by individuals with traits that are best adapted to their ENVIRONMENTS.

Nautical mile. Standard MEASUREMENT at SEA, equalling 6,076.12 feet (1.85 km.). The mile used for land measurements is called a statute mile and measures 5,280 feet (1.6 km.).

Neap tide. TIDE with the minimum RANGE, or when the level of the high tide is at its lowest.

Near-polar orbit. Earth ORBIT that lies in a plane that passes close to both the north and south POLES.

Nekton. PELAGIC organisms that can swim freely, without having to rely on OCEAN CURRENTS or WINDS. Nekton includes shrimp; crabs; oysters; MARINE reptiles such as turtles, crocodiles, and snakes; and even sharks; porpoises; and whales.

Net migration. Net balance of a COUNTRY or REGION'S IMMIGRATION and EMIGRATION.

Nomadism. Lifestyle in which pastoral people move with grazing animals along a defined route, ensuring adequate pasturage and water for their flocks or herds. This lifestyle has decreased greatly as countries discourage INTERNATIONAL MIGRATION. A more restricted form of nomadism is TRANSHUMANCE.

North geographic pole. Northernmost REGION of the earth, located at the northern point of the PLANET'S AXIS of ROTATION.

North magnetic pole. Small, nonstationary area in the Arctic Circle toward which a COMPASS needle points from any LOCATION on the earth.

Notch. Erosional feature found at the base of a SEA CLIFF as a result of undercutting by WAVE EROSION, bioabrasion from MARINE organisms, and dissolution of ROCK by GROUNDWATER seepage. Also known as a nip.

Nuclear energy. ENERGY produced from a naturally occurring isotope of uranium. In the process of nuclear FISSION, the unstable uranium isotope absorbs a neutron and splits to form tin and molybdenum. This releases more neurons, so a chain reaction proceeds, releasing vast amounts of heat energy. Nuclear energy was seen in the 1950s as the energy of the future, but safety fears and the problem of disposal of radioactive nuclear waste have led to public condemnation of nuclear power plants.

Nuée ardente. Hot cloud of ROCK fragments, ASH, and gases that suddenly and explosively erupt from some VOLCANOES and flow rapidly down their slopes.

Nunatak. Isolated MOUNTAIN PEAK or RIDGE that projects through a continental ICE SHEET. Found in Greenland and Antarctica.

Obduction. Tectonic collisional process, opposite in effect to SUBDUCTION, in which heavier OCEANIC CRUST is thrust up over lighter CONTINENTAL CRUST.

Oblate sphere. Flattened shape of the earth that is the result of ROTATION.

Occultation. ECLIPSE of any astronomical object other than the SUN or the MOON caused by the Moon or any PLANET, SATELLITE, or ASTEROID.

Ocean basins. Large worldwide depressions that form the ultimate RESERVOIR for the earth's water supply.

Ocean circulation. Worldwide movement of water in the SEA.

Ocean current. Predictable circulation of water in the OCEAN, caused by a combination of WIND friction, Earth's ROTATION, and differences in TEMPERATURE and density of the waters. The five great oceanic circulations, known as GYRES, are in the North Pacific, North Atlantic, South Pacific, South Atlantic, and Indian Oceans. Because of the CORIOLIS EFFECT, the direction of circulation is CLOCKWISE in the NORTHERN HEMISPHERE and COUNTERCLOCKWISE in the SOUTHERN HEMISPHERE, except in the Indian Ocean, where the direction changes annually with the pattern of winds associated with the Asian MONSOON. Currents flowing toward the EQUATOR are cold currents; those flowing away from the equator are warm currents. An important current is the warm Gulf Stream, which flows north from the Gulf of Mexico along the East Coast of the United States; it crosses the North Atlantic, where it is called the North Atlantic Drift, and brings warmer conditions to the western parts of Europe. The West Coast of the United States is affected by the cool, south-flowing California Current. The cool Humboldt, or Peru, Current, which flows north along the South American coast, is an important indicator of whether there will be an El Niño event. Deep currents, below 300 feet (100 meters), are extremely complicated and difficult to study.

Oceanic crust. Portion of the earth's crust under its ocean basins.

Oceanic island. Islands arising from seafloor volcanic eruptions, rather than from continental shelves. The Hawaiian Islands are the best-known examples of oceanic islands.

Off-planet. Pertaining to regions off the earth in orbital or planetary space.

Ore deposit. Natural accumulation of mineral matter from which the owner expects to extract a metal at a profit.

Orogeny. Mountain-building episode, or event, that extends over a period usually measured in tens of millions of years; also termed a revolution.

Orographic precipitation. Phenomenon caused when an air mass meets a topographic barrier, such as a mountain range, and is forced to rise; the air cools to saturation, and orographic precipitation falls on the windward side as rain or snow. The lee side is a rain shadow. This effect is noticeable on the West Coast of the United States, which has rainforest on the windward side of the mountains and deserts on the lee.

Orography. Study of mountains that incorporates assessment of how they influence and are affected by weather and other variables.

Oscillatory flow. Flow of fluid with a regular back-and-forth pattern of motion.

Overland flow. Flow of water over the land surface caused by direct precipitation.

Oxbow lake. Lake created when floodwaters make a new, shorter channel and abandon the loop of a meander. Over time, water in the oxbow lake evaporates, leaving a dry, curving, low-lying area known as a meander scar. Oxbow lakes are common on floodplains. Another name for this feature is a cut-off.

Ozone hole. Decrease in the abundance of Antarctic ozone as sunlight returns to the pole in early springtime

Ozone layer. Narrow band of the stratosphere situated near 18 miles (30 km.) above the earth's surface, where molecules of ozone are concentrated. The average concentration is only one in 4 million, but this thin layer protects the earth by absorbing much of the ultraviolet light from the Sun and reradiating it as longer-wavelength radiation. Scientists were disturbed to discover that the ozonosphere was being destroyed by photochemical reaction with chlorofluorocarbons (CFCs). The ozone holes over the South and North Poles negatively affect several animal species, as well as humans; skin cancer risk is increasing rapidly as a consequence of depletion of the ozone layer. Stratospheric ozone should not be confused with ozone at lower levels, which is a result of photochemical smog. Also called the ozonosphere.

P wave. Fastest elastic wave generated by an earthquake or artificial energy source; basically an acoustic or shock wave that compresses and stretches solid material in its path.

Pangaea. Name used by Alfred Wegener for the supercontinent that broke apart to create the present continents.

Parasitic cone. Small volcanic cone that appears on the flank of a larger volcano, or perhaps inside a caldera.

Particulate matter. Mixture of small particles that adversely affect human health. The particles may come from smoke and dust and are in their highest concentrations in large urban areas, where they contribute to the "dust dome." Increased occurrences of illnesses such as asthma and bronchitis, especially in children, are related to high concentrations of particulate matter.

Pastoralism. Type of agriculture involving the raising of grazing animals, such as cattle, goats, and sheep. Pastoral nomads migrate with their domesticated animals in order to ensure sufficient grass and water for the animals.

Paternoster lakes. Small circular LAKES joined by a STREAM. These lakes are the result of glacial EROSION. The name comes from the resemblance to rosary beads and the accompanying prayer (the Our Father).

Pedestal crater. A CRATER that has assumed the shape of a pedestal as a result of unique shaping processes caused by WIND.

Pedology. Scientific study of SOILS.

Pelagic. Relating to life-forms that live on or in open SEAS, rather than waters close to land.

Peneplain. In the geomorphic CYCLE, or cycle of LANDFORM development, described by W. M. Davis, the final stage of EROSION led to the creation of an extensive land surface with low RELIEF. Davis named this a peneplain, meaning "almost a plain." It is now known that tectonic forces are so frequent that there would be insufficient time for such a cycle to complete all stages required to complete this landform.

Percolation. Downward movement of part of the water that falls on the surface of the earth, through the upper layers of PERMEABLE SOIL and ROCKS under the influence of GRAVITY. Eventually, it accumulates in the zone of SATURATION as GROUNDWATER.

Perforated state. STATE whose territory completely surrounds another state. The classic example of a perforated state is South Africa, within which lies the COUNTRY of Lesotho. Technically, Italy is perforated by the MICROSTATES of San Marino and Vatican City.

Perihelion. Point in Earth's REVOLUTION when it is closest to the SUN (usually on January 3). At perihelion, the distance between the earth and the Sun is 91,500,000 miles (147,255,000 km.). The opposite of APHELION.

Periodicity. The recurrence of related phenomena at regular intervals.

Permafrost. Permanently frozen SUBSOIL. The condition occurs in perennially cold areas such as the ARCTIC. No trees can grow because their roots cannot penetrate the permafrost. The upper portion of the frozen SOIL can thaw briefly in the summer, allowing many smaller plants to thrive in the long daylight. Permafrost occurs in about 25 percent of the earth's land surface, and the condition even hampers construction in REGIONS such as Siberia and ARCTIC Canada.

Perturb. To change the path of an orbiting body by a gravitational force.

Petrochemical. Chemical substance obtained from NATURAL GAS or PETROLEUM.

Petrography. Description and systematic classification of ROCKS.

Photochemical smog. Mixture of gases produced by the interaction of sunlight on the gases emanating from automobile exhausts. The gases include OZONE, nitrogen dioxide, carbon monoxide, and peroxyacetyl nitrates. Many large cities suffer from poor AIR quality because of photochemical smog. Severe health problems arise from continued exposure to photochemical smog.

Photometry. Technique of measuring the brightness of astronomical objects, usually with a photoelectric cell.

Phylogeny. Study of the evolutionary relationships among organisms.

Phylum. Major grouping of organisms, distinguished on the basis of basic body plan, grade of anatomical complexity, and pattern of growth or development.

Physiography. The PHYSICAL GEOGRAPHY of a PLACE—the LANDFORMS, water features, CLIMATE, SOILS, and VEGETATION.

Piedmont glacier. GLACIER formed when several ALPINE GLACIERS join together into a spreading glacier at the base of a MOUNTAIN or RANGE. The Malaspina glacier in Alaska is a good example of a piedmont glacier.

Place. In geographic terms, space that is endowed with physical and human meaning. Geographers study the relationship between people, places, and ENVIRONMENTS. The five themes that geographers use to examine the world are LOCATION, place, human/environment interaction, movement, and REGIONS.

Placer. Accumulation of valuable MINERALS formed when grains of the minerals are physically deposited along with other, nonvaluable mineral grains.

Planetary wind system. Global atmospheric circulation pattern, as in the BELT of prevailing westerly WINDS.

Plantation. Form of AGRICULTURE in which a large area of agricultural land is devoted to the production of a single cash crop, for export. Many plantation crops are tropical, such as bananas, sugarcane, and rubber. Coffee and tea plantations require cooler CLIMATES. Formerly, slave labor was used on most plantations, and the owners were Europeans.

Plate boundary. REGION in which the earth's crustal PLATES meet, as a converging (SUBDUCTION ZONE), diverging (MID-OCEAN RIDGE), TRANSFORM FAULT, or collisional interaction.

Plate tectonics. Theory proposed by German scientist Alfred Wegener in 1910. Based on extensive study of ancient geology, STRATIGRAPHY, and CLIMATE, Wegener concluded that the CONTINENTS were formerly one single enormous LANDMASS, which he named PANGAEA. Over the past 250 million years, Pangaea broke apart, first into LAURASIA and GONDWANALAND, and subsequently into the present continents. Earth scientists now believe that the earth's CRUST is composed of a series of thin, rigid PLATES that are in motion, sometimes diverging, sometimes colliding.

Plinian eruption. Rapid ejection of large volumes of VOLCANIC ASH that is often accompanied by the collapse of the upper part of the VOLCANO. Named either for Pliny the Elder, a Roman naturalist who died while observing the ERUPTION of Mount Vesuvius in 79 CE, or for Pliny the Younger, his nephew, who chronicled the eruption.

Plucking. Term used to describe the way glacial ice can erode large pieces of ROCK as it makes its way downslope. The ice penetrates JOINTS, other openings on the floor, or perhaps the side wall, and freezes around the block of stone, tearing it away and carrying it along, as part of the glacial MORAINE. The rocks contribute greatly to glacial ABRASION, causing deep grooves or STRIATIONS in some places. The jagged torn surface left behind is subject to further plucking. ALPINE GLACIERS can erode steep VALLEYS called glacial TROUGHS.

Plutonic. IGNEOUS ROCKS made of MINERAL grains visible to the naked eye. These igneous rocks have cooled relatively slowly. GRANITE is a good example of a plutonic rock.

Pluvial period. Episode of time during which rains were abundant, especially during the last ICE AGE, from a few million to about 10,000 years ago.

Polar stratospheric clouds. CLOUDS of ice crystals formed at extremely low TEMPERATURES in the polar STRATOSPHERE.

Polder. Lands reclaimed from the SEA by constructing DIKES to hold back the sea and then pumping out the water retained between the dikes and the land. Before AGRICULTURE is possible, the SOIL must be specially treated to remove the SALT. Some polders are used for recreational land; cities also have been built on polders. The largest polders are in the Netherlands, where the northern part, known as the Low Netherlands, covers almost half of the total area of this COUNTRY.

Polygenetic. Pertaining to volcanism from several physically distinct vents or repeated ERUPTIONS from a single vent punctuated by long periods of quiescence.

Polygonal ground. Distinctive geological formation caused by the repetitive freezing and thawing of PERMAFROST.

Possibilism. Concept that arose among French geographers who rejected the concept of ENVIRONMENTAL DETERMINISM, instead asserting that the relationship between human beings and the ENVIRONMENT is interactive.

Potable water. FRESH WATER that is being used for domestic consumption.

Potholes. Circular depressions formed in the bed of a RIVER when the STREAM flows over BEDROCK. The scouring of PEBBLES as a result of water TURBULENCE wears away the sides of the depression, deepening it vertically and producing a smooth, rounded pothole. (In modern parlance, the term is also applied to holes in public roads.)

Primary minerals. Minerals formed when magma crystallizes.

Primary wave. Compressional type of earthquake wave, which can travel in any medium and is the fastest wave.

Primate city. City that is at least twice as large as the next-largest city in that country. The "law of the primate city" was developed by U.S. geographer Mark Jefferson, to analyze the phenomenon of countries where one huge city dominates the political, economic, and cultural life of that country. The size and dominance of a primate city is a pull factor and ensures its continuing dominance.

Principal parallels. The most important lines of latitude. Parallels are imaginary lines, parallel to the equator. The principal parallels are the equator at zero degrees, the tropic of Cancer at 23.5 degrees North, the tropic of Capricorn at 23.5 degrees south, the Arctic Circle at 66.5 degrees north, and the Antarctic Circle at 66.5 degrees south.

Protectorate. Country that is a political dependency of another nation; similar to a colony, but usually having a less restrictive relationship with its overseeing power.

Proterozoic eon. Interval between 2.5 billion and 544 million years ago. During this period in the geologic record, processes presently active on Earth first appeared, notably the first clear evidence for plate tectonics. Rocks of the Proterozoic eon also document changes in conditions on Earth, particularly an apparent increase in atmospheric oxygen.

Pull factors. Forces that attract immigrants to a new country or location as permanent settlers. They include economic opportunities, educational facilities, land ownership, gold rushes, climate conditions, democracy, and similar factors of attraction.

Push factors. Forces that encourage people to migrate permanently from their homelands to settle in a new destination. They include war, persecution for religious or political reasons, hunger, and similar negative factors.

Pyroclasts. Materials that are ejected from a volcano into the air. Pyroclastic materials return to Earth at greater or lesser distances, depending on their size and the height to which they are thrown by the explosion of the volcano. The largest pyroclasts are volcanic bombs. Smaller pieces are volcanic blocks and scoria. These generally fall back onto the volcano and roll down the sides. Even smaller pyroclasts are lapilli, cinders, and volcanic ash. The finest pyroclastic materials may be carried by winds for great distances, even completely around the earth, as was the case with dust from the Krakatoa explosion in 1883 and the early 1990s explosions of Mount Pinatubo in the Philippines.

Qanat. Method used in arid regions to bring groundwater from mountainous regions to lower and flatter agricultural land. A qanat is a long tunnel or series of tunnels, perhaps more than a mile long. The word *qanat* is Arabic, but the first qanats are thought to have been constructed in Farsi-speaking Persia more than 2,000 years ago. Qanats are still used there, as well as in Afghanistan and Morocco.

Quaternary sector. Economic activity that involves the collection and processing of information. The rapid spread of computers and the Internet caused a major increase in the importance of employment in the quaternary sector.

Radar imaging. Technique of transmitting radar toward an object and then receiving the reflected radiation so that time-of-flight measurements provide information about surface topography of the object under study.

Radial drainage. The pattern of stream courses often reveals the underlying geology or structure of a region. In a radial drainage pattern, streams radiate outward from a center, like spokes on a wheel, because they flow down the slopes of a volcano.

Radioactive minerals. Minerals combining uranium, thorium, and radium with other elements. Useful for nuclear technology, these minerals furnish the basic isotopes necessary not only for nuclear reactors but also for advanced medical treatments, metallurgical analysis, and chemicophysical research.

Rain gauge. Instrument for measuring RAINFALL, usually consisting of a cylindrical container open to the sky.

Rain shadow. Area of low PRECIPITATION located on the LEEWARD side of a topographic barrier such as a mountain RANGE. Moisture-laden WINDS are forced to rise, so they cool ADIABATICALLY, leading to CONDENSATION and precipitation on the WINDWARD side of the barrier. When the AIR descends on the other side of the MOUNTAIN, it is dry and relatively warm. The area to the east of the Rocky Mountains is in a rain shadow.

Range, mountain. Linear series of MOUNTAINS close together, formed in an OROGENY, or mountain-building episode. Tall mountain ranges such as the Rocky Mountains are geologically much younger than older mountain ranges such as the Appalachians.

Rapids. Stretches of RIVERS where the water flow is swift and turbulent because of a steep and rocky CHANNEL. The turbulent conditions are called WHITE WATER. If the change in ELEVATION is greater, as for small WATERFALLS, they are called CATARACTS.

Recessional moraine. Type of TERMINAL MORAINE that marks a position of shrinkage or wasting or a GLACIER. Continued forward flow of ice is maintained so that the debris that forms the moraine continues to accumulate. Recessional moraines occur behind the terminal moraine.

Recumbent fold. Overturned fold in which the upper part of the fold is almost horizontal, lying on top of the nearest adjacent surface.

Reef (geology). VEIN of ORE, for example, a reef of gold.

Reef (marine). Underwater ridge made up of sand, rocks, or coral that rises near to the water's surface.

Refraction of waves. Bending of waves, which can occur in all kinds of waves. When OCEAN WAVES approach a COAST, they start to break as they approach the SHORE because the depth decreases. The wave speed is retarded and the WAVE CREST seems to bend as the wavelength decreases. If waves are approaching a coast at an oblique angle, the crest line bends near the shore until it is almost parallel. If waves are approaching a BAY, the crests are refracted to fit the curve of the bay.

Regression. Retreat of the SEA from the land; it allows land EROSION to occur on material formerly below the sea surface.

Relative humidity. Measure of the HUMIDITY, or amount of moisture, in the ATMOSPHERE at any time and place compared with the total amount of moisture that same AIR could theoretically hold at that TEMPERATURE. Relative humidity is a ratio that is expressed as a percentage. When the air is saturated, the relative humidity reaches 100 percent and rain occurs. When there is little moisture in the air, the relative humidity is low, perhaps 20 percent. Relative humidity varies inversely with temperature, because warm air can hold more moisture than cooler air. Therefore, when temperatures fall overnight, the air often becomes saturated and DEW appears on grass and other surfaces. The human COMFORT INDEX is related to the relative humidity. Hot temperatures are more bearable when relative humidity is low. Media announcers frequently use the term "humidity" when they mean relative humidity.

Replacement rate. The rate at which females must reproduce to maintain the size of the POPULATION. It corresponds to a FERTILITY RATE of 2.1.

Reservoir rock. Geologic ROCK layer in which OIL and gas often accumulate; often SANDSTONE or LIMESTONE.

Retrograde orbit. ORBIT of a SATELLITE around a PLANET that is in the opposite sense (direction) in which the planet rotates.

Retrograde rotation. ROTATION of a PLANET in a direction opposite to that of its REVOLUTION.

Reverse fault. Feature produced by compression of the earth's CRUST, leading to crustal shortening. The UPTHROWN BLOCK overhangs the downthrown block, producing a FAULT SCARP where the overhang is prone to LANDSLIDES. When the movement is mostly horizontal, along a low angle FAULT, an overthrust fault is formed. This is commonly associated with extreme FOLDING.

Reverse polarity. Orientation of the earth's MAGNETIC FIELD so that a COMPASS needle points to the SOUTHERN HEMISPHERE.

Ria coast. Ria is a long narrow ESTUARY or RIVER MOUTH. COASTS where there are many rias show the effects of SUBMERGENCE of the land, with the SEA now occupying former RIVER VALLEYS. Generally, there are MOUNTAINS running at an angle to the coast, with river valleys between each RANGE, so that the ria coast is a succession of estuaries and promontories. The submergence can result from a rising SEA LEVEL, which is common since the melting of the PLEISTOCENE GLACIERS, or it can be the result of SUBSIDENCE of the land. There is often a great TIDAL RANGE in rias, and in some, a tidal BORE occurs with each TIDE. The eastern coast of the United States, from New York to South Carolina, is a ria coast. The southwest coast of Ireland is another. The name comes from Spain, where rias occur in the south.

Richter scale. SCALE used to measure the magnitude of EARTHQUAKES; named after U.S. physicist Charles Richter, who, together with Beno Gutenberg, developed the scale in 1935. The scale is a quantitative measure that replaced the older MERCALLI SCALE, which was a descriptive scale. Numbers range from zero to nine, although there is no upper limit. Each whole number increase represents an order of magnitude, or an increase by a factor of ten. The actual MEASUREMENT was logarithm to base 10 of the maximum SEISMIC WAVE amplitude (in thousandths of a millimeter) recorded on a standard SEISMOGRAPH at a distance of 60 miles (100 km.) from the earthquake EPICENTER.

Rift valley. Long, low REGION of the earth's surface; a VALLEY or TROUGH with FAULTS on either side. Unlike valleys produced by EROSION, rift valleys are produced by tectonic forces that have caused the faults or fractures to develop in the ROCKS of Earth's CRUST. TENSION can lead to the block of land between two faults dropping in ELEVATION compared to the surrounding blocks, thus forming the rift valley. A small LANDFORM produced in this way is called a GRABEN. A rift valley is a much larger feature. In Africa, the Great Rift Valley is partially occupied by Lake Malawi and Lake Tanganyika, as well as by the Red Sea.

Ring dike. Volcanic LANDFORM created when MAGMA is intruded into a series of concentric FAULTS. Later EROSION of the surrounding material may reveal the ring dike as a vertical feature of thick BASALT rising above the surroundings.

Ring of Fire. Zone of volcanic activity and associated EARTHQUAKES that marks the edges of various TECTONIC PLATES around the Pacific Ocean, especially those where SUBDUCTION is occurring.

Riparian. Term meaning related to the BANKS of a STREAM or RIVER. Riparian VEGETATION is generally trees, because of the availability of moisture. RIPARIAN RIGHTS allow owners of land adjacent to a river to use water from the river.

River terraces. LANDFORMS created when a RIVER first produces a FLOODPLAIN, by DEPOSITION of ALLUVIUM over a wide area, and then begins downcutting into that alluvium toward a lower BASE LEVEL. The renewed EROSION is generally because of a fall in SEA LEVEL, but can result from tectonic UPLIFT or a change in CLIMATE pattern due to increased PRECIPITATION. On either side of the river, there is a step up from the new VALLEY to the former alluvium-covered floodplain surface, which is now one of a pair of river terraces. This process may occur more than once, creating as many as three sets of terraces. These are called depositional terraces, because the terrace is cut into river deposits. Erosional terraces, in contrast, are formed by lateral migration of a river, from one part of the valley to another, as the river creates a floodplain. These terraces are cut into BEDROCK, with only a thin layer of alluvium from the point BAR deposits, and they do not occur in matching pairs.

River valleys. VALLEYS in which STREAMS flow are produced by those streams through long-term EROSION and DEPOSITION. The LANDFORMS produced by FLUVIAL action are quite diverse, ranging from spectacular CANYONS to wide, gently sloping valleys. The patterns formed by stream networks are complex and generally reflect the BEDROCK geology and TERRAIN characteristics.

Rock avalanche. Extreme case of a rockfall. It occurs when a large mass of ROCK moves rapidly down a steeply sloping surface, taking everything that lies in its path. It can be started by an EARTHQUAKE, rock-blasting operations, or vibrations from thunder or artillery fire.

Rock cycle. Cycle by which ROCKS are formed and reformed, changing from one type to another over long PERIODS of geologic time. IGNEOUS ROCKS are formed by cooling from molten MAGMA. Once exposed at the surface, they are subject to WEATHERING and EROSION. The products of erosion are compacted and cemented to form SEDIMENTARY ROCKS. The heat and pressure accompanying a volcanic intrusion causes adjacent rocks to be altered into METAMORPHIC ROCKS.

Rock slide. Event that occurs when water lubricates an unconsolidated mass of weathered ROCK on a steep slope, causing rapid downslope movement. In a RIVER VALLEY where there are steep SCREE slopes being constantly carried away by a swiftly flowing STREAM, the undercutting at the base can lead to constant rockslides of the surface layer of rock. A large rockslide is a ROCK AVALANCHE.

S waves. Type of SEISMIC disturbance of the earth when an EARTHQUAKE occurs. In an S wave, particles move about at right angles to the direction in which the wave is traveling. S waves cannot pass through the earth's CORE, which is why scientists believe the INNER CORE is liquid. Also called transverse wave, shear wave, or secondary wave.

Sahel. Southern edge of the Sahara Desert; a great stretch of semiarid land extending from the Atlantic Ocean in Senegal and Mauritania through Mali, Burkina Faso, Nigeria, Niger, Chad, and Sudan. Northern Ethiopia, Eritrea, Djibouti, and Somalia usually are included also. This transition zone between the hot DESERT and the tropical SAVANNA has low summer RAINFALL of less than 8 inches (200 millimeters) and a natural VEGETATION of low grasses with some small shrubs. The REGION traditionally has been used for PASTORALISM, raising goats, camels, and occasionally sheep. Since a prolonged DROUGHT in the 1970s, DESERTIFICATION, SOIL EROSION, and FAMINE have plagued the Sahel. The narrow band between the northern Sahara and the Mediterranean North African COAST is also called Sahel. "Sahel" is the Arabic word for edge.

Saline lake. LAKE with elevated levels of dissolved solids, primarily resulting from evaporative concentration of SALTS; saline lakes lack an outlet to the SEA. Well-known examples include Utah's Great Salt Lake, California's Mono Lake and Salton Sea, and the Dead Sea in the Middle East.

Salinization. Accumulation of SALT in SOIL. When IRRIGATION is used to grow crops in semiarid to arid REGIONS, salinization is frequently a problem. Because EVAPORATION is high, water is drawn upward through the soil, depositing dissolved salts at or near the surface. Over years, salinization can build up until the soil is no longer suitable for AGRICULTURE. The solution is to maintain a plentiful flow of water while ensuring that the water flows through the soil and is drained away.

Salt domes. Formations created when deeply buried salt layers are forced upwards. SALT under pressure is a plastic material, one that can flow or move slowly upward, because it is lighter than surrounding SEDIMENTARY ROCKS. The salt forms into a plug more than a half mile (1 km.) wide and as much as 5 miles (8 km.) deep, which passes through overlying sedimentary rock layers, pushing them up into a dome shape as it passes. Some salt domes emerge at the earth's surface; others are close to the surface and are easy to mine for ROCK SALT. OIL and NATURAL GAS often accumulate against the walls of a salt dome. Salt domes are numerous around the COAST of the Gulf of Mexico, in the North Sea REGION, and in Iran and Iraq, all of which are major oil-producing regions.

Sand dunes. Accumulations of SAND in the shape of mounds or RIDGES. They occur on some COASTS and in arid REGIONS. Coastal dunes are formed when the prevailing WINDS blow strongly onshore, piling up sand into dunes, which may become stabilized when grasses grow on them. DESERT sand dunes are a product of DEFLATION, or wind EROSION removing fine materials to leave a DESERT PAVEMENT in one region and sand deposits in another. Sand dunes are classified by their shape into barchans, or crescent-shaped dunes; seifs or LONGITUDINAL DUNES; TRANSVERSE DUNES; star dunes; and sand drifts or sand sheets.

Sapping. Natural process of EROSION at the bases of HILL slopes or CLIFFS whereby support is removed by undercutting, thereby allowing overlying layers to collapse; SPRING SAPPING is the facilitation of this process by concentrated GROUNDWATER flow, generally at the heads of VALLEYS.

Saturation, zone of. Underground REGION below the zone of AERATION, where all pore space is filled with water. This water is called GROUNDWATER; the upper surface of the zone of saturation is the WATER TABLE.

Scale. Relationship between a distance on a MAP or diagram and the same distance on the earth. Scale can be represented in three ways. A linear, or graphic, scale uses a straight line, marked off in equally spaced intervals, to show how much of the map represents a mile or a kilometer. A representative fraction (RF) gives this scale as a ratio. A verbal scale uses words to explain the relationship between map size and actual size. For example, the RF 1:63,360 is the same as saying "one inch to the mile."

Scarp. Short version of the word "ESCARPMENT," a short steep slope, as at the edge of a PLATEAU. EARTHQUAKES lead to the formation of FAULT SCARPS.

Schist. METAMORPHIC ROCK that can be split easily into layers. Schist is commonly produced from the action of heat and pressure on SHALE or SLATE. The rock looks flaky in appearance. Mica-schists are shiny because of the development of visible mica. Other schists include talc-schist, which contains a large amount of talc, and hornblende-schist, which develops from basaltic rocks.

Scree. Broken, loose ROCK material at the base of a slope or CLIFF. It is often the result of FROST WEDGING of BEDROCK cliffs, causing rockfall. Another name for scree is TALUS.

Sedimentary rocks. ROCKS formed from SEDIMENTS that are compressed and cemented together in a process called LITHIFICATION. Sedimentary rocks cover two-thirds of the earth's land surface but are only a small proportion of the earth's CRUST. SANDSTONE is a common sedimentary rock. Sedimentary rocks form STRATA, or layers, and sometimes contain FOSSILS.

Seif dunes. Long, narrow RIDGES of SAND, built up by WINDS blowing at different times of year from two different directions. Seif dunes occur in parallel lines of sand over large areas, running for hundreds of miles in the Sahara, Iran, and central Australia. Another name for seif dunes is LONGITUDINAL DUNES. The Arabic word means sword.

Seismic activity. Movements within the earth's CRUST that often cause various other geological phenomena to occur; the activity is measured by SEISMOGRAPHS.

Seismology. The scientific study of EARTHQUAKES. It is a branch of GEOPHYSICS. The study of SEISMIC WAVES has provided a great deal of knowledge about the composition of the earth's interior.

Shadow zone. When an EARTHQUAKE occurs at one LOCATION, its waves travel through the earth and are detected by SEISMOGRAPHS around the world. Every earthquake has a shadow zone, a band where neither P nor S WAVES from the earthquake will be detected. This shadow zone leads scientists to draw conclusions about the size, density, and composition of the earth's CORE.

Shale oil. SEDIMENTARY ROCK containing sufficient amounts of hydrocarbons that can be extracted by slow distillation to yield OIL.

Shallow-focus earthquakes. EARTHQUAKES having a focus less than 35 miles (60 km.) below the surface.

Shantytown. URBAN SQUATTER SETTLEMENT, usually housing poor newcomers.

Shield. Large part of the earth's CONTINENTAL CRUST, comprising very old ROCKS that have been eroded to REGIONS of low RELIEF. Each CONTINENT has a shield area. In North America, the Canadian Shield extends from north of the Great Lakes to the Arctic Ocean. Sometimes known as a CONTINENTAL SHIELD.

Shield volcano. VOLCANO created when the LAVA is quite viscous or fluid and highly basaltic. Such lava spreads out in a thin sheet of great radius but comparatively low height. As flows continue to build up the volcano, a low DOME shape is created. The greatest shield volcanoes on Earth are the ISLANDS of Hawaii, which rise to a height of almost 30,000 feet (10,000 meters) above SEA LEVEL.

Shock city. CITY that typifies disturbing changes in social and cultural conditions or in economic conditions. In the nineteenth century, the shock city of the United States was Chicago.

***Sierra*.** Spanish word for a mountain RANGE with a serrated crest. In California, the Sierra Nevada is an important range, containing Mount Whitney, the highest PEAK in the continental United States.

Sill. Feature formed by INTRUSIVE volcanic activity. When LAVA is forced between two layers of ROCK, it can form a narrow horizontal layer of BASALT, parallel with the adjacent beds. Although it resembles a windowsill in its flatness, a sill may be hundreds of miles long and can range in thickness from a few centimeters to considerable thickness.

Siltation. Build-up of SILT and SAND in creeks and waterways as a result of SOIL EROSION, clogging water courses and creating DELTAS at RIVER MOUTHS. Siltation often results from DEFORESTATION or removal of tree cover. Such ENVIRONMENTAL DEGRADATION causes loss of agricultural productivity, worsening of water supply, and other problems.

Sima. Abbreviation for SILICA and *ma*gnesium. These are the two principal constituents of heavy ROCKS such as BASALT, which forms much of the OCEAN floor. Lighter, more abundant rock is SIAL.

Sinkhole. Circular depression in the ground surface, caused by WEATHERING of LIMESTONE, mainly through the effects of SOLUTION on JOINTS in the ROCK. If a STREAM flows above ground and then disappears down a sinkhole, the feature is called a swallow hole. In everyday language, many events that cause the surface to collapse are called sinkholes, even though they are rarely in limestone and rarely caused by weathering.

Sinking stream. STREAM or RIVER that loses part or all of its water to pathways dissolved underground in the BEDROCK.

Situation. Relationship between a PLACE, such as a TOWN or CITY, and its RELATIVE LOCATION within a REGION. A situation on the COAST is desirable in terms of overseas TRADE.

Slip-face. LEEWARD side of a SAND DUNE. As the WIND piles up sand on the WINDWARD side, it then slips down the rear or slip-face. The angle of the slip-face is gentler than the angle of the windward slope.

Slump. Type of LANDSLIDE in which the material moves downslope with a rotational motion, along a curved slip surface.

Snout. Terminal end of a GLACIER.

Snow line. The height or ELEVATION at which snow remains throughout the year, without melting away. Near the EQUATOR, the snow line is more than 15,000 feet (almost 5,000 meters); at higher LATITUDES, the snow line is correspondingly lower, reaching SEA LEVEL at the POLES. The actual snow line varies with the time of year, retreating in summer and coming lower in winter.

Soil horizon. SOIL consists of a series of layers called horizons. The uppermost layer, the O horizon, contains organic materials such as decayed leaves that have been changed into HUMUS. Beneath this is the A horizon, the TOPSOIL, where farmers plow and plant seeds. The B HORIZON often contains MINERALS that have been washed downwards from the A horizon, such as calcium, iron, and aluminum. The A and B horizons together comprise a solum, or true soil. The C horizon is weathered BEDROCK, which contains pieces of the original ROCK from which the soil formed. Another name for the C horizon is REGOLITH. Beneath this is the R horizon, or bedrock.

Soil moisture. Water contained in the unsaturated zone above the WATER TABLE.

Soil profile. Vertical section of a SOIL, extending through its horizon into the unweathered parent material.

Soil stabilization. Engineering measures designed to minimize the opportunity and/or ability of EXPANSIVE SOILS to shrink and swell.

Solar energy. One of the forms of ALTERNATIVE or RENEWABLE ENERGY. In the late 1990s, the world's largest solar power generating plant was located at Kramer Junction, California. There, solar energy heats huge OIL-filled containers with a parabolic shape, which produces steam to drive generating turbines. An alternative is the production of energy through photovoltaic cells, a TECHNOLOGY that was first developed for space exploration. Many individual homes, especially in isolated areas, use this technology.

Solar system. SUN and all the bodies that ORBIT it, including the PLANETS and their SATELLITES, plus numerous COMETS, ASTEROIDS, and METEOROIDS.

Solar wind. Gases from the SUN'S ATMOSPHERE, expanding at high speeds as streams of charged particles.

Solifluction. Word meaning flowing SOIL. In some REGIONS of PERMAFROST, where the ground is permanently frozen, the uppermost layer thaws during the summer, creating a saturated layer of soil and REGOLITH above the hard layer of frozen ground. On slopes, the material can flow slowly downhill, creating a wavy appearance along the hillslope.

Solution. Form of CHEMICAL WEATHERING in which MINERALS in a ROCK are dissolved in water. Most substances are soluble, but the combination of water with CARBON DIOXIDE from the ATMOSPHERE means that RAINFALL is slightly acidic, so that the chemical reaction is often a combination of solution and carbonation.

Sound. Long expanse of the SEA, close to the COAST, such as a large ESTUARY. It can also be the expanse of sea between the mainland and an ISLAND.

Source rock. ROCK unit or bed that contains sufficient organic carbon and has the proper thermal history to generate OIL or gas.

Spatial diffusion. Notion that things spread through space and over time. An understanding of geographic change depends on this concept. Spatial diffusion can occur in various ways. Geographers distinguish between expansion diffusion, relocation diffusion, and hierarchical diffusion.

Spheroidal weathering. Form of ROCK WEATHERING in which layers of rock break off parallel to the surface, producing a rounded shape. It results from a combination of physical and CHEMICAL WEATHERING. Spheroidal weathering is especially common in GRANITE, leading to the creation of TORS and similar rounded features. Onion-skin weathering is a term sometimes used, especially when this is seen on small rocks.

Spring tide. TIDE of maximum RANGE, occurring when lunar and solar tides reinforce each other, a few DAYS after the full and new MOONS.

Squall line. Line of vigorous THUNDERSTORMS created by a cold downdraft that spreads out ahead of a fast-moving COLD FRONT.

Stacks. Pieces of ROCK surrounded by SEA water, which were once part of the mainland. WAVE EROSION has caused them to be isolated. Also called sea stacks.

Stalactite. Long, tapering piece of calcium carbonate hanging from the roof of a LIMESTONE CAVE or cavern. Stalactites are formed as water containing the MINERAL in solution drips downward. The water evaporates, depositing the dissolved minerals.

Stalagmite. Column of calcium carbonate growing upward from the floor of a LIMESTONE CAVE or cavern.

Steppe. Huge REGION of GRASSLANDS in the midlatitudes of Eurasia, extending from central

Europe to northeast China. The region is not uniform in ELEVATION; most of it is rolling PLAINS, but some mountain RANGES also occur. These have not been a barrier to the migratory lifestyle of the herders who have occupied the steppe for many centuries. The Asian steppe is colder than the European steppe, because of greater elevation and greater continentality. The best-known rulers from the steppe were the Mongols, whose empire flourished in the thirteenth and fourteenth centuries. Geographers speak of a steppe CLIMATE, a semiarid climate where the EVAPORATION rate is double that of PRECIPITATION. South of the steppe are great DESERTS; to the north are midlatitude mixed FORESTS. In terms of climate and VEGETATION, the steppe is like the short-grass PRAIRIE vegetation west of the Mississippi River. Also called steppes.

Storm surge. General rise above normal water level, resulting from a HURRICANE or other severe coastal STORM.

Strait. Relatively narrow body of water, part of an OCEAN or SEA, separating two pieces of land. The world's busiest SEAWAY is the Johore Strait between the Malay Peninsula and the island of Sumatra.

Strata. Layers of SEDIMENT deposited at different times, and therefore of different composition and TEXTURE. When the sediments are laid down, strata are horizontal, but subsequent tectonic processes can lead to tilting, FOLDING, or faulting. Not all SEDIMENTARY ROCKS are stratified. Singular form of the word is stratum.

Stratified drift. Material deposited by glacial MELTWATERS; the water separates the material according to size, creating layers.

Stratigraphy. Study of sedimentary STRATA, which includes the concept of time, possible correlation of the ROCK units, and characteristics of the rocks themselves.

Stratovolcano. Type of VOLCANO in which the ERUPTIONS are of different types and produce different LAVAS. Sometimes an eruption ejects cinder and ASH; at other times, viscous lava flows down the sides. The materials flow, settle, and fall to produce a beautiful symmetrical LANDFORM with a broad circular base and concave slopes tapering upward to a small circular CRATER. Mount Rainier, Mount Saint Helens, and Mount Fuji are stratovolcanoes. Also known as a COMPOSITE CONE.

Streambed. Channel through which a STREAM flows. Dry streambeds are variously known as ARROYOS, DONGAS, WASHES, and WADIS.

Strike. Term used when earth scientists study tilted or inclined beds of SEDIMENTARY ROCK. The strike of the inclined bed is the direction of a horizontal line along a bedding plane. The strike is at right angles to the dip of the rocks.

Strike-slip fault. In a strike-slip fault, the surface on either side of the fault moves in a horizontal plane. There is no vertical displacement to form a FAULT SCARP, as there is with other types of faults. The San Andreas Fault is a strike-slip fault. Also called a transcurrent fault.

Subduction zone. CONVERGENT PLATE BOUNDARY where an oceanic PLATE is being thrust below another plate.

Sublimation. Process by which water changes directly from solid (ice) to vapor, or vapor to solid, without passing through a liquid stage.

Subsolar point. Point on the earth's surface where the SUN is directly overhead, making the Sun's rays perpendicular to the surface. The subsolar point receives maximum INSOLATION, compared with other PLACES, where the Sun's rays are oblique.

Sunspots. REGIONS of intense magnetic disturbances that appear as dark spots on the solar surface; they occur approximately every eleven years.

Supercontinent. Vast LANDMASS of the remote geologic past formed by the collision and amalgamation of crustal PLATES. Hypothesized supercontinents include PANGAEA, GONDWANALAND, and LAURASIA.

Supersaturation. State in which the AIR'S RELATIVE HUMIDITY exceeds 100 percent, the condition necessary for vapor to begin transformation to a liquid state.

Supratidal. Referring to the SHORE area marginal to shallow OCEANS that are just above high-tide level.

Swamp. WETLAND where trees grow in wet to waterlogged conditions. Swamps are common close to the RIVER on FLOODPLAINS, as well as in some coastal areas.

Swidden. Area of land that has been cleared for SUBSISTENCE AGRICULTURE by a farmer using the technique of slash-and-burn. A variety of crops is planted, partly to reduce the risk of crop failure. Yields are low from a swidden because SOIL fertility is low and only human labor is used for CLEARING, planting, and harvesting. See also INTERTILLAGE.

Symbolic landscapes. LANDSCAPES centered on buildings or structures that are so visually emblematic that they represent an entire CITY.

Syncline. Downfold or TROUGH shape that is formed through compression of ROCKS. An upfold is an ANTICLINE.

Tableland. Large area of land with a mostly flat surface, surrounded by steeply sloping sides, or ESCARPMENTS. A small PLATEAU.

Taiga. Russian name for the vast BOREAL FORESTS that cover Siberia. The marshy ground supports a tree VEGETATION in which the trees are CONIFEROUS, comprising mostly pine, fir, and larch.

Talus. Broken and jagged pieces of ROCK, produced by WEATHERING of steep slopes, that fall to the base of the slope and accumulate as a talus cone. In high MOUNTAINS, a ROCK GLACIER may form in the talus. See also SCREE.

Tarn. Small circular LAKE, formed in a CIRQUE, which was previously occupied by a GLACIER.

Tectonism. The formation of MOUNTAINS because of the deformation of the CRUST of the earth on a large scale.

Temporary base level. STREAMS or RIVERS erode their beds down toward a BASE LEVEL—in most cases, SEA LEVEL. A section of hard ROCK may slow EROSION and act as a temporary, or local, base level. Erosion slows upstream of the temporary base level. A DAM is an artificially constructed temporary base level.

Tension. Type of stress that produces a stretching and thinning or pulling apart of the earth's CRUST. If the surface breaks, a NORMAL FAULT is created, with one side of the surface higher than the other.

Tephra. General term for volcanic materials that are ejected from a vent during an ERUPTION and transported through the AIR, including ASH (volcanic), BLOCKS (volcanic), cinders, LAPILLI, scoria, and PUMICE.

Terminal moraine. RIDGE of unsorted debris deposited by a GLACIER. When a glacier erodes it moves downslope, carrying ROCK debris and creating a ground MORAINE of material of various sizes, ranging from big angular blocks or boulders down to fine CLAY. At the terminus of the glacier, where the ice is melting, the ground moraine is deposited, building the ridge of unsorted debris called a terminal moraine.

Terrain. Physical features of a REGION, as in a description of rugged terrain. It should not be confused with TERRANE.

Terrane. Piece of CONTINENTAL CRUST that has broken off from one PLATE and subsequently been joined to a different plate. The terrane has quite different composition and structure from the adjacent continental materials. Alaska is composed mostly of terranes that have accreted, or joined, the North American plate.

Terrestrial planet. Any of the solid, rocky-surfaced bodies of the inner SOLAR SYSTEM, including the PLANETS Mercury, Venus, Earth, and Mars and Earth's SATELLITE, the MOON.

Terrigenous. Originating from the WEATHERING and EROSION of MOUNTAINS and other land formations.

Texture. One of the properties of SOILS. The three textures are SAND, SILT, and CLAY. Texture is measured by shaking the dried soil through a series of sieves with mesh of reducing diameters. A mixture of sand, silt, and clay gives a LOAM soil.

Thermal equator. Imaginary line connecting all PLACES on Earth with the highest mean daily TEMPERATURE. The thermal equator moves south of the EQUATOR in the SOUTHERN HEMISPHERE summer, especially over the CONTINENTS of South America, Africa, and Australia. In the northern summer, the thermal equator moves far into Asia, northern Africa, and North America.

Thermal pollution. Disruption of the ECOSYSTEM caused when hot water is discharged, usually as a thermal PLUME, into a relatively cooler body of water. The TEMPERATURE change affects the aquatic ecosystem, even if the water is chemically pure. Nuclear power-generating plants use large volumes of water in the process and are important sources of thermal pollution.

Thermocline. Depth interval at which the TEMPERATURE of OCEAN water changes abruptly, separating warm SURFACE WATER from cold, deep water.

Thermodynamics. Area of science that deals with the transformation of ENERGY and the laws that govern these changes; equilibrium thermodynamics is especially concerned with the reversible conversion of heat into other forms of energy.

Thermopause. Outer limit of the earth's ATMOSPHERE.

Thermosphere. Atmospheric zone beyond the MESOSPHERE in which TEMPERATURE rises rapidly with increasing distance from the earth's surface.

Thrust belt. Linear BELT of ROCKS that have been deformed by THRUST FAULTS.

Thrust fault. FAULT formed when extreme compression of the earth's CRUST pushes the surface into folds so closely spaced that they overturn and the ROCK then fractures along a fault.

Tidal force. Gravitational force whose strength and direction vary over a body and thus act to deform the body.

Tidal range. Difference in height between high TIDE and low tide at a given point.

Tidal wave. Incorrect name for a TSUNAMI.

Till. Mass of unsorted and unstratified SEDIMENTS deposited by a GLACIER. Boulders and smaller rounded ROCKS are mixed with CLAY-sized materials.

Timberline. Another term for tree line, the BOUNDARY of tree growth on MOUNTAIN slopes. Above the timberline, TEMPERATURES are too cold for tree growth.

Tombolo. Strip of SAND or other SEDIMENT that connects an ISLAND or SEA stack to the mainland. Mont-Saint-Michel is linked to the French mainland by a tombolo.

Topography. Description of the natural LANDSCAPE, including LANDFORMS, RIVERS and other waters, and VEGETATION cover.

Topological space. Space defined in terms of the connectivity between LOCATIONS in that space. The nature and frequency of the connections are measured, while distance between locations is not considered an important factor. An example of topological space is a transport network diagram, such as a bus route or a MAP of an underground rail system. Networks are most concerned with flows, and therefore with connectivity.

Toponyms. PLACE names. Sometimes, names of features and SETTLEMENTS reveal a good deal about the history of a REGION. For example, the many names starting with "San" or "Santa" in the Southwest of the United States recall the fact that Spain once controlled that area. The scientific study of place names is toponymics.

Tor. Rocky outcrop of blocks of ROCK, or corestones, exposed and rounded by WEATHERING. Tors frequently form in GRANITE, where three series of JOINTS often developed as the rock originally cooled when it was formed.

Transform faults. FAULTS that occur along DIVERGENT PLATE boundaries, or SEAFLOOR SPREADING zones. The faults run perpendicular to the spreading center, sometimes for hundreds of miles, some for more than five hundred miles. The motion along a transform fault is lateral or STRIKE-SLIP.

Transgression. Flooding of a large land area by the SEA, either by a regional downwarping of continental surface or by a global rise in SEA LEVEL.

Transmigration. Policy of the government of Indonesia to encourage people to move from the densely overcrowded ISLAND of Java to the sparsely populated other islands.

Transverse bar. Flat-topped body of SAND or gravel oriented transverse to the RIVER flow.

Trophic level. Different types of food relations that are found within an ECOSYSTEM. Organisms that derive food and ENERGY through PHOTOSYNTHESIS are called autotrophs (self-feeders) or producers. Organisms that rely on producers as their source of energy are called heterotrophs (feeders on others) or consumers. A third trophic level is represented by the organisms known as decomposers, which recycle organic waste.

Tropical cyclone. STORM that forms over tropical OCEANS and is characterized by extreme amounts of rain, a central area of calm AIR, and spinning WINDS that attain speeds of up to 180 miles (300 km.) per hour.

Tropical depression. STORM with WIND speeds up to 38 miles (64 km.) per hour.

Tropopause. BOUNDARY layer between the TROPOSPHERE and the STRATOSPHERE.

Troposphere. Lowest and densest of Earth's atmospheric layers, marked by considerable TURBULENCE and a decrease in TEMPERATURE with increasing ALTITUDE.

Tsunami. SEISMIC SEA WAVE caused by a disturbance of the OCEAN floor, usually an EARTHQUAKE, although undersea LANDSLIDES or volcanic ERUPTIONS can also trigger tsunami.

Tufa. LIMESTONE or calcium carbonate deposit formed by PRECIPITATION from an alkaline LAKE. Mono Lake is famous for the dramatic tufa towers exposed by the lowering of the level of lake water. Also known as TRAVERTINE.

Tumescence. Local swelling of the ground that commonly occurs when MAGMA rises toward the surface.

Tunnel vent. Central tube in a volcanic structure through which material from the earth's interior travels.

U-shaped valley. Steep-sided VALLEY carved out by a GLACIER. Also called a glacial TROUGH.

Ubac slope. Shady side of a MOUNTAIN, where local or microclimatic conditions permit lower TIMBERLINES and lower SNOW LINES than occur on a sunny side.

Ultimate base level. Level to which a STREAM can erode its bed. For most RIVERS, this is SEA LEVEL. For streams that flow into a LAKE, the ultimate base level is the level of the lakebed.

Unconfined aquifer. AQUIFER whose upper BOUNDARY is the WATER TABLE; it is also called a water table aquifer.

Underfit stream. STREAM that appears to be too small to have eroded the VALLEY in which it flows. A RIVER flowing in a glaciated valley is a good example of underfit.

Uniformitarianism. Theory introduced in the early nineteenth century to explain geologic processes. It used to be believed that the earth was only a few thousand years old, so the creation of LANDFORMS would have been rapid, even catastrophic. This theory, called CATASTROPHISM, explained most landforms as the result of the Great Flood of the Bible, when Noah, his family, and animals survived the deluge. Uniformitarian- ism, in contrast, stated that the processes in operation today are slow, so the earth must be immensely older than a mere few thousand years.

Universal time (UT). See GREENWICH MEAN TIME.

Universal Transverse Mercator. Projection in which the earth is divided into sixty zones, each six DEGREES of LONGITUDE wide. In a traditional Mercator projection, the earth is seen as a sphere with a cylinder wrapped around the EQUATOR. UTM can be visualized as a series of six-degree side strips running transverse, or north-south.

Unstable air. Condition that occurs when the AIR above rising air is unusually cool so that the rising air is warmer and accelerates upward.

Upthrown block. When EARTHQUAKE motion produces a FAULT, the block of land on one side is displaced vertically relative to the other. The higher is the upthrown block; the lower is the downthrown block.

Upwelling. OCEAN phenomenon in which warm SURFACE WATERS are pushed away from the coast and are replaced by cold waters that carry more nutrients up from depth.

Urban heat island. Cities experience a different microclimate from surrounding regions. The city temperature is typically higher by a few degrees, both day and night, because of factors such as surfaces with higher heat absorption, decreased wind strength, human heat-producing activities such as power generation, and the layer of air pollution (dust dome).

Vadose zone. The part of the soil also known as the zone of aeration, located above the water table, where space between particles contains air.

Valley train. Fan-shaped deposit of glacial moraine that has been moved down-valley and redeposited by meltwater from the glacier.

Van Allen radiation belts. Bands of highly energetic, charged particles trapped in Earth's magnetic field. The particles that make up the inner belt are energetic protons, while the outer belt consists mainly of electrons and is subject to day-night variations.

Varnish, desert. Shiny black coating often found over the surface of rocks in arid regions. This is a form of oxidation or chemical weathering, in which a coating of manganese oxides has formed over the exposed surface of the rock.

Varve. Pair of contrasting layers of sediment deposited over one year's time; the summer layer is light, and the winter layer is dark.

Ventifacts. Pebbles on which one or more sides have been smoothed and faceted by abrasion as the wind has blown sand particles.

Volcanic island arc. Curving or linear group of volcanic islands associated with a subduction zone.

Volcanic rock. Type of igneous rock that is erupted at the surface of the earth; volcanic rocks are usually composed of larger crystals inside a fine-grained matrix of very small crystals and glass.

Volcanic tremor. Continuous vibration of long duration, detected only at active volcanoes.

Volcanology. Scientific study of volcanoes.

Voluntary migration. Movement of people who decide freely to move their place of permanent residence. It results from pull factors at the chosen destination, together with push factors in the home situation.

Warm temperate glacier. Glacier that is at the melting temperature throughout.

Water power. Generally means the generation of electricity using the energy of falling water. Usually a dam is constructed on a river to provide the necessary height difference. The potential energy of the falling water is converted by a water turbine into mechanical energy. This is used to power a generator, which produces electricity. Also called hydroelectric power. Another form of water power is tidal power, which uses the force of the incoming and outgoing tide as its source of energy.

Water table. The depth below the surface where the zone of aeration meets the zone of saturation. Above the water table, there may be some soil moisture, but most of the pore space is filled with air. Below the water table, pore space of the rocks is occupied by water that has percolated down through the overlying earth material. This water is called groundwater. In practice, the water table is rarely as flat as a table, but curved, being far below the surface in some places and even intersecting the surface in others. When groundwater emerges at the surface, because it intersects the water table, this is called a spring. The depth of the water table varies from season to season, and with pumping of water from an aquifer.

Watershed. The whole surface area of land from which rainfall flows downslope into a stream. The watershed comprises the streambed or channel, together with the valley sides, extending up to the crest or interfluve, which separates that watershed from its neighbor. Each watershed is separated from the next by the drainage divide. Also called a drainage basin.

Waterspout. Tornado that forms over water, or a tornado formed over land which then moves over water. The typical funnel cloud, which reaches down from a CUMULONIMBUS CLOUD, is a narrow rotating STORM, with WIND speeds reaching hundreds of miles per hour.

Wave crest. Top of a WAVE.

Wave-cut platform. As SEA CLIFFS are eroded and worn back by WAVE attack, a wave-cut platform is created at the base of the cliffs. ABRASION by ROCK debris from the cliffs scours the platform further, as waves wash to and fro and TIDES ebb and flow. The upper part of the wave-cut platform is exposed at high tide. These areas contain rockpools, which are rich in interesting MARINE life-forms. Offshore beyond the platform, a wave-built TERRACE is formed by DEPOSITION.

Wave height. Vertical distance between one WAVE CREST and the adjacent WAVE TROUGH.

Wave length. Distance between two successive WAVE CRESTS or two successive WAVE TROUGHS.

Wave trough. The low part of a WAVE, between two WAVE CRESTS.

Weather analogue. Approach to WEATHER FORECASTING that uses the WEATHER behavior of the past to predict what a current weather pattern will do in the future.

Weather forecasting. Attempt to predict WEATHER patterns by analysis of current and past data.

Wilson cycle. Creation and destruction of an OCEAN BASIN through the process of SEAFLOOR SPREADING and SUBDUCTION of existing ocean basins.

Wind gap. Abandoned WATER GAP. The Appalachian Mountains contain both wind gaps and water gaps.

Windbreak. Barrier constructed at right angles to the prevailing WIND direction to prevent damage to crops or to shelter buildings. Generally, a row of trees or shrubs is planted to form a windbreak. The feature is also called a shelter belt.

Windchill. MEASUREMENT of apparent TEMPERATURE that quantifies the effects of ambient WIND and temperature on the rate of cooling of the human body.

World Aeronautical Chart. International project undertaken to map the entire world, begun during World War II.

World city. CITY in which an extremely large part of the world's economic, political, and cultural activity occurs. In the year 2018, the top ten world cities were London, New York City, Tokyo, Paris, Singapore, Amsterdam, Seoul, Berlin, Hong Kong, and Sydney.

Xenolith. Smaller piece of ROCK that has become embedded in an IGNEOUS ROCK during its formation. It is a piece of older rock that was incorporated into the fluid MAGMA.

Xeric. Description of SOILS in REGIONS with a MEDITERRANEAN CLIMATE, with moist cool winters and long, warm, dry summers. Since summer is the time when most plants grow, the lack of SOIL MOISTURE is a limiting factor on plant growth in a xeric ENVIRONMENT.

Xerophytic plants. Plants adapted to arid conditions with low PRECIPITATION. Adaptations include storage of moisture in tissue, as with cactus plants; long taproots reaching down to the WATER TABLE, as with DESERT shrubs; or tiny leaves that restrict TRANSPIRATION.

Yardangs. Small LANDFORMS produced by WIND EROSION. They are a series of sharp RIDGES, aligned in the direction of the wind.

Yazoo stream. TRIBUTARY that flows parallel to the main STREAM across the FLOODPLAIN for a considerable distance before joining that stream. This occurs because the main stream has built up NATURAL LEVEES through flooding, and because RELIEF is low on the floodplain. The yazoo stream flows in a low-lying wet area called backswamps. Named after the Yazoo River, a tributary of the Mississippi.

Zero population growth. Phenomenon that occurs when the number of deaths plus EMIGRATION is matched by the number of births plus IMMIGRATION. Some European countries have reached zero population growth.

Bibliography

The Nature of Geography

Adams, Simon, Anita Ganeri, and Ann Kay. *Geography of the World*. London: DK, 2010. Print.

Harley, J. B., and David Woodward, eds. *The History of Cartography: Cartography in the Traditional Islamic and South Asian Societies*. Vol. 2, book 1. Chicago: University of Chicago Press, 1992. Offers a critical look at maps, mapping, and mapmakers in the Islamic world and South Asia.

_______, eds. *The History of Cartography: Cartography in the Traditional East and Southeast Asian Societies*. Vol. 2, book 2. Chicago: University of Chicago Press, 1994. Similar in thrust and breadth to volume 2, book 1.

Marshall, Tim, and John Scarlett. *Prisoners of Geography: Ten Maps That Tell You Everything You Need to Know About Global Politics*. London : Elliott and Thompson Limited, 2016. Print

Nijman, Jan. *Geography: Realms, Regions, and Concepts*. Hoboken, NJ : Wiley, 2020. Print.

Snow, Peter, Simon Mumford, and Peter Frances. *History of the World Map by Map*. New York: DK Smithsonian, 2018.

Woodward, David, et al., eds. *The History of Cartography: Cartography in the Traditional African, American, Arctic, Australian, and Pacific Societies*. Vol. 2, book 3. Chicago: University of Chicago Press, 1998. Investigates the roles that maps have played in the wayfinding, politics, and religions of diverse societies such as those in the Andes, the Trobriand Islanders of Papua-New Guinea, the Luba of central Africa, and the Mixtecs of Central America.

Physical Geography

Christopherson, Robert W, and Ginger H. Birkeland. *Elemental Geosystems*. Hoboken, NJ : Wiley, 2016. Print.

Lutgens, Frederick K., and Edward J. Tarbuck. *Foundations of Earth Science*. Upper Saddle River, N.J.: Prentice-Hall, 2017. Undergraduate text for an introductory course in earth science, consisting of seven units covering basic principles in geology, oceanography, meteorology, and astronomy, for those with little background in science.

McKnight, Tom. Physical Geography: A Landscape Appreciation. 12th ed. New York: Prentice Hall, 2017. Now-classic college textbook that has become popular because of its illustrations, clarity, and wit. Comes with a CD-ROM that takes readers on virtual-reality field trips.

Robinson, Andrew. *Earth Shock: Climate Complexity and the Force of Nature*. New York: W. W. Norton, 1993. Describes, illustrates, and analyzes the forces of nature responsible for earthquakes, volcanoes, hurricanes, floods, glaciers, deserts, and drought. Also recounts how humans have perceived their relationship with these phenomena throughout history.

Weigel, Marlene. *UxL Encyclopedia of Biomes*. Farmington Hills, Mich.: Gale Group, 1999. This three-volume set should meet the needs of seventh grade classes for research. Covers all biomes such as the forest, grasslands, and desert. Each biome includes sections on development of that particular biome, type, and climate, geography, and plant and animal life.

Woodward, Susan L. *Biomes of Earth*. Westport, CT: Greenwood Press, 2003. Print.

Human Geography

Blum, Richard C, and Thomas C. Hayes. *An Accident of Geography: Compassion, Innovation, and the Fight against Poverty*. Austin, TX: Greenleaf Book Group, 2016. Print.

Dartnell, Lewis. *Origins: How the Earth Made Us*. New York: Hachette Book Group, 2019. Print.

Glantz, Michael H. *Currents of Change: El Niño's Impact on Climate and Society*. New York: Cambridge University Press, 1996. Aids readers in understanding the complexities of the earth's weather pattern, how it relates to El Niño, and the impact upon people around the globe.

Morland, Paul. *The Human Tide: How Population Shaped the Modern World*. New York: PublicAffairs, 2019. Print.

Novaresio, Paolo. *The Explorers: From the Ancient World to the Present*. New York: Stewart, Tabori and Chang, 1996. Describes amazing journeys and exhilarating discoveries from the earliest days of seafaring to the first landing on the moon and beyond.

Rosin, Christopher J, Paul Stock, and Hugh Campbell. *Food Systems Failure: The Global Food Crisis and the Future of Agriculture*. New York: Routledge, 2014. Print.

Economic Geography

Diamond, Jared M. *Guns, Germs, and Steel: The Fates of Human Societies*. New York: Norton, 2011. Print.

Esping-Andersen, Gosta. *Social Foundations of Postindustrial Economies*. New York: Cambridge University Press, 1999. Examines such topics as social risks and welfare states, the structural bases of postindustrial employment, and recasting welfare regimes for a postindustrial era.

Michaelides, Efstathios E. S*Alternative Energy Sources*. Berlin: Springer Berlin, 2014. Print. This book offers a clear view of the role each form of alternative energy may play in supplying energy needs in the near future. It details the most common renewable energy sources as well as examines nuclear energy by fission and fusion energy.

Robertson, Noel, and Kenneth Blaxter. *From Dearth to Plenty: The Modern Revolution in Food Production*. New York: Cambridge University Press, 1995. Tells a story of scientific discovery and its exploitation for technological advance in agriculture. It encapsulates the history of an important period, 1936-86, when government policy sought to aid the competitiveness of the agricultural industry through fiscal measures and by encouraging scientific and technical innovation.

REGIONAL GEOGRAPHY

Biger, Gideon, ed. *The Encyclopedia of International Boundaries*. New York: Facts on File, 1995. Entries for approximately 200 countries are arranged alphabetically, each beginning with introductory information describing demographics, political structure, and political and cultural history. The boundaries of each state are then described with details of the geographical setting, historical background, and present political situation, including unresolved claims and disputes.

Leinen, Jo, Andreas Bummel, and Ray Cunningham. *A World Parliament: Governance and Democracy in the 21st Century*. Berlin Democracy Without Borders, 2018. Print.

Pitts, Jennifer. *Boundaries of the International: Law and Empire*. Cambridge, Mass: Harvard University Press, 2018. Print.

Index

B

D

E

G

H

N

Q

R

T